Heinz K. Flack · Georg Möllerke
Illustrated Engineering Dictionary

Springer

Berlin
Heidelberg
New York
Barcelona
Budapest
Hongkong
London
Mailand
Paris
Santa Clara
Singapur
Tokio

Heinz K. Flack · Georg Möllerke

Illustrated Engineering Dictionary

Bildwörterbuch Maschinenbau und Elektrotechnik
Englisch/Deutsch - Deutsch/Englisch

Mit 619 Abbildungen

Heinz K. Flack
Trottackerstraße 5
CH-5507 Mellingen

Georg Möllerke
Kornweg 5
CH-5415 Nussbaumen

Die Deutsche Bibliothek - CIP-Einheitsaufnahme
Flack, Heinz K.:
Illustrated engineering dictionary = Bildwörterbuch Maschinenbau und Elektrotechnik /
Heinz K. Flack; Georg Möllerke.
Berlin; Heidelberg; New York; Barcelona; Budapest; Hongkong; London; Mailand; Paris;
Santa Clara; Singapur; Tokio: Springer, 1997
ISBN 3-540-62441-4
NE: Möllerke, Georg: HST

ISBN 3-540-62441-4 Springer-Verlag Berlin Heidelberg New York

Einband-Entwurf: MEDIO, S. Vitale, Berlin
Satz: Camera ready Vorlage durch Autor
SPIN: 10566286 68/3020 - Gedruckt auf säurefreiem Papier

Vorwort

Manchem Techniker ist es sicher auch schon so ergangen: Man tritt eine Dienstreise an und fragt sich, welches technische Wörterbuch sich für eine Reise eignet. Die Bände im Büro sind zu schwer für das Fluggepäck oder werden gerade benötigt; so reist man letzten Endes mit einem kleinen allgemeinsprachlichen Wörterbuch. Im Ausland stellt man bald fest, daß ein technisches Wörterbuch - und sei es noch so klein - eine Notwendigkeit ist.

Damit aber nicht genug: Wie oft kommt es vor, daß die reine Übersetzung mit Hilfe des Wörterbuches nicht ausreicht, einen Ausdruck zu verstehen. Bilder sind eine große Hilfe. Aus dieser Überlegung heraus ist unser Illustrated Engineering Dictionary entstanden. Um z.B. Unterschiede zwischen *machine jack, machinist's jack, screw packing* und *screw jack* zu erkennen, sind Zeichnungen unerläßlich. Daher enthält der englisch-deutsche Vokabelteil über 400 technische Abbildungen.

Einem reichen Wortschatz zuliebe wurde auf solche Hauptwörter verzichtet, die sich ohne weiteres aus den bekannten Zeit- und Eigenschaftswörtern bilden lassen. Von den Wortzusammensetzungen sind die regelmäßig gebildeten sparsam aufgenommen, um mehr Platz für abnormal gebildete zu erhalten. Viele Zusammensetzungen dieser Art wurden nicht berücksichtigt, z.B. *bit..., cam..., carbide..., chip..., face..., holding..., milling..., turning..., work...*usw.

Eine Besonderheit bildet der Fachteil *Pictorial Machine Tool Vocabulary*, der innerhalb des Buchblocks durch gerasterten Balken am rechten Seitenrand leicht aufzufinden ist. Speziell für Ingenieure und Techniker der Fertigungstechnik ist hier, ebenfalls unterstützt durch mehr als 180 Bilder, das Grundvokabular des Werkzeugmaschinenbaus zusammengestellt. Der Schwerpunkt liegt dabei auf den spanenden Werkzeugmaschinen und deren Steuerungen; ein maschinenbauliches Grundvokabular ist jedoch mit einbezogen. Überschneidungen mit dem maschinenbaulich-elektrotechnischen Hauptteil des Wörterbuches können vorkommen und sind beabsichtigt, um möglichst nicht in zwei verschiedenen Vokabelteilen nachschlagen zu müssen. Ergänzt wird dieser Fachteil durch fachliche Redewendungen des Werkzeugmaschinenbaus (*machine tool phrases*) im Anhang. Wertvolle Unterstützung hierbei haben Bedienungsanleitungen der Joseph Binkert AG, Glattbrugg/Schweiz, geleistet, wofür die Verfasser zu besonderem Dank verpflichtet sind.

Von hohem Nutzen für den Praktiker dürfte schließlich – für ein Taschenwörterbuch dieser Art sicher ein Novum – der umfangreiche Anhang sein. Auf rd. 48 Seiten kann man, nach Themen gegliedert, spezielle technische Redewendungen, aber auch allgemeinsprachliche (die im Berufsalltag häufig vorkommen), nachschlagen. Die Bandbreite reicht von der Ausdrucksweise bei *Formelzeichen und Größengleichungen* über *Spannungslegung und -messung, Montage und Inbetriebnahme* sowie *Microcomputer Glossary* bis hin zu Fach- und Alltagsgesprächen (*"At the meeting"* und *"Everyday phrases"*).

Wir hoffen sehr, daß Sie recht viel Nutzen aus diesem Wörterbuch ziehen können.

Mellingen/Schweiz
und Nussbaumen/Schweiz

Ing. Heinz K. Flack
Ing. Georg Möllerke

im Januar 1997

Inhaltsverzeichnis/Table of Contents

How to pronounce

Es handelt sich hier um eine Zusammenstellung von Wörtern, die oft nicht richtig ausgesprochen werden.

accessories [æk·sesɔris] Zubehör(teile)
alignment [ə·lainmənt] Ausrichtung, Anpassung
alloy [·ælɔi] Legierung
alternative [ɔːl·təːnətiv] Alternative
analysis [ə·næləsis] Analyse
ancillary [æn·siləri] Hilfs . . ., Neben . . .
armature [·ɑːmətjuə] Anker (elektr.)
auxiliaries [ɔːg·ziljəris] Hilfsbetriebe
bearing [·bεəriŋ] Lager
buoy [bɔi] Boje, Bake
cascade [kæs·keid] Kaskade
catastrophe [kə·tæstrəfi] Katastrophe
cathode [·kæθoud] Kat(h)ode
centrifugal force [sen·trifjugəl] Zentrifugalkraft
ceramics [si·ræmiks] Keramik
chaos [·keiɔs] Chaos
circumference [sə·kʌmfərəns] Umkreis, Umfang
cleanliness [·klenlinis] Sauberkeit, Reinlichkeit
coefficient [koui·fiʃənt] Koeffizient
comment [·kɔment] Stellungnahme, Kommentar
comparison [kəm·pærisn] Vergleich
component [kəm·pounənt] (Bestand-)Teil
concrete [·kɔnkriːt] Beton
console [·kɔnsoul] Konsole, Pult
crystal [kristl] Kristall
data [·deitə] Daten, Angaben
debris [·debriː] Trümmer, Schutt
deprecation [depri·keiʃən] Ablehnung, Missbilligung
depreciation [dipriːʃi·eiʃən] Abschreibung, Herabsetzung
detail [·diːteil] Einzelheit
diagram [·daiəgræm] Plan, Schema, Diagramm
discrepancy [dis·krepənsi] Diskrepanz
discretion [dis·kreʃən] Geschick, Umsicht, Belieben
divergence [dai·vəːdʒəns] Divergenz, Auseinanderlaufen
electrolyte [i·lektroulait] Elektrolyt
emphasis [·emfəsis] Betonung, Gewicht, Schwerpunkt
example [ig·zɑːmpl] Beispiel
excerpt [·eksəːpt] Auszug
executive [ig·zekjutiv] leitender Angestellter, Beamter
exhaust gas [ig·zɔːst] Abgas
facsimile [fæk·simili] Faksimile, Reproduktion
fatigue limit [fə·tiːg] Ermüdungsgrenze
filament [·filəmənt] (Glüh-)Faden
floodlight [·flʌdlait] Flutlicht
frequency [·friːkwənsi] Frequenz

galvanometer [gælvə'nɔmitə] Galvanometer
gasometer [gæ'sɔmitə] Gasbehälter
gauge [geidʒ] Eichmass, Messgerät
goniometer [gouni'ɔmitə] Winkelmesser
hazard ['hæzəd] Gefahr, Risiko
hexagonal bolt [hek'sægənl boult] Sechskantschraube
hydraulics [hai'drɔ:liks] Hydraulik
hydrogen ['haidridʒən] Wasserstoff
hygrometer [hai'grɔmitə] Luftfeuchtemesser
hyperbola [hai'pə:bələ] Hyperbel
impedance [im'pi:dəns] Impedanz
indictment [in'daitmənt] Anklage
incandescent [inkæn'desnt] Glühlampe
increment ['inkrimənt] Zuwachs, Zunahme
instantaneous [instən'teinjəs] Sofort . . ., Moment . . .
knowledge ['nɔlidʒ] Wissen, Kenntnisse
laboratory [lə'bɔrət(ə)ri] Laboratorium
legend ['ledʒənd] Legende, erläuternder Text
lever ['li:və] Hebel
maintenance ['meintinəns] Instandhaltung, Wartung
mercury ['mə:kjuri] Quecksilber
miscellaneous [misi'leinjəs] Verschiedenes
molydenum [mə'libdənəm] Molybdän
neutron ['nju:trɔn] Neutron
nomenclature [nou'menklətʃə] Nomenklatur, Bezeichnungen
oxygen ['ɔksidʒən] Sauerstoff
parabola [pə'ræbələ] Parabel
parameter [pə'ræmitə] Parameter
parentheses [pə'renθisis] (runde) Klammern
personnel [pə:sə'nel] Personal
phenomenon [fi'nɔminən] Phänomen
pivot ['pivət] Drehpunkt, (Dreh-)Zapfen
porcelain ['pɔ:slin] Porzellan
potentiometer [pətenʃi'ɔmitə] Potentiometer
preference ['prefərəns] Vorrang, Vorzug
process ['prouses] Prozess
prototype ['proutətaip] Prototyp
psychiatrist [sai'kaiətrist] Psychiater
record ['rekəd] Niederschrift, Rekord
receipt [ri'si:t] Empfang, Quittung
sample [sɑ:mpl] Muster
schedule brit.: ['ʃedju:l] am.: ['skedʒu:l] Aufstellung, Plan
scheme [ski:m] Schema, Anlage
species ['spi:ʃi:z] Art, Spezies
specimen ['spesimin] Exemplar, Muster
speedometer [spi:'dɔmitə] Geschwindigkeitsmesser
telegraphy [ti'legrəfi] Telegraphie
thread [θred] Gewinde, Faden
thermometer [θə:'mɔmitə] Thermometer
trigonometry [trigə'nɔmitri] Trigonometrie
trough [trɔf] Trog, Mulde, Rinne
uranium [juə'reinjəm] Uranium
voltage ['voultidʒ] Spannung
weight [weit] Gewicht

Guide to pronunciation

[ɑː]	plant [plɑːnt] cast [kɑːst]	Reines, langes «a» wie in Lager.
[ʌ]	current [ˈkʌrənt] multiple [ˈmʌltipl]	Kurzes, dunkles «a», bei dem die Lippen nicht gerundet sind.
[æ]	reactor [riˈæktə] tapping [ˈtæpiŋ]	Heller, ziemlich offener, nicht zu kurzer Laut; ungefähr wie «Fähre».
[ɛə]	air [ɛə] bearing [ˈbɛəriŋ]	Nicht zu offenes halblanges «ä«, nur vor r.
[ai]	provide [prəˈvaid] derive [diˈraiv]	Helles «a» und schwächeres, offenes «i».
[au]	power [ˈpauə] out [aut]	Helles «a» und schwächeres, offenes «u».
[ei]	make [meik] decay [diˈkei]	Halboffenes «e», nach «i» auslautend.
[e]	effect [iˈfekt] stem [stem]	Halboffenes, kurzes «e», etwas geschlossener als das in «Fett».
[ə]	about [əˈbaut] connect [kəˈnekt]	Flüchtiger Gleichlaut, ähnlich wie das «e» in «Anlage».
[iː]	peak [piːk] wheel [wiːl]	Langes «i» wie in «Liter», etwas offener einsetzend als im Deutschen.
[i]	deliver [diˈlivə] mechanic [miˈkænik]	Kurzes, offenes «i» wie in «mit».
[iə]	here [hiə] inferior [inˈfiəriə]	Halboffenes, halblanges «i» mit nachhallendem «a».
[ɔː]	order [ˈɔːdə] ore [ɔː]	Offener, langer Laut zwischen «a» und «o», ungefähr wie in «Wort».
[ɔ]	volume [ˈvɔljum] stop [stɔp]	Offener, kurzer Laut zwischen «a» und «o».
[o]	molest [moˈlest, mouˈlest]	Flüchtiges, geschlossenes «o».
[ɔi]	hoist [hɔist] exploit [iksˈplɔit]	Offenes «o» und schwächeres, offenes «i».
[uː]	rule [ruːl] shoe [ʃuː]	Langes «u» ohne Lippenrundung, jedoch ungefähr wie in «Buch».
[uə]	sure [ʃuə] pure [pjuə]	Halboffenes, halblanges «u» mit nachhallendem «ə».

[u]	careful [ˈkɛəful] put [put]	Flüchtiges «u».
[r]	radar [ˈreidə] there is [ðɛərˈiz]	Nur vor Vokalen gesprochen und am Ende eines Wortes bei Bindungen mit dem Anlautvokal des folgenden Wortes.
[ʒ]	gyrate [dʒaiəˈreit] large [lɑ:dʒ]	Stimmhaftes «sch» wie in «Journal».
[ʃ]	shape [ʃeip] share [ʃɛə]	Stimmloses «sch» wie in «Schacht».
[θ]	method [ˈmeθəd] thin [θin]	Lispellaut, entsteht durch Anlegen der Zunge an die oberen Schneidezähne.
[ð]	there [ðɛə] breathe [bri:ð]	Wie «θ», jedoch mit Stimmton.
[s]	see [si:] decide [diˈsaid]	Stimmloser Zischlaut wie in «Fass».
[z]	zero [ˈziərou] horizon [həˈraizn]	Stimmhafter Zischlaut wie in «sauber».
[ŋ]	heating [ˈhi:tiŋ] ring [riŋ]	Nasallaut wie in «fangen».
[w]	water [ˈwɔ:tə] will [wil]	Flüchtiges, mit Lippe an Lippe gesprochenes «w», aus der Mundstellung für «u» gebildet.
[f]	effort [ˈefət] tough [tʌf]	Stimmloser Lippenlaut wie in «Pfeife».
[v]	vice [vais] vessel [ˈvesl]	Stimmhafter Lippenlaut wie in «Ventil».
[j]	filial [ˈfiljəl] yacht [jɔt]	Flüchtiger Laut zwischen «j» und «i».

Illustrated Engineering Vocabulary

Englisch - Deutsch

A

abbreviate, to kürzen, abkürzen
abbreviation Abkürzung
aberration Abweichung, Abweichungsfehler
abrasion Abrieb, Abschleifen, Abschmirgeln
abscissa Abzisse
absorb, to absorbieren, aufnehmen
absorber Absorber, nichtreflektierender Körper
absorption Absorption, Aufnahme
absorption type refrigerator Absorptionskältemaschine
abstract, to entziehen, abziehen; (Adj.) abstrakt, theoretisch
abundance Überfluss
abuse, to übermässig beanspruchen, missbrauchen
abut, to anstossen
abutment Widerlager, Gegenpfeiler
abyss Kluft, Abgrund
a.c., AC (alternating current) Wechselstrom
accelerate, to beschleunigen
acceleration Beschleunigung
accelerator Beschleunigungsvorrichtung
accentuate, to hervorheben, akzentuieren
accentuation Anhebung, Akzentuierung
accept, to abnehmen
acceptance Abnahme, Annahme
acceptance certificate Abnahmeprotokoll
acceptance test Abnahmeprüfung
access Zugang
access cycle Zugriffzyklus
access time Suchzeit, Zugriffzeit
accessible zugänglich
accessibility Zugänglichkeit
accessories Zubehör
accessory Zubehörteil
accident Zwischenfall, Unfall
accommodate, to aufnehmen
accommodation Unterbringung
accomplish, to bewerkstelligen, erreichen
accord, to bewilligen, bewähren, übereinstimmen
account Bericht, Darstellung

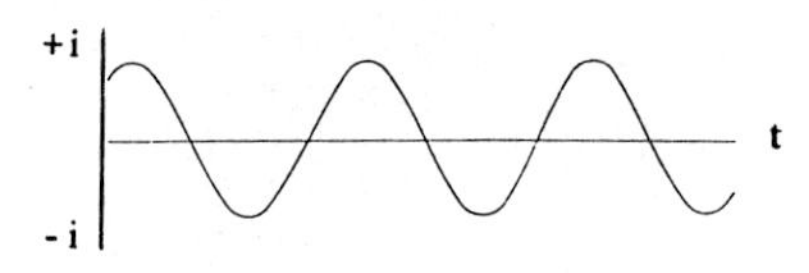

a. c. curve
Wechselstrom-Kurve

+i **positive current direction** positive Stromrichtung
- i **negative current direction** negative Stromrichtung
t **time** Zeit

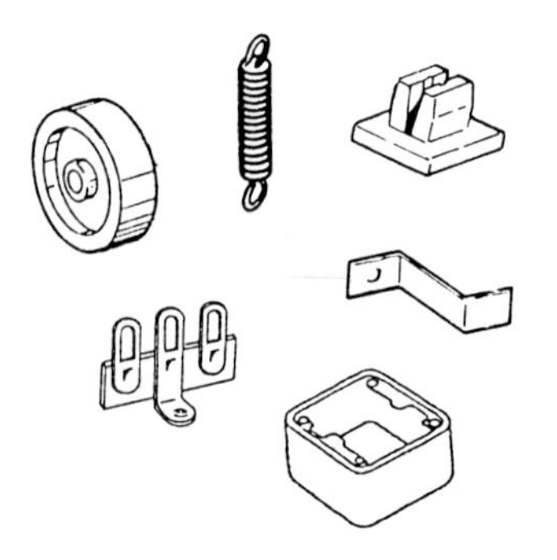

Accessories Zubehör

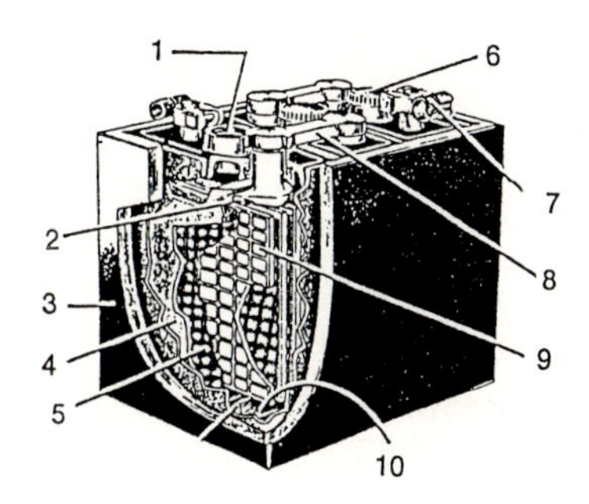

Accumulator (battery)
Akkumulator, Sammler (Batterie)

1 **filler opening** Füllöffnung
2 **plate connector** Plattenverbindung
3 **container** Behälter
4 **negative plate** Minusplatte
5 **separator** Trennplatte
6 **vent plug** Entlüftungsstopfen
7 **terminal** Klemme
8 **cell link** Zellenverbindung
9 **positive plate** Plusplatte
10 **sediment** Ablagerung

accretion Zuwachs
accumulate, to speichern
accumulation Speicherung
accumulator Akkumulator, Sammler
accuracy Genauigkeit, Richtigkeit
accurate genau
acetylene Azetylen
acetylene welding Autogen-schweissung
achieve, to erreichen, erzielen
acicular nadelförmig
acid Säure
acidless säurefrei
acid-proof säurefest, -beständig
acknowledge, to bestätigen, anerkennen
acknowledgement Bestätigung, Anerkennung
acoustical akustisch
acoustics Akustik
acquire, to erwerben, gewinnen
acquisition Erfassung, Aufnahme
acrid ätzend, scharf, beissend
across parallel zu, im Durchmesser
across-the-line starting Anlassen mit voller Spannung
act (air-cooled triode) luftgekühlte Triode
act, to (ein)wirken auf
action Einwirkung, Wirken, Tätigkeit
action turbine Aktionsturbine (auch: impulse turbine)
action wheel Aktionsrad
activate, to wirksam machen, anregen, aktivieren
activation Aktivierung, Anschaltung
active aktiv, wirksam, tätig
active current Wirkstrom
active component Wirkanteil
active power Wirkleistung
activity Wirksamkeit, Tätigkeit, Aktivität
actual wirklich, tatsächlich
actual size Istmass
actual value Istwert
actuate, to betätigen
actuation Betätigung
actuator Betätigungsglied
acuity Schärfe
acute angle spitzer Winkel
ACV (air cushion vehicle) Luftkissenfahrzeug

acyclic unperiodisch, azyklisch
A-D converter (analog-digital) Analog-Digital-Umsetzer
adamantean diamanthart
adapt, to anpassen
adaption Anpassung
adapter Adapter, Zwischenstück
adapter plug Zwischenstecker
adapter sleeve Spannbüchse
adapting apparatus Vorsatzgerät
addendum Kopfhöhe (Zahnrad)
addendum circle Kopfkreis (Zahnrad)
add, to beimengen, hinzufügen
add up, to summieren, zusammenrechnen
added filter Zusatzfilter
addend Summand, Addend
adding machine Additionsmaschine
addition Beimengung, Hinzufügung
additional zusätzlich
addition agent Zusatz
additive additiv, Wirkstoff
additivity Additivität
address Adresse
addressable adressierbar
addressee Adressat
ADE (audible Doppler enhancer) akustisches Dopplergerät
adept sachverständig
adequate entsprechend, angemessen, adäquat
ADF (automatic direction finder) automatischer Funkpeiler
adhere, to (an)haften, kleben
adherence Einhaltung, Beachtung
adherent anhaftend
adhesion Adhäsion, Haftvermögen
adhesion coefficient Reibungskoeffizient
adhesive Klebstoff, anhaftend
adhesive tape Klebstreifen
adhesiveness Haftfähigkeit
adiabatic adiabatisch
adiabatic curve Adiabate
adjacent benachbart
adjacent attenuation Trennschärfe
adjoining nebenstehend
adjoint-piece Verlängerungsstück
adjunct Zubehör
adjunction Anschluss
adjust, to einstellen, nachstellen, abgleichen

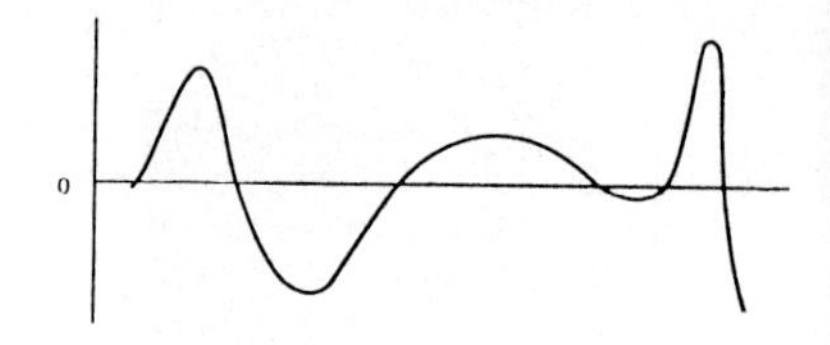

Acyclic waveform
azyklische Wellenform

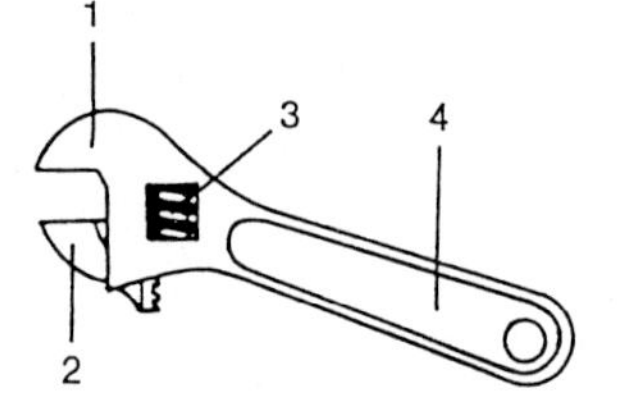

Adjustable spanner (wrench)
Verstellbarer Schraubenschlüssel

1 **fixed jaw** feste Schraubbacke
2 **adjustable jaw** verstellbare Schraubbacke
3 **adjusting screw** Verstellschraube
4 **handle** Handgriff

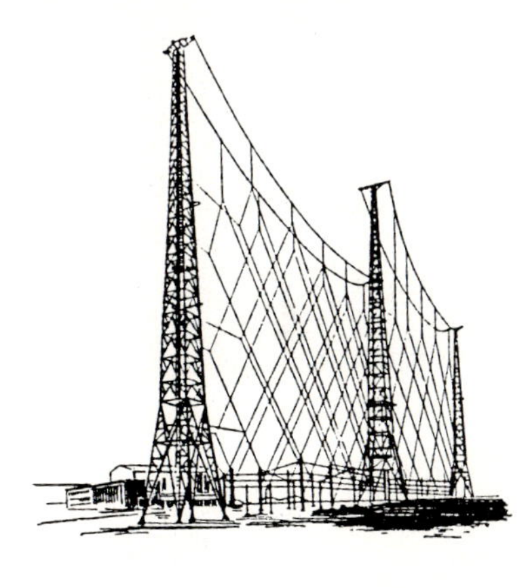

Aerials (Antennae) for high-frequency transmission
Antennen für Hochfrequenz-Übertragung

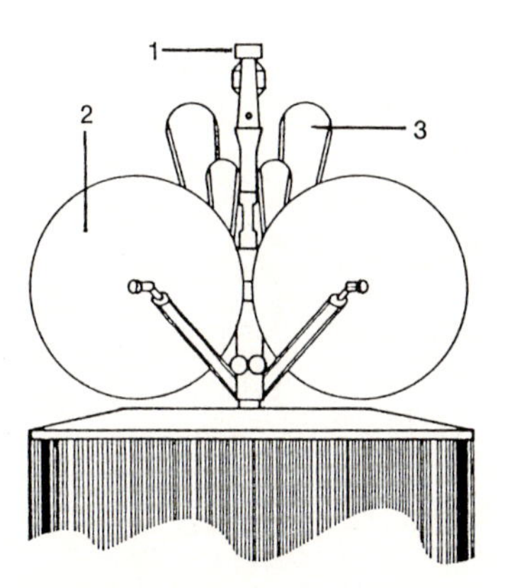

Aerials (Antennas) on Intelsat IV
Antennen an Intelsat IV

1 **omnidirectional aerial for receiving** Rundstrahlantenne für Empfang
2 **directional aerials** Richtstrahlantennen
3 **wide-band aerials (earth coverage)** breitbündige Antennen (Erd-deckend)

adjustability Einstellbarkeit, Nachstellbarkeit
adjust, to einstellen, nachstellen,
adjusted abgeglichen
adjuster Verstellvorrichtung
adjusting Verbesserung, Herrichtung
adjusting ear Nachspannöse
adjusting device Justiereinrichtung
adjusting nut Stellmutter
adjusting wedge Anzugkeil
adjustment Einstellung, Nachstellung, Abgleichung
admissible zulässig
admission Zulassung, Einströmung
admission valve Einlassventil
admit, to zulassen, zuführen
admittance Admittanz, Scheinleitwert
admitting port Einströmungsöffnung
admix, to beimengen
admixture Beimengung, Zusatz
ADP (automatic data processing) automatische Datenverarbeitung
ADU method (powder metallurgy) ADU-Methode
advance, to vorrücken, fortschreiten
advance Fortschreiten, Voreilen
advance angle Voreilwinkel
advancement Vorrücken, Beförderung
advantageous vorteilhaft
advertise, to annoncieren
advertising department Werbeabteilung
advise, to beraten, informieren, mitteilen
advisory Beratungs . . .
aerate, to lüften, der Luft aussetzen
aeration Belüftung
aerial Antenne
aerial circuit Antennenkreis
aerial loading coil Pupinspule für Luftkabel
aerodynamic aerodynamisch
aerodynamics Aerodynamik
aerofoil Tragfläche, -flügel
aerologist Aerologe
aeronautical aeronautisch
aeronautics Flugwesen
aeroplane Flugzeug
aerospace industry Luft- und Raumfahrtindustrie

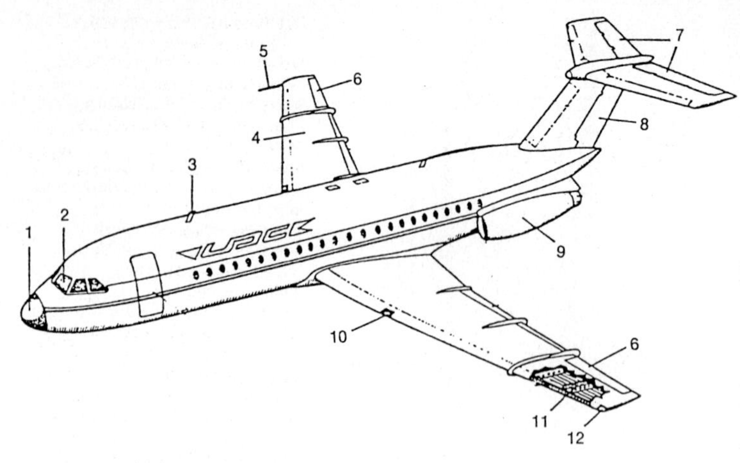

Aeroplane/Aircraft (Airliner)
Flugzeug (Verkehrsflugzeug)

1 **radar dom** Radarkuppel
2 **flight deck (cockpit)** Pilotenkanzel
3 **aerial** Antenne
4 **aerofoil** Tragfläche
(**wing** Flügel)
5 **pitot tube for measuring speed** Staurohr zur Geschwindigkeitsmessung
6 **aileron** Querruder
7 **elevators** Höhenruder
8 **rudder** Seitenruder
9 **jet engine** Turbo-Maschine
10 **landing light** Landeleuchte
11 **ribs** Spanten
12 **navigation light** Kennleuchte

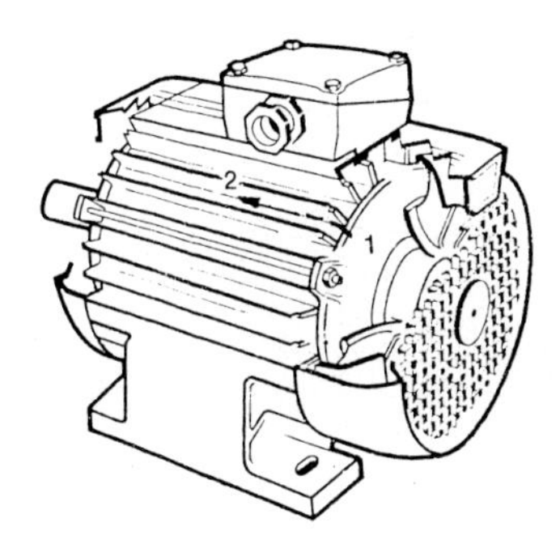

Air-cooled motor
Luftgekühlter Motor

1 **fan** Lüfter
2 **flow of air** Luftfluß

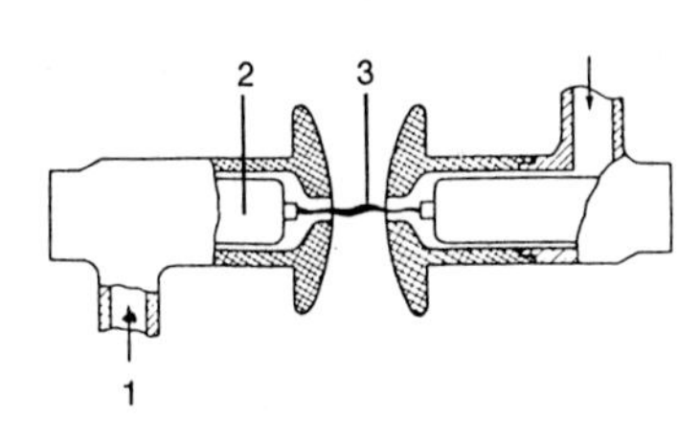

Air-blast circuit-breaker (contact arrangement)
Druckluftschalter (Kontaktanordnung)

1 **compressed air supply** Druckluftzuführung
2 **contact** Kontakt
3 **arc extinguished by blast of air** durch Luftstrahl gelöschter Lichtbogen

AF (audio frequency) Niederfrequenz, Tonfrequenz
affect, to beeinflussen, einwirken (auf), beeinträchtigen
affection Beeinflussung, Einwirkung, Beeinträchtigung
AFC (automatic frequency control) automatische Frequenznachsteuerung
affidavit Affidavit (eidesstattliche Erklärung)
affiliate Zweigorganisation
affinity Affinität
affix, to aufkleben, anheften
afflux Zufluss, Zustrom
afloat schwimmend, im Wasser befindlich
afterglow Nachleuchten
age, to altern, veredeln, tempern
agency (Administrations-)Stelle
aging Alterung, Tempern
agglomerate, to anhäufen, zusammenballen
agglomeration Anhäufung, Zusammenballung
aggravate, to erschweren
aggravation Erschwerung
aggregate Aggregat
aggregation Zusammensetzung, Aggregation
agitation heftige Bewegung
agree with, to übereinstimmen mit
agreement Übereinstimmung, Zustimmung
aileron Querruder
aim Ziel
aim at, to zielen auf, anvisieren
air-blast circuit-breaker Druckluft(leistungs)schalter
airborne im Flugzeug eingebaut
airborne radar Flugzeugradar
air brake Landeklappe
air bubble Luftblase
air chamber Windkessel
air circulation Luftumlauf, -zirkulation
air conditioning plant Klimaanlage
air contamination Luftverunreinigung
air cooling Luftkühlung
air-cored coil eisenfreie Spule
aircraft Flugzeug
air current Luftströmung
air cushion vehicle Luftkissenfahrzeug

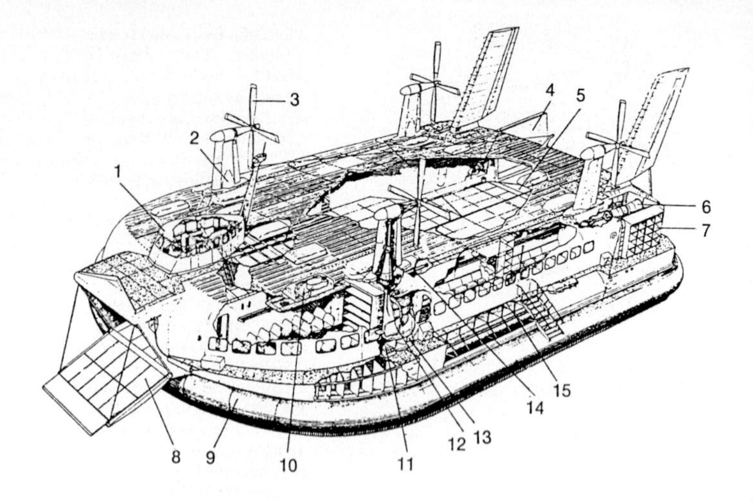

Air-cushion vehicle (hovercraft)
Luftkissen(Schwebe)fahrzeug

1 **control cabin** Steuerkabine (Brücke)
2 **pivoted pylon** drehbare Tragsäule
3 **propeller for directional control** Propeller für Richtungssteuerung
4 **stern loading doors** Heck-Ladeklappen
5 **loading ramp** Laderampe
6 **gas turbine engines** Gasturbinen-Antriebsmaschinen
7 **engine air intakes** Lufteintritt für Antriebsmaschine
8 **bow unloading ramp** Bug-Entladerampe
9 **skirting trunk** Schlauchgürtel
10 **passenger cabin ventilator** Ventilator für Passagierkabine
11 **refuelling point** Treibstoff-Nachfüllung
12 **transmission main gear** Hauptgetriebe für Kraftübertragung
13 **centrifugal lift fan** Zentrifugal-Hebegebläse
14 **transmission shafting** Wellengestänge für Kraftübertragung
15 **plenum chamber** Druckkammer

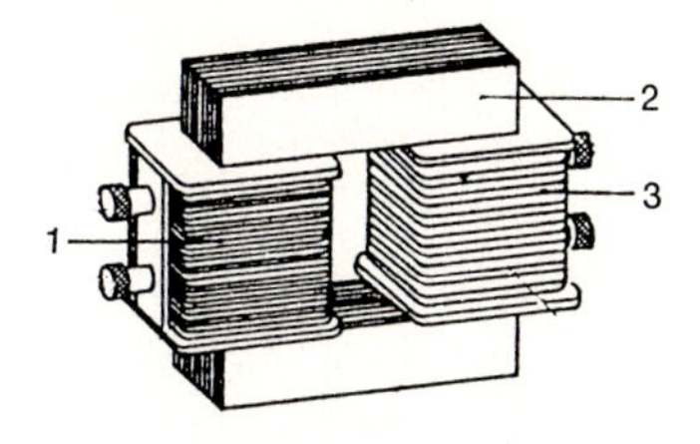

Air (-cooled) transformer
Trockentransformator
(luftgekühlter Transformator)

1 **primary winding** Primärwicklung
2 **magnet core** Magnetkern
3 **secondary winding** Sekundärwicklung

air draught Luftzug
air draught blower (Luft-)Gebläse
air duct Luftkanal
air ejector Saugluftförderer, Luftejektor
air escape Entlüftung
air exhauster Absaugventilator
air-flow indicator Luftströmungsanzeiger
air-flow monitor Luftströmungswächter
air gap Luftspalt
air hardening steel Lufthärtungsstahl
air heater Lufterhitzer
air hose Luftschlauch
air humidity Luftfeuchtigkeit
air inlet (air intake) Lufteintritt
air lock Luftschleuse
air outlet Luftaustritt
air nozzle Luftdüse
air path Luftweg
air pocket Luftblase
air pollution Luftverunreinigung
air pressure Luftdruck
air switch Luftdruckschalter
air resistance Luftwiderstand
air shaft Luftschacht
air shutter Luftklappe
air supply Luftzufuhr
airtight luftdicht
air transducer Luftschallwandler
air transformer Trockentransformator
air vessel Luftflasche
airway beacon Luftstreckenfeuer
airway lighting Streckenbefeuerung
airway marking Streckenerkennung
aisle Mittelgang (Flugzeug)
alarm, to melden, alarmieren, Alarm geben
alc (automatic level control) automatische Pegelregelung
alcatron Alkatron (Feldeffekt-Transistor)
alcohol Alkohol
alcyde resin Alkydharz
Aldis lamp Handmorselampe
alert notice Alarmmeldung
algebraic algebraisch
alien frequencies Fremdtöne, Fremdfrequenz
alight, to landen, aussteigen
alighting gear Fahrwerk

align, to abgleichen, ausfluchten, ausrichten
alignment Abgleichung, Ausfluchtung, Ausrichtung, Aneinanderreihung
alike on both sides seitengleich
alive unter Spannung befindlich, stromführend, unter Dampf, dampfführend (besser: live)
alkaline alkalisch
all-metal Ganzmetall
all ready signal lamp Klarmeldelampe
allege, to behaupten, angeben
Allen key Imbus-Schlüssel
alleviate, to erleichtern, mildern
alley(way) Gang, Laufgang
alligator clip Krokodilklemme
allocable zuteilbar, anrechenbar
allocate, to zuweisen
allocation Zuteilung
allot, to zuteilen, zuweisen
allotment Zuteilung, Zuweisung, Anteil
allow, to bewilligen, erlauben, berücksichtigen
allowable zulässig
allowance Kleinstspiel, Grösstübermass
alloy Legierung
alnico Alnico (Aluminium-Nickel-Kobalt-Legierung)
alphanumeric alphanumerisch
altazimuth Höhenazimut
alter, to abändern, umändern
alteration Abänderung, Umänderung
alternate, to wechseln, abwechseln
alternating current Wechselstrom
alternative Alternative, alternativ
alternation Wechsel, Wechselfolge
alternator Wechselstromgenerator
altigraph Höhenschreiber
altimeter Höhenmesser
altitude Höhe
aluminium (am.: aluminum) Aluminium
AM (amplitude modulation) Amplitudenmodulation
amalgate, to amalgamieren, verschmelzen
amalgamation Amalgamierung, Verschmelzung
amber Bernstein
ambient temperature Umgebungstemperatur

Alignment check of two axes
Fluchtungsprüfung von zwei Achsen

1 **connecting rod** Pleuelstange
2 **scribing block** Parallelreißer
A **axis through small-end bushing** Achse durch obere Pleuelbuchse
B **axis through big-end bushing** Achse durch untere Pleuelbuchse

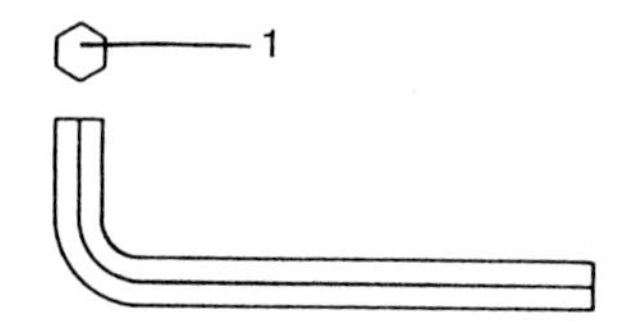

Allen key
Imbus-Schlüssel, Schraubenschlüssel für Innensechskant

1 **hexagonal cross-section** Sechskant-Querschnitt

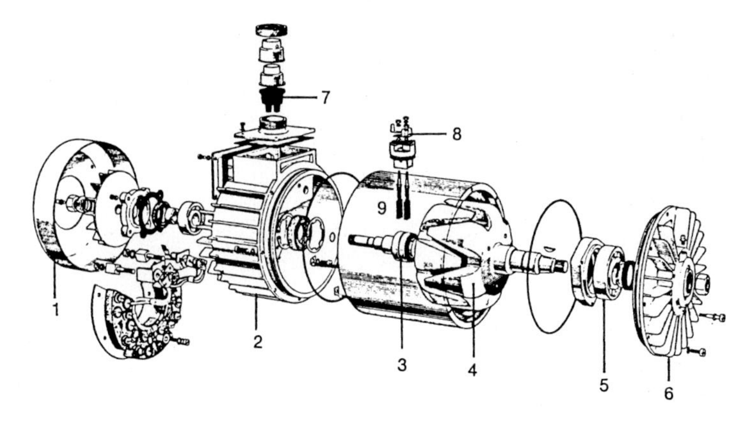

Alternator for heavy trucks (exploded view)
Wechselstromgenerator für Lastkraftwagen (aufgelöste Darstellung)

1 **end shield** Lagerschild
2 **stator** Ständer
3 **slip rings** Schleifringe
4 **claw poles** Klauenpole
5 **bearing** Lager
6 **fan wheel** Lüfter rad
7 **terminal bushing** Anschlußdurchführung
8 **terminals** Anschlüsse
9 **coal brushes** Kohlebürsten

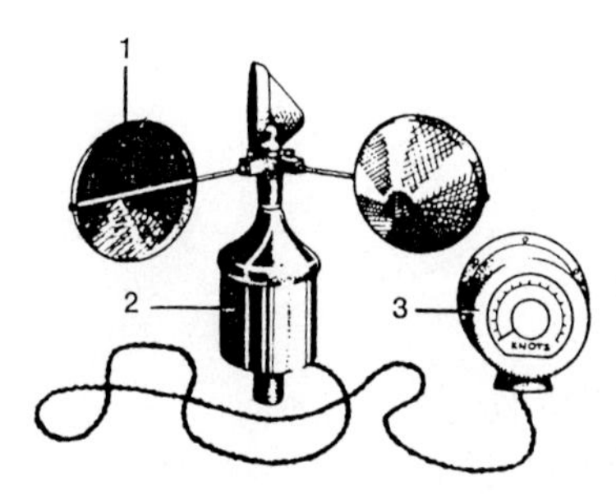

Anemometer
Windmesser

1 **metal cups** Metallschalen
2 **generator** Generator
3 **wind speed gauge** Windgeschwindigkeitsanzeige

ambiguity Unklarheit, Zweideutigkeit
ambiguous unklar, zweideutig
ammeter Amperemeter
ammonia Ammoniak
ammonium Ammonium
ammunition Munition
amorphous gestaltlos, amorph, unkristallinisch
amount, to sich belaufen, beziffern (auf)
amount Höhe, Betrag
amp, ampere Ampere
ampere-turn Amperewindung
amp-hour Amperestunde (Ah)
amperage Amperezahl
amplidyne Amplidyne (el. Masch.)
amplification Verstärkung
amplifier Verstärker
amplify, to verstärken
amplifying circuit Verstärkerschaltung
amplifying tube Verstärkungsröhre
amplitude Amplitude, Schwingungsweite
ampoule (ampulla) Ampulle
AMX (automatic message exchange) automatische Speichervermittlung
analog computer Analogrechner
analogous analog, entsprechend
analyse, to analysieren, zergliedern
analyser Analysator
analytic(al) analytisch
anchor Anker
anchor, to verankern
anchoring Verankerung
ancillaries Zusatz-, Hilfsgeräte
ancillary apparatus Zusatzgerät
anemograph Windschreiber
anemometer Windmesser
anemometry Windstärkenmessung
aneroid Dosenbarometer
angle Winkel, Ecke
angle drive Winkeltrieb
angle joint Winkelgelenk
angle of climb Steigwinkel
angle piece Winkelstück
angled eckig
angular winkelförmig, Winkel . . .
angular contact ball bearing Radialschräglager
angularity Winkelstellung, Winkelform

Angles
Winkel

1 **right angle** rechter Winkel
2 **acute angle** spitzer Winkel
3 **obtuse angle** stumpfer Winkel

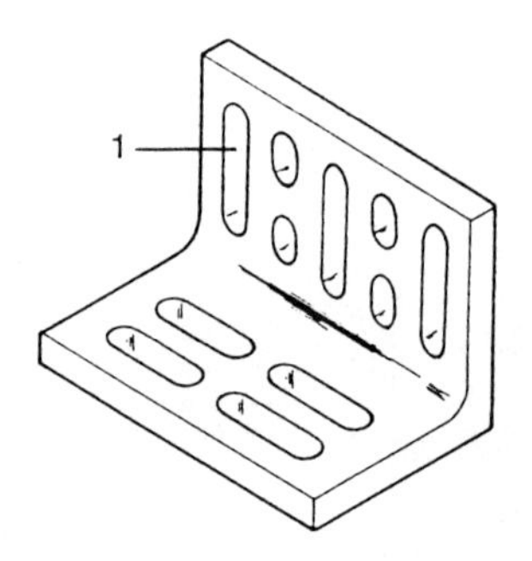

Angle iron
Winkeleisen

1 **fixing slits** Befestigungsschlitze

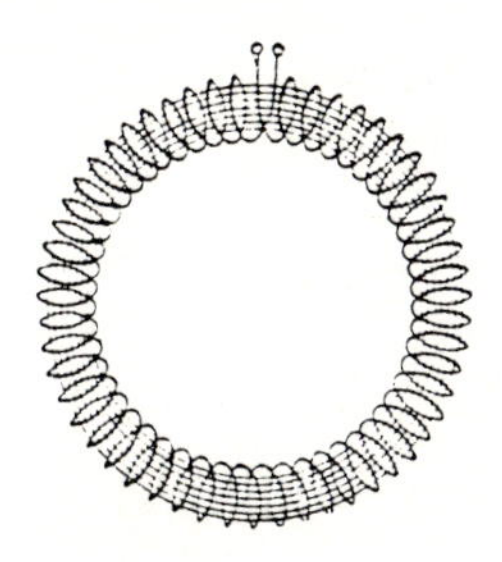

Annular coil
Ringspule

aniline Anilin
anneal, to glühen (Metall)
annealing Glüh . . .
announce, to ankündigen, bekanntgeben, ansagen
announcement Ankündigung, Bekanntgabe, Ansage
announcer Sprecher, Ansager
annoyance Lästigkeit, Störung, Belästigung
annual jährlich
annul, to ausser Kraft setzen, annullieren
annulment Annullierung, Aufhebung
annular ringförmig
annulus (Kreis-)Ring, Ringraum
annunciator Signaltafel
anode Anode
anodic anodisch
anodize, to eloxieren
antagonistic entgegengesetzt wirkend
answering equipment Abfrage-einrichtung
antenna Antenne
anti-clockwise gegen den Uhrzeigersinn
anticipate, to voraussehen, erwarten, zuvorkommen
anticipatory control Vorsteuerung
anti-detonent Mittel zur Erhöhung der Klopffestigkeit
anti-fouling paint anwuchs-behindernde Farbe
anti-friction bearing Wälzlager
anti-friction fuel klopffreier Brennstoff
anti-vibration mounting Schwing-metall-Lagerung
anvil Amboss
apart, to be auseinanderliegen
aperture Öffnung, Strahl-eröffnung, Apertur
apex Scheitel(punkt), Spitze, Gipfel
apparatus Apparat(ur), Gerät
apparent output Scheinleistung
appearance Erscheinen, Aussehen
appendage Anhängsel
appendix Anhang, Ergänzung
appliance Gerät, Vorrichtung, Einrichtung
applicable anwendbar, verwendbar
applicant Antragsteller, Be-werber, Patentanmelder

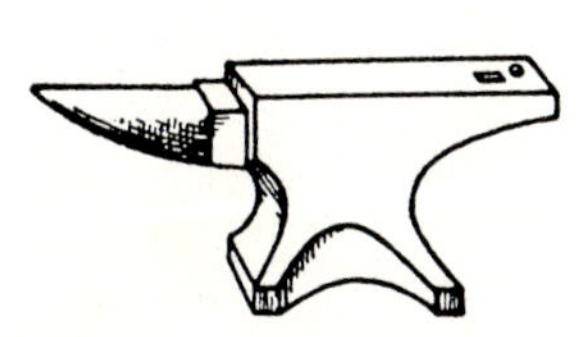

Anvil
Amboß

application Anwendung, Verwendung, Gebrauch
application for patent Patentanmeldung
apply, to anwenden, anlegen, verwenden
apply the brake, to bremsen, abbremsen
appoint, to ernennen, berufen, verabreden
appointment Ernennung, Berufung, Verabredung, Festsetzung
appraisal Schätzung, Abschätzung
appraise, to schätzen, abschätzen
appreciable abschätzbar, nennenswert, beträchtlich
appreciation Schätzung, Würdigung
appreciate, to schätzen, würdigen
apprentice Lehrling
apprenticeship Lehre
approach, to nähern
approach Annäherung, Lösungsweg, Zugang, Anflug
appropriate geeignet, sachgemäss
appropriation Aneignung, Bereitstellung, Zuteilung

Arc furnace
Lichtbogenofen

1 **power tubes** Stromrohre
2 **cable connections** Kabelanschlüsse
3 **electrodes carrying arm** Elektroden-Tragarm
4 **electrode clamps** Elektroden-Klemmbacken
5 **pouring spout** Abgußschnauze
6 **arched roof** Deckelgewölbe
7 **furnace shell** Ofenkessel
8 **slagging door** Abschlacktür
9 **rocker** Wälzwiege
10 **roof-lifting cylinder** Deckelhebezylinder
11 **roof-swing structure** Deckelschwenkkonstruktion
12 **electrodes raising cylinder** Elektrodenhebezylinder
13 **pivot pin** Drehzapfen
14 **tilting cylinder** Kippzylinder

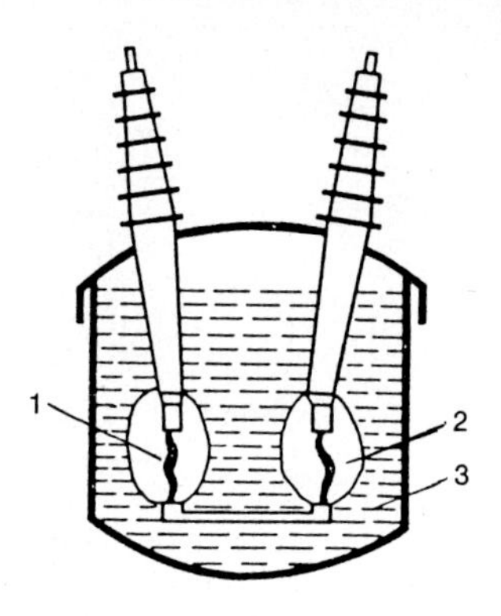

Arc extinction in an oil circuit-breaker
Lichtbogenlöschung in einem Ölkesselschalter

1 **arc** Lichtbogen
2 **gas bubble** Gasblase
3 **oil** Öl

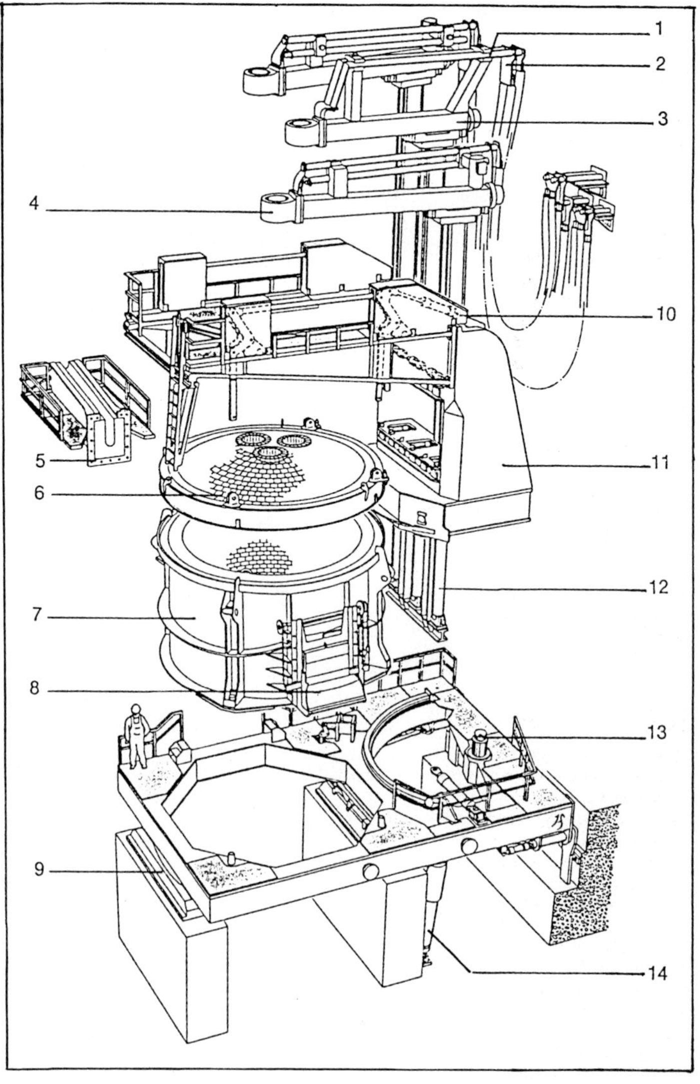

ABB Asea Brown Boveri AG

approval Genehmigung, Billigung
approve, to genehmigen, billigen
approximate angenähert
approximately ungefähr, zirka
approximation Annäherung, Näherung
appurtenances Zubehör
apron Abweisblech, Schlossplatte, Schürze
apt geeignet, tauglich, passend
aptness Eignung, Tauglichkeit, Befähigung
aqua ammonia Salmiakgeist
aquanaut Froschmann
aqueous wasserhaltig, wässerig
arabic figure arabische Zahl
arbitrary willkürlich, beliebig
arbitration schiedsrichterliches Verfahren
arbo(u)r Spindel, Dorn, Achse
arc Bogen, Lichtbogen
arc, to Lichtbogen bilden, funken
arc duration Lichtbogendauer
arc extinguishing medium Funkenlöschmittel
arc furnace Lichtbogenofen
arc welding Lichtbogenschweissung
arch Wölbung
arcing Lichtbogenbildung
area Bereich, Bezirk, Fläche
areometer Aräometer, Tauchwaage
argentic silbrig
argentiferous silberhaltig
argument Beweisführung, Begründung, Streit
arithmetics Arithmetik
arithmetical arithmetisch
armament Rüstung, Bewaffnung
armature Anker (el.)
armo(u)r, to armieren, panzern, bewehren (Kabel)
armo(u)ring Armierung, Panzerung, Bewehrung
arrange, to anordnen
arrangement Anordnung
array Anordnung, Reihe
arrears Rückstände (Gebühren)
arrest, to festhalten, anhalten, abstellen, blockieren, zum Stillstand bringen
arrester Überspannungsableiter
arresting Stillsetzung, Blockierung, Feststellung

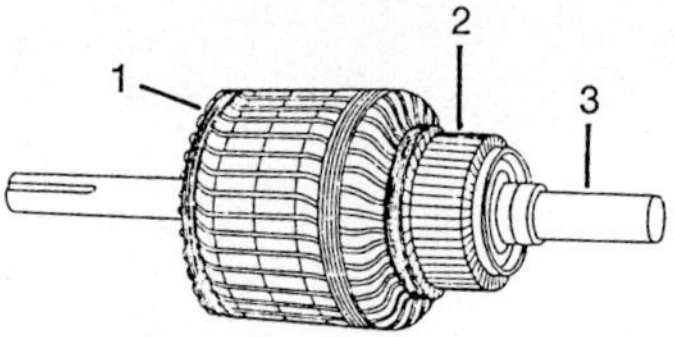

Armature of a d.c. machine
Anker einer Gleichstrommaschine

1 **winding** Wicklung
2 **commutator** Kollektor
3 **shaft** Welle

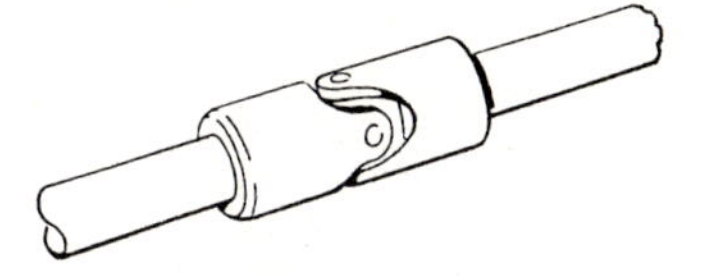

Articulated coupling
Gelenkkupplung

arresting gear Sperrvorrichtung
arrival Ankunft, Eintreffen
arrive, to ankommen, eintreffen, einlaufen
arrival curve Empfangskurve (teleph.)
arrow Pfeil
arsenic Arsen
articulate deutlich, artikuliert
articulated arm Gelenkausleger
articulated coupling Gelenkkupplung
articulated shaft Gliederwelle
articulation Deutlichkeit, (Sprach-)Verständlichkeit
artificial künstlich
artisan Handwerker
artist's impression künstlerische Darstellung
ASA (American Standards Association) Amerikanischer Normenverband
asbestos Asbest
ascent Aufstieg, Anstieg
ascent, to aufsteigen, ansteigen
ascension Aufsteigung
ascertain, to feststellen, ermitteln
ascertainable feststellbar
ash chest Aschenkasten
aspect Aspekt, Gesichtswinkel
aspirate, to ansaugen, aufsaugen (Pumpe)
aspirator Strahlpumpe
assemblage Zusammenbau, Montage
assemble, to zusammenbauen, -setzen
assembler Monteur
assemblies zusammengesetzte Bauteile
assembling jig Montagelehre
assembly Zusammenbau, Apparatesatz, Gerätegruppe
assent Zustimmung, Einwilligung, Genehmigung
assertion Behauptung, Erklärung
assess, to schätzen, taxieren
assessment Schätzung, Einschätzung, Bewertung
assign, to zuweisen, zuteilen
assignment Zuteilung, Verteilung
assist, to helfen, mitwirken, beistehen
assistant Assistent, Gehilfe
associate Mitarbeiter
associated zugehörig, zugeordnet

A frightened assistant?
Ein erschreckter Assistent?

association Verband, Vereinigung
assort, to sortieren
assume the value, to den Wert annehmen
assurance Versicherung
assure, to versichern
assumption Annahme, Voraussetzung
astatic astatisch
astatine Astatin
asterisk Sternchen (Fussnotensymbol)
astern achtern
astronautics Raumfahrt (Wissenschaft)
astronomical astronomisch
astronomy Astronomie
assunder, to be far weit auseinanderliegen
asylum switch verschlossener Schalter
asymmetric(al) asymmetrisch
asymmetry Asymmetrie
asymptotic asymptotisch
asynchronism Ungleichlauf
asynchronous asynchron
athwart dwars querab
atmosphere Atmosphäre
atmospheric atmosphärisch
atmospherics atmosphärische Störung(en)
atomic atomisch, atomar, Atom ...
atomic energy Kern-, Atomenergie
atomize, to zerstäuben
atomizer Zerstäuber
atomizing Zerstäubung
attach, to anbringen, befestigen, anheften
attaching (attachment) Befestigung
attack, to angreifen
attain, to erreichen, erzielen, erlangen
attainment Erreichung, Erlangung
attend, to warten, betreuen, bedienen, pflegen
attendance Bedienung, Wartung, Pflege
attendant Wärter, Bedienungsmann, Pfleger
attendant phenomenon Begleiterscheinung
attention Achtung, Beachtung
attention getter Blickfänger
attentive achtsam, aufmerksam
attenuate, to (ab)schwächen, dämpfen, verdünnen

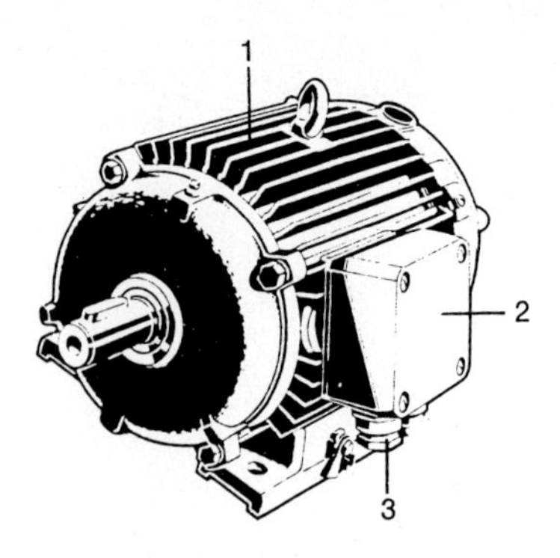

Asynchronous (a.c.) motor
Asynchron- (Wechselstrom-) Motor

1 **cooling fins** Kühlrippen
2 **terminal box** Klemmenkasten
3 **cable supply** Kabelzuführung

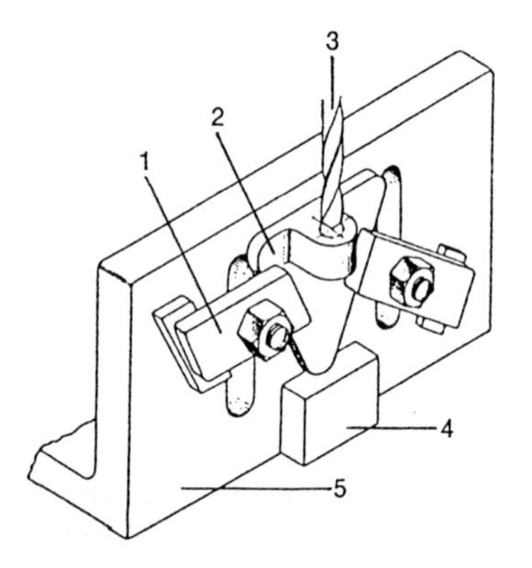

Attachment of a workpiece for drilling
Befestigung eines Werkstücks zum Bohren

1 **attachments** Befestigungsteile
2 **workpiece to be drilled** zu bohrendes Werkstück
3 **drill** Bohrer
4 **packing** Packungsunterlage
5 **holding plate** Aufspannplatte

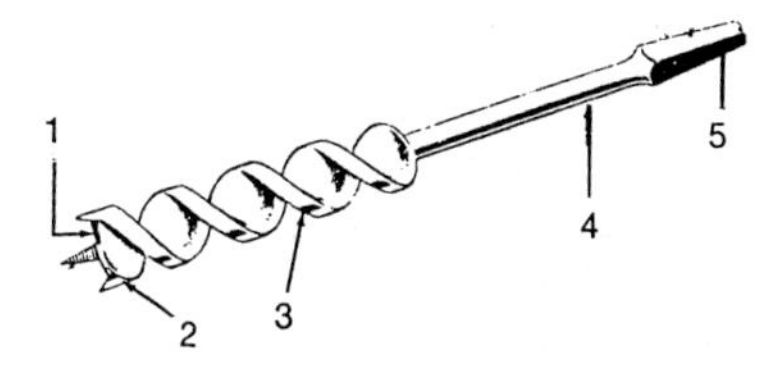

Auger bit
Holzbohrer-Einsatz

1 **cutter** Schneider
2 **spur** Sporn
3 **twist** Windung
4 **shank** Schaft
5 **tang** Griffzapfen

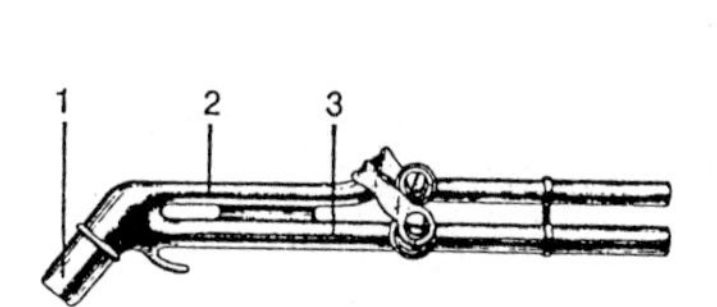

Autogenous blow pipe
Autogen-Schweißbrenner
(**blow cutter** Schneidbrenner)

1 **nozzle** Düse
2 **oxygen pipe** Sauerstoffrohr
3 **acetylene gas pipe** Azetylengasrohr

attenuation Dämpfung, Schwächung, Spannungsteilung
attenuator Dämpfungsglied, Spannungsteiler
attitude Lage, Haltung
attract, to anziehen
attraction Anziehung
attributable zuschreibbar
attribute, to zuschreiben, anrechnen
attrition Abnutzung, Verschleiss
audibility Hörbarkeit, Vernehmbarkeit, Verständlichkeit (Tel.)
audible hörbar
audience Zuhörerschaft
audio Ton . . .
audio amplifier Tonfrequenzverstärker
audio engineer Toningenieur, Tonmeister
audio engineering Tontechnik
audio frequency Hörfrequenz
audio pressure Schalldruck
auditory accuity Hörschärfe
augment, to vermehren
augmentation Erhöhung, Vermehrung
auger Holzbohrer, Erdbohrer, Schneckenbohrer
aural alarm akustisches Alarmsignal
aural receiving Hörempfang
author Verfasser, Urheber
autoclean filter Spaltfilter, selbstreinigender Filter
autogenous welding Autogenschweissung
automate, to automatisieren
automatic automatisch, selbsttätig, Selbst . . .
automatic dial service Selbstwählbetrieb
automatic exchange Wählervermittlung, Vermittlungsstelle
automatic winding machine Wickelautomat
automation Automatisierung, Automation
automize, to automatisieren
automorphous automorph
automotive selbstgetrieben, mit Eigenantrieb
automotive industry Kraftfahrzeugindustrie
autopilot Selbststeueranlage

auto-transformer Spartransformator, Autotransformator
auxiliary Hilfs . . . , Zusatz . . .
auxiliary apparatus Zusatzapparat
auxiliary circuit Hilfsstromkreis
auxiliaries Hilfsmaschinen, Hilfsbetriebe, Hilfsfahrzeuge
availability Verfügbarkeit, Vorhandensein
available verfügbar, lieferbar, vorrätig, zur Verfügung
avalanche diode Avalanche-Diode (avalanche: Lawine)
average Durchschnitt, Mittel(wert)
average, to auf einen Mittelwert bringen, den Durchschnitt nehmen
averaged gemittelt
aviation Flugwesen, Luftfahrt, Fliegerei
avoid, to vermeiden, verhindern
avoidable vermeidbar
award, to (Auftrag) zuerkennen, zuteilen
A. W. G. (American wire ga[u]ge) amerikanische Drahtlehre
awl Ahle
axial axial, achsrecht
axiom Grundsatz, Axiom
axis (Pl.: axes) Achse
axle Spindel, Welle, Achse
axle bearing Achslager
azimuth Azimut

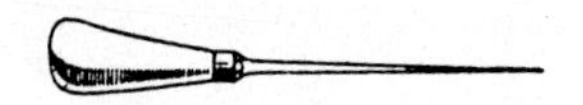

Awl
Ahle

Back axle of a motor-car
Hinterachse eines Kraftfahrzeugs

1 **driving shaft** Antriebswelle
2 **axle** Achse
3 **differential gear** Differentialgetriebe

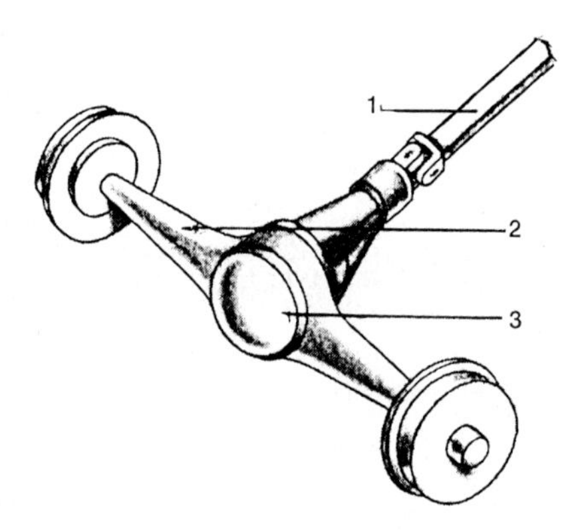

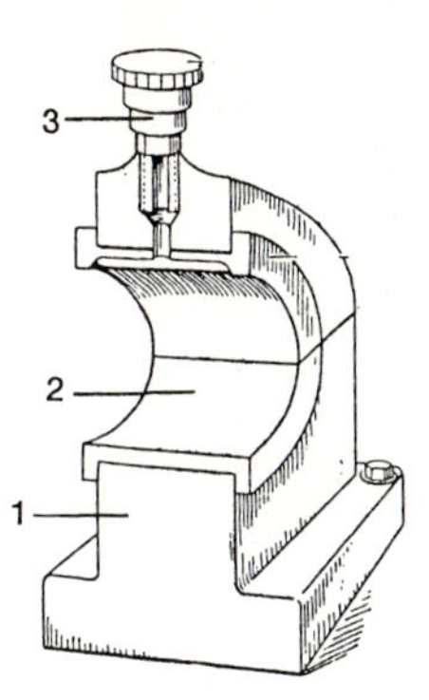

Babbit
Lagermetall

1 **bearing block (section)** Lagerbock (Schnitt)
2 **babbit** Lagermetall
3 **lubricator** Schmiernippel

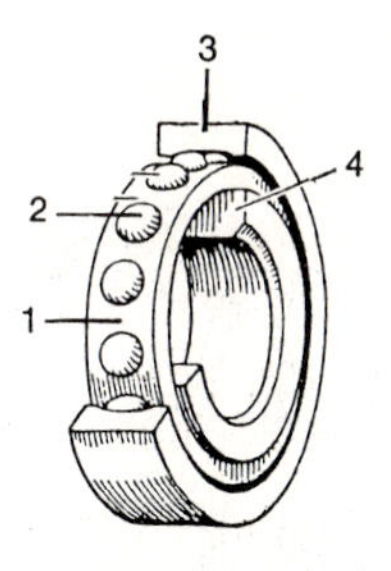

Ball bearing
Kugellager

1 **cage** Käfig
2 **ball** Kugel
3 **outer race** Außenring
4 **inner race** Innenring

B

babbit (metal) Lagermetall, Babbitmetall
back Rücken, Rückseite
back elevation Hinteransicht, Rückansicht
back nut Gegenmutter, Kontermutter
back up, to unterstützen
backfire grid Flammenrückschlagsieb
backfiring Zurückschlagen der Verbrennungsflamme (z. B. in den Vergaser)
background Hintergrund
backlash toter Gang, Lose, Flankenspiel
back-to-back connection gegensinnig gepolte Schaltung
back view Rückansicht, Hinteransicht
backwards rückwärts
baffle Prallblech, Resonanzwand, Schallwand
bag Tasche, Beutel, Tüte
baggage Gepäck
baggy bauchig, sackartig
bakelite Bakelit
balance Ausgleich, Ausgewogenheit
balance, to (Brücke) abgleichen
balanced amplifier Gegentaktverstärker
balancer Ausgleichvorrichtung
balancing Ausgleich ...
bale Ballen, Packen
ball bearing Kugellager
ball joint Kugelgelenk
ball journal Kugelzapfen
ball lightning Kugelblitz
ball-shaped kugelförmig
balloon Ballon
banana jack Bananenbuchse
banana plug Bananenstecker
band brake Bandbremse
bandage Bandage
bandage, to bandagieren
bang Knall
bang, to knallen
bank of capacitors Kondensatorenbatterie
bar Stange (Metall), Profilstahl
bar, to abriegeln

barbed hook Widerhaken
barbed wire Stacheldraht
bare blank, nackt (Draht)
bare, to Isolation entfernen, blosslegen
bare wire Blankdraht
barge Schleppkahn, Schute
barred versperrt, vergittert
barrel Tonne, Fass, Trommel
barrel-shaped walzenförmig
barrier Schranke, Grenze, Schutz
base Grundfläche, Grundplatte, Fundament
base line Nullinie, Nullachse, Grundlinie
basement Untergeschoss, Fundament
basic Grund . . . , fundamental
basic pig iron basisches Roheisen
basic size Sollmass
batch Menge, Schub, Masse, Haufen
bath Bad (Schmelze)
batten Latte
battery Batterie
bay Fach, Rahmen, Gestell, Bucht
bayonet cap Bajonettsockel
bayonet socket Bajonettfassung
bauxite Bauxit
beacon Bake, Leitstrahlsender
beacon fire Leuchtfeuer
bead, to falzen
bead Perle
bead weld Wulstnaht
beam Balken, Strahl, Träger
beam, to mit Richtstrahler senden
bear, to tragen, stützen
bearer wire Tragseil, Tragdraht
bearing Lager
beat Takt, Schlag
beat, to schlagen, klopfen
beat, in im Takt
become airborne, to abheben
bed Bett, Lage, Schicht
bed, to betten, einbetten
behave, to sich verhalten
behaviour Verhalten
bell Glocke, Klingel
bell-shaped glockenförmig
bell signal Glocken-, Klingelsignal
bellied bauchig
belling Aufweitung (eines Rohres)
bellows Blasebalg
below, to go below a value einen Wert unterschreiten

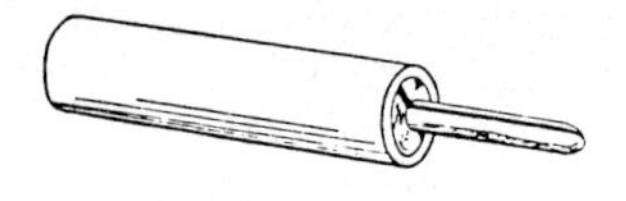

Banana plug
Bananenstecker

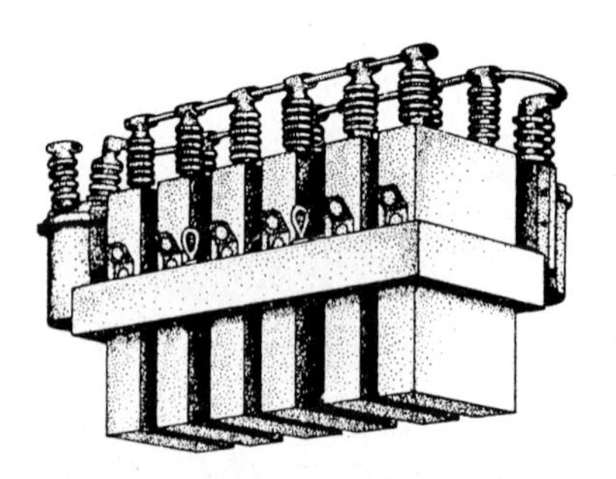

Bank of capacitors
Kondensatorenbatterie (-gruppe)

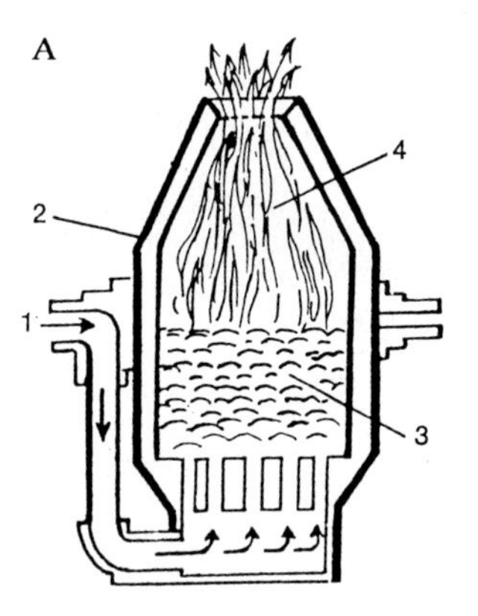

Bessemer converter
Bessemerbirne

A **Process principle**
Arbeitsprinzip

1 **compressed air inlet** Druckluftzuführung
2 **pear-shaped brick-lined steel shell** birnenförmiges, mit Ziegel ausgekleidetes Stahlgefäß
3 **molten pig iron** geschmolzenes Roheisen
4 **escaping CO_2 and SO_2** entweichendes CO_2 und SO_2

B **Tilted Bessemer converter pouring the converted steel**

Gekippte Bessemer Birne beim Abguß des gewonnenen Stahls

belt Riemen
belt conveyor Förderband, Fliessband, laufendes Band
belt drive Riemenantrieb
belt pulley Riemenscheibe
belted insulation cable Gürtelkabel
belted motor drive Antrieb durch Motor über Riemen
bench (work bench) Werkbank, Arbeitstisch
bench lathe Mechanikerdrehbank
bench mounting Tischaufbau
bend, to biegen, beugen, krümmen
bend off, to abbiegen
bend out, to nach aussen biegen, ausbiegen
bend over, to umbiegen
bend to breaking umknicken
bend up, to aufbiegen
bend Biegung, Abbiegung, Krümmung
bendable (ver)biegbar
bending Biegung, Verbiegung
bending endurance Dauerbiegefestigkeit
bending moment Biegemoment
bending strength Biegefestigkeit
bending stress Biegespannung
bendix driver starter Schraubtriebanlasser
benefit, to Nutzen ziehen
bent lever Winkelhebel
bent pipe gebogenes Rohr
benzene Benzol
benzin(e) Waschbenzin
benzol(e) Rohbenzol
Berlin blue Preussischblau
berth Koje, Bett
beryllium Beryllium
Bessemer converter Bessemerbirne
Bessemer steel Bessemerstahl
beta rays Betastrahlen
betatron Betatron

betray, to verraten
better-than-average überdurchschnittlich
bevel, to abkanten, abschrägen
bevel Schrägkante
bevel driving pinion Kegelantriebsritzel
bevel gear Kegel(zahn)rad
bevel gear tooth system Kegelradverzahnung
bevel joint Schrägverbindung
bevelled abgeschrägt, schräg
bevelling Abschrägung
beware! Obacht!
b/f (brought forward) Übertrag
biannual Halbjahres . . . , halbjährlich
bias, to vormagnetisieren, vorspannen
bias Vormagnetisierungsstrom, Vorspannung
bias voltage Vorspannung
biatomic doppelatomig, zweiatomig
biax(ial) zweiachsig
bid Angebot, Offerte
bidder Bieter
bidirectional doppelseitig gerichtet, doppelsinnig
bifilar doppelfädig, bifilar
bifilar winding Bifilarwicklung
bilateral zweiseitig, doppelseitig, bilateral
bilge Bilge, Schiffsbauch
bilge and fire pump Lenz- und Feuerlöschpumpe
bill Rechnung
bill of cost Kostenrechnung
billet (Stahl-)Knüppel
bimetal Bimetall
bimotored zweimotorig
bin Behälter
binary binär
binary digit Binärziffer
binary notation Binärschreibweise
bind, to (an)binden, schnüren
binder Bindemittel
binding Bindung, Schnürung
binding upon the vendor bindend für den Verkäufer
binnacle Kompasshaus
binocular zweiäugig, binokular
binominal binomisch
binomial series Binomialreihe
biophysics Biophysik
biphase zweiphasig

Bevel pinion (bevel gear)
Kegelzahnrad

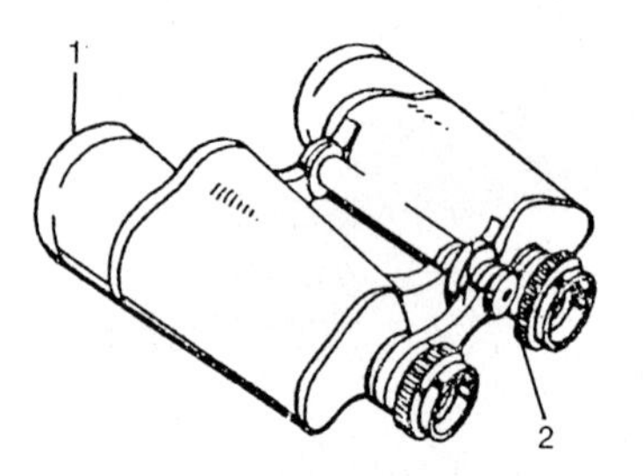

Binoculars
Fernglas

1 **lens** Objektiv
2 **focussing adjustment ring** Okularring

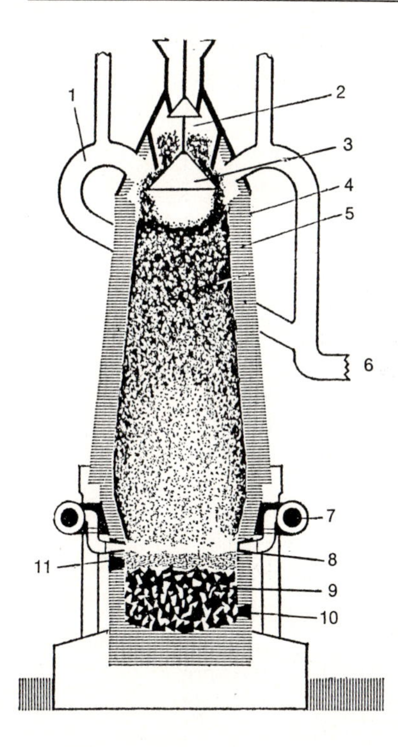

Blast furnace (section)
Hochofen (Schnitt)

1 **throat gas pipe** Gichtgasleitung
2 **charge (iron ore, coke, limestone)** Einsatz (Eisenerz, Koks, Kalkstein)
3 **bell** Gichtverschluß
4 **stack** Ofenschacht
5 **firebrick lining** Schamottestein-Auskleidung
6 **dust collection** Staubabzug
7 **bustle pipe** Windringleitung
8 **Tuyères** Düsenstock
9 **melt** Schmelze
10 **tap hole** Abstichöffnung
11 **slag notch** Schlackenloch

biphase connection Zweiphasenschaltung
biplug Doppelstecker
bipolar doppelpolig
bird's eye view Vogelperspektive
bisect, to halbieren (Fläche)
bisecting (bisection) Halbierung, Zweiteilung
bismuth Wismut
bismuthiferous wismuthaltig
bistable bistabil
bit Nachrichteneinheit
bits/s Bits pro Sekunde (NE/s)
bitumen Bitumen, Erdpech, Asphalt
bivalence Zweiwertigkeit
bivalent zweiwertig
B/L (bill of loading) Konnossement
black enamelled schwarz feuerlackiert
black-finishing Brünieren
black finish mild steel Schwarzblech
black lead Graphit
black-out Stromausfall, Verdunkelung
blacken, to (ein)schwärzen
blacksmith Schmied
blade Blatt, Flügel, Klinge, Messer (bei Messerschaltern)
blank leer, weiss, unbeschrieben
blanket Mantel, Brutzone
blast Stoss (Wind)
blast furnace Hochofen
blast pressure Gebläsedruck
bleary verwaschen (Bild)
bled steam Abzapfdampf
bleeding Dampfanzapfung
bleep, to piepsen (Funksignale)
blend, to verschmelzen
blend Mischung
blind, to blenden
blind Lichtblende, Lichtvorhang
blind approach beacon Blindlandefeuer
blind flange Blindflansch
blind flying Blindfliegen, Blindflug
blinding Blendung
blink, to blinken
blinker signal Blinkzeichen
blip Echosignal (Radar)
blister, to Blasen bilden (Lack)
blister Blase
block, to blockieren, blocken
block Block, Klotz

Blast furnace plant
Hochofenanlage

1 **charging crane** Chargierkran
2 **charging bin** Beschickungskübel
3 **charging platform** Gichtbühne
4 **throat gas purifier** Gichtgasreiniger
5 **furnace framework** Ofengerüst
6 **hot air pipe** Heißwindleitung
7 **tapping spout** Abgußrinne
8 **bucket taking molten pig-iron** Kübel zur Aufnahme des geschmolzenen Roheisens

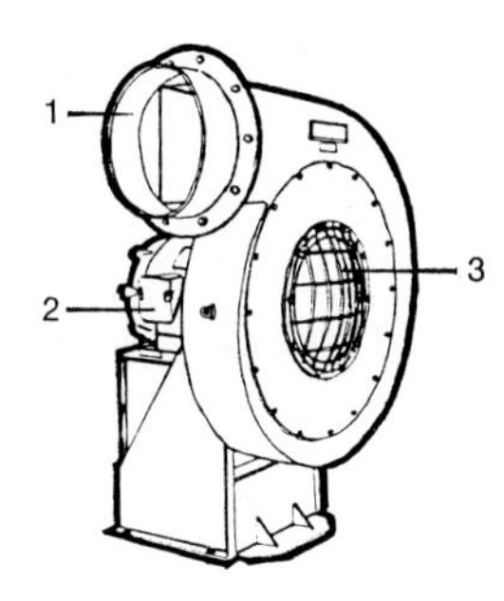

Blower
Gebläse

1 **air exhaust** Luftausstoß
2 **driving motor** Antriebsmotor
3 **air intake** Lufteintritt

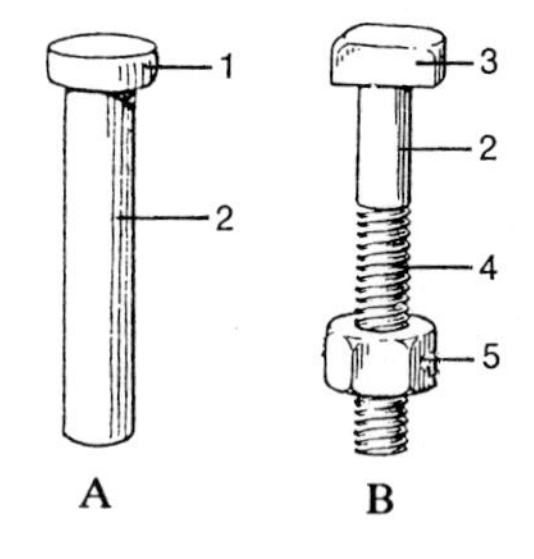

Bolts
Bolzen

A **Plain bolt**
Einfacher Bolzen

1 **round head** Rundkopf
2 **shank** Schaft

B **Threaded bolt**
Gewindebolzen

3 **square head** Vierkantkopf
4 **thread** Gewinde
5 **nut** Mutter

block diagram Blockschaltbild
blockage Blockierung
blocking Blockieren, Sperrung, Verriegelung
block voltage Sperrspannung
blow, to durchbrennen (Sicherung), blasen
blow in, to einblasen
blow out, to ausblasen
blow up, to explodieren
blow Hieb, Schlag
blow magnet Blasmagnet
blower Lüfter, Gebläse
blower cooled engine durch Gebläse gekühlter Motor
blower cooling Gebläsekühlung
blowing engine Gebläsemaschine
blueprint (Blau-)Pause
blunder grober Fehler
blunt stumpf
blur, to trüben, verwischen, verschwimmen
blurred verwischt, verschwommen, unscharf
B.M.E. (Bachelor of Mechanical Engineering) etwa: Maschinenbauingenieur
board Tafel, Brett, Diele
boarding Täfelung, Verkleidung
bobbin Beutel
body plan Spantenriss (Schiff)
bogie, bogy Drehgestell (Wagen und Anhänger)
boil, to sieden, kochen
boiled oil Leinölfirnis
boiler Kessel, Boiler, Heizkessel
boiler feed pump Kesselspeisepumpe
boiling-water reactor Siedewasserreaktor
bold-face (type) Fettdruck
bold-faced fettgedruckt
bolt, to verriegeln, abriegeln
bolt together, to verschrauben
bolt Riegel, Schraube (mit Mutter)
bolted connection Schraubverbindung
bond, to (ver)binden
bond together, to zusammenkitten
bond Verbindungsstelle, Bindemittel
bonding Bindung
bonnet Motorhaube
booklet Broschüre
Boole's function Boolesche Funktion

boost, to erhöhen (Spannung, Druck)
boost charge Schnelladung
boost pressure Ladedruck (bei Motoren)
booster Spannungserhöher, Zusatzverstärker
booth Stand, Zelle
border, to umranden
border Kante, Grenze
bore, to bohren
bore Bohrung, Ausbohrung
bore bit Bohrspitze
borer Bohrer
boron Bor
bottle, to abfüllen (in Flaschen)
bottle Flasche
bottom Boden, Grund, Unterteil
bottom face Grundfläche
bottom side Unterseite
bottom view Untenansicht
bounce, to springen
bounce Prellen (Schalter), Rückprall, plötzlicher Sprung
bound Grenze
boundary Abgrenzung, Grenze
boundary layer Grenzschicht
boundary line Begrenzungslinie
boundary region Grenzgebiet
bounded by eingegrenzt von
boundless unbegrenzt
bow Bügel, Bug
bow jet system (or: rudder) Bugstrahlruder
bowl Schale, Napf, Pfanne
box, to in Kisten verpacken
box Gehäuse, Schachtel, Kiste, Kasten
boxed verpackt
boxing Einpacken in Kisten
brace, to versteifen, abstützen, verstreben
brace geschweifte Klammer (Math.) Strebe, Abstützung, Versteifung
braced versteift, verstrebt
bracket eckige Klammer (Math.) Stütze, Konsole, Befestigungsarm
bracket, to remove the die Klammer auflösen
bracket term Ausdruck in eckigen Klammern
braid, to umflechten, umklöppeln
braid Umflechtung, Umspinnung, Geflecht

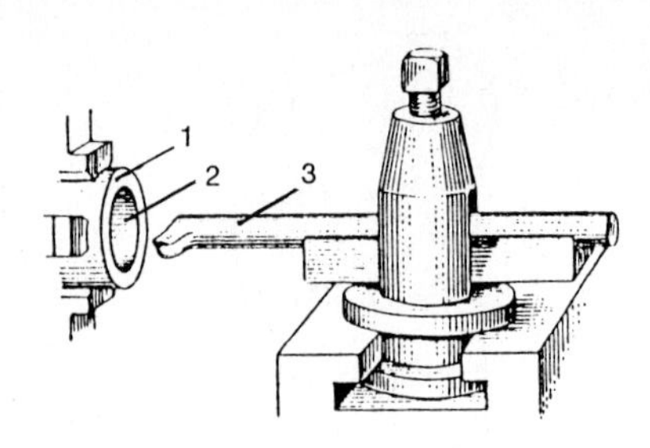

Boring set-up
Einrichtung zum Aufbohren

1 **turning workpiece** sich drehendes Werkstück
2 **bore** Bohrung
3 **boring tool** Bohrwerkzeug

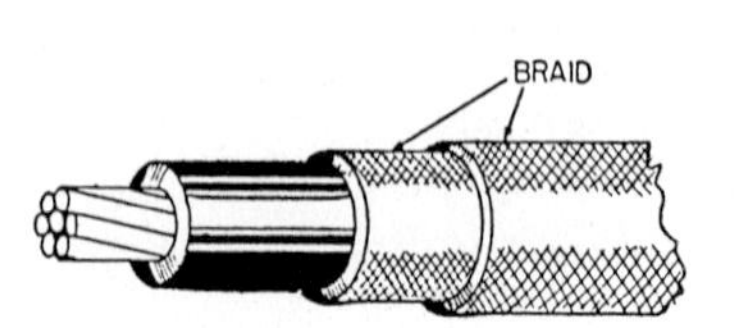

Fibrous braid covering of a cable
Textil-Umflechtung eines Kabels

braid wire Flechtdraht
brake, to (ab)bremsen
brake, to apply the (ab)bremsen
brake Bremse
brake action Bremswirkung
brake lifting magnet Bremslüftmagnet
brake light Bremslicht
brake shoe Bremsbacke
braking Bremsen, Bremsung, Hemmung
braking by short-circuiting armature Ankerkurzschlussbremsung
braking torque Bremsmoment
branch, to verzweigen
branch off, to abzweigen
branch Verzweigung, Zweig
branch box Abzweigdose
branch circuit Zweigstromkreis
branch distribution centre Leitungsverteiler
branch current Zweigstrom, Teilstrom
branch joint Abzweigpunkt
branch office Zweigstelle
branch pipe Abzweigrohr
branch tee of a pipe Rohrabzweigstück
branched verzweigt, gegabelt
brand Handels(marke) von Waren
branding iron Brennstempel
brand-new nagelneu

Brake system of a motor car
Auto-Bremssystem

1 **brake drum** Bremstrommel
2 **brake lining** Bremsbelag
3 **brake shoe** Bremsbacke
4 **bleeder screw** (Bremsflüssigkeits-) Ablaßschraube
5 **adjustment** Einstellung

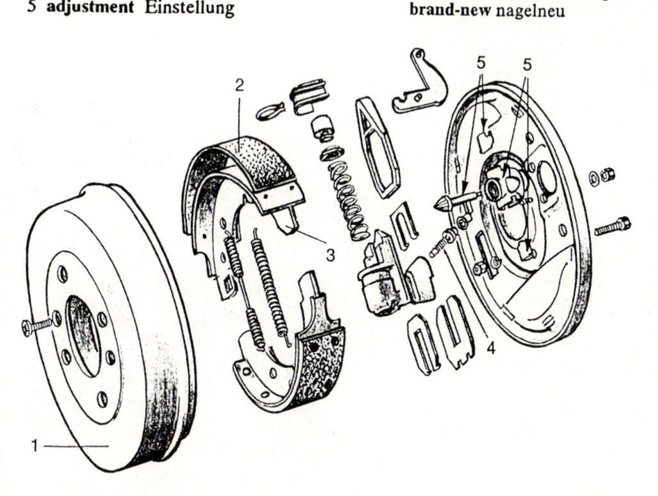

brass Messing
brass box Messingbüchse
brass case Messinggehäuse
brasses (Pl.) Lagerschalen
brass sheet Messingblech
brass plate Messingblech, -platte
brass wire Messingdraht
brass work Messingbeschläge (Pl.)
Braun tube Braunsche Röhre, Kathodenstrahlröhre
braze, to hartlöten, verlöten
brazier Klempner, Kupferschmied
braze-welding Schweisslöten
brazing Hartlöten
brazing solder Hartlot
breadth Breite, Weite
break, to brechen, unterbrechen, zerschlagen
break asunder, to auseinanderbrechen
break down, to zusammenbrechen, abbrechen
break open, to aufbrechen, aufplatzen
break Unterbrechung, Bruch, Bruchlinie
break contact Öffnungskontakt, Ruhekontakt
break-even point Unkostendeckungspunkt
break of current Stromunterbrechung
break-off Abbruch
break period Öffnungsdauer
break-proof bruchfest
break stress Bruchspannung
breakability Zerbrechlichkeit
breakable zerbrechlich
breakable connection lösbare Verbindung
breakdown Panne, Versagen, Zusammenbruch, Aufgliederung, Zerlegung (Chem.)
breakdown of service Betriebsstörung
breakdown voltage Zündspannung, Funkenstrecke
breaker Unterbrecher
breaking Zerbrechen, Knickung
breaking of a code Entschlüsselung
breaking capacity Ausschaltvermögen, Ausschaltleistung
breaking elongation Bruchdehnung
breaking limit (or: point) Zerreissgrenze
break strength Bruchfestigkeit Zerreissfestigkeit

A breakdown of service?
Eine Betriebsstörung?

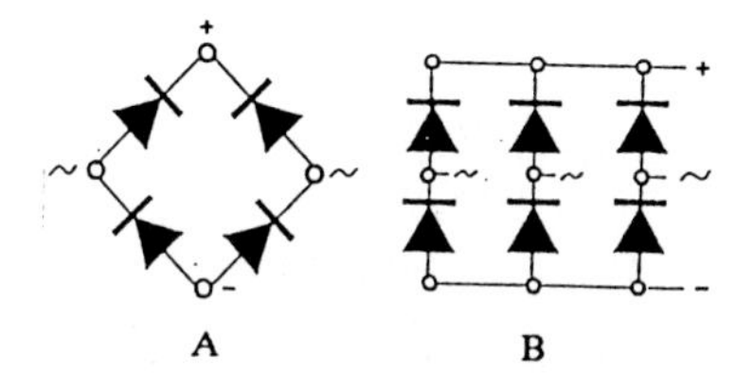

Bridge rectifier circuits
Brücken-Gleichrichterschaltungen
A **single-phase circuit** Einphasen-Schaltung
B **three-phase circuit** Dreiphasen-Schaltung

break-out Durchbruch
breakover voltage Kippspannung
breakup Zerlegung
breathe, to atmen
breather Entlüfter, Entlüftungsöffnung
breeder reactor Brüter, Brutreaktor
breeding Brüten
breeding ratio Konversionsgrad, Umwandlungsgrad
breeze Guss
breeze oven Gussofen
brick Ziegel, Mauerstein
brick layer Maurer
brick lining Ausmauerung
bricking Mauerwerk
bridge, to überbrücken
bridge across, to durchschleifen, einschleifen
bridge Brücke
bridge amplifier Messbrückenverstärker
bridge balance Brückenabgleich
bridge circuit Brückenschaltung
bridge controlled machinery von der Brücke gefahrene Maschinenanlage
bridge rectifier circuit Graetz-Schaltung, Brückengleichrichter
bridge type diplexer Brückenweiche
bridging Überbrückung
Brigg's logarithm Briggscher Logarithmus, dekadischer Logarithmus
bright hell, lichtstark, glänzend
bright adaption Hellanpassung
bright chromium plating Glanzverchromung
bright-anneal, to blankglühen
bright-polished glanzpoliert
bright wire blanker Draht
brighten up, to aufhellen, erhellen
brightener Glanzmittel, Glanzzusatz
brightness Helligkeit, Glanz
brilliance Wiedergabebrillianz, Glanz, Helligkeit
brilliant strahlend, leuchtend
brilliant varnish Glanzlack
brimstone yellow schwefelgelb
brine Sole, Salzwasser
Brinell hardness Brinellhärte
bring about, to ermöglichen, bewirken, bewerkstelligen

bring into position, to in Stellung bringen
brining Versalzen, Versalzung
brittle spröde, brüchig
brittleness Sprödigkeit, Brüchigkeit
broach, to räumen
broach Räumnadel, Reibahle
broaching Räumen
broad breit, weit
broadband Breitband
broad-base tower Hochspannungsmast mit zwei Füssen
broadcast Rundfunk, Funksendung
broadcast receiver Rundfunkempfänger
broadcast transmitter Rundfunksender, -station
broadcasting Rundfunkübertragung
brochure Druckschrift, Broschüre
broil, to (am.) auf dem Rost braten
broker Makler, Agent
bromic Brom . . .
bromide Bromid
bronze Bronze
broom Besen
bruise, to quetschen, einbeulen
bruise Quetschung
brush Bürste
brush holder Bürstenhalter
brush lifter Bürstenabheber
brushing discharge Büschelentladung
B.S.G. (British Standard Gauge) Britische Normallehre
B.Th.U. (British Thermal Unit) Britische Wärmeeinheit
bubble, to brodeln, Blasen bilden
bubble Blase
bubbling Blasenbildung
Buchholz relay Buchholz-Relais
buck, to in Lauge einweichen, entgegenwirken
buck Lauge
bucket Eimer
bucket wheel dredge Schaufelradbagger
bucking voltage Kompensationsspannung
buckle, to krümmen
buckle Knick, Spange, Schnalle
buckled krumm, verbeult
buckling Verbeulung, Verkrümmung

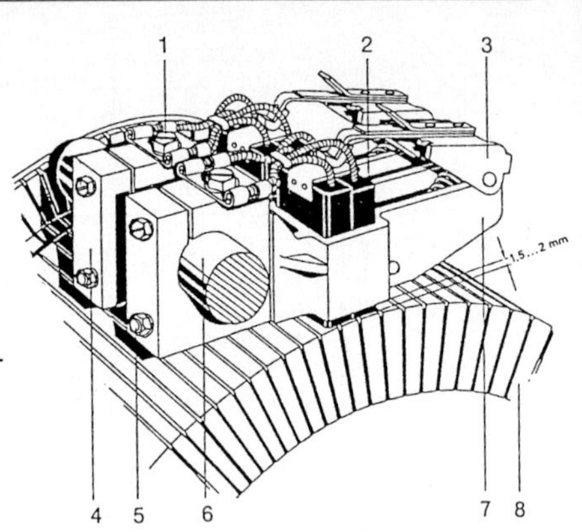

Brush gear
Bürstenhaltevorrichtung

1 **brush connection** Bürstenanschluß
2 **coal brush** Kohlebürste
3 **press finger** Druckfinger
4 **clamp** Klemmstück
5 **fixing screw** Befestigungsschraube
6 **holder bolt** Haltebolzen
7 **brush holder** Bürstenhalter
8 **commutator** Kollector

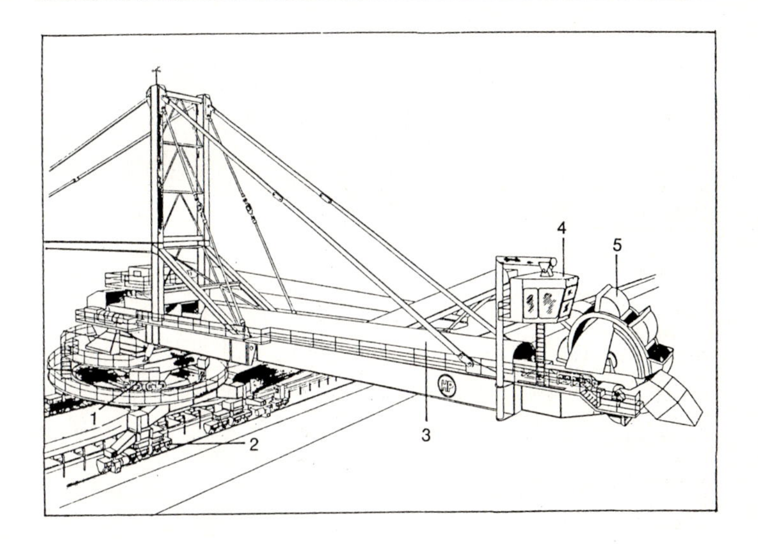

Bucket-wheel dredger
Schaufelrad-Bagger

1 **turn table** Drehgestell
2 **travel truck** Fahrgestell
3 **covered conveyor belt** abgedecktes Förderband
4 **operator cab** Steuerkabine
5 **bucket wheel** Schaufelrad

buckram Steifleinen
budget year Rechnungsjahr, Haushaltsjahr
buff, to polieren, glanzschleifen
buff Polierscheibe
buffer Puffer
buffering Puffern, Zwischenspeichern
buffer battery Pufferbatterie
bug, to abhören
bug Abhörvorrichtung, unsteter Fehler, zeitweises Versagen
buggy Lore
build, to bauen, herstellen, fertigen, erzeugen
build-up Aufbau, Zuwachs
builder Erbauer, Bauherr
building Bauwerk, Aufbau, Gebäude
built-in eingebaut
built-in unit Einbauteil, -aggregat
bulb Birne, Röhrenkolben
bulbous bow Wulstbug
bulge, to ausbeulen, ausbuchten
bulge Beule
bulge of the earth Erdkrümmung
bulk Masse, Umfang, Grösse, Hauptteil
bulk cargo Schüttgut, Massengut
bulkhead Schott
bulk manufacture Massenfertigung
bulk oil circuit-breaker Kesselölschalter
bulkiness Sperrigkeit
bulky sperrig, umfangreich
bull, to bullern, ballern
bulletin Tagesbericht
bull's eye Bull(en)auge
bulwark Schanzkleid
bumper Stoßstange
bunch, to bündeln, zusammenballen
bunch Bündel, Bund
bunched gebündelt, geballt
bundle, to bündeln, binden
bundle Bündel
bungle, to verpfuschen
Bunsen burner Bunsenbrenner
buoy Boje, Bake, Tonne
buoyage Betonnung
buoyancy Auftrieb, Schwimmfähigkeit
buoyant schwimmfähig

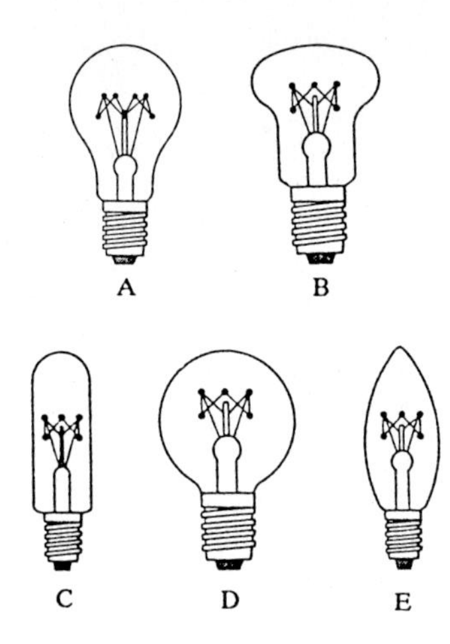

Forms of light bulbs
Glühbirnen-Formen

A **pear shape** Birnenform
B **mushroom shape** Pilzform
C **tube shape** Röhrenform
D **ball shape** Kugelform
E **candle shape** Kerzenform

Bushing-type insulators on a transformer lid
Durchführungsisolatoren an einem Transformatorendeckel

burden Bürde
burette Bürette, Messglas
burglar alarm Einbruchswarnanlage
burglar-proof einbruchssicher, diebessicher
burial ground Abfallgrube, Abfallager
buried erdverlegt, unter Putz, versenkt, eingebaut
buried cable Erdkabel
buried wiring Leitungsverlegung unter Putz
burn, to brennen, verbrennen
burn off, to abbrennen
burn out (or: through), to durchbrennen
burner Brenner
burner nozzle Brennerdüse
burner tip Brennerkopf
burning Verbrennung, Verbrennen
burning of contact Kontaktabbrand
burnish, to glanzschleifen, presspolieren
burnisher Polierstahl, Poliermaschine
burnishing Presspolierung
burr, to entgraten, abgraten
burr Gussnaht, Grat
burring machine Abgratmaschine
burst, to bersten, aufreissen, platzen
burst Impuls, Stoss
bursting Platzen, Bersten, Zerspringen
bus (bar) Sammelschiene, Stromschiene
bush Buchse, Lagerbuchse
bushing Durchführung, Ausfütterung
bushing guide Führungslager
bushing transformer Durchsteckwandler
bushing type capacitor Durchführungskondensator
busy besetzt, belegt (Tel.)
busy signal Besetztzeichen
butane Butan
butt Stoss, Verbindungsstoss, Stumpfstoss
butt-joint, to stumpf aneinanderfügen
butt-weld, to stumpfschweissen
butt-joint Stumpfverbindung, Stossnaht

butt-riveting Laschennietung
butt weld Stumpfschweissung
butterfly nut Flügelmutter
Butterworth system Butterworth-system
button Druckknopf, Taste, Knopf
butylene Butylen
buyer Einkäufer, Käufer
buzz, to summen
buzzer Summer
buzzing Summen
BWR (boiling-water reactor) Siedewasser-Reaktor
by-pass, to überbrücken, umgehen
by-pass Beipass, Nebenleitung Umgehung
by-pass engine Mantelstrom-triebwerk
by-passing Überbrückung, Umgehung
by-path Parallelweg, Nebenweg
by-product Nebenerzeugnis, -produkt
by-reaction Nebenwirkung
by virtue of kraft
by way of mittels
by way of expedient behelfsmässig

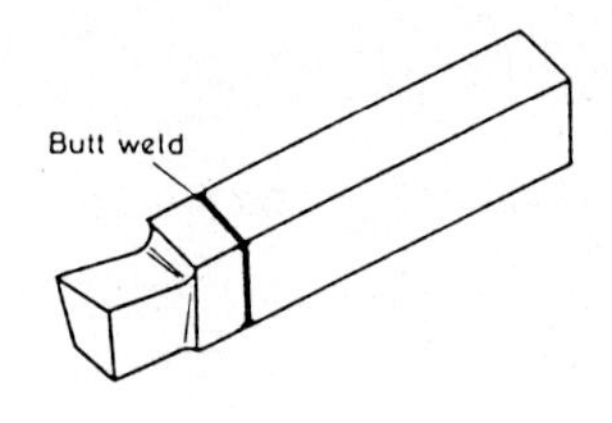

Butt-welded lathe tool
Stumpfgeschweißter Drehmeißel

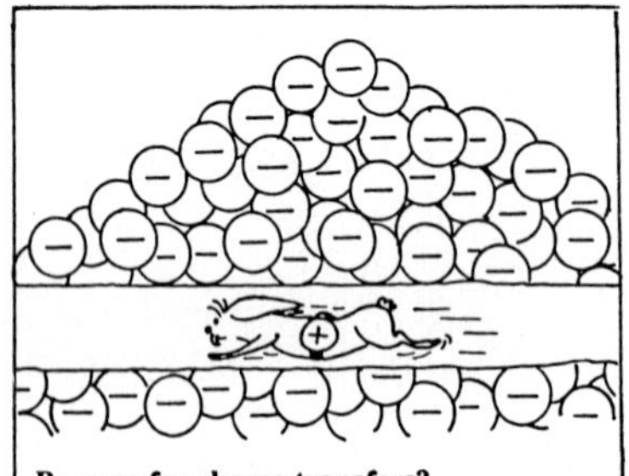

By-pass for charge transfers?
Nebenleitung für Ladungsverschiebungen?

C

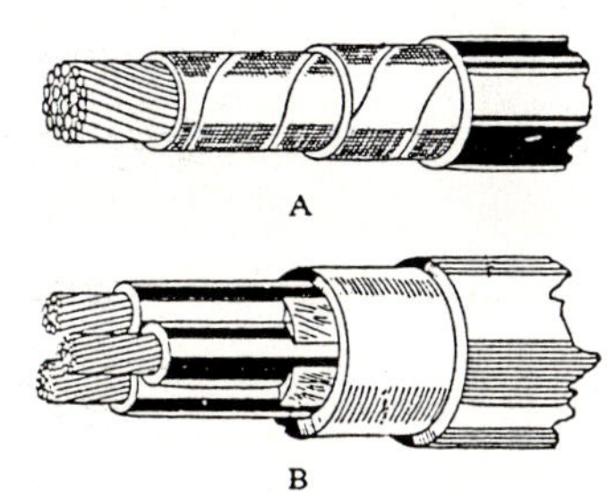

Cables
Kabel

A **Single-core cable** Einleiter-Kabel

B **3-core cable** 3-Leiter-Kabel

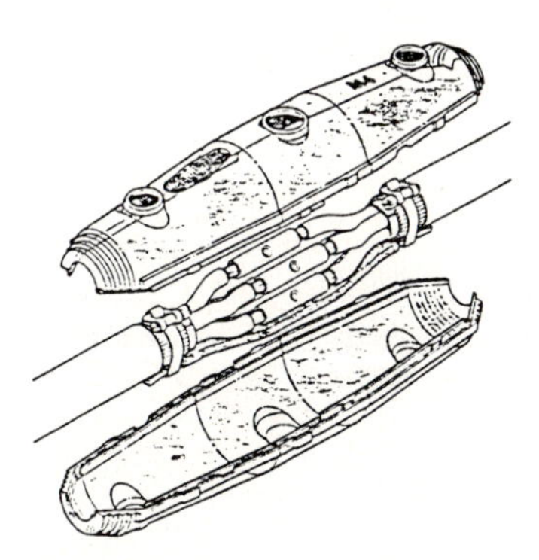

Cable joint
Kabelverbindung

c-shaped washer geschlitzter Federring
c to c (centre to centre) von Mitte zu Mitte
c & f (cost and freight) Kosten und Fracht
c & i (cost and insurance) Kosten und Versicherung
cabin Zelle, Häuschen, Kabine
cabinet Schrank(kästchen), Werkzeugschrank
cabinet drawing Möbelzeichnung
cabinetmaker Tischler, Schreiner
cable, to kabeln, verkabeln, mit Kabel verbinden
cable-connect, to anschliessen
cable Kabel, Drahtseil, Drahtnachricht, Kabeltrosse, Telegramm
cable armour(ing) Kabelbewehrung
cablegram Kabelgramm
cable bearer Kabelträger, -halter
cable box Kabelkasten
cable bracket Kabelträger, -konsole
cable break Kabelbruch
cable chute Kabelschacht
cable clamp Kabelschelle
cable fittings Kabelgarnitur
cable screw Kabelklemmschraube
cable head Kabelabschluss, Kabelkopf
cable jacket Kabelmantel
cable joint Kabellötstelle, Kabelverbindungsstelle
cable laying Kabellegung
cable layout Kabelplan
cable pothead Kabelabschlussmuffe
cable rack Kabelgestell
cable reel Kabeltrommel
cable runway Kabelschacht
cable compound Kabelvergussmasse
cable shelf Kabelrost, -gestell
cable strand Kabellitze
cable suspension Kabelaufhängung
cable terminal Kabelanschluss
cable winch Kabelwinde
cabling Verkabelung, Kabellegung
caboose Kombüse

cadet Kadett, Zögling, Offiziersanwärter
cadmium Kadmium
caesium Zäsium
cage Käfig
cage rotor Käfigläufer
cage winding Käfigwicklung
calcareous kalkhaltig
calcination Kalzinierung
calcium Kalzium
calcium carbide Kalziumkarbid
calculable berechenbar
calculate, to rechnen, kalkulieren
calculating disc Rechenscheibe
calculating roughly überschlägig berechnet
calculation Rechnung
calculator Rechenmaschine, Kalkulator
calculus Rechnung (bes. Infinitesimalrechnung)
caldron grosser Kessel
calendar Kalender
caliber (calibre) Kaliber, Rohrweite
calibrate, to eichen
calibrating potentiometer Eichpotentiometer
calibration Eichung
calibration instrument Eichgerät, -instrument
calibrator Eichgerät
caliper, to mit Tastlehre messen
caliper rule Schublehre
caliper rules Tastzirkel
call (up), to anrufen
call signal (or: sign) Rufzeichen
caller Besucher
calm down, to (Wind) abflauen
calm Windstille
caloric kalorisch
calorie (calory) Kalorie
calorific wärmeerzeugend
calorific power Heizwert, -kraft
calorimeter Kalorimeter, Wärmemesser
calorimetry Wärmemengenmessung
calotte Kalotte
cam, to mit Nocken versehen
cam Nocke(n), Steuerkurve
cam controlled (operated) nockenbetätigt
cam controller Nockenfahrschalter
cam disc Nockenscheibe

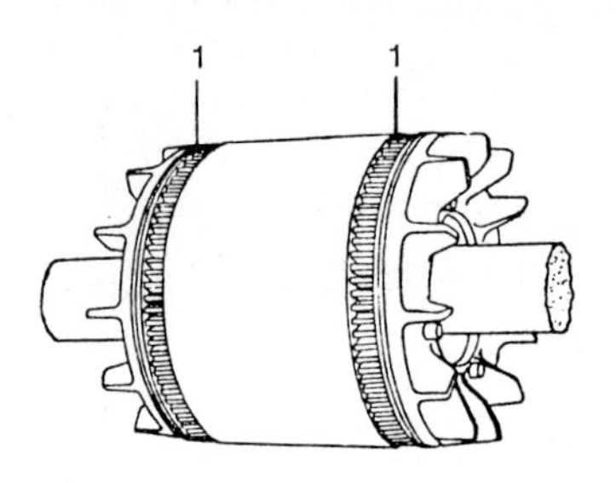

Cage rotor
Käfigläufer

1 **short-circuited conductor bars** kurzgeschlossene Leiterstäbe

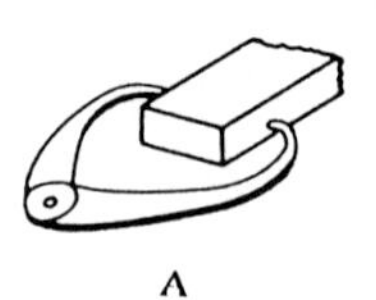

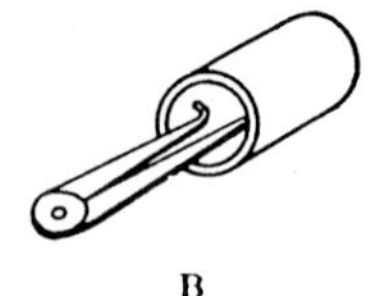

Calipers
Taster, Tastlehre

A **Outside calipers** Außentaster

B **Inside calipers** Innentaster

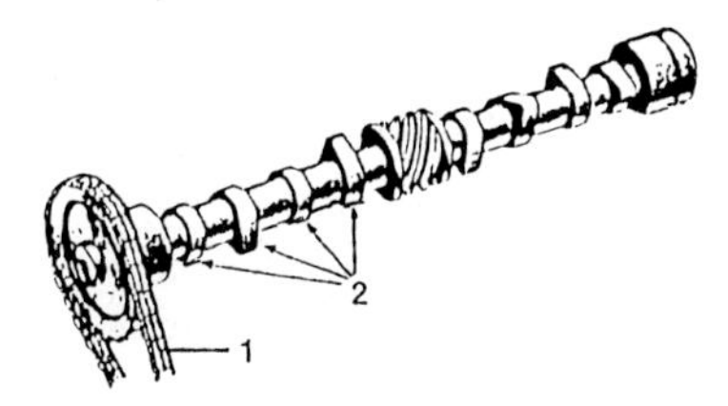

Camshaft
Nockenwelle

1 **drive** Antrieb
2 **cams** Nocken

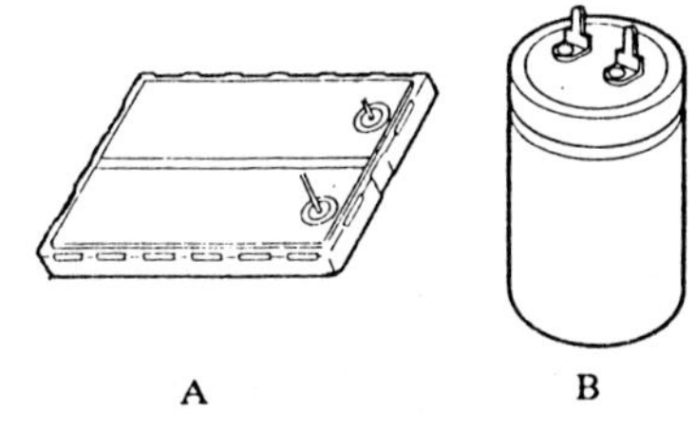

Capacitors
Kondensatoren

A **Ceramic low-profile type** Keramik-Flachausführung

B **Electrolytic type** Elektrolyt-Ausführung

cam drive Nockenantrieb
cam follower Nockenstössel
cam switch Nockenschalter
camber Wölbung, Ausbauchung
cambric Batist
cambric tape Band aus Oeltuch
camouflage, to tarnen
camouflage Tarnung
camouflage material Tarnstoff
camshaft Nockenwelle
can Becher, Dose, Kanne
canal (künstlicher) Kanal
cancel, to streichen, annullieren, für ungültig erklären
cancel key Löschtaste
cancellation Kündigung, Aufhebung
cancelling Löschung, Streichung
candela Candela (Lichtstärkeeinheit)
candle Kerze
canned assembly Becherkonstruktion
cannibalisation Ausschlachtung (Teile)
cannibalise, to ausschlachten (Teile)
canning Tonaufzeichnung, Umhüllung
cant, to verkanten
cant Verkantung, Schräge
CANTAT Transatlantic cable laid 1961; 80 circuits, 3kc-spaced
canted shot verkantete Aufnahme
cantilever freitragender Ausleger
cantilever beam Auslegerbalken, Freiträger
caolin Kaolin, Porzellanerde
CANUSE Canada–US Eastern Interconnection (ein Verbundnetz)
canvas Segeltuch
canyon Tunnel (für Brennstoffstäbe)
caoutchouc Kautschuk
cap, to sockeln
cap Sockel, Kappe, Haube
cap nut Hutmutter
capability Fähigkeit
capacitance Kapazität (Farad)
capacitive kapazitiv
capacitor Kondensator
capacity Fähigkeit, Leistungsfähigkeit, Fassungsvermögen
capillarity Kapillarität
capillary kapillar

capitalise, to gross schreiben
capstan Spill, Ankerwinde
capsule Kapsel
caption Überschrift, Titel, Bildunterschrift
capture, to fangen, einfangen
carat Karat
carbide Karbid
carbon Kohlenstoff
carbon compound Kohlenstoffverbindung
carbon dust Kohlenstaub
carbon filament Kohlenfaden
carbon steel Kohlenstoffstahl
carbonaceous kohlenstoffhaltig
carbonate Karbonat
carbonic acid Kohlensäure
carbonic oxide Kohlenoxyd
carbonisation Verkohlung
carbonise, to verkohlen
carborundum Karborund
carboy Korbflasche, Ballon
carburet, to vergasen (Benzin)
carburettor Vergaser
carburettor adjustment Vergasereinstellung
carburettor engine Vergasermotor
carcassing Rohbau
card Karte, Lochkarte
cardboard Karton
cardboard box Pappschachtel, -karton
Cardan drive Kardanantrieb
Cardan joint Kardangelenk
Cardan shaft Kardanwelle
Cardew voltmeter Hitzdrahtvoltmeter
cardinal number Grundzahl
cardinal point Hauptpunkt (einer Skala)
cardiogram Kardiogramm
cardiograph Kardiograph
cardioid herzförmig
carelessness Unachtsamkeit
cargo Fracht
cargo and passenger vessel Fracht- und Fahrgastschiff
Carnot working cycle Carnotscher Kreisprozess
carpenter Zimmermann
carpenter bench Hobelbank
carriage Wagen, Frachtgebühren
carriage body Karosserie
carriage paid frachtfrei
carrier Träger, Mitnehmer, Trägermaterial

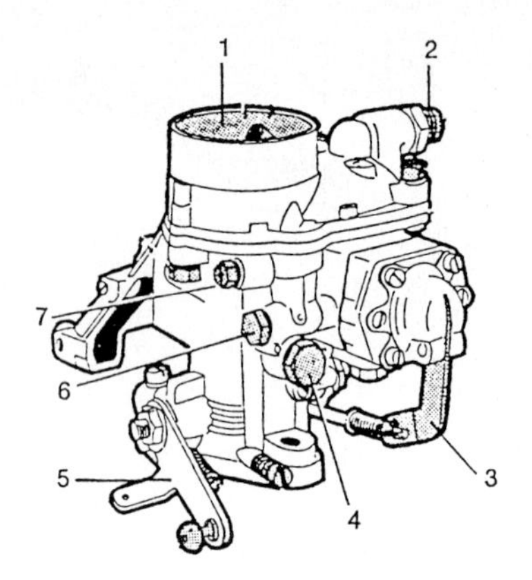

Carburettor (conventional type)
Vergaser (herkömmliche Ausführung)

1 **air intake** Lufteintritt
2 **petrol supply** Kraftstoffzuführung
3 **acceleration pump lever** Beschleunigungspunmpenhebel
4 **main nozzle** Hauptdüse
5 **speed lever** Gashebel
6 **acceleration pump nozzle** Beschleunigungspumpendüse
7 **idling nozzle** Leerlaufdüse

Cartridge fuse
Patronensicherung

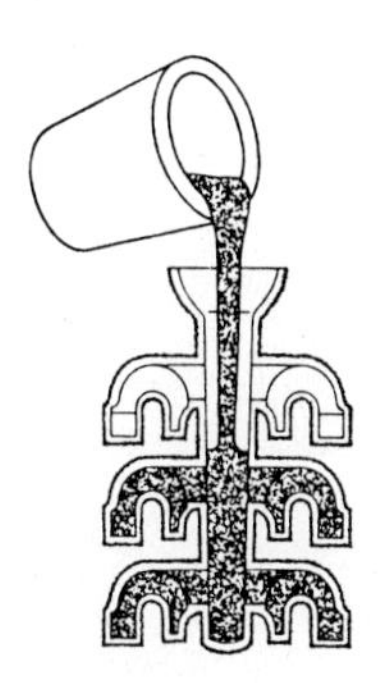

Casting
Gußherstellung

carrier telegraphy Träger-telegraphie
carrier telephony Trägertelephonie
carrier frequency Trägerfrequenz
carrier wave Trägerwelle
carry, to tragen, übertragen
carry over Übertrag
carrying capacity Tragfähigkeit, Tragvermögen, Belastbarkeit
carrying force Tragkraft
carrying handle Traggriff
cart Karre, kleiner Wagen
Cartesian coordinates kartesische Koordinaten
carton Karton (Behälter)
cartridge Patrone
cartridge fuse Sicherungspatrone
carve, to einschneiden, schnitzen, ausstechen
cascade, to in Kaskade schalten
cascade Kaskade
cascade tube Kaskadenröhre, Ionenröhre
case Gehäuse, Kasten, Futteral
case at hand der vorliegende Fall
case of damage Schadensfall
cash box Münzbehälter
casing Gehäuse, Ummantelung
cask Tonne, Fass, Fässchen
cassette Kassette
cast, to giessen (Metall)
cast integral, to angiessen
cast Guss, Gussform, Abguss
cast iron Gusseisen
castability Vergiessbarkeit
caster wheel schwenkbare Rolle
casting Guss(stück)
casting die Metallgussform
casting flaw Gussfehler
castle nut Kronenmutter
castor schwenkbare Rolle
cat's eye Katzenauge
catalog (catalogue) Katalog
catalogue, to katalogisieren
catalysis Katalyse
catalyst Katalysator
catalytic katalytisch
catapult, to katapultieren
category Kategorie
catenary Kettenlinie, Seilkurve
catenary wire Tragdraht
caterpillar drive Raupenantrieb
caterpillar tractor Raupen-schlepper

cathode Kat(h)ode
cathode-ray tube Kat(h)odenstrahlröhre
cathodic kat(h)odisch
catholyte Kat(h)olyt
cation Kation
cation migration Kationenwanderung
caulk, to verstopfen, abdichten
causal ursächlich, kausal
causal connection Kausalzusammenhang
cause, to bewirken, verursachen, hervorrufen
caustic ätzend
caustic lime gebrannter, ungelöschter Kalk
caustic potash Kalihydrat
caustic soda Aetznatron
caustic cell Alkalielement
caustic lye Natronlauge
causticity Aetzkraft
caution Warnung, Vorsicht
cave (heisse) Zelle
cavitation Kavitation
cavity Hohlraum, Kavität, Höhlung
cavity resonance Gehäuseresonanz
cavity resonator Hohlraumresonator
cease, to aufhören
cedar wood Zedernholz
ceiling Decke
ceiling duct Deckendurchführung
ceiling lamp Deckenlampe
ceiling suspension Deckenmontage
ceiling voltage Spitzenspannung
celestial axis Himmelsachse
celestial body Himmelskörper
celestial coordinates Himmelskoordinaten
celestial navigation Astronavigation
cell Zelle, Element
cell cover Zellendeckel
cell terminal Klemme eines Elements
cellophane Zellophan
cellular zellenartig, zellenförmig
celluloid Zelluloid
cellulose Zellulose, Zellstoff
cement, to zementieren, abdichten, kitten
cement Zement
cementation Zementierung
cemented lens verkittete Linse

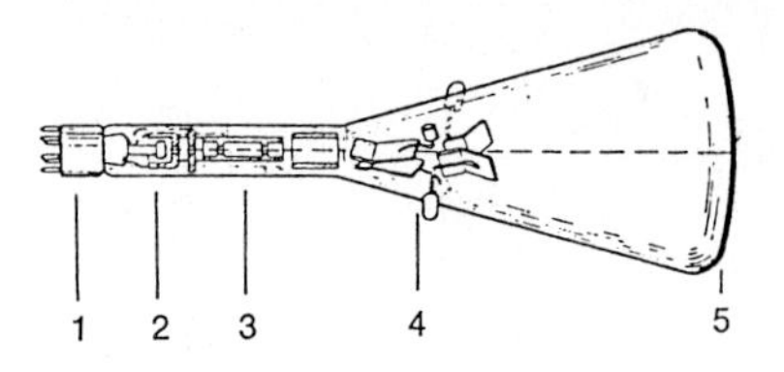

Cathode-ray tube
Kathodenstrahlröhre

1 **plug-in socket** Steckfassung
2 **cathode** Kathode
3 **acceleration electrodes** Beschleunigungselektroden
4 **deflection plates** Ablenkplatten
5 **screen** Bildschirm

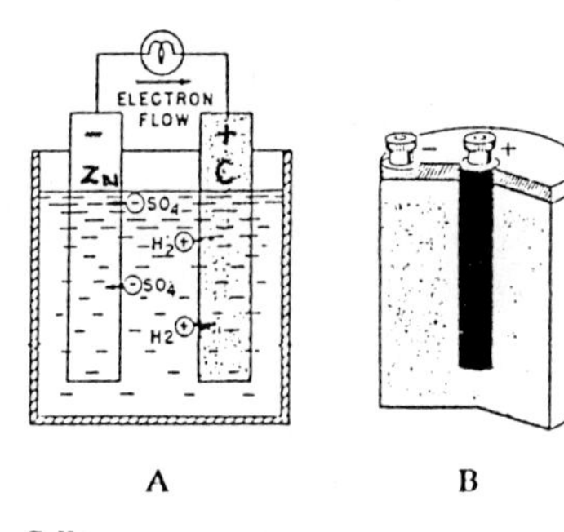

Cells
Zellen, Elemente (elektr.)

A **Voltaic cell** Voltasches Element

B **Battery dry cell** Batterie-Trockenelement

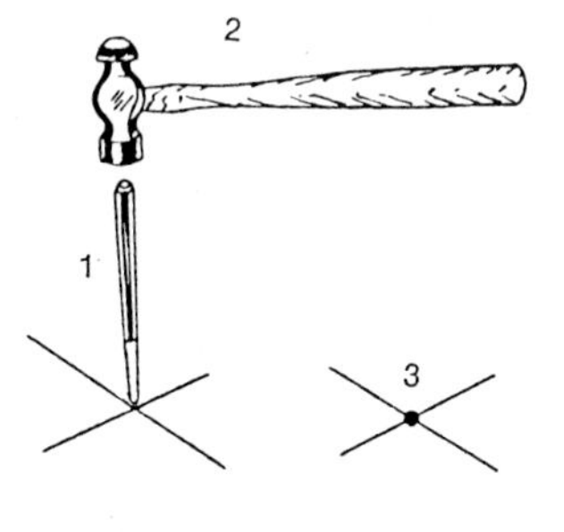

Centre punch
Körner (1)

2 **hammer** Hammer
3 **centre mark** Mittelpunkt

cementing Zementierung
c.e.m.f. (counter electromotive force) gegenelektromotorische Kraft, Gegen-EMK
census Totalerhebung
center (centre), to zentrieren, mittig einstellen
center (centre) Zentrum, Mittelpunkt, Mitte
center pin Zentrierstift
center tapping Mittelanzapfung
centered zentriert
centering Zentrierung
centesimal hundertteilig, zentesimal
centibar Zentibar
centibel Zentibel
centigrade Celsius (hundertteilige Skala)
centigram Zentigramm
centimetre graduation Zentimeterteilung
centimetre radio link Zentimeterrichtverbindung
central mittig, zentrisch, zentral
central exchange Fernsprechvermittlung
centralization Zentralisation, Zentralisierung
centralize, to zentralisieren, zusammenlegen, konzentrieren
centrally situated zentral gelegen
centre hole Zentrierloch
centre lathe Spitzendrehbank
centre line Mittellinie
centre of anticyclone Hochdruckzentrum
centre of attraction Anziehungsmittelpunkt
centre gravity Schwerkraftzentrum, Schwerkraft
centre of pressure Hochdruckzentrum
centre inertia Trägheitsmittelpunkt
centre low (barometric) pressure Tiefdruckzentrum
centre point Mittelpunkt
centre position Totlage
centre punch Körner
centre to centre von Mitte zu Mitte
centre-to-centre spacing Abstand von Mitte zu Mitte
centre scale Skala mit Nullpunkt in der Mitte
centric zentrisch

centrifugability Schleuderverhalten
centrifugal zentrifugal, mittelpunktflüchtig
centrifugal force Zentrifugalkraft, Fliehkraft
centrifugal impeller Zentrifugalgebläse
centrifugal machine Zentrifuge
centrifugal mass Schwungmasse
centrifugal moment Schwungmoment
centrifugal starter Fliehkraftanlasser
centrifugalize, to zentrifugieren
centrifuge, to zentrifugieren
centrifuge Zentrifuge
centring Zentrierung, Zentrieren
centripetal zum Mittelpunkt hinstrebend, zentripetal
ceramic keramisch
ceramics Keramik
cerium Zerium
cermet fuel element Cermet-Brennelement
certificate Zertifikat, Zeugnis
certificate of apprenticeship Lehrzeugnis
certify, to bescheinigen
chad Stanzblättchen
chads Schnitzel
chafe, to scheuern, reiben, schaben
chain Kette
chain of atoms Atomkette
chain reaction Kettenreaktion
chalcocite Kupferglanz
chalk Kreide
challenge, to abfragen (Radar)
challenger Abfragesender
chamfer, to fasen, abkanten, abschrägen
chamfer angle Abfasungswinkel
chamfering Abschrägung
change, to ändern, umstellen, umwechseln, abändern
change over, to umschalten
change Aenderung, Abänderung, Umstellung
change in cross section Querschnittübergang
change in direction Richtungsänderung
change lever Umschalthebel
change-over Übergang, Umschaltung, Zeichenwechsel

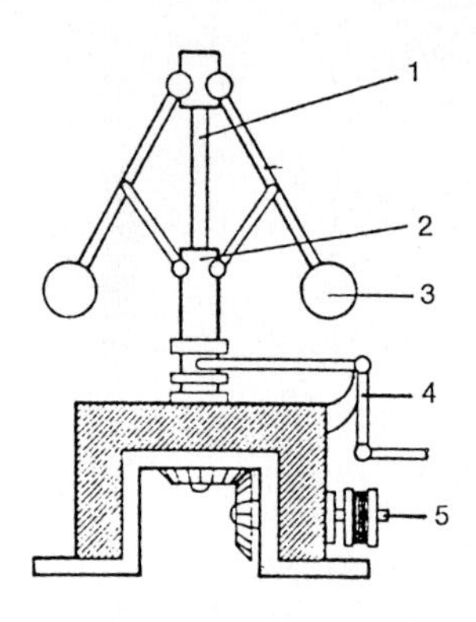

Centrifugal speed controller (principle of operation)
Zentrifugal-Geschwindigkeitsregler (Arbeitsprinzip)

1 **rotating spindle** rotierende Spindel
2 **sleeve** Muffe
3 **ball weights** Kugelgewichte
4 **activating levers** Betätigungshebel
5 **speed-controlled shaft** geschwindigkeitsgesteuerte Welle

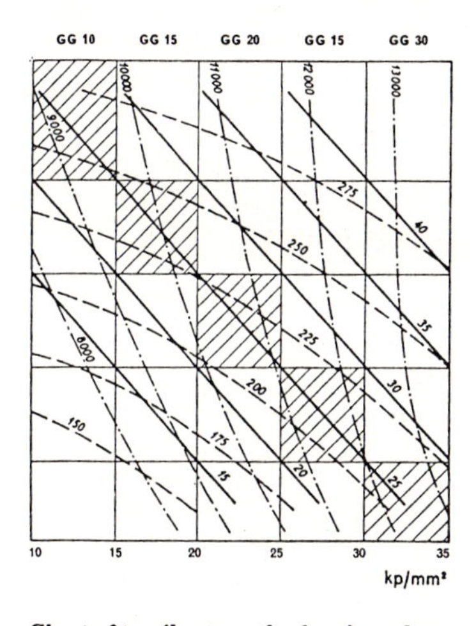

Chart of tensile strength of various classes of cast iron
Schaubild der Zugfestigkeit von verschiedenen Gußeisensorten

change-over contact Wechselkontakt
change-over switch Umschalter
change point Wechselpunkt
change-pole motor polumschaltbarer Motor
change-speed motor Motor mit Drehzahländerung
changing cycle Wechselvorgang
channel, to auskehlen
channel Rinne, Kanal, Auskehlung, Riefe, Fahrwasser
channelled kanneliert
channelled (steel) plate Riffelblech
characteristics Kennwerte, Kenndaten, Eigenschaften
charcoal Holzkohle
charge, to laden
charge Ladung, Charge
charge carrier Ladungsträger
charge for cancellation Streichungsgebühr
chargeable gebührenpflichtig
charged geladen
charger Ladeeinrichtung, Ladegerät
charges Kosten, Gebühren
charging Berechnung, Verrechnung, Belastung, Aufladung
charging board Ladetafel
charging circuit Ladestromkreis
charging equipment Ladevorrichtung
charging panel Ladetafel
charging rectifier Ladegleichrichter
charging switchboard Ladeschalttafel
charging voltage Ladespannung
chariot Schlitten
chart, to graphisch darstellen
chart Karte, Tafel, Kurvenblatt, Übersicht
charting Registrierung
chase, to treiben
chase a screw thread, to ein Gewinde nachschneiden
chaser type thread snap gauge Gewinde-Rachenlehre mit festem Gewindeprofilstück
chassis Fahrgestell, Chassis, Aufbauplatte
chassis ground Masse
chatter, to klappern, prellen
chatter of switch Schalterprellen

check, to überprüfen, nachsehen, kontrollieren
check Überprüfung, Durchsicht, Kontrolle
check nut Gegenmutter
checker Abnahmebeamter
checking Prüfung, Kontrolle, Nachprüfung
checkout Funktionsprüfung, Startvorbereitungen
chemical chemisch
chemically pure chemisch rein
chemicals Chemikalien
chemist Chemiker
chemistry Chemie
cherry red heat volle Rotglut
chest Kiste
chief hauptsächlich
chief designer Chefkonstrukteur
chief engineer Chefingenieur
chill, to abschrecken (Metall)
chill Kokille
chill casting Hartguss
chimney Kamin, Schornstein
china clay Kaolin, Porzellanerde
chip off, to abblättern, abspringen
chip Plättchen, Span, Schnitzel
chip board Holzfaserplatte, Spanholzplatte
chips Stanzabfälle, Schnitzel
chirp, to zwitschern (Radio)
chirping Zwitschern (Radio)
chisel, to meisseln
chisel off, to abmeisseln, wegmeisseln
chisel Meissel, Stemmeisen
chiseling-out Ausstemmung
chlorate Chlorat
chloride Chlorid
chlorine Chlor
chloroprene Chloropren
choice Wahl, Auswahl
choke, to drosseln
choke out, to abdrosseln
choke Drossel, Drosselspule
choking Drosselung
choking coil Drosselspule
choose, to wählen, auswählen, aussuchen
chop, to zerhacken
chop up into pulses, to zu Impulsen zerhacken
chopped zerhackt
chopper Zerhacker, Chopper
Christmas-tree antenna Tannenbaumantenne

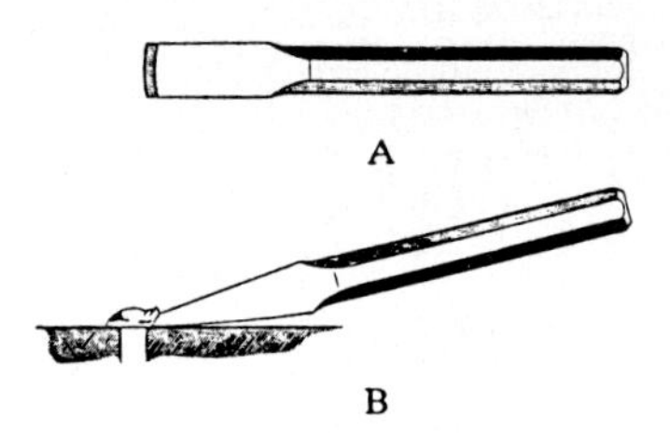

Chisel
Meißel

A **Plan view** Draufsicht

B **Side-view showing chisel cutting off a rivet head** Seitenansicht eines Meißels beim Abhauen eines Nietenkopfes

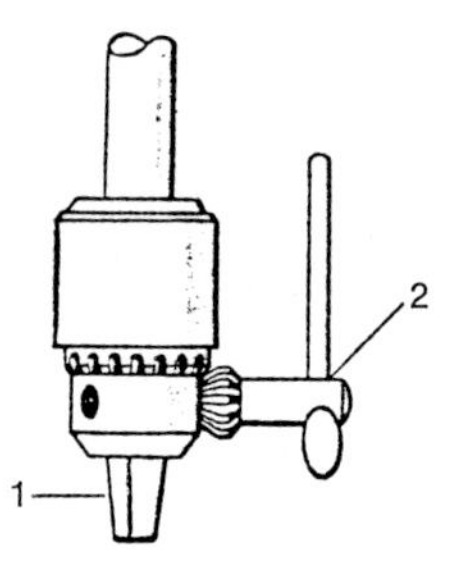

Drill chuck
Bohr-Spannfutter

1 **jaws** Spannbacken
2 **chuck key** Spannfutter-Schlüssel

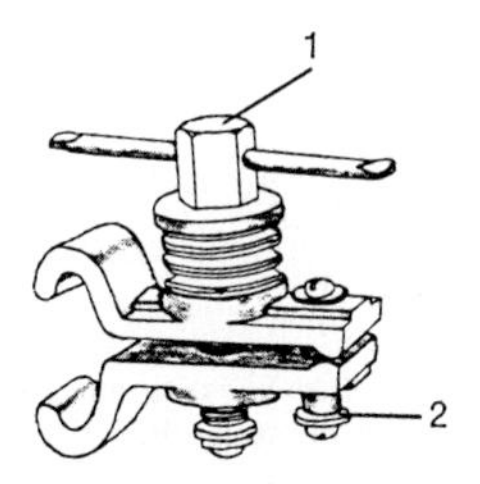

Clamp for earthing electric equipment
Klemme zum Erden elektrischer Ausrüstungen

1 **clamping screw** Spannschraube
2 **earthing terminal** Erdungsanschluß

chromate chromsaures Salz, Chromat
chromatic chromatisch, farbig
chromatics Farbenlehre
chrome, to verchromen
chrome Chrom
chromium Chrom
chronometer Chronometer, Zeitmesser
chuck (Auf-)Spannfutter, Einspannvorrichtung
chuck jaws Einspannbacken
chuck tool Spannwerkzeug
chug, to knattern
c.i.f. (cost, insurance, freight) Kosten, Versicherung, Fracht inkl.
cinematograph camera Filmkamera
cinematographic kinematographisch
cinematography Kinematographie
cineradiography Röntgenkinematographie
cinnabar Zinnober
cipher, to verschlüsseln
circle Kreis
circle cutter Kreisbohrer
circle minute Kreisminute
circle of longitude Längenkreis
circle segment Kreisabschnitt
circuit Stromkreis, Schaltung, Leitung (Tel.)
circuit algebra Schaltungsalgebra
circuit arrangement Schaltungsaufbau
circuit-breaker Leistungsschalter
circuit-breaking capacity Schaltleistung
circuit closing connection Arbeitsstromschaltung
circuit design Schaltungsaufbau
circuit designer Schaltungsfachmann
circuitry Schaltung, Schaltungsaufbau
circuitwise schaltungsmässig
circular kreisförmig, kreisrund, umlaufend, zirkulierend, wiederkehrend, Kreis . . . , Rund . . .
circulate, to zirkulieren, umlaufen, kreisen
circulate around, to umströmen
circulation Zirkulation, Umlauf, Kreislauf

circumference Umkreis, Peripherie, Umfang
circumferential am Umfang verlaufend, Umfangs . . .
circumferential velocity Umfangsgeschwindigkeit
citation Entgegenhaltung (Pat.)
cite, to entgegenhalten (Pat.)
clad, to plattieren
clad Überzug
cladding Verkleidung
claim Anspruch
clamp, to einspannen
clamp in place, to festklemmen
clamp Zwinge, Klammer, Schelle
clamp bolt Spannschraube
clamp collar Klemmring
clamp coupling Schalenkupplung
clamp chuck Klemmfutter
clamp device Einspannvorrichtung
clamp roller Klemmrolle
clamp jaw Klemmbacke
clamp ring Klemmring, Schelle
clamp sleeve Spannhülse
clamp tool Spannwerkzeug
clapper Klöppel, Schwengel
clarifier Feinabstimmvorrichtung
clarify, to klären
clarity Deutlichkeit
clasp Stange
classification Einteilung, Klassifizierung
classified directory Branchentelefonbuch
classify, to einteilen, klassieren
clatter, to rattern
clause Klausel
claw Greifer, Klaue
clay Ton (Erde)
clean, to reinigen, säubern, putzen
clean rein, sauber
cleaning Reinigung
cleanse, to reinigen, säubern
clear, to beheben, räumen, freigeben
clear a fault, to eine Störung beseitigen
clear deutlich, klar, hell
clear height lichte Höhe
clearance Freigabe
clearness Deutlichkeit, Klarheit, Schärfe, Reinheit
cleat, to befestigen
cleat Klampe, Leiste
cleavability Spaltbarkeit
cleavable spaltbar

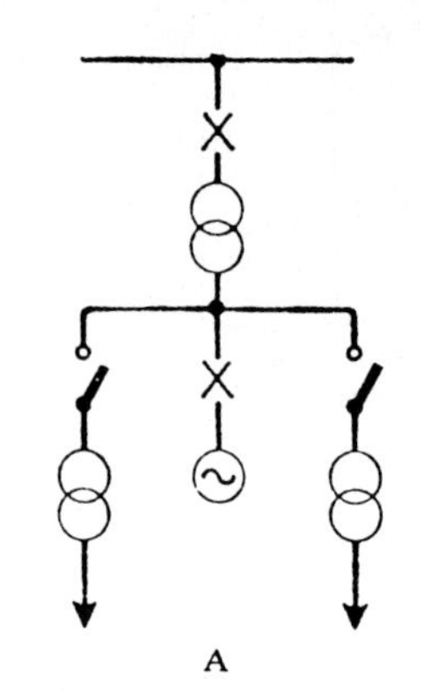

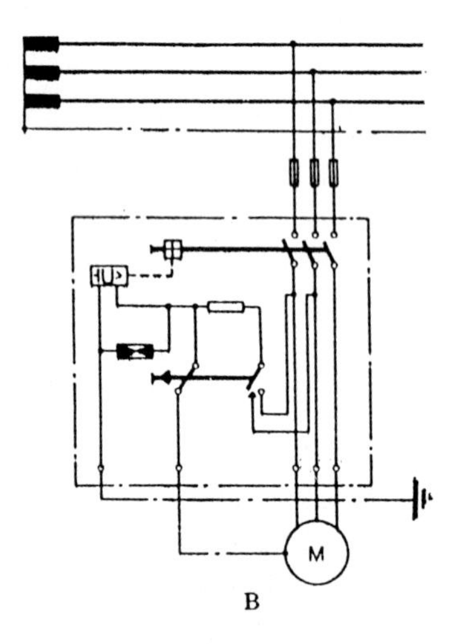

Circuit diagrams
Schaltpläne

A **High-voltage main-line diagram** Hochspannungs-Übersichtsschaltplan

B **Control schematic diagram** Steuerschaltplan

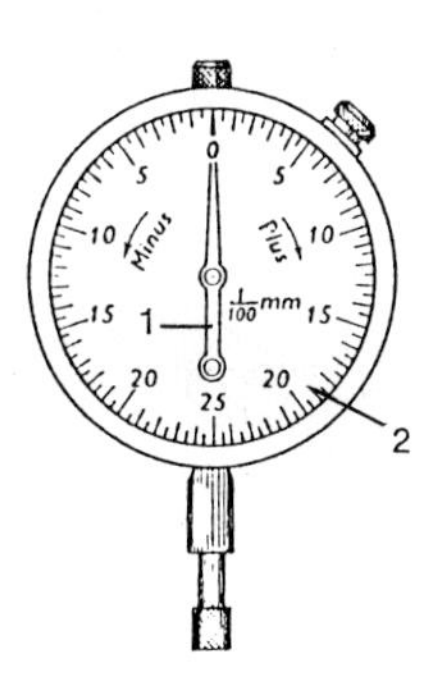

Clock indicator
Meßuhr

1 **pointer** Zeiger
2 **dial** Skala

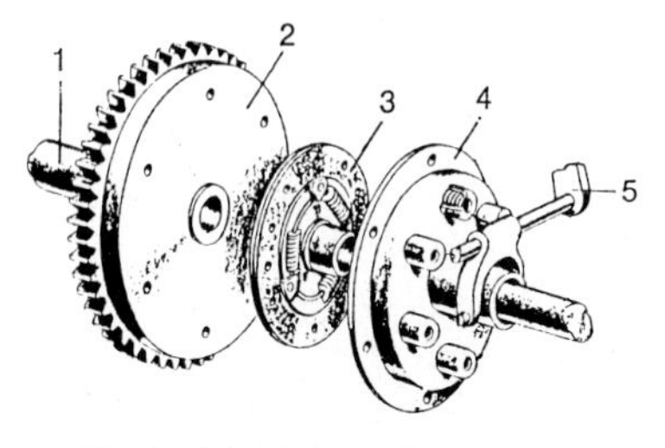

Clutch of a motor vehicle
Kupplung eines Motorfahrzeuges

1 **crank shaft** Kurbelwelle
2 **fly-wheel** Schwungrad
3 **clutch plate** Kupplungscheibe
4 **pressure plate** Druckplatte
5 **activating shaft** Betätigungswelle

cleavage Spaltung
cleave, to spalten, zerspalten
cleft Sprung, Riss
clerical work Büroarbeit
clew Knäuel
click, to knacken, klicken
click Klicken, Knacken
climate Klima
climatology Klimalehre
climb, to klettern
climbers Steigeisen
cling, to anhaften
clinker Klinker
clinometer Neigungswinkelmesser
clip, to abschneiden, stutzen
clip Klammer
clip-on ammeter Zangenstrommesser
clock, to die Zeit stoppen, nehmen
clock Uhr, Taktgeber
clockwise im Uhrzeigersinn, nach rechts drehend
clock rotation Rechtslauf
close, to schliessen, zugehen
close nahe, dicht, eng, genau
closed geschlossen
closed circuit geschlossener Stromkreis, Ruhestromkreis
closing Schliessung, Schliessen
closure Schliessen, Schliessung
clot, to sich zusammenballen, Klumpen bilden, gerinnen
cloth Stoff, Tuch
cloud, to verdecken
cloud Wolke
clouded bewölkt
cloudy bewölkt
cluster, to sich zusammenballen, anhäufen
cluster Haufen, Anhäufung
clutch, to kuppeln
clutch (ausrückbare) Kupplung
clutter Störflecke
coal Kohle
coarse grob
coarse adjustment Grobeinstellung
coat, to vergüten, (an)streichen
coat Lage, Schicht, Überzug
coating Überzug, Anstrich
coaxial koaxial, gleichachsig konzentrisch
coaxial cable Koaxialkabel
coefficient Koeffizient, Beiwert
coercive force Koerzitivkraft
cogged ingot Vorblock (Giesserei)

cohere, to zusammenhalten
coherence Kohärenz
coherer Fritter, Frittröhre
cohesion Kohäsion
cohesive power Kohäsionskraft
coil, to spulen, wickeln, wendeln
coil Spule, Windung
coil-loaded bespult, pupinisiert, spulenbelastet
coin box Münzfernsprecher
coincide, to zusammenfallen, übereinstimmen
coincidence Zusammenfallen, Zusammentreffen, Gleichzeitigkeit
coincident gleichzeitig (eintretend)
coke, to verkoken
coke Koks
coking Verkokung
cold Kälte
cold-bend, to (Metall) kalt biegen
cold cathode kalte Kat(h)ode
cold chamber Kälteschrank
cold-forge, to kalt schmieden
cold-rolled kaltgewalzt
collaborator Mitarbeiter
collapse, to zusammenfallen, zusammenbrechen
collapse Zusammenfall, Zusammensturz
collapsible zerlegbar, auseinandernehmbar, zusammenlegbar
collar Stellring, Manschette, Bund, Kragen
colleague Fachgenosse, Kollege
collect, to sammeln, auffangen
collection Sammlung
collector Kollektor, Stromabnehmer
collet Klemmring
collide, to kollidieren, zusammenstossen, zusammentreffen
collision Anprall, Kollision, Stoss, Zusammenprall
colour (color) Farbe
colour picture Farbbild
coloured farbig, gefärbt
colouring Färben, Kolorierung
column Säule, Spalte
column of water Wassersäule
comb Kamm
comb collector Kammstromabnehmer
combination Kombination, Vereinigung, Zusammensetzung

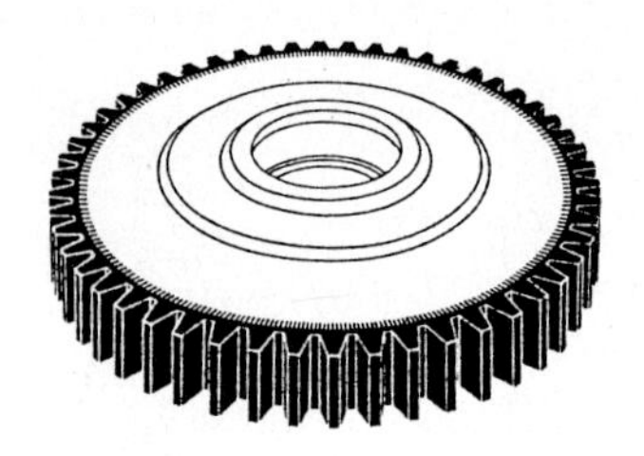

Cog wheel
Zahnrad

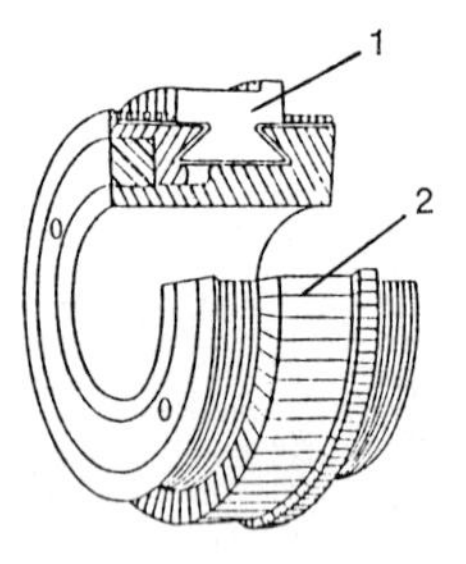

Collector of a direct-current machine
Kollektor (Stromwender) einer Gleichstrommaschine

1 **copper lamella with dove tail** Kupferlamelle mit Schwalbenschwanz
2 **mecanite separation** Glimmerzwischenlage

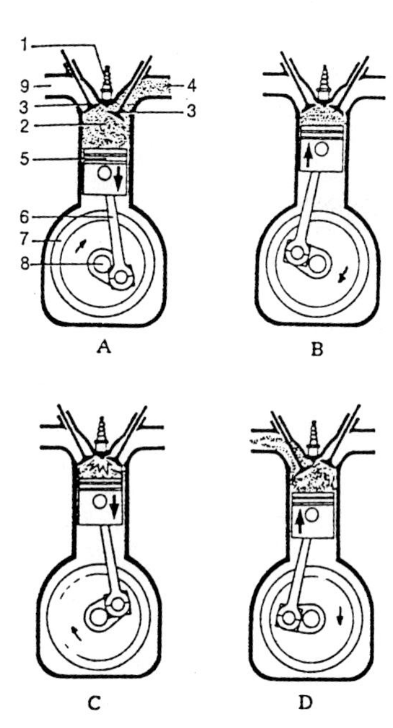

Four-stroke combustion
Viertakt-Verbrennung

A **Induction** Ansaugung

B **Compression** Verdichtung

C **Firing** Zündung

D **Exhaust** Ausstoß

1 **sparcing plug** Zündkerze
2 **cylinder** Zylinder
3 **valve** Ventil
4 **inlet** Einlaß
5 **piston** Kolben
6 **connecting rod** Pleuelstange
7 **fly-wheel** Schwungrad
8 **crankshaft** Kurbelwelle
9 **outlet** Ausgang

combine, to kombinieren, vereinigen, zusammenfassen
combustibility Brennbarkeit
combustible verbrennbar, brennbar
combustion Verbrennung, Verbrennen
combustion engine Verbrennungskraftmaschine
comfortable grip bequemer Griff
command Befehl
comment Stellungnahme, Kommentar
commentator Kommentator
commercial wirtschaftlich, kommerziell
commission Vermittlungsgebühr, Provision
committee Kommission, Ausschuss
common gemeinsam
common aerial Gemeinschaftsantenne
communicate, to in Verkehr stehen, mitteilen, in Verbindung stehen
communication Nachricht, Verkehr, Verbindung
communications Fernmeldewesen
community antenna Gemeinschaftsantenne
commutate, to kommutieren, umpolen
commutating field Wendefeld
commutating pole Wendepol
commutating winding Wendefeldwicklung
commutator Stromwender, Kommutator, Kollektor
compact dicht, gedrängt, kompakt
company standard Werknorm
comparable vergleichbar
comparator Komparator, Vergleicher
compare, to vergleichen
comparison Vergleich
compartment Fach
compass Kompass, Bussole
compasses (pair of) Zirkel
compensate, to kompensieren, ausgleichen
compensating (compensation) Ausgleich, Kompensieren, Kompensation
competent fachmännisch

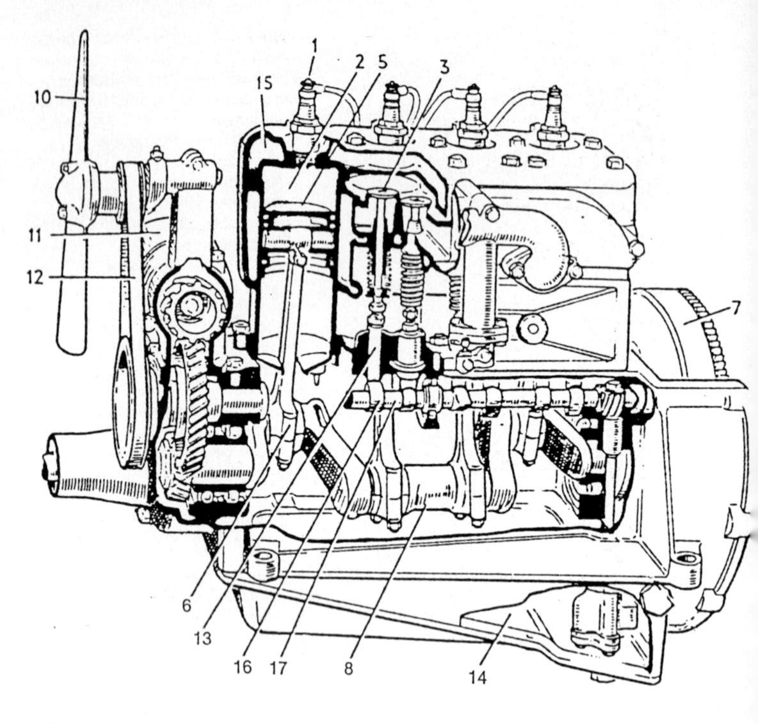

Combustion engine
Verbrennungsmotor

(1 **to** 9 **see opposite page** siehe gegenüberliegende Seite)
10 **fan** Lüfter
11 **dynamo** Lichtmaschine
12 **fan belt** Lüftertreibriemen
13 **tappet** Ventilstößel
14 **sump** Ölwanne
15 **cooling water jacket** Kühlwassermantel
16 **cam** Nocke
17 **camshaft** Nockenwelle

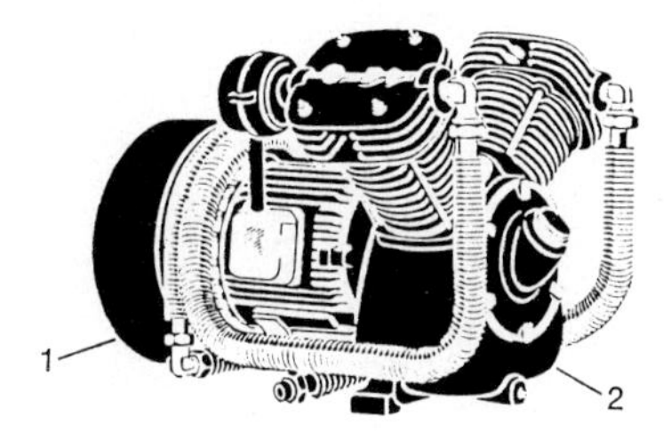

Compressor
Kompressor

1 **driving motor** Antriebsmotor
2 **compressor** Luftverdichter

Computerized menu order?
Rechnergesteuerte Menubestellung?

competitor Konkurrent
compile, to zusammenstellen
complaint Beanstandung, Reklamation
complement, to komplementieren, ergänzen
complement Ergänzung, Komplement
complementary komplementär, ergänzend
complete, to vervollständigen, komplettieren, fertigstellen, ausbauen
complete komplett, vollständig, ganz
completion Vervollständigung, Herstellung
complex verwickelt, kompliziert
complicated kompliziert
composition Zusammensetzung, Zusammenstellung
compound, to kompoundieren
compound zusammengesetzt, gemischt, Verbindung
compound connection Verbundschaltung
compound generator Doppelschlussgenerator
compound motor Verbundmotor
compound winding Verbundwicklung
compounding Kompoundierung, Verbundbetrieb
comprehensive umfassend
compress, to zusammenpressen, -drücken, komprimieren, verdichten
compressed air Pressluft, Druckluft
compressed air gauge Druckluftmanometer
compressible zusammendrückbar, verdichtbar
compression Verdichtung, Kompression
compressive force Druckkraft
compressor Kompressor, Verdichter, Lader
computable berechenbar
computation Errechnung, Berechnung
compute, to berechnen, errechnen, ausrechnen
computer Rechner, Rechenanlage
computer control unit Befehlswerk, Leitwerk

computer operation Rechnerbetrieb
computing centre Rechenzentrale
concave konkav, hohl
concave cutter Hohlfräser
concave mirror Hohlspiegel
concealed wire Unterputzleitung
concealed wiring Unterputzverlegung
concentrate, to konzentrieren, zusammenlegen
concentrate Konzentrat
concentration Konzentration, Sammlung
concentric konzentrisch
concept Begriff
concession Zulassung, Konzession
concessionary Konzessionsinhaber
conclude, to folgern, den Schluss ziehen
conclusion Folgerung, Schlussfolgerung
concrete, to betonieren
concrete Beton
concreting Betonierung
condensable verdichtbar
condensate, to kondensieren
condensation Kondensierung
condense, to kondensieren, zusammenpressen, verdichten
condenser Kondensator
condition Bedingung, Zustand, Beschaffenheit
conditions of acceptance Abnahmebedingungen
conduct, to leiten, führen
conductance Leitwert, Konduktanz
conducting leitend
conduction Leitung
conductive leitend
conductive material elektrischer Leiter, leitendes Material
conductivity Leitfähigkeit
conductor Leiter, Ader
conduit Rohr, Schutzrohr, Installationsrohr
conduit box Abzweigdose, Rohrleitungskasten
cone Konus
cone of dispersion Streukegel
cone of rays Strahlenbündel
confidential vertraulich
configuration Anordnung, Gestaltung

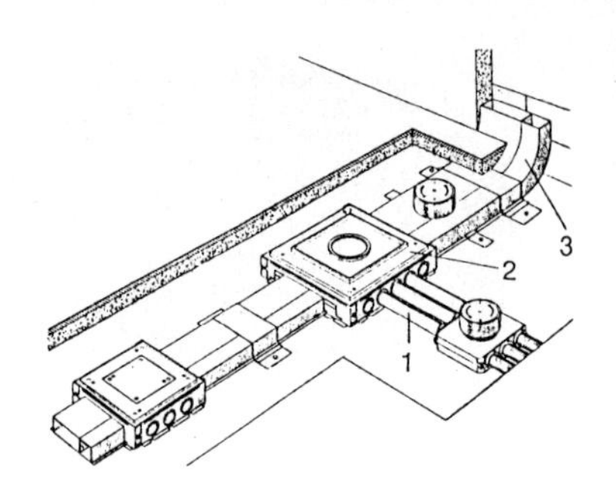

Conduits
Rohrleitungen

1 **conduit pipe** Leitungsrohr
2 **junction box** Abzweigkasten
3 **channel** Kanal

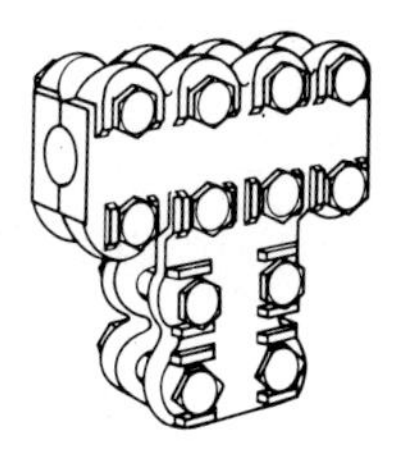

Connector for steel ropes
Verbindungsklemme für Stahlseile

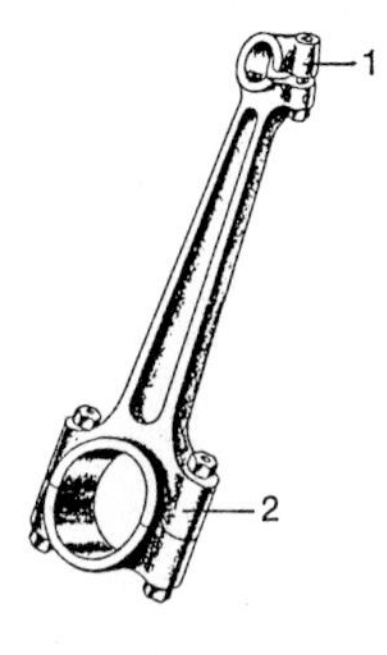

Connecting rod
Pleuelstange

1 **small end** oberes Ende
2 **big end** unteres Ende

confine, to beschränken
confirm in writing, to schriftlich bestätigen
confirmation Bestätigung
conform with, to übereinstimmen mit
conforming to specifications den Vorschriften entsprechend
confuse, to verwechseln
confusion Verwechslung, Verwirrung
congestion Verstopfung, Besetztsein
conical konisch
connect, to zusammenschalten, verbinden, anklemmen, anschliessen
connect in parallel, to parallel schalten
connect in series, to in Reihe schalten
connection Anschluss, Schaltung, Verbindung
connector Stecker, Leitungswähler
consecutive aufeinanderfolgend
consecutive number laufende Nummer
conservation Erhaltung
consider, to berücksichtigen, in Betracht ziehen
consistence Konsistenz
console Steuerpult, Kontrollpult
conspicuous auffallend, hervortretend
constant konstant
constant load Dauerlast, Dauerbelastung
constant speed konstante Drehzahl
constantan Konstantan
constituent Bestandteil
constraining force Zwangskraft
constrict, to zusammenziehen
constriction Zusammenziehung, Verengung, Einschnürung
construct, to bauen
construction Bau, Aufbau
constructional baulich
consult, to konsultieren, um Rat fragen
consulting engineer beratender Ingenieur, Konsulent
consulting engineers Ingenieurbüro, beratende Ingenieure
consume, to verbrauchen

Container crane
Containerkran

1 **grab** Greifer
2 **crane bridge** Kranbrücke
3 **travel carriage** Fahrgetell
4 **container** Container
5 **loading platform** Ladefläche

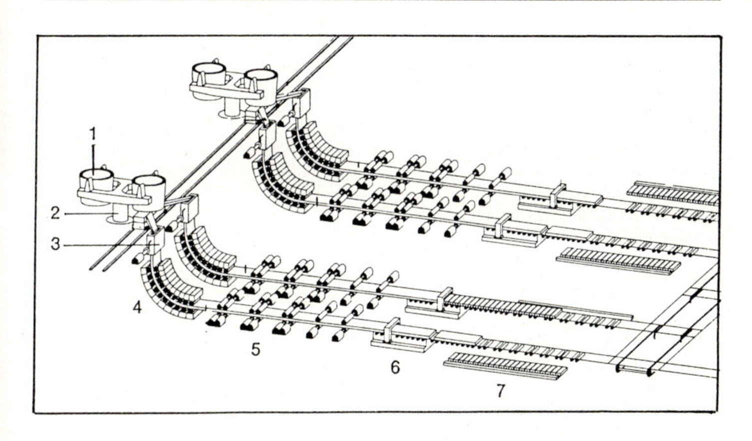

Continuous casting plants (two-strand)
Stranggußanlagen (zweisträngig)

1 **pouring ladle** Gießpfanne
2 **ladle turning standard** Pfannendrehturm
3 **mould** Kokille
4 **casting machine** Gießmaschine
5 **straighteners** Treibrichtrollen
6 **cutting machine (shears)** Schneidmaschine
7 **billets** Knüppel

consumer Verbraucher, Konsument
consumption Verbrauch
contact, to berühren
contact Kontakt, Berührung
contact clearance Kontaktabstand
contact making Kontaktgabe
contact piece Schaltstück
contactor Schütz, Kontaktgeber
contain, to enthalten
container Behälter
containment shell Sicherheitsbehälter, Schutzbehälter
contaminate, to verunreinigen
contamination Verseuchung, Vergiftung (eines Reaktors), Kontamination
content Gehalt, Inhalt
contents Inhalt
contex, in this in diesem Zusammenhang
contingencies (Pl.) unvorhergesehene Ausgaben
continuance Andauern, Fortdauer
continuation Fortsetzung
continue, to fortdauern, fortfahren, fortsetzen
continuity Kontinuität, Stetigkeit
continuous dauernd, fortdauernd, fortlaufend, stetig, kontinuierlich
continuous duty Dauerbetrieb
continuous welding Nahtschweissung
contour Kontur, Umgrenzungslinie, Umrisslinie
contract, to zusammenziehen
contract Vertrag, Auftrag, Kontrakt
contraction Zusammenziehung, Verengung
contractor Auftragnehmer (Firma)
contractual vertraglich
contraction Widerspruch
contradictory widersprüchlich
contrail Kondensstreifen
contrariwise umgekehrt
contrary gegenteilig, entgegengesetzt
contrast, to kontrastieren, Kontrast bilden, sich absetzen
contrast Kontrast
contrasting sample Gegenprobe

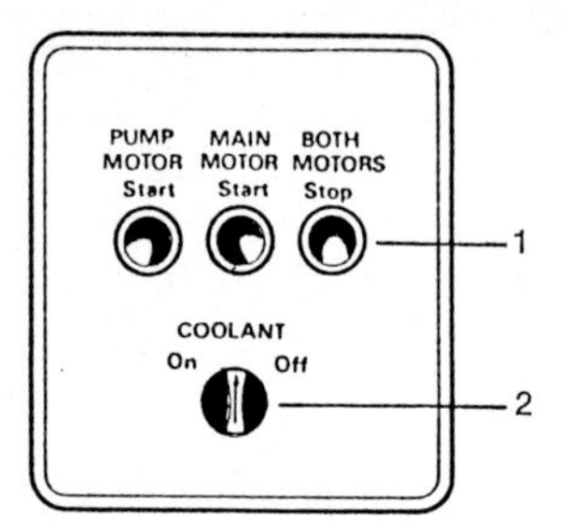

Control panel
Steuertafel

1 **push-button** Druckknopf
2 **rotary switch** Drehschalter

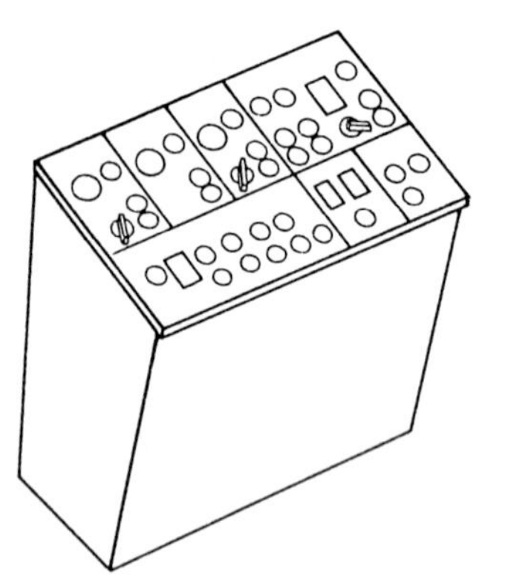

Control console
Steuerpult

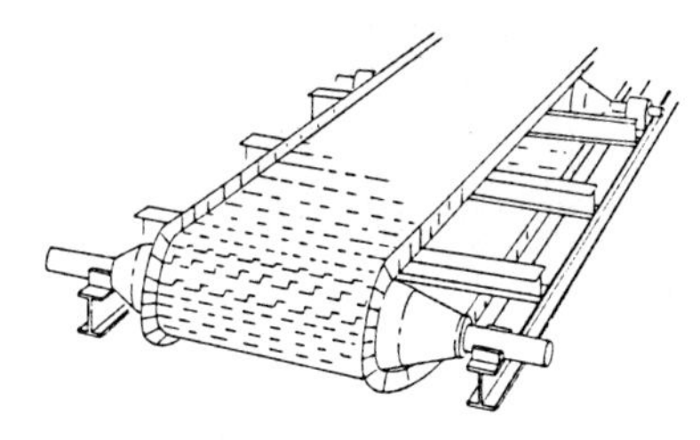

Conveyor belt
Förderband

contribute, to beitragen
contrivance Einrichtung, Vorrichtung
contrive, to bewerkstelligen
control, to steuern, regeln, kontrollieren
control Steuerung, Regelung, Kontrolle
control circuit Steuerkreis
control desk Steuerpult, Schaltpult
control device Steuervorrichtung
control station Leitstation, Kontrollstelle
controllable steuerbar, regelbar
controlled gesteuert
controller Kontroller, Regler, Regelgerät
controls Bedienungselemente, Steuerzüge
convector, electric elektrischer Konvektionsofen
conventional gebräuchlich, üblich
converge, to konvergieren
convergence Konvergenz, Zusammenlaufen
convergency Konvergenz
convergent konvergent, zusammenlaufend
conversion Umwandlung, Wandlung
convert, to umwandeln
converter Konverter, Umsetzer, Umformer
convertible verwandelbar, umwandelbar
convey, to befördern, übertragen
conveyance Beförderung
conveyer (conveyor) Fördergerät
conveyor belt Transportband, Fliessband
cooker, electric Elektrohaushaltsherd
cool, to kühlen, abkühlen
coolant Kühlmittel, Kühlflüssigkeit
cooling Kühlung
co-operate, to zusammenarbeiten, mitwirken, mitarbeiten
co-operation Zusammenwirken, Zusammenarbeit
co-ordinate, to koordinieren
co-ordinate axis Koordinatenachse
co-ordination Koordinierung

copal Kopal, Kopalharz
copper, to verkupfern
copper Kupfer
coppered verkupfert
coppering Verkupferung
copy, to kopieren, vervielfältigen
copy Kopie, Pause, Durchschlag
cord Schnur
cordless schnurlos
core Kern, Spaltraum
coreless kernlos
corner Ecke, Eckpunkt
cornered eckig
corona Korona
correct, to korrigieren, verbessern, ausbessern
correction Korrektion, Verbesserung, Ausbesserung, Richtigstellung
corrective verbessernd, korrigierend
corresponding entsprechend
corrode, to korrodieren
corrosion Korrosion
corrosive korrodierend
corrugated gewellt
corrugated metal Wellblech
cost Kosten
cost increase index Steigerungskostenindex
cost per unit Stückpreis
costing office Kalkulationsbüro
cotter, to versplinten
cotter Querkeil
cotton Baumwolle
cough, to zeitweise aussetzen (Motor)
count, to zählen
counter Zähler, Schaltertisch
counterbalance, to Gegengewicht bilden, auswuchten
counterbore Flachsenker
counterbore, to flachsenken
counterclockwise gegen den Uhrzeigersinn
counter-electromotive force gegenelektromotorische Kraft (Gegen-EMK)
countersunk bolt versenkter Bolzen
couple, to koppeln, kuppeln
coupler Kopplungsspule, Muffe
coupling Kupplung, Kopplung
course Kurs, Verlauf
cover, to verdecken, bedecken, zudecken

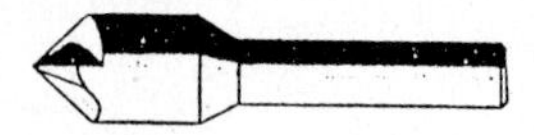

Counter sink
Senkbohrer

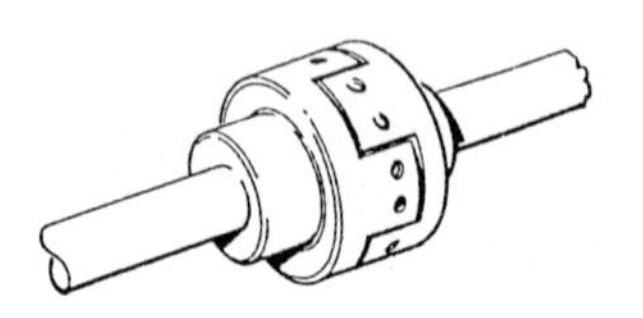

Coupling (Spider)
Kupplung (Finger-, Greif-)

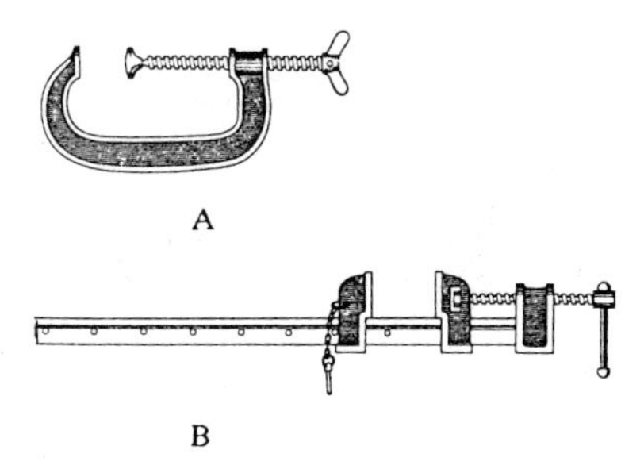

Cramps
Schraubzwingen

A **'G' cramp** (normale) Schraubzwinge

B **Sash cramp** (verschiebbare) Schienenschraubzwinge

Crank shaft
Kurbelwelle

cover Kappe, Schutzhaube, Hülle
coverage Reichweite
crack, to platzen, aufspringen, rissig werden
crack Sprung, Riss, Knall, Krach
craft Handwerk
craftsman Handwerker
crane Kran
crank, to ankurbeln, anwerfen
crank Kurbel, Kröpfung
cranked gekröpft
crank shaft Kurbelwelle
crate, to verpacken (in Kratten)
cream solder Lötpaste
crease, to falten, kniffen
create, to hervorbringen, produzieren
creep, to kriechen, nacheilen
creepage Kriechstrom
crest value Spitzenwert
crevice Spalt, Riss
criterion Kriterium
crocodile clip Krokodilklemme
cross, to durchkreuzen, durchqueren
cross Kreuz
crossbar Querschiene, Querarm, Querstange
cross piece Kreuzstück
cross section Querschnitt
crossing Überquerung, Kreuzung
crowbar Brecheisen, Brechstange
crowd, to zusammendrängen
crucible Schmelztiegel
crumble, to zerbröckeln
crumple, to zerknittern
crush, to zerdrücken, brechen, zerquetschen
crust, to verkrusten
crust Kruste
cryogenic kryogen, kälteerzeugend
crystal Kristall
cube Würfel, Kubikzahl, dritte Potenz
cubicle Zelle, Kabine, (Schalt-)Feld
cup Napf, Schale
curdle, to gerinnen
curdle Gerinnen
cure, to Fehler beseitigen, beheben
cure Abhilfe
curing of defects Fehlerbeseitigung
curl, to Schlingen bilden, sich ringeln, rollen
curl Gussblase
curly bracket geschweifte Klammer

current Strom
current-carrying stromführend
curvature Krümmung, Kurvenform
curve, to sich krümmen
curve Kurve
curved gekrümmt, gebogen, krummlinig, geschweift
cushion, to auspolstern
cushion Kissen, Polster
cusp Spitze (Math.)
custom-make, to auf Bestellung machen
custom-made speziell hergestellt, auf Bestellung gemacht
customer Kunde, Auftraggeber
cut, to schneiden
cut Schnitt
cut-out Ausschnitt, Ausschalter
cutter Fräser
cybernetics Kybernetik
cycle, to periodisch wiederholen
cycle Zyklus
cycles per second Hertz (Hz)
cycling Durchlaufen
cylinder Zylinder, Trommel, Walze
cylindrical zylinderförmig, walzenförmig

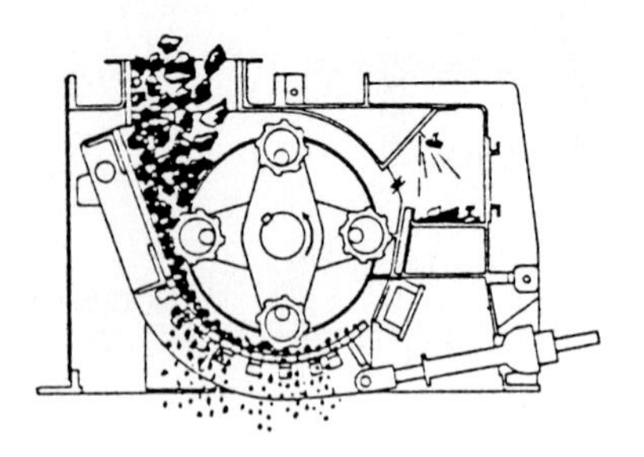

Crusher
Brecher (Brechmaschine)

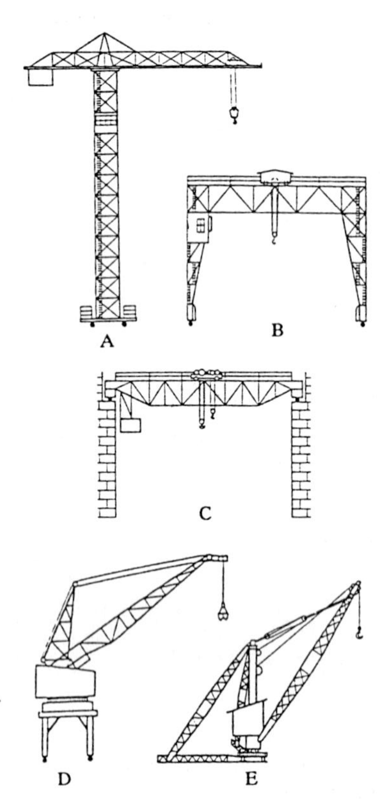

Cranes
Krâne

A **Tower crane** Hochkran
B **Gantry crane** Portalkran
C **Overhead travelling crane** Hallen-Laufkran
D **Wharf crane** Hafenkran
E **Derrick** Drehkran

Damped waveform
gedämpfte Wellenform

A data carrier? Ein Datenträger?

D

dab, to betupfen
dam, to stauen
dam Staudamm, Wehr
damage, to beschädigen
damage Schaden, Beschädigung
damp, to dämpfen
damp feuchter Dunst, Dampf
dampen, to befeuchten
damper Dämpfer
damping Dämpfung
dampness Feuchtigkeit
danger Gefahr
dash, to stricheln
dash Strich
data Angaben, Daten
daub, to verschmieren
dazzle, to blenden
dazzling Blendung
d.c. (DC, dc) (direct current) Gleichstrom
dead spannungslos, verbraucht, tot
dead centre Totpunkt
de-aerate, to entlüften
deaf taub
debugging Fehlerbeseitigung
decay, to zerfallen
decay Verfall
decelerate, to verlangsamen
deceleration Verlangsamung
deception Täuschung
decipher, to entziffern, entschlüsseln
decode, to dekodieren, entschlüsseln
decoder Dekodierer, Dekodiergerät
decompose, to zersetzen, zerfallen
decomposition Zerfall, Zerstreuung
decouple, to entkoppeln
decrease, to abnehmen, sich vermindern, herabsetzen
decrease Abnahme, Verringerung, Verminderung
decree a standard, to eine Norm festlegen
decrement, to verringern
decrement Dekrement
deduce, to ableiten, herleiten (Formel)
deduct, to abziehen
deduction Ableitung, Folgerung

de-energise, to abschalten, ausser Spannung setzen, energielos machen
defect Defekt, Mangel, Schaden
defective defekt, schadhaft, fehlerhaft
defer, to verschieben, verzögern
definable definierbar
define, to definieren, eindeutig festlegen
definite abgegrenzt, bestimmt
deflagration explosionsartige Verbrennung
deflect, to umlenken, ablenken
deflection Umlenkung, Ablenkung
deflector Leitblech
deform, to deformieren
deformable verformbar
degas, to entgasen
degassing Entgasen
degauss, to entmagnetisieren
degaussing Entmagnetisierung
degrade, to herabsetzen
degree Grad
dehydrate, to Wasser entziehen
dehydration Wasserentziehung
deicer Enteisungsgerät
deion circuit-breaker Leistungsschalter mit magnetischer Bogenlöschung, Deionisationsschalter
delay, to verzögern
delay Verzögerung, Laufzeit
deliver, to liefern, abliefern, (Vortrag) halten
delivery Lieferung, Zustellung
delta connection Dreieckschaltung
demand, to erfordern, fordern
demand Bedarf
demesh, to ausrücken (Getriebe)
demijohn Korbflasche
demodulate, to demodulieren
demodulation Entmodelung, Demodelung
demolish, to zertrümmern, zerstören
demolition Zerstörung, Zertrümmerung
demonstrate, to vorführen
demonstration Vorführung, Beweisführung
demount, to ausbauen
demountable demontierbar
denotation Bezeichnung
dense dicht
density Dichte

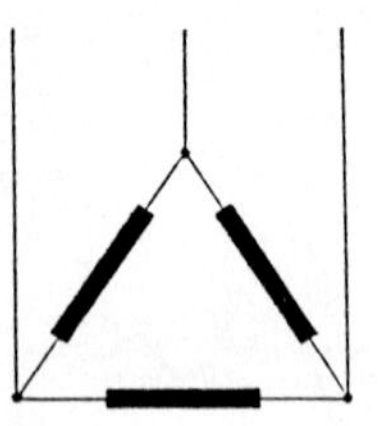

Delta connection (e. g. of a transformer winding)
Dreieckschaltung (z. B. einer Transformatorwicklung)

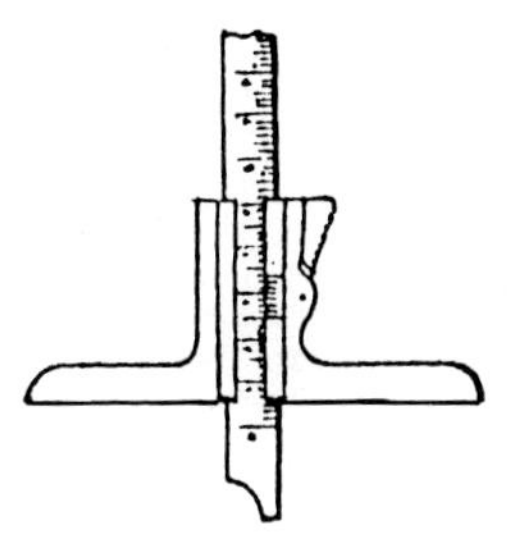

Depth gauge
Tiefen-Meßschieber

dent, to einbeulen
dent Beule, Einbeulung
denude, to (Isolation) entfernen
depart from, to abweichen von
departure Abweichung, Abgang, Abfahrt
dependable verlässlich, betriebssicher, zuverlässig
dependence Abhängigkeit
dependent abhängig
depict, to bildlich darstellen
depletion Verarmung, Sperrschicht
deposit, to sich niederschlagen, ablagern, bilden
deposit Niederschlag
deposition Abscheidung, Niederschlag
depreciation allowance Abschreibungssatz
depress, to niederdrücken, herabdrücken
depression Tief, Depression
depth Tiefe
derail, to entgleisen
derating Lastminderung
derivable ableitbar
derivative Ableitung
derive, to ableiten, herleiten
derrick Ladebaum, Kran
descale, to entzundern
descend, to absteigen
describe, to beschreiben
description Beschreibung, Darstellung
descriptive geometry darstellende Geometrie
descriptive literature technische Unterlagen
design, to konstruieren, entwerfen
design Entwurf, Konstruktion
designate, to bezeichnen, betiteln
designation Bezeichnung, Benennung
designer Konstrukteur
designing department Konstruktionsabteilung
desk Pult, Tisch
deslag, to entschlacken
despatch Absendung
de-spin system Stabilisierungssystem
destination Zielpunkt, Bestimmungsort
destroy, to zerstören, vernichten

destruction Zerstörung
detach, to lösen, entfernen, losmachen, abmachen
detachable abnehmbar, trennbar, lösbar
detail Einzelheit
detailed ausführlich
detailer Teilkonstrukteur
detect, to demodulieren, entdecken, auffinden
detection Demodulation, Auffindung, Entdeckung
detector Demodulator, Detektor
detent Sperrglied
detention Sperrung
deteriorate, to verschlechtern
deterioration Verschlechterung
determinable feststellbar
determination Ermittlung, Bestimmung, Feststellung
determine, to bestimmen, ermitteln, feststellen
develop, to entwickeln
development Entwicklung
deviate, to abweichen
deviation Abweichung, Ablenkung
diagram, to als Schema zeichnen
diagram Diagramm, Schema, graphische Darstellung
dial, to wählen (eine Nummer)
dial Wählscheibe, Skala, Zifferblatt
dialling Wählen
diameter Durchmesser
diamond Diamant
diaphragm Membran, Blende
die Gesenk, Matritze
die-cast Spritzgußstück
dielectric dielektrisch
dielectric transducer dielektrischer Wandler
differ, to sich unterscheiden, differieren, abweichen
difference Unterschied, Differenz, Abweichung
diffluent zerfliessend
diffuse, to zerlaufen, diffundieren
diffuse zerstreut, diffus
diffusion Zerstreuung, Diffusion
digit Ziffer, Stelle
digital digital, ziffermässig
dilatation Dehnung, Ausdehnung
dilate, to dehnen
dilute, to verdünnen
dilution Verdünnung

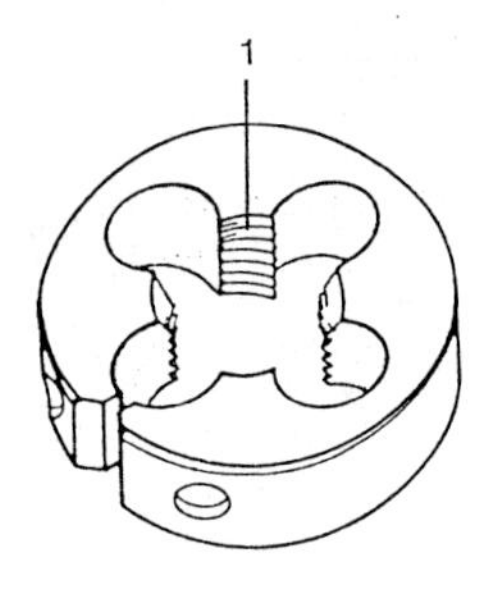

Threading die
(Gewinde-) Schneideisen

1 **cutting thread** Schneidgewinde

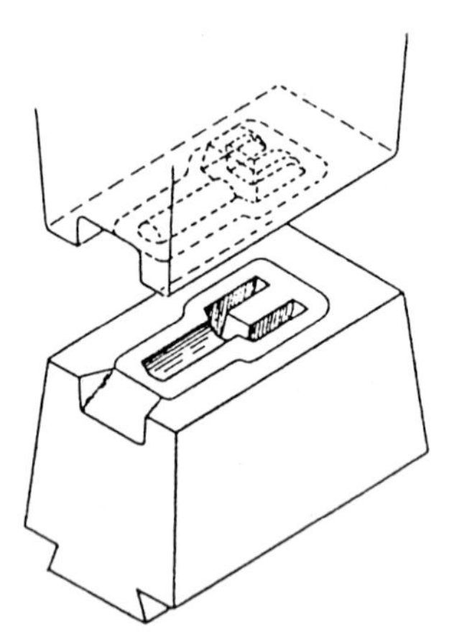

Stamping dies
(Stanz-) Preßmatrizen

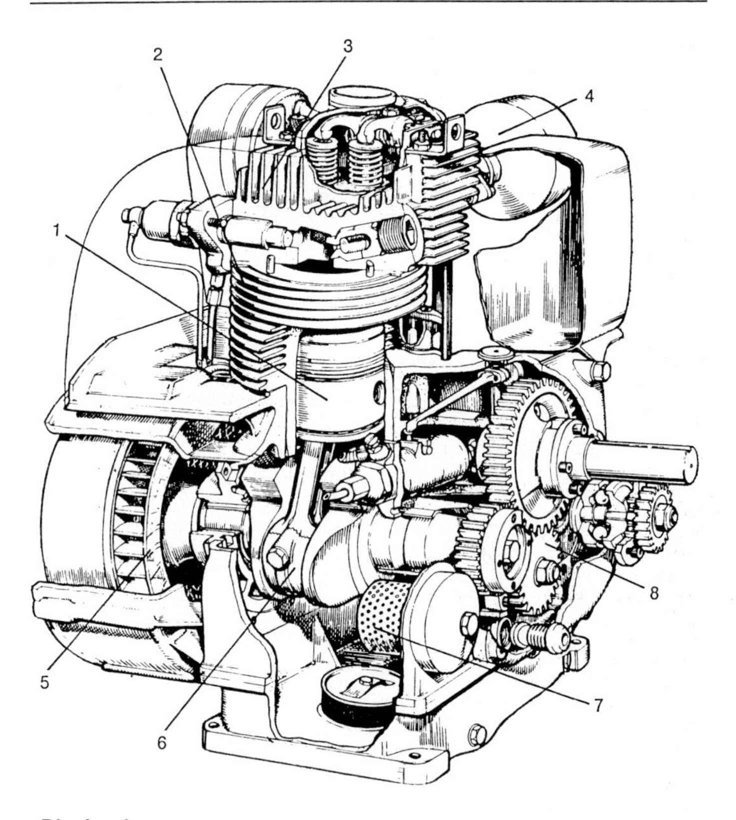

Diesel engine
Dieselmotor

1 **piston** Kolben
2 **cylinder** Zylinder
3 **finned cylinder head** mit Kühlrippen versehener Zylinderkopf
4 **fuel system** Kraftstoffsystem
5 **fan** Ventilator
6 **crank shaft** Kurbelwelle
7 **lubricating system** Schmiersystem
8 **gear** Getriebe

dim, to abblenden
dim lichtschwach, trübe
diminish, to verringern, vermindern, abschwächen
diminution Verminderung, Verkleinerung, Abnahme
dimmer Abblendregler, -schalter
dioxide Dioxyd
dip, to tauchen, eintauchen
diplexer Frequenzweiche, Diplexer
dipstick Meßstab zum Eintauchen
direct current Gleichstrom
direction Führung, Steuerung, Richtung
direction finder Peiler, Peilgerät
direction finding Peilung, Peilen
directional gerichtet, richtungsempfindlich
directional radio Richtfunk
directions for use Gebrauchsanweisung
directivity Richtwirkung, Richtvermögen
directory Telephonbuch
disable, to unwirksam machen, abschalten
disabled unbrauchbar, abgeschaltet
disappear, to verschwinden, entweichen
disassemble, to zerlegen, auseinandernehmen, demontieren
disassembly Zerlegung, Demontage
disc Scheibe, Platte
disc-type alternator Generator mit Scheibenanker
discharge, to entladen
discharge Entladung, Stromentnahme, Abfluss
discharger Entladewiderstand
disconnect, to abtrennen, abschalten, abklemmen, Verbindung lösen
disconnecting link Trennlasche
disconnecting switch Trennschalter
disconnection Trennung, Abschaltung, Ausrückung (Kupplung)
discontinue, to unterbrechen
discontinued nicht mehr gebaut (Bez. auf Datenblättern)
discontinuity Unstetigkeit, Diskontinuierlichkeit, Sprungstelle
discontinuous diskontinuierlich, unstetig
discover, to entdecken

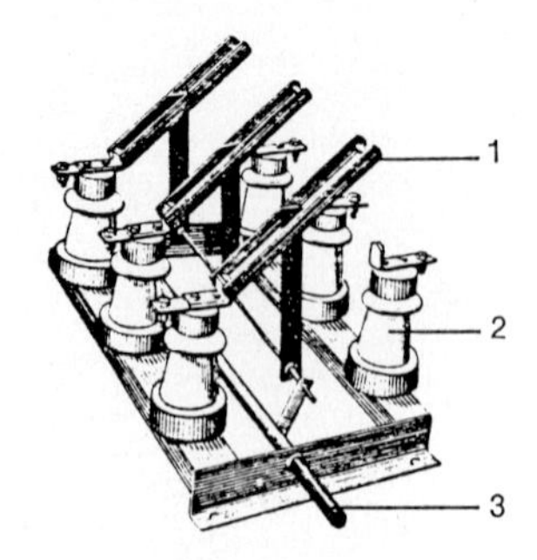

Disconnecting switch (high voltage)
Trennschalter (Hochspannung)

1 **switching blade** Schaltmesser
2 **insulator** Isolator
3 **operating rod** Betätigungsstange

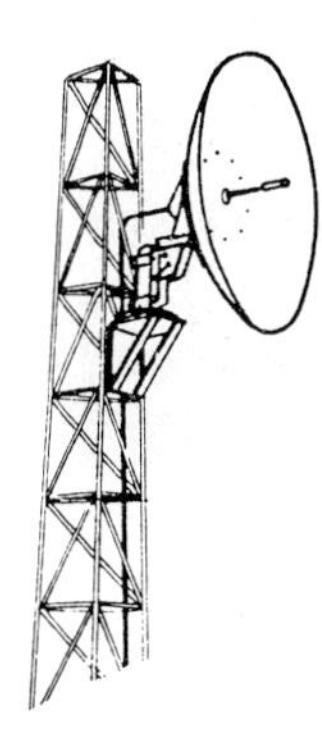

Dish (antenna)
Antennenschale, -schüssel, Parabolreflektor

discriminate, to unterscheiden, diskriminieren
discrimination Unterscheidung
discuss, to erörtern, besprechen, behandeln
discussion Erörterung, Besprechung
disengage, to lösen, entfernen, loskuppeln, ausrücken
disengaged frei, unbesetzt
disentangle, to entwirren
dish Parabolreflektor, Schale
dished schalenförmig
disintegrate, to (sich) zersetzen, auflösen, zerfallen
disintegration Zerfall, Zersetzung
dislocate, to verdrängen
dislocation Versetzung
dislodge, to verdrängen
dismantle, to demontieren, abbauen, auseinandernehmen
dismantling Demontage
dismember, to auseinandernehmen
dismount, to zerlegen, auseinandernehmen, ausbauen, demontieren
dismounting Demontage, Abmontierung
dispatch, to schicken, abfertigen
dispatch Beförderung, Abfertigung
disperse, to streuen, zerstreuen
dispersion Zerstreuung, Verteilung, Streuung, Dispersion
displace, to verdrängen, verrücken, verschieben
displaceable verschiebbar
displaced versetzt, verschoben
displacement Verdrängung, Verschiebung, Trennungsentschädigung
display, to darstellen
display Darstellung, Oszillogramm, Schirmbild
disposal plant Abfallbeseitigungsanlage
disruption Zerreissen
dissect, to zerlegen, zergliedern aufgliedern
dissection Zergliederung, Zerlegung
dissimilar unähnlich
dissipate, to verbrauchen, verzichten, verzehren, vergeuden
dissipated heat Verlustwärme

dissipation Zerstreuung, Verbrauch, Vergeudung
dissipator Kühlkörper
dissociate, to zerfallen, dissoziieren
dissociation Zerfall, Aufspaltung, Dissoziation, Auflösung
dissoluble auflösbar
dissolution Auflösung
dissolve, to zergehen, sich auflösen
distant entlegen, weit, fern
distance Abstand, Entfernung, Distanz
distant reading Fernablesung
distil, to destillieren
distillate Destillat
distilling apparatus Destillierapparat
distinct deutlich, klar
distinction Unterscheidung
distinctive auffallend, charakteristisch
distort, to entstellen, verzerren
distortion Entstellung, Verzerrung
distress Notlage
distress at sea Seenot
distribute, to verteilen
distribution Verteilung
distribution panel Verteilertafel
distributor Verteiler
district Bereich, Distrikt
disturb, to stören
disturbance Störung
ditch, to notwassern
dive, to tauchen
diverge, to auseinandergehen, divergieren
divergence Divergenz
diversified vielfältig, verschiedenartig
diversity Mannigfaltigkeit, Verschiedenheit
divert, to umleiten, umlenken
divide, to teilen, unterteilen
divining rod Wünschelrute
divisibility Teilbarkeit
divisible teilbar, unterteilbar
division Teilung, Einteilung, Unterteilung, Abteilung
document Beleg, Dokument
dog Klaue, Mitnehmer (Masch.)
domain Bereich
dome Deckel
domestic appliances Haushaltgeräte

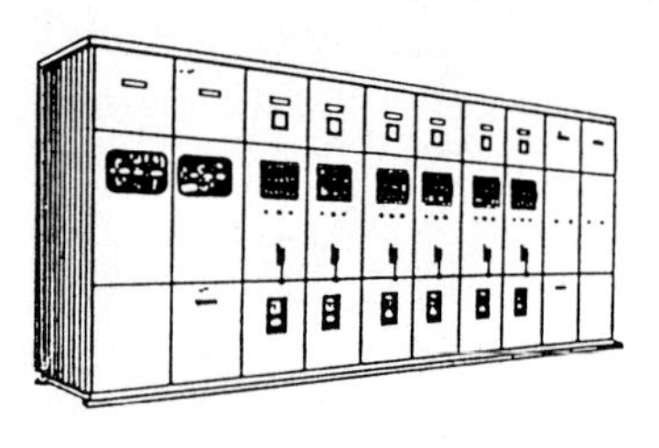

Distribution switchboard
Verteiler-Schalttafel

Dividers
Teil-, Stechzirkel

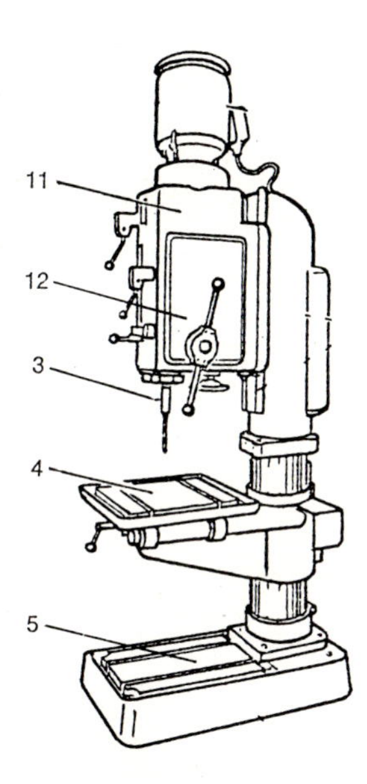

Drilling machine
Bohrmaschine

1 **machine column** Maschinenständer
2 **feed mechanism** Vorschubgetriebe
3 **spindle with twist drill** Spindel mit Spiralbohrer
4 **work table (slewing and adjustable for height)** Bohrtisch (schwenkbar und höhenverstellbar)
5 **slotted base plate for large workpieces** Grundplatte mit Spanneinrichtung für große Werkstücke

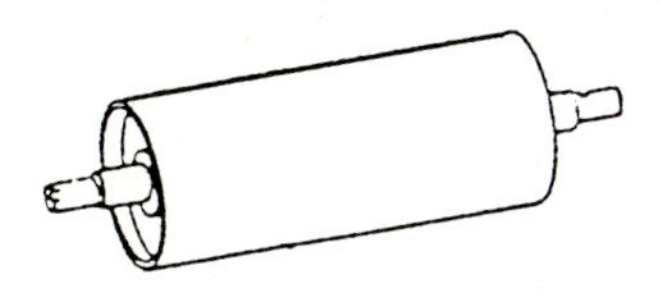

Drum
Trommel

(reversing drum of a conveyor belt Umkehrtrommel eines Förderbandes)

doped gedopt, angereichert
doping Dopen, Dotieren
dosage Dosis, Dosierung
dose, to dosieren
dose Dosis
dosed dosiert
dosing Dosierung
dot, to punktieren
dot Punkt, Fleck
double, to verdoppeln
double doppelt
double bridge Thomson-Brücke, Doppelbrücke
doubler Frequenzverdoppler
doubling Verdopplung
dovetail, to zinken
dovetail Schwalbenschwanz
dowel Dübel
dowelling Verdübelung
down tools, to Arbeit niederlegen
draft, to zeichnen
draft Zug, Entwurfsskizze, Tratte, Tiefgang
draftsman Zeichner
drag, to schleppen, ziehen, zerren
dragging Schleppen, Ziehen
drain (off), to entleeren
drain Ablauf, Abfluss
draining off Entleerung
draughtsman Zeichner
draw, to zeichnen, aufnehmen, abnehmen, ziehen
draw bolt Spannbolzen
dredger Bagger
dress, to nach(be)arbeiten
drier Trockenapparat
drill, to bohren
drill Bohrer
drill gauge Bohrlehre
drill jig Bohrschablone
drip, to tropfen
drive, to fahren, führen, aussteuern
drive Antrieb, Trieb
driver Treiber, Mitnehmer, Antriebsrad
drop, to fallen, sinken, abnehmen
drop Fall, Tropfen
droplet Tröpfchen
drum Trommel, Walze, Zylinder
dry, to trocknen
dry trocken
drying agent Trockenmittel
duct Kabelkanal, Kanal, Kanalzug, Dukt

duplex Duplex, Gegensprechverkehr, Duplexverkehr
duplexer Duplexer, Sende-Empfang-Schalter
duplicate, to vervielfältigen
duplicate Doppel, Kopie
duplicator Vervielfältiger
durability Haltbarkeit, Dauerfestigkeit
durable haltbar, widerstandsfähig, dauerhaft
duration Dauer
durobronze Hartbronze
dust, to Staub abwischen
dust Staub
duty Betriebsart, Arbeitszyklus
dwindle, to schrumpfen
dye Farbstoff
dyeing Färbung
dying-out Ausschwingen
dynamic range Lautstärkeumfang, Aussteuerungsbereich
dynamic dynamisch
dynamics Dynamik
dynamo Dynamo
dynamometer Dynamometer, Zugkraftgeber
dyne Dyn

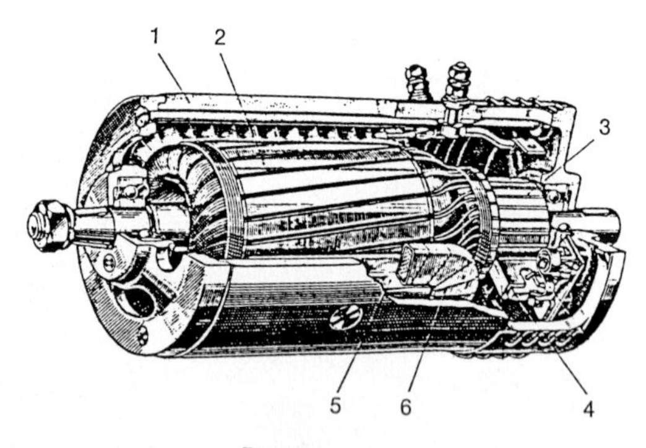

Dynamo
Gleichstrommaschine (Lichtmaschine)

1 **stator poles** Ständerpole
2 **armature** Anker
3 **collector** Kollektor
4 **coal brushes** Kohlebürsten
5 **exciter winding** Erregerwicklung
6 **pole shoe** Polschuh

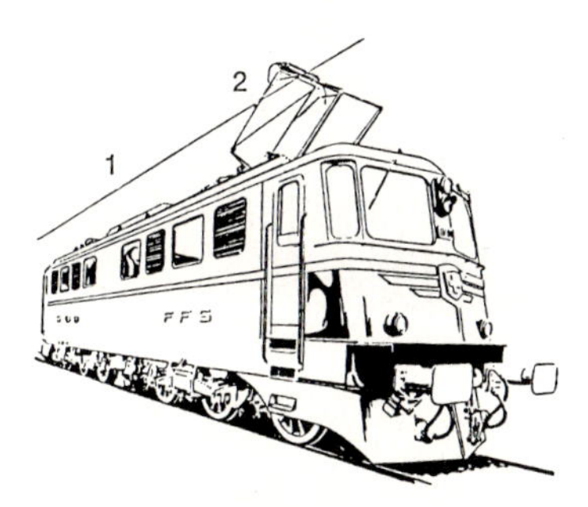

Electric locomotive
Elektro-Lokomotive

1 **overhead contact line** Fahrleitung
2 **current collector** Stromabnehmer

E

E-bend E-Krümmer, E-Bogen
E-coupling E-Kupplung
E-layer E-Schicht
ear Ohr, Zipfel, Oese
early stage Frühstadium
earnings card Lohnkarte
earth current Erdstrom
earth detector Erdschlussprüfer, Erdschlussanzeiger
earth return Erdrückleitung
earthed geerdet
earthenware Steingut
earthing Erdung
easing of work Arbeitserleichterung
ebonite Ebonit
ebulliometer Siedepunktmesser
ebullition Aufwallen
eccentric exzentrisch, Exzenter
echo, to nachhallen, widerhallen
echo Widerhall, Nachhall
echoing area Rückstrahlfläche
economical rationell, sparsam, wirtschaftlich
economy Wirtschaftlichkeit, Sparsamkeit
eddy Strudel, Wirbel
eddy current Wirbelstrom
eddy flow Wirbelströmung
edge Rand, Kante, Schneide
edgewise hochkant, hochkantig
educate, to ausbilden, schulen
education Ausbildung, Schulung
education film Lehrfilm
educational television Schulfernsehen
effect, to bewirken, erwirken, hervorrufen
effect Wirkung, Folge, Resultat
effective wirkungsvoll, wirksam
effectiveness Wirksamkeit, Leistungsfähigkeit
either-or circuit Oder-Oder-Schaltung
eject, to herausschleudern, auswerfen, ablegen
ejector pin Ausrückstift
elaborate, to ausarbeiten
elaborate aufwendig
elapse, to verstreichen (Zeit)
elastic elastisch, federnd

elasticity Elastizität, Federkraft
elbow Knie, Kniestück, Winkelstück
electric elektrisch
electrical elektrisch
electrified elektrisiert
electrification Elektrifizierung
electrify, to elektrifizieren, elektrisieren
electrodynamics Elektrodynamik
electrolysis Elektrolyse
electrolyte Elektrolyt
electrolyser Elektrolyseur
electromechanics Elektromechanik
electromotive elektromotorisch
electron beam Elektronenstrahl
electronic elektronisch
electronics Elektronik
electronogenic elektronenerzeugend
electrostatic elektrostatisch
elementary elementar
elevated railway Hochbahn
elevation Höhe, Erhöhung
elevator Höhenruder
eliminate, to beseitigen, eliminieren
elimination Beseitigung, Elimination
ellipse Ellipse
elliptical ellipsenförmig, elliptisch
elongate, to verlängern, dehnen
embed, to einbetten, betten
emboss, to erhaben ausarbeiten
embossed erhaben, plastisch
embrace, to umfassen
embrittlement Sprödewerden, Versprödung
emergency Notlage, Notfall, Notstand
emergency plant Notstromanlage
emery Schmirgel
emission Ausgabe, Strahlung, Emission
emissivity Emissionsvermögen
emitter Strahler, Emitter
emitting area Emissionsfläche, Strahlungsfläche
emphasis Betonung, Hervorhebung, Anhebung
emphasize, to betonen, hervorheben
empirical empirisch, erfahrungsmässig

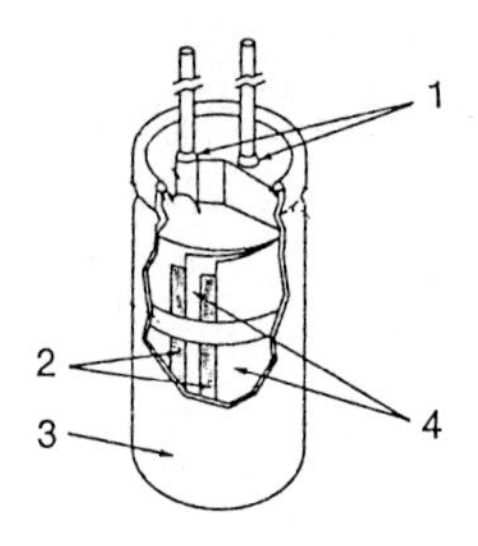

Electrolytic capacitor
Elektrolyt-Kondensator

1 **connections to electrodes** Elektrodenanschlüsse
2 **electrode foils** Elektrodenfolien
3 **aluminium case** Aluminiumkapsel
4 **paper separators saturated in electrolyte** in Elektrolyt getränkte Papierabtrennungen

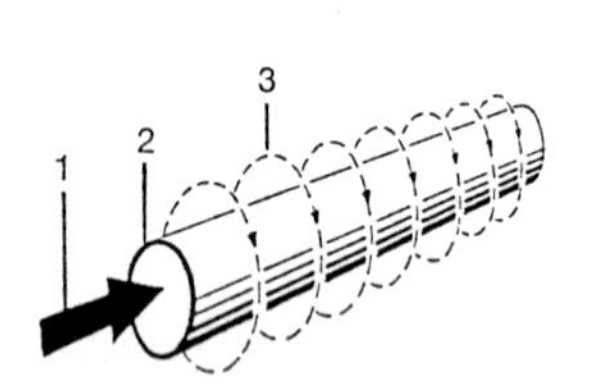

Electromagnetic field
Elektromagnetisches Feld

1 **current flow** Stromfluß
2 **electric conductor** elektrischer Leiter
3 **lines of force** Kraftlinien

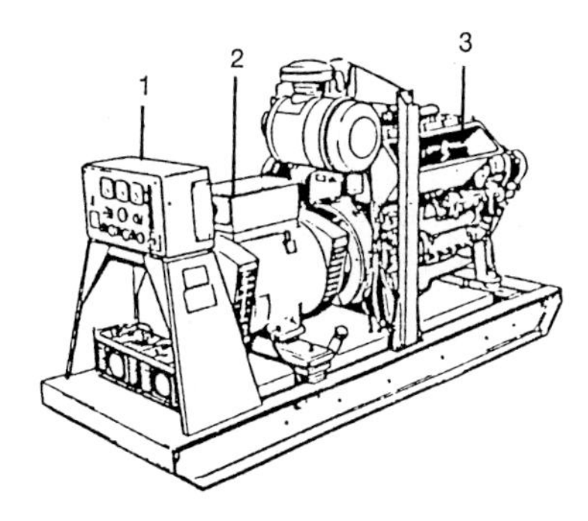

Emergency power plant (diesel-generator set)
Notstromanlage (Diesel-Generatorsatz)

1 **control box** Steuerkasten
2 **generator** Generator
3 **diesel engine** Dieselmotor

employ, to anwenden, verwenden, gebrauchen, beschäftigen
employee Arbeitnehmer, Angestellter
employer Arbeitgeber
employment Benutzung, Gebrauch, Beschäftigung, Anstellung
empty, to leeren, entleeren
empty leer
emulsify, to emulgieren
emulsion Emulsion
enamel, to emaillieren
enamel Emaille
enamelled wire Emailledraht
encapsulate, to einkapseln
encapsulation Kapselung, Einkapselung
encapsulated eingekapselt, gekapselt
encase, to mit einem Gehäuse versehen, einschliessen, einkapseln
encipher, to verschlüsseln
encircle, to umranden (Kreis)
enclosed umschlossen
enclosion Kapselung
enclosure Hülle, Kapsel, Einhüllung
encoder Kodierer, Kodiergerät
encrypt, to verschlüsseln
encryption Verschlüsselung
end Stirnfläche, Stirnseite
endurance strength Dauerfestigkeit
endurance test Dauerversuch
enduring dauerhaft, widerstandsfähig
energetics Energetik
energise, to Energie zuführen, (Relais) erregen
energisation Erregung, Erregen, Energiezuführung
energy Energie
enforce, to erzwingen
engage, to schalten, eingreifen
engaged eingreifend, belegt, besetzt
engaging lever Einrückhebel
engagement Eingriff, Einklinken
engine Motor, Triebwerk, Maschine
engineer Ingenieur
engineering Ingenieurwesen, Ingenieurwissenschaft, Technik
engrave, to gravieren
enhance, to steigern, vergrössern

enlarge, to erweitern, aufweiten, vergrössern
enlarged vergrössert
enlarger Vergrösserungsapparat
enlargement Vergrösserung, Erweiterung, Aufweitung
enlarging Vergrössern
enquire, to abfragen, anfragen
enquiry Abfrage, Rückfrage, Anfrage
enrich, to anreichern
enriched angereichert
enrichment Anreicherung
ensure, to sicherstellen, dafür sorgen
entangle, to sich verwickeln
enter, to eintreten, einlaufen, einfliessen
enterprise Unternehmen
entrance Einlass, Zugang
entrepreneur Unternehmer
entry Eintrag(ung), Eintritt
entwine, to verflechten
enumerate, to aufzählen
envelope, to einhüllen, umhüllen, einwickeln
envelope Hülle, Umhüllung, Umschlag
environment Umwelt
environmental Umwelt . . .
enwrap, to einwickeln
epoxy Epoxyd
equal gleich, gleichwertig
equality Gleichheit
equalisation Gleichmachen, Gleichmachung
equalise, to gleichmachen
equaliser Ausgleichsverbindung, Entzerrer
equalising Ausgleich, Entzerrung
equate, to gleichsetzen
equating Gleichsetzung
equation Gleichung
equilibrium Gleichgewicht
equip, to ausrüsten, einrichten, ausstatten
equipment Ausrüstung, Gerät(e), Einrichtung
equipped ausgerüstet
equivalent Aequivalent, äquivalent
erasable löschbar
erase, to löschen
erasing Löschen
erasure Löschung
erect, to errichten, aufstellen

A straight way to use electrical energy?
Eine direkte Methode, elektrische Energie zu nutzen?

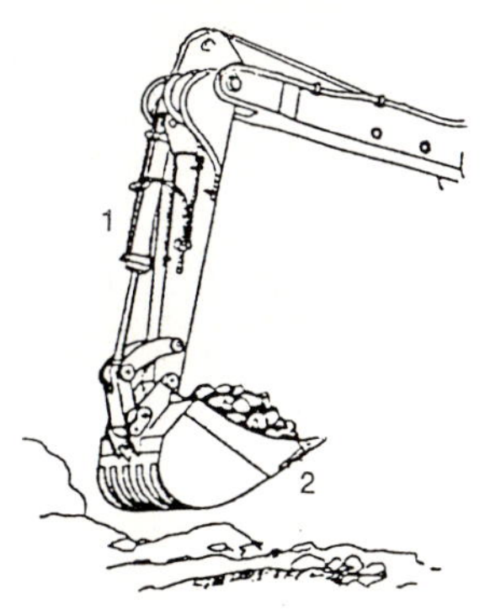

Excavator
(Trocken-)Bagger

1 **hydraulic cylinder** Hydraulikzylinder
2 **bucket** Eimer

erection Aufstellung, Errichtung
erg Erg
ergodic ergodisch
erosion Erosion
erratic sprunghaft
erroneous fehlerhaft
error Fehler, Irrtum
escalator Rolltreppe
escape, to entweichen, entströmen
escape Ausströmen
establish, to herstellen, errichten, festsetzen
establishment Errichtung, Herstellung, Festsetzung
estimate, to schätzen, veranschlagen, überschlagen, grob rechnen
estimate Abschätzung, Schätzung, Schätzwert, Voranschlag
estimation Schätzung
etch, to ätzen
etching method Ätzmethode
ethane Äthan
ether Äther
ethyl Äthyl
ethylene Äthylen
evacuate, to evakuieren, entleeren
evacuation Evakuierung, Entleerung
evaluate, to auswerten
evaluation Auswertung
evaporate, to verdampfen, verdunsten
evaporation Verdampfung, Verdunstung
evaporator Verdampfer
even, to glätten
even number gerade Zahl
evenly distributed gleichmässig verteilt
evenness Glätte, Ebenheit
event Vorgang, Ereignis
evidence Beweis
evolution Entwicklung
evolve, to entwickeln
exact genau
examination Untersuchung, Prüfung, Durchsicht, Beobachtung
examine, to prüfen, untersuchen
excavation Ausgrabung, Aushub
excavator Trockenbagger
exceed, to überschreiten, übersteigen
exceeding Überschreitung
excess Überschuss

exchange, to auswechseln, austauschen
exchange Auswechselung, Umtausch, Austausch
exchangeable auswechselbar, ersetzbar, austauschfähig
excitable erregbar
excitant Erregermasse (des Trockenelements)
excitation Erregung
excitation electrode Zündelektrode
excite, to erregen, anregen
exciter Erreger
exciter circuit Erregerkreis
exciting anode Zündanode
excursion Auslenkung, Auswanderung
execution time Ausführungszeit
executive leitender Angestellter, Chef, Leiter, Direktor
executive aircraft Geschäftsflugzeug
executive engineer etwa Oberingenieur
exempt from duty zollfrei
exemption Ausschliessung, Befreiung
exercise, to üben
exercise Übung, Ausführung
exert, to ausüben, anstrengen
exertion Anspannung, Anstrengung
exfoliate, to abblättern, abplatzen
exfoliation Abblätterung, Abschieferung
exhalation Ausdünstung, Brodem
exhale, to ausdünsten, aushauchen
exhaust, to auslassen, ausblasen, abbauen, absaugen, entleeren, erschöpfen
exhaust Abdampf, Abgas, Auspuff, Abströmgas, Austritt
exhaust stroke Auslasshub
exhausted abgebaut, abgejagt, verbraucht, vergriffen
exhauster Entlüfter, Exhaustor, Sauggebläse, Saugventilator
exhaustion vollkommene Abtrennung, Aufspaltung, Entleerung, Erschöpfung
exhaustive eingehend, erschöpfend
exhibit, to aufweisen, ausstellen, zur Schau stellen, vorzeigen

An experiment gone wrong?
Ein mißglücktes Experiment?

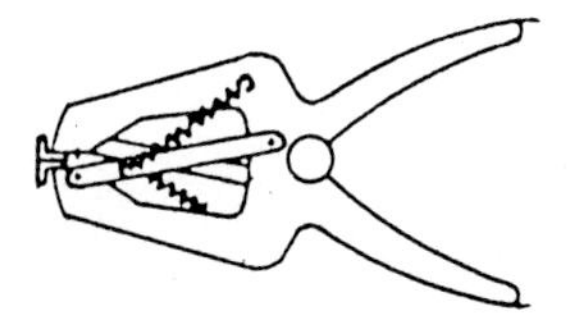

Expander for piston rings
Kolbenringzange

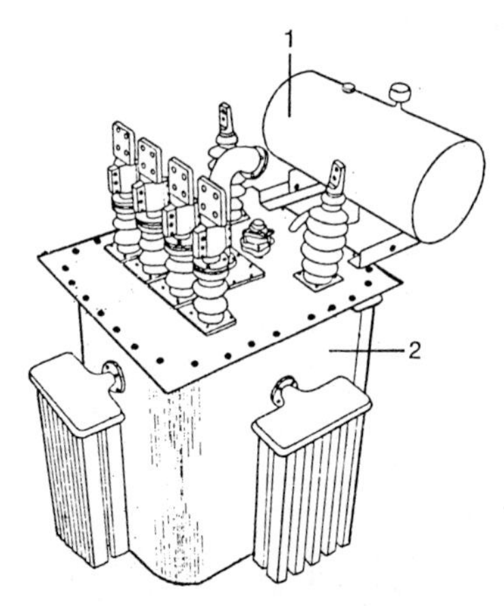

Expansion tank of an oil transformer
Expansionsgefäß eines Öltransformators

1 **expansion tank** Expansionsgefäß
2 **oil-filled transformer tank** ölgefüllter Transformatortank

exhibit Ausstellungsstück
exhibition Ausstellung, Darstellung, Schau
exhibitor Aussteller
exhort, to mahnen
exigency Anforderung, Bedarf, Erfordernis
exist, to bestehen
existence Bestand, Bestehen, Vorhandensein, Existenz
existent, to be vorliegen
existing vorhanden
exit, to hinausgehen
exit Ausgang, Auslass, Austritt
exoergic exotherm
exogenous nach aussen hin wachsend
exograph Röntgenbild
exorbitant übermässig, übertrieben
exothermal wärmeabgebend, wärmeliefernd
exothermic exotherm(isch), Wärme abgebend oder erzeugend
expand, to aufblähen, ausweiten, ausbreiten, ausdehnen
expansible ausdehnbar, dehnbar
expansion Aufblähung, Ausweitung, Ausbreitung, Ausdehnung
expansive ausdehnbar, expansiv
expatiate, to sich auslassen
expect, to abwarten, entgegensehen, erwarten
expediency Zweckmässigkeit
expedient Aushilfe, Ausweg, Behelf, Kunstgriff
expedite, to beschleunigen
expediting Versand
expel, to ausstossen, abtreiben, austreiben, herausspülen
expend, to aufwenden
expenditure Aufwand, Aufwendung, Ausgabe, Verbrauch
expense Abgabe, Ausgabe, Unkosten
expenses Aufwand, Auslagen, Kosten, Spesen
expensive kostbar, kostspielig, teuer
experience, to empfinden, erfahren
experience Erfahrung, Sachkenntnis
experienced erfahren, geschäftskundig, routiniert
experiment, to experimentieren

experiment Erprobung, Versuch, Experiment
experimental experimentell, versuchsmässig
experimenter Forscher
expert Fachmann, Sachbearbeiter, Experte
expertness Geschicklichkeit
expiration Ablauf, Ablaufen, Erlöschen, Verfall
expiration date Fristablauf
expire, to ablaufen, erlöschen, ausatmen, verstreichen
explain, to erklären, deuten, klar machen
explainable erklärlich
explanation Aufschluss, Auslegung, Deutung, Erklärung
explicit ausdrücklich, ausführlich, explizit
explodable explosibel, explodierbar
explode, to explodieren, bersten, detonieren
exploded view Darstellung in auseinandergezogener Anordnung
exploit, to ausbeuten, ausnutzen, auswerten, gewinnen
exploitable abbauwürdig, aufschliessbar
exploitation Abbau, Ausbeutung, Nutzung, Gewinnung
exploiter Ausbeuter
exploration Erforschung, Untersuchung, Abtastung, Schürfung
explore, to erforschen, untersuchen, abtasten, schürfen
explorer Entdecker, Erforscher, Taster
explosion Detonation, Explosion, Sprengung, Knall
explosion hazard Explosionsgefahr
explosive Sprengstoff, explodierbar, explosiv
explosiveness Explosionsfähigkeit
exponent Exponent, Anzeiger, Potenz
exponential exponentiell, exponential
exporter Exporteur
export, to ausführen, exportieren
exports Aussenhandel, Exportartikel
expose, to aussetzen, auslegen, beschicken, entblössen, belichten

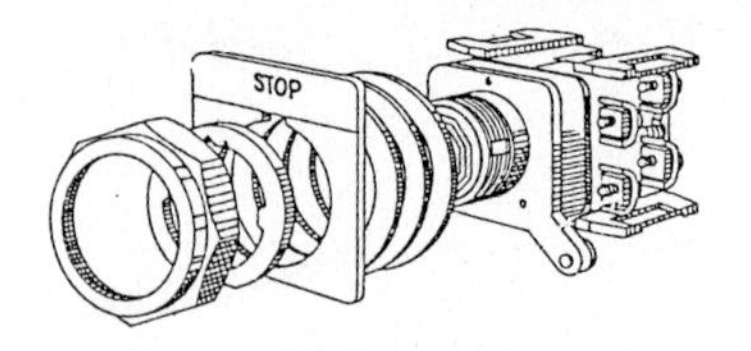

Exploded view of a control push-button switch
Auseinandergezogene Darstellung eines Steuer-Druckknopfschalters

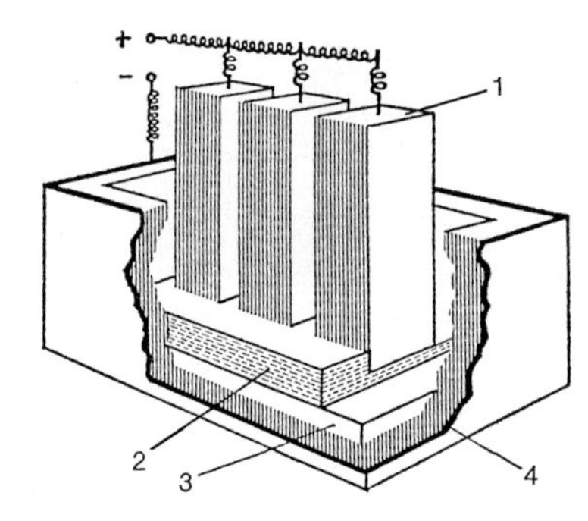

Extraction of aluminium by electrolysis
Gewinnung von Aluminium durch Elektrolyse

1 **carbon electrode** Kohle-Elektrode
2 **molten solution of alumina in cryolite** Tonerde-Kryolith-Schmelze
3 **carbon lining** Kohle-Auskleidung

Extractor (for removing pulleys and pinions)
Abziehvorrichtung (zum Abziehen von Scheiben und Ritzeln)

exposition Schaustellung
exposure Aussetzen, Aussetzung, Aufnahme, Belichtung
express, to ausdrücken, äussern
expression Ansatz, Ausdruck, Bezeichnung
expressions Formelausdrücke
expressive ausdrucksvoll
expropriate, to enteignen
expropriation Enteignung, Beschlagnahme
expulsion Ausstoss, Ausstossung, Ausweisung
expunged ausgestrichen
exquisite auserlesen
exsiccate, to austrocknen
exsiccation Austrocknung
extend, to ausbauen, ausbreiten, ausdehnen, auseinanderbiegen, ausfahren
extension Ansatz, Ausbau, Ausbreitung, Ausdehnung, Verlängerung, Zuwachs
extensive ausgedehnt, umfangreich, umfassend, vielseitig
extent Ausdehnung, Aushalten, Umfang
exterior äusserlich, Aussen . . .
external aussenliegend, äusserlich
exterpolate, to extrapolieren
extinct ausgestorben, erloschen
extinction Auslöschung, Ablöschung, Tilgung
extinguish, to auslöschen, zum Erlöschen bringen
extinguishing Löschung, Ablöschung
extract, to abreissen, abscheiden, ausfördern, ausscheiden, herausziehen
extract Auszug
extraction Gewinnung, Ausziehung
extractor Abziehvorrichtung
extradite, to ausliefern
extraordinary ausserordentlich, ungewöhnlich
extrapolate, to extrapolieren
extrapolation Extrapolation
extreme ausserordentlich, übermässig, übertrieben
extremity Ende, Endpunkt
extricate, to freimachen, losmachen, herauswickeln
extrinsic semiconductor Störhalbleiter

extrovert, to herausstülpen
extrude, to strangpressen, ausstossen, herauspressen
extruder Strangpresse, Spritzmaschine
extrusion Aufdornen, Ausstossung, Strangpressen, Pressen
exudation Ausscheidung, Ausschwitzung
exude, to ausscheiden, ausschwitzen, sickern
ex works supplies Werkablieferungen
eye Auge, Loch, Oese, Oehr
eyebolt Augbolzen, Gewindeöse, Hebeauge, Ringschraube
eyepiece Lupe, Einblick, Okular
eyelet Anschlussöse, Oesenloch, Oehr

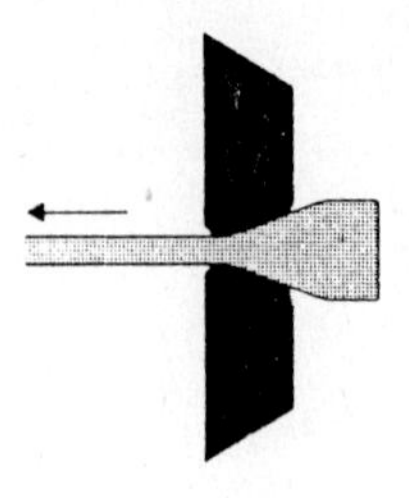

Extrusion
Strangpressen

F

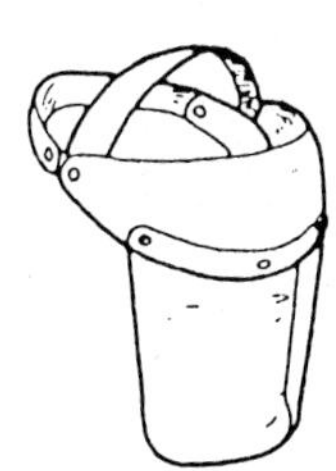

Face screen
Gesichtsschutz

fabric Bespannstoff, Erzeugnis, Tuch
fabricate, to fabrizieren, verarbeiten
fabricating operation Fabrikationsvorgang
fabrication Fabrikation, Fertigung
face, to abfasen, abflachen, planen
face Front, Fläche, Kranz
face-plate Frontplatte, Planscheibe
facet Abschrägung, Schrägung, Seitenfläche
facilitate, to erleichtern
facilities Einrichtungen, Erleichterungen
facility Einrichtung, Erleichterung
facing Aufschlag, Belagring, Fräsen, Planfräsen, Schlichten
facings Plandrehspäne
facsimile Bildtelegraphie, Fernphotographie
factorial function Fakultät (Math.)
factory Betrieb, Fabrik
facts Daten
factual tatsächlich
factual report Tatbestandaufnahme
faculty Fakultät
fade, to erblassen, verbleichen, verschwinden
fader Mischer, Lautstärkeregler, Überblender
fadings Lautstärkeschwankungen Überblendung
fail, to aussetzen, fehlgehen, misslingen, versagen
fail-safe ausfallsicher, Selbstschutz
failure Ausfall, Aussetzen, Aussetzer, Misserfolg, Versagen
faint, to abbauen, ohnmächtig werden
faintness Schwäche
fair Mustermesse, Ausstellung
fairing Füllstück, Umkleidung
fairlead Halterung
fake echo Täuschecho

A facsimile transmission of the pilot's wife?
Eine Bildsendung der Frau des Piloten?

fallacious argument Trugschluss
fallacy Trugschluss
fallow fahl
false falsch, irrig, nachgemacht
falsification Fälschung
falsify, to fälschen
falter, to stocken
familiar vertraulich, vertraut
familiarise, to vertraut machen
fan Gebläse, Lüfter, Ventilator, Fächer
fanned beam antenna Fächerstrahlantenne
fast access storage Schnellspeicher
fasten, to anheften, anschrauben, befestigen, festmachen
fastener Befestigungselement, Halter
fastening Befestigung, Verschluss, Anknüpfung
fastidious anspruchsvoll, wählerisch
fathom, to eine Tiefe abmessen, ergründen
fathom Faden, Fadon, Klafter
fatigue, to altern, ermüden
fatigue Abschwächung, Erlahmung, Ermüdung
fatigue bend test Dauerbiegeversuch
fatigue test Dauerprüfung
faucet Hahn, Absperrvorrichtung, Zapfen
fault Fehler, Störung, Störstelle, Sprung
faultless fehlerfrei, tadellos
faulty fehlerhaft, gestört, mangelhaft, rissig, schadhaft
favour, to begünstigen, bevorzugen, fördern, unterstützen
favour Gefälligkeit, Vergünstigung
fawn fahl
fay, to zum Fluchten bringen
faying surface Dichtungsfläche, anpassende Oberfläche
feasability Ausführbarkeit, Durchführbarkeit
feasible durchführbar, ausführbar
feat of endurance Dauerleistung
feather Feder (für Keilnut), Federkeil
feathering Schlingern (bei Turbinenschaufeln)
feature Besonderheit, Gesichtspunkt, Hauptartikel

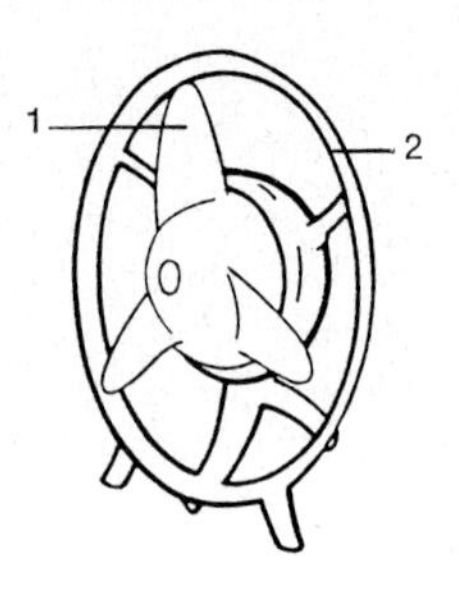

Fan
Lüfter, Ventilator

1 **blade** Flügel
2 **frame** Rahmen

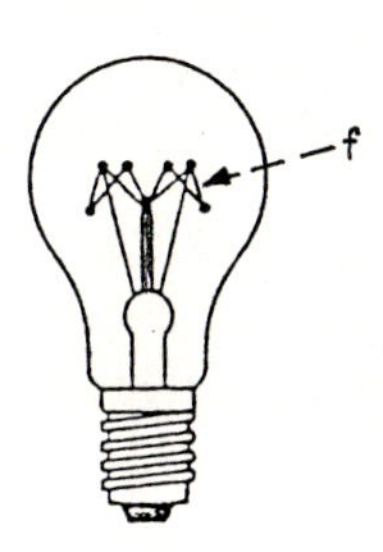

Filament
Glühfaden

fee Abgabe, Gebühr, Honorar
feeble dünn, matt, schwach
feed, to speisen, beschicken, beliefern, Vorschub geben, zuführen
feed Einspeisung, Beschickung, Förderung, Vorschub, Zufluss
feedback Rückführung, Rückkopplung, Rückwirkung
feeder Speiseleitung, Füllvorrichtung, Abzweig, Vorschubeinrichtung
feeler Füllehre, Taster
feldspar Feldspat
feldspathic feldspathaltig
felly Felge
felt Filz, Pappe
felted verfilzt
ferment, to fermentieren, aufgehen, treiben
ferment Ferment, Gärstoff
ferric eisenhaltig
ferrite Ferrit
ferrous eisenhaltig, eisern
ferrule Ring, Kapsel
ferry, to überführen
ferry Fähre
fertile ergiebig, fett, fruchtbar
fertility Ausgiebigkeit, Ergiebigkeit
fertilise, to düngen
fertiliser Düngmittel, Dünger
festoon Gehänge
fibre (fiber) Faser, Fiber
fibery faserig
fibrous fadenförmig, faserförmig, fibrös
fidelity Genauigkeit (Treue der Wiedergabe)
field Fach, Gebiet
figurative bildlich
figure Abbildung, Bild, Figur, Form, Zahl, Ziffer
figures Bezifferung, Werte
filament Draht, Glühfaden, Leuchtdraht
filamentary drahtförmig, fadenförmig
file, to ablegen, einordnen, feilen, registrieren
file Ablage, Bündel, Ordner, Feile
files, in gliederweise
fileting Ebnen
filigree Drahtarbeit
filing Anmeldung, Zustellung

filings Feilspäne, Feilstaub
fill, to auffüllen, aufschütten, beladen, strecken
filler Einlage, Formstück, Fördermann, Füller
filling Auffüllen, Anfüllung, Füllung
fillister Falzhobel
film, to beschlagen, filmen, überziehen
film Schicht, Überzug, Anstrichfilm
filth Dreck, Kot, Schmutz
filthy schmutzig
filtrate, to durchfiltern
filtrate Filtrat
fin Flosse, Grat, Finne, Bart
find Fund
finder Sucher, Finder
finding Bestimmen, Suchen
findings Ermittlungen, Befunde
fine Strafe
fined iron Frischfeuereisen
fineness Feinheit, Feingehalt
fines Abrieb, Feinschlag
fine adjustment Feineinstellung
fining Frischen
finish, to aufarbeiten, bearbeiten, beendigen, glätten, schlichten
finish Anstrich, Bearbeitung, Politur, Ende
finisher Ausrüster, Polierwalze
finite begrenzt, endlich
finiteness Endlichkeit
finned gerippt
fins cast integral angegossene Rippen
fire, to feuern, zünden, abdrücken, schiessen
firebox Brennkammer, Feuerbüchse, Feuerraum
fire-brick Chamottenstein, Brandziegel
fire-clay Schamotte, Feuerton
fire department Feuerwehr
fire-fighting Brandbekämpfung
firm dicht, derb, fest, stabil
fiscal year Rechnungs-, Berichtsjahr
fishplate, to verlaschen
fishplate Knotenblech, Fischplatte, Stossblech
fission Spaltung
fissionable spaltbar
fissure Bruch, Naht, Sprung

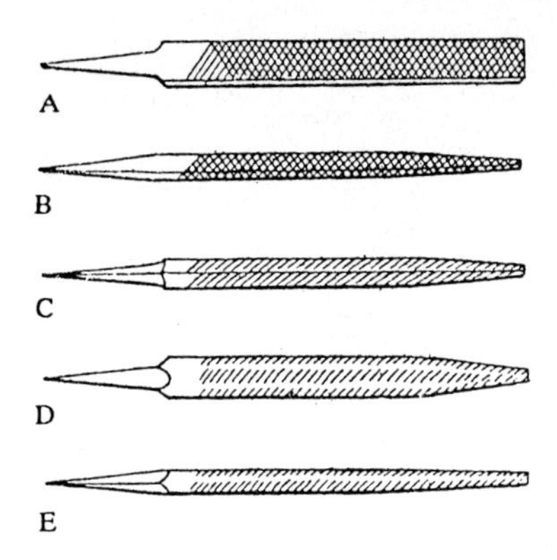

Files
Feilen

A **flat file** Flachfeile
B **square file** Vierkantfeile
C **three-square file** Dreikantfeile
D **half-round file** Halbrundfeile
E **round file** Rundfeile

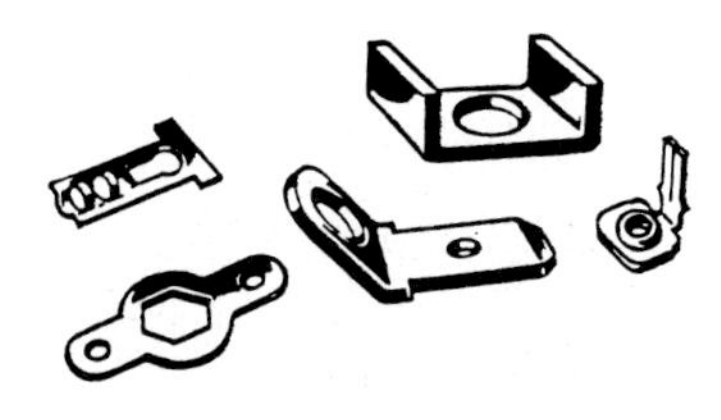

Fittings
Beschlagteile

fit, to anpassen, anbringen, aufmontieren, zurichten, zusammensetzen
fit Sitz, Passung
fitness Brauchbarkeit, Eignung
fitter Monteur, Maschinenschlosser, Rüster
fittings Apparatur, Fittings, Garnitur, Zubehör, Beschlagteile
fix, to befestigen, einsetzen, bestimmen, abgrenzen
fix Beizen, Festpunkt, Standortbestimmung
fixable feststellbar
fixture Aufsatz, Vorrichtung, Aufsatz
flabby schlaff, lappig
flag Fahne, Flagge, Fliese
flagged floor Fliesboden
flake, to abblättern
flammability Feuergefährlichkeit
flammable entflammbar, entzündlich
flange, to anflanschen, bördeln
flange Flansch, Bördel, Gurtung
flank, to umgehen, begrenzen, flankieren
flank Flanke, Schenkel, Ende
flapper Prallplatte
flapping Schlagen
flash, to blinken, flackern, auflodern, schiessen
flash Aufblitzen, Blitz, Flackern
flashlight Taschenlampe, Blitzlicht
flashover Funkenüberschlag
flashings Abweisbleche
flask Flasche, Fläschchen, Kastengussform
flat Reifenpanne, Schwimmer, Anflächung
flat flach, eben, platt
flatness Flachung, Ebenheit, Fläche
flats Flacheisen
flatten, to abflachen, abplatten, ebnen, strecken, flachschlagen
flattening Abflachung, Abplattung, Dämpfung
flaw Riss, Blase, Flinse, Sprung
flaw in casting Gussfehler
flawless fehlerlos, rissfrei, tadellos
flection Biegung
fleece roller Aufroller

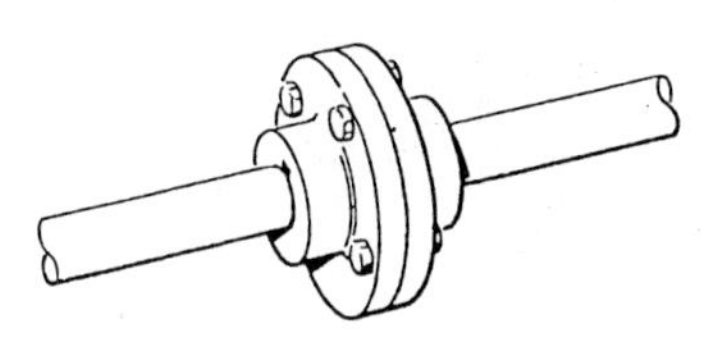

Flange coupling
Flanschkupplung

flex, to biegen
flexible biegsam, elastisch, anpassungsfähig
flexural fatigue strength Biegungsdauerschwingfestigkeit
flexure Abbiegung, Beugung, Biegebeanspruchung
flicker, to flackern, flimmern, blinken
flickerless flimmerfrei
flight Flug, Plattfuss, Flucht
fling, to schlenkern, schleudern
fling Schleudern
flint Kiesel, Flint(stein)
flintiness Glasigkeit
flinty kieselartig, kieselig
flip coil Feldinduktionsspule
flip-flop circuit Flip-Flop-Schaltung, binäre Zählstufe
flitch plate Verstärkungslasche
float, to schwimmen, aufschwemmen, (Batterie) puffern, schweben
float Floss, Schwimmkörper
floatplane Schwimmerflugzeug
floatability Schwimmfähigkeit
floatable flössbar, flotierbar
floater Pegel, Schwimmer
flocculate, to flocken, flockig machen
flock, to strömen, zulaufen
flood, to anstauen, überfluten, überschwemmen
flood Flut, Hochwasser
floodlight, to anstrahlen
floodlight Flutleuchte, Flutlicht, Tiefstrahler
floor Boden, Arbeitsbühne, Stockwerk
flooring Fussboden, Belag
flow, to fliessen, strömen, laufen
flow Ausfluss, Ausströmung, Fluss
flowmeter Durchfluss-, Durchlauf-Mengenmesser
flowing Fliessen, Verlauf
fluctuate, to fluktuieren, schwanken, streuen
fluctuation Fluktuation, Schwankung
flue Feuerkanal, Feuerrohr, Flammenrohr, Ofenzug
flue boiler Flammrohrkessel
flue dust Flugasche
fluid Flüssigkeit, Fluidum
fluidify, to flüssig machen

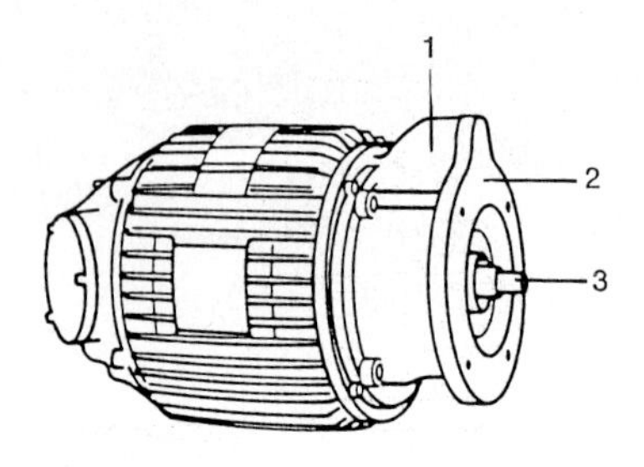

Flange motor
Flanschmotor

1 **housing** Gehäuse
2 **fitting flange** Befestigungsflansch
3 **shaft end** Wellenende

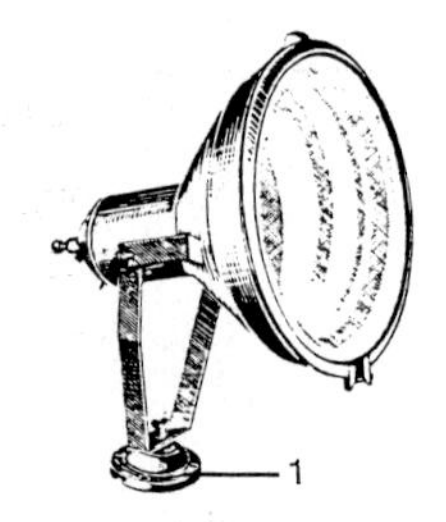

Floodlight
Flutlichtleuchte

1 **fixture** Befestigungssockel

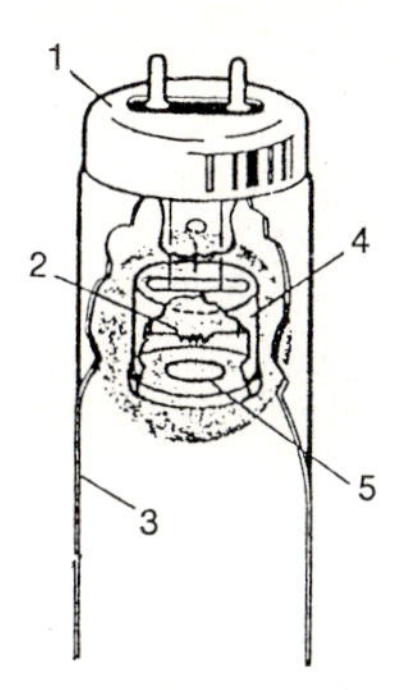

Fluorescent tube
Leuchtstoffröhre

1 **base** Sockel
2 **cathode of coiled tungsten** Kathode aus Wolframwendel
3 **inside coating of fluorescent powder** inseitiger Leuchtstoffüberzug
4 **reflective anode screen** reflektierender Anodenschirm
5 **mica disc with hole** gelochte Glimmerscheibe

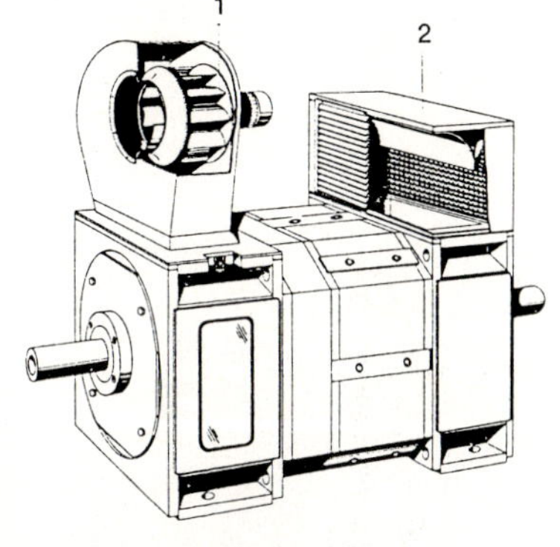

Forced-air cooled motor
Motor mit Luftumwälzkühlung

1 **air-circulating fan** Luftumwälzventilator
2 **air-exhaust housing** Luftaustrittsgehäuse

fluke Ausräumer, Klaue, Schaufel
fluoresce, to fluoreszieren
fluorenscence Fluoreszenz
fluorescent lamp Leuchtstoffröhre
fluoride Fluorid, Fluorsalz
fluorine Fluor
fluorite Fluorkalzium, Flußspat
flush, to bündig machen, bündig abschneiden
flush Baufluchtlinie, bündig, glatt, in gleicher Ebene
flush pin gauge Tiefenlehre
flushing Ausspülen, Wasserspülung
flute cutter Gewindenutenfräser
fluted ausgekehlt, genutet, gerillt
flutter, to schwanken, flattern, vibrieren
flutter Flattern, Vibrieren, Zittern
flux Fluss, Kraftströmung (el.), Magnetfluss
fluxing Flussmittel
fly, to schiessen lassen, fliegen
fly ash Flugasche
fly-wheel Schwungrad
foam, to schäumen
foam Schaum
foamer Schaumbildner
foci (Pl.) Brennpunkte
focus, to scharf einstellen, in den Brennpunkt stellen
focus Brennpunkt, Fokus
focussing Einstellung, Bündelung
fog, to sich verschmieren, verschleiern, anlaufen
foil Blatt, Blattmetall, Folie
fold, to falten, falzen, knicken
fold Falte, Biegewulst
foldable faltbar, zusammenklappbar
folder Broschüre, Faltprospekt, Bördelmaschine
folio Blatt, Kolumnenziffer
follow, to folgen, nacheilen, sich anhängen
follow-up Rückführung
follower Anhänger, Mitnehmer, Gewindebacke
foment, to blähen
foment action Blähung
foolproof narrensicher, betriebssicher
foot Fuss, Unterteil
foot-operated fussgeschaltet
force, to treiben, zwingen, forcieren

force Kraft, Energie, Gewalt
forced air cooling Gebläsekühlung
forceful gewaltsam
forceps Kluppzange, Pinzette, Zange
forcible heftig, kräftig
fore shorten, to perspektivisch zeichnen
forecast, to vorhersagen
forecast Prognose, Vorhersage
forecastle Back, Vordeck
foreman Vorarbeiter, Werkmeister, Bauführer
forfeit, to aufgeben, einbüssen
forge, to schmieden
forge Schmiede
forgeable schmiedbar
forger Fälscher
forgery Fälschung
fork, to sich gabeln
fork Gabel, Maulschlüssel
form, to bilden, formen
formed gestaltet
former Bildner, Gestalter, Formblock
former's tools Formerwerkzeug
formless formlos
forms of joints Stossverbindungsart
formula Formel
forward, to befördern, transportieren, zustellen
forward vorn, voraus, vorwärts
forwarder Transporteur
forwarding agent Transportmakler
foul, to verschmutzen, verschleimen
found, to giessen, gründen
foundation Fundament, Gründung
founder, to scheitern
founder Gründer, Giesser, Schmelzarbeiter
founding Giessereiwesen, Abguss, Guss
foundry Giesserei
fount Guss
fountain Brunnen, Fontäne
fraction Bruch
fractional gebrochen, Bruch ...
fractionise, to brechen
fractions Bruchziffern
fracture, to brechen, zerbrechen
fractured brüchig
fracturing Zerstückelung
fragile brechbar, zerbrechlich, brüchig

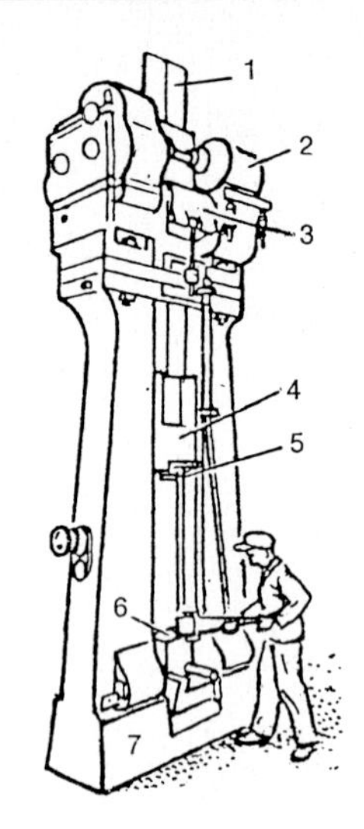

Forge drop hammer
Schmiede-Fallhammer

1 **lifting board** Hubbrett
2 **driving motor** Antriebsmotor
3 **winch** Aufzugsrolle
4 **hammer block** Hammerbär
5 **hammer die** Ober-(Hammer-) Gesenk
6 **bottom die** Untergesenk
7 **anvil block** Amboßblock (Schabotte)

Fork-lift truck
Gabelstapler

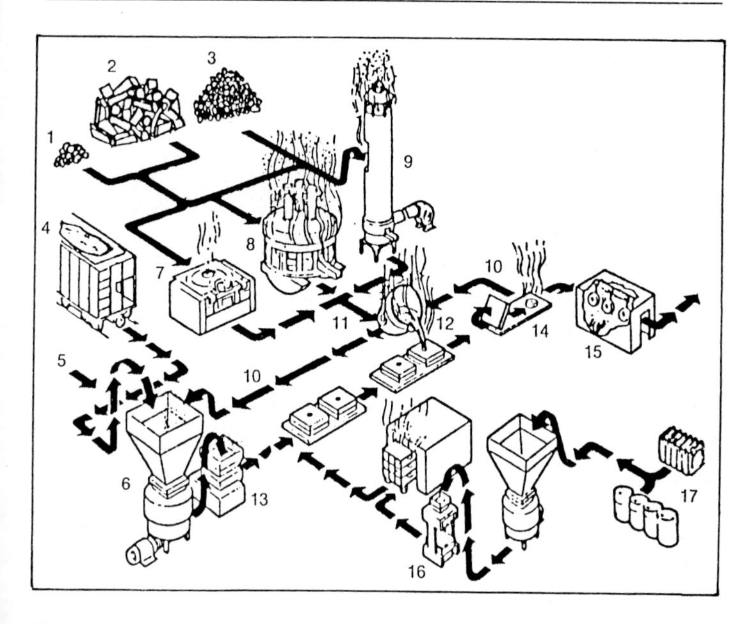

Foundry processes and equipment
Gießerei-Prozesse und Ausrüstungen

1 **additives** Zuschläge
2 **scrap metal** Schrott
3 **coke** Koks
4 **moulding sand** Formsand
5 **binder** Binder
6 **sand preparation** Sandaufbereitung
7 **induction furnace** Induktionsofen
8 **arc furnace** Lichtbogenofen
9 **cupola** Kupolofen
10 **used (burnt) sand** Altsand
11 **liquid iron** Flüssigeisen
12 **pouring** gießen
13 **moulding** formen
14 **shake-out** auspacken
15 **after-treatment of castings** Gußnachbehandlung
16 **core making** Kernmacherei
17 **core sand and binder** Kernsand und Binder

fragment Bruchteil, Brocken, Stück
fragments (Pl.) Bruchstücke, Trümmer
frail morsch, hinfällig
frame, to einfassen, umrahmen
frame Rahmen, Gestell, Ständer, Gehäuse
frangible brechbar
fray, to abfasern, sich abnutzen
freak value Zufallswert
free-wheel ungebremst, Walzenfreilauf
free-wheeling Freilauf, freilaufend
freeze, to gefrieren, frieren
freeze Frieren
freezer Eismaschine
frequency Frequenz
frequent, to frequentieren
frequent häufig, vielfach
fret, to durch Reibung abnutzen
fret Abnutzung durch Reibung
friable bröckelig, brüchig
friction Reibung
friction coupling Reibungskupplung
frictionless reibungslos
frigid kalt
frill, to kräuseln
fringe Franse, Streifen, Saum
frit, to fritten, schmelzen, sintern
frit Fritte, Glasmasse
front-end Stirnseite
frontier Grenze, Grenzgebiet
frost, to gefrieren, mattieren
froth, to gären, aufschäumen
frustum Kegelstumpf
frying Knallgeräusche, Prasseln
fuel, to Brennstoff tanken
fuel Benzin, Kraftstoff, Betriebsstoff
fuel-pump Brennstoffpumpe
fueling Tanken
fulcrum, to schwenken, anlenken
fulcrum Drehpunkt, Hebelpunkt
full-scale maßstäblich
fume, to dampfen, rauchen
fume Abgas, Dunst, Rauch
fumers Brodem, Dämpfe
function, to funktionieren, arbeiten
function Funktion, Gang
functional funktionell, betrieblich
funds Fonds
fungicide Konservierungsmittel
funicular Berg- und Talbahn

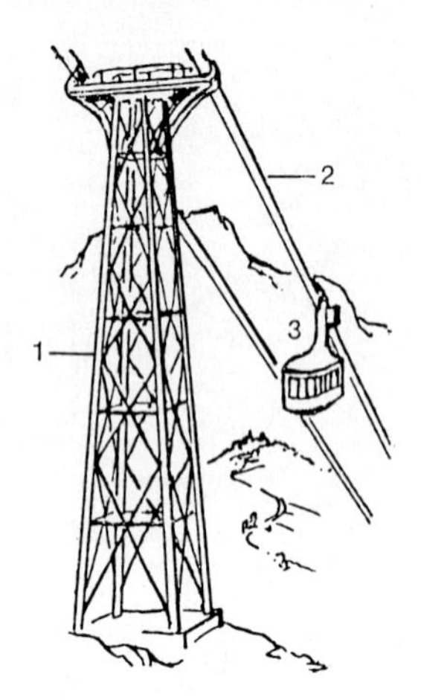

Funicular
Bergseilbahn

1 **pylon** Tragmast
2 **traction and carrying rope** Zug- und Tragseil
3 **cabin** Kabine

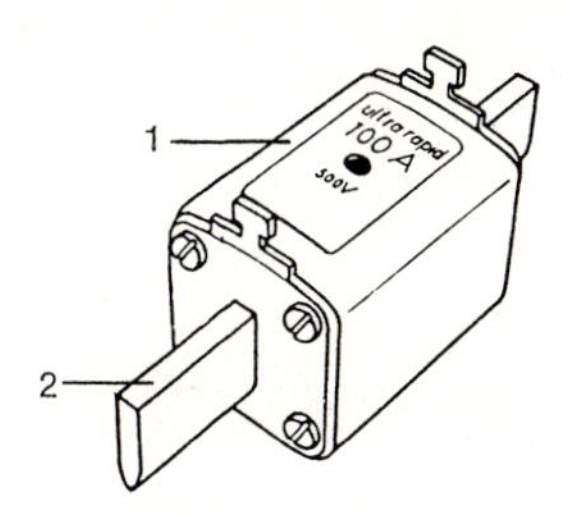

(electric) fuse
(elektrische) Sicherung

1 **fuse cartridge** Sicherungspatrone
2 **plug-in contact** Einsteckkontakt

funnel Schornstein
fur Kesselstein
furbish, to abschleifen, polieren
furlough Urlaub
furnace Brennofen, Feuerungsanlage
furnish, to ausfüllen, liefern, versehen
furrow, to riffeln, durchfurchen
furrow Furche, Nute, Rille
further, to fördern
furtherance Vorschub
furthering Förderung
fuse, to durchbrennen (Sicherung), aufschmelzen, aufschweissen
fuse Sicherung, Brennzünder, Lunte
fuselage Flugzeugrumpf
fusetron Elektronensicherung
fusible schmelzbar, Schmelzsicherung
fusing Abschmelzen, Ansprechen, Durchbrennen
fusion Anschluss, Bindung
fuzz Flaum, Franse
fuzz, to abfasern

G

gag Knebel
gag press Richtpresse
gain, to gewinnen, erzielen, verstärken
gain Ausbeute, Gewinn, Pegel
gainfully employed erwerbstätig
gale Sturmwind, Kühle
gall, to durch Reibung abnutzen, beschädigen
gall Abnutzung durch Reibung
galleries Förderstollen
gallery Bühne, Galerie, Stollen
galley Kombüse
gallon Gallone
galvanic galvanisch
galvanism Galvanismus
galvanisation Galvanisierung, Verzinkung
galvanise, to galvanisieren, verzinken
galvanometer Galvanometer, Strommesser
gangway Durchgang, Laufbrücke
gantry Bockkran, Krangerüst
gap Fuge, Lücke, Leerstelle
gape, to klaffen
garage, to Fahrzeuge unterstellen
garbage Müll, Schund
garble Fremdstörung, Verstümmelung
garnet Granat
garnish, to ausstaffieren
garniture Beschlag, Garnitur
gas-tight gasdicht
gaseous gasartig, gasförmig
gash, to mit Einschnitten versehen
gasification Gasbildung, -erzeugung
gasket Dichtring, Packung, Abdichtung
gasket board Dichtungspappe
gasoline Benzin, Gasolin, Leichtbenzin
gassy gasführend
gate, to austasten, einblenden, abblenden
gate Tor, Eingußstelle, Ablaufpunkt
gate valve Absperrschieber
gather, to sammeln, entwickeln, auflesen

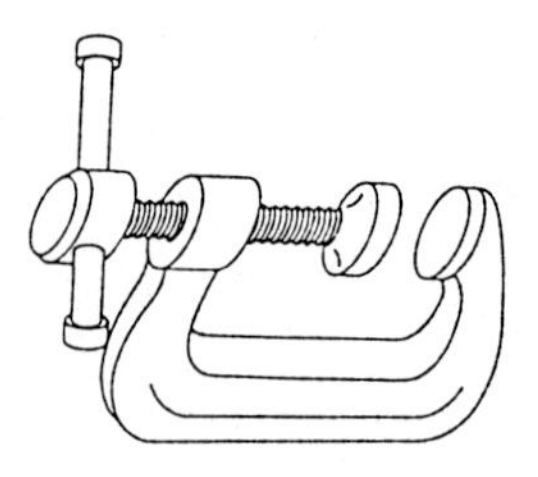

G-clamp
Schraubzwinge (G-förmig)

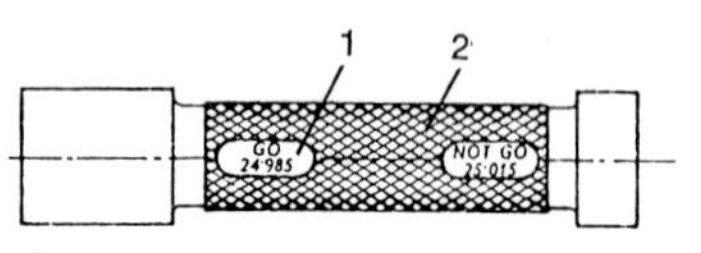

A

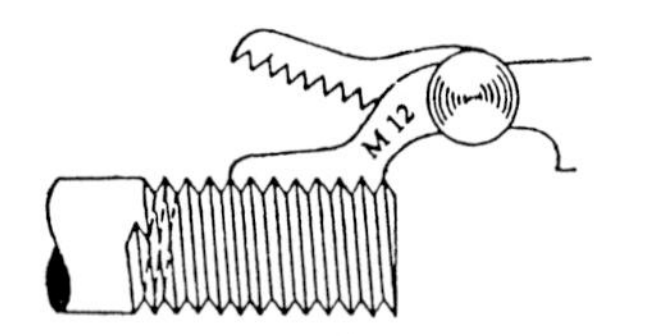

B

Ga(u)ges
Lehren

A **drill gauge** Bohrlehre

1 **indication** Anzeige
2 **knurled handle** gerändelter Griff

B **thread gauge** Gewindelehre

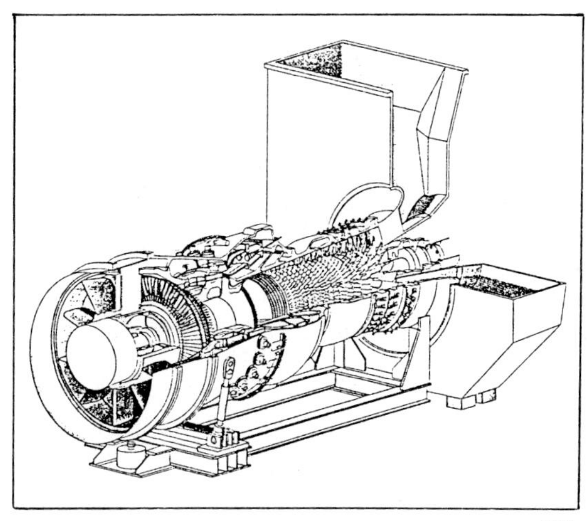

A

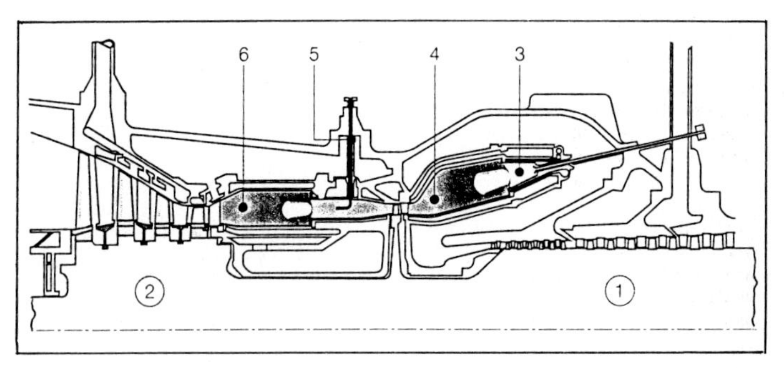

B

Gas turbine
Gasturbine

A **three-dimensional cut-away view**
drei-dimensionale aufgeschnittene Ansicht

B **sectional drawing** Schnittzeichnung

1 **compressor portion** Verdichter-Teil
2 **turbine portion** Turbinen-Teil
3 **burner** Brenner
4 **combustion chamber** Brennkammer
5 **fuel injection** Brennstoff-Einspritzung
6 **circular combustion chamber** Ringbrennkammer

gathering Versammlung, Zubrand
gauge, to abmessen, eichen, zurichten
gauge Kaliber, Lehre, Spurweite, Stichmass
gauging Abtasten, Eichung, Messung, Normierung
gear, to eingreifen, verzahnen, mit Getriebe versehen
gear Gang, Gerät, Getriebe, Verzahnung
gear-box Getriebe-, Räderkasten
geared verzahnt, übersetzt
gears Getrieberäder, Räderwerk
gem Edelstein
general-purpose Allzweck . . .
generate, to erzeugen, abgeben, entwickeln
generation Erzeugung, Entwicklung, Ausscheidung
generator Generator, Dynamo
genuine echt, gediegen, wirklich
genuineness Echtheit
geodesy Geodäsie, Vermessungskunde
geognosy geognostische Wissenschaften, Gesteinskunde
geologist Geologe
geology Geologie
geomagnetic field erdmagnetisches Feld
geometric geometrisch
geometrician Geometer
geometry Geometrie
geophysics Geophysik
geothermal geothermisch
getaway speed Abfluggeschwindigkeit
getter Fangstoff, Füllung
geyser Geiser, Geyser
ghost Doppelbild, Achterleitung
gib Beilagekeil, Führungslineal
gibs Leisten
gild, to vergolden
gilding Vergoldung
gilt vergoldet
gimbal Kardanrahmen, Tragbügel
gimlet Bohrer, Vorbohrer
gin, to entkörnen
giratory breaker Walzenbrecher
gird, to gürten
girder Balken, Träger
girdle, to gürten
girdle Gurt

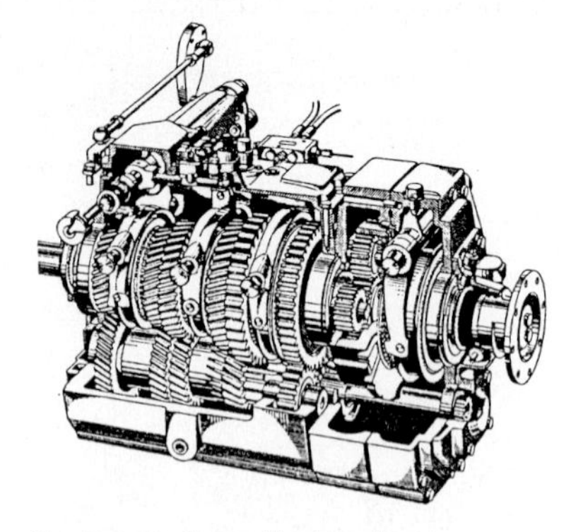

Gearbox of a heavy diesel truck
Getriebe eines Diesel-Lastkraftwagens

lowest gear : speed-crawler (change ratio 13 : 1)
niedrigster Gang: Kriechgeschwindigkeit (Übersetzungsverhältnis 13 : 1)

ninth gear : top speed (change ratio 1 : 1)
neunter Gang: Spitzengeschwindigkeit (Übersetzungsverhältnis 1 : 1)

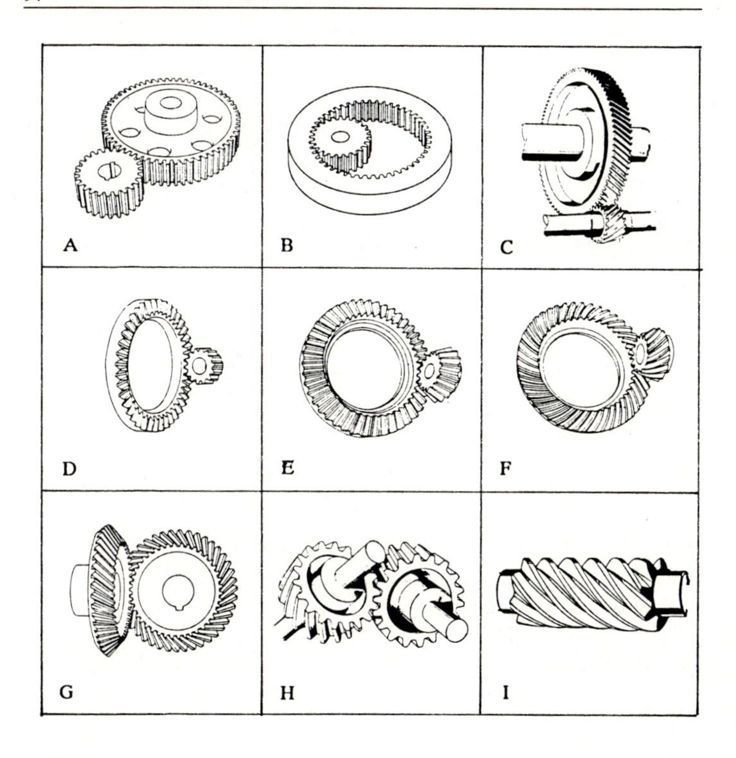

Gears
Getriebearten

A **spur gear** Stirnradgetriebe

B **internal spur gear** Stirnradgetriebe mit Innenverzahnung

C **helical spur gear** Stirnradgetriebe mit Spiralverzahnung

D **face gear** Getriebe mit Seitenverzahnung

E **straight bevel gear** Kegelradgetriebe mit Geradverzahnung

F **spiral bevel gear** Kegelradgetriebe mit Spiralverzahnung

G **scew bevel gear** Kegelradgetriebe mit Schrägverzahnung

H **crossed-axis helical gear** Spiralzahnradgetriebe mit gekreuzten Achsen

I **worm gear** Schneckenradgetriebe

git Eingusstrichter
glacial Eis . . ., eisig
gland Kappe, Stopfbüchse
gleam, to leuchten
gleam Schimmer
glimmer, to flimmern
global communication network Weltverkehrsnetz
globe Ball, Globus, Glocke, Kugel
globose kugelförmig, kugelig
gloss Glanz, Glasur
glossary Nomenklatur, Spezialwörterbuch
glow, to glimmen, glühen, aufleuchten
glow Glühen, Glimmen, Schein
glow-lamp Glimmlampe, Gasentladungsröhre
glue, to kitten, kleben, leimen
glue Leim, Kleister
glycerin Glyzerin
gnarled knotig
goal Ziel, Zielpunkt
gobo Blendenschirm, Schalltilgungsmittel
goliath crane Schwerlastkran
goniometer Peiler, Winkelmesser
goodness Arbeitssteilheit (Röhre)
goods Artikel, Werkstoff, Güter
goods traffic Güterverkehr
gooseneck Anschlusstück, Schwanenhals
gorge Kehle, Rinne, Schlucht
govern, to herrschen, leiten, regeln regulieren
governing Kontrolle
governor Drehzahl-, Geschwindigkeitsregler, Regulator
grab Greifer, Exkavator
grace Frist, Fristverlängerung
gradation Abstimmbarkeit, Abstufung, Grad, Stufenfolge
grade, to abstufen, einteilen, sortieren
grade Grad, Klasse, Marke
grader Planierer, Sortierer, Trenner
gradient Anstieg, Gefälle, schiefe Ebene
grading Anreicherung, Klassifizierung, Einteilung, Sortierung
gradual stufenartig, fortschreitend
gradually stufenweise, allmählich, absatzweise
graduate, to abstufen, einteilen

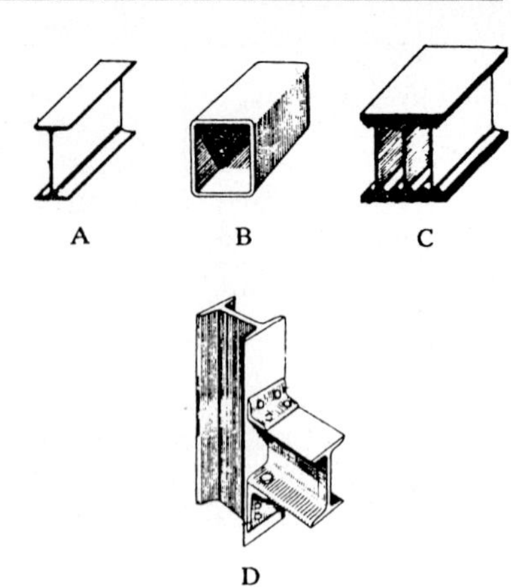

Girders
(Stahl-) Träger

A **double-T girder** Doppel-T-Träger

B **box girder** Kastenträger

C **compound girder** Kombi-Träger

D **girder bolted or riveted with angle iron to a stanchion** Träger mit Winkeleisen an einen Stützpfeiler geschraubt oder genietet

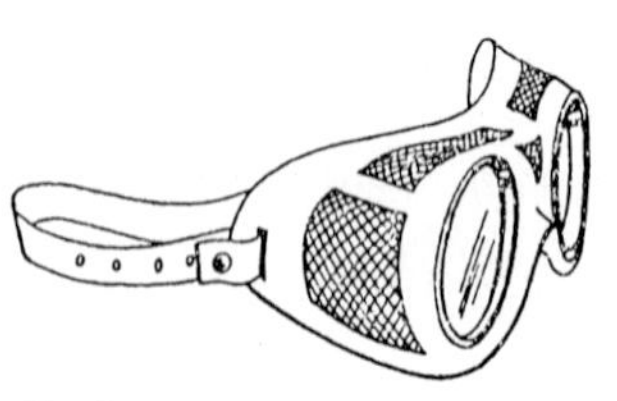

Goggles
Schutzbrille

Grab crane
Greiferkran

graduate Absolvent
grain Faden, Faser, Faserung, Kern, Korn
graining Kornbildung, Maserung
gram Gramm
granite Granit
grant, to bewilligen, erteilen gewähren, zusprechen
grant Beihilfe, Beisteuer, Bewilligung
grant-in-aid Hilfeleistung, Zuschuss, Subvention
granted erteilt
grantee Patentinhaber
granting a license Erteilung einer Konzession
granular gekörnt, körnig
granulate, to aufrauhen, granulieren, körnen
granulation Kornbildung, Granulierung
graph, to grafisch darstellen
graph grafische Darstellung, Diagramm, Schaubild
graph paper Millimeterpapier, Koordinatenpapier
graphic arts Grafik
graphical grafisch, bildlich
graphite Graphit
grapple, to anklammern, dreggen
grappling Verankerung, Dreggen
grasp, to anfassen, begreifen, fassen, greifen
grasp Griff, Klaue
grate, to gittern, kratzen, reiben schaben
grate Feuerrost, Fangrechen, Netz
grater Raspel, Reibe
graticule Fadenkreuz
grating Raster
gravel, to aufschottern
gravel Flusskies, Kiessand, Schotter
gravitate, to sinken, sich durch Schwerkraft fortbewegen
gravity Erdenschwere, Schwerkraft
graze, to bestreichen, streifen
grease, to fetten, ölen, schmieren
grease Fett, Talg
greaser Schmiervorrichtung
greasy fettig, ölig, speckig
greedy gierig, gefrässig
grid Gitter, Rost, Reuse
grind, to schleifen, reiben, zermahlen

grind Schleifen, Mahlen
grinder Schleifmaschine, Mahlstein, Mühle
grindery Schleiferei
grinds Schleifschlamm
grip Greifer, Griff, Greifklaue
grip, to fassen, klemmen, packen
gripper Greifer, Halter, Mitnehmer
gripping-device Einspannvorrichtung, Greifwerkzeug
grommet Durchführungshülse, Öse, Auge
groom, to putzen
groove, to aushöhlen, auskehlen kerben
groove Aushöhlung, Aussparung, Auskehlung
ground, to erden, an Erde legen kurzschliessen
ground Erde, Boden
group, to gruppieren, in Gruppen zusammenfassen, anordnen
group Gruppe, Konzern, Konsortium
grout Vergussmaterial, Mörtelschlamm
guidance Führung, Anleitung, Richtschnur
guide, to anleiten, führen, lenken
guide-beam Leitstrahl
gullet Rinne, Nische
gum, to aufkleben
gum Harz
gun Kanone, Geschütz
gusset Eckblech, Anschluss
guttapercha Guttapercha
gyrate, to sich schnell drehen, kreisen, umlaufen
gyration Drehung, Wirbel
gyro Kreisel, Drehung
gyroscope Kreisel, Gyroskop
gyro compass Kreiselkompass

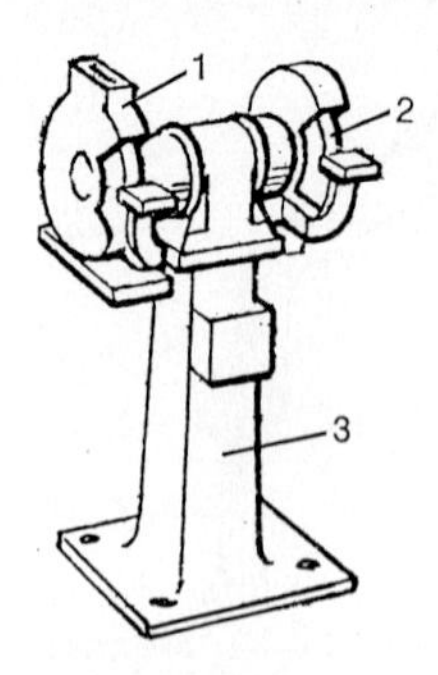

(Pedestal) grinder
(Ständer-) Schleifmaschine

1 **protective cover** Schutzhaube
2 **grinding wheel** Schleifscheibe
3 **pedestal** Maschinenständer

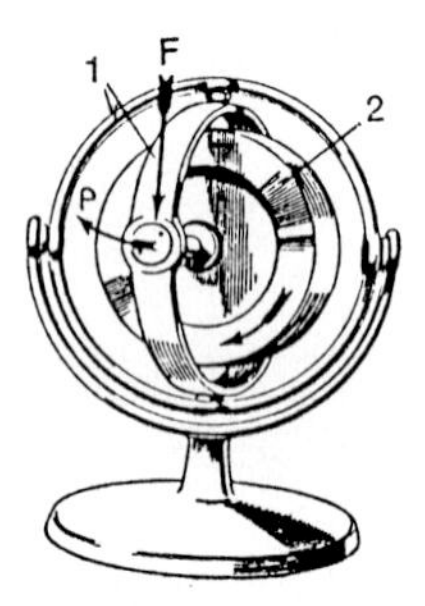

Gyroscope
Gyroskop

1 **gimbals** Kardanringe
2 **fly-wheel** Schwungrad
F **downward force** abwärtsgerichtete Kraft
P **direction of motion** Bewegungsrichtung

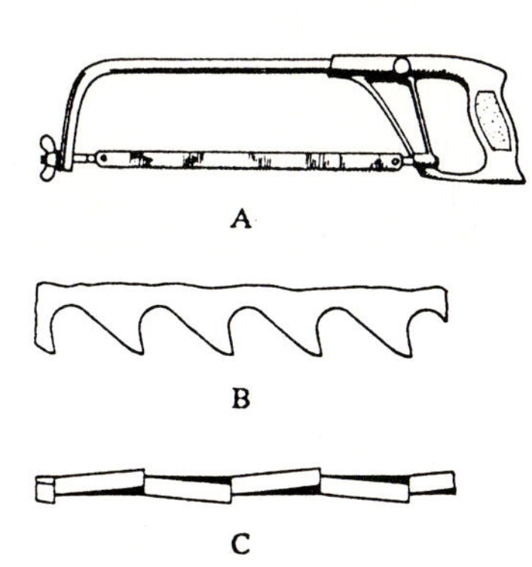

Hack-saw (iron saw)
Bügelsäge (Eisensäge)

A saw Säge

B tooth shape Form der Sägezähne

C offset of teeth Zahnverschränkung

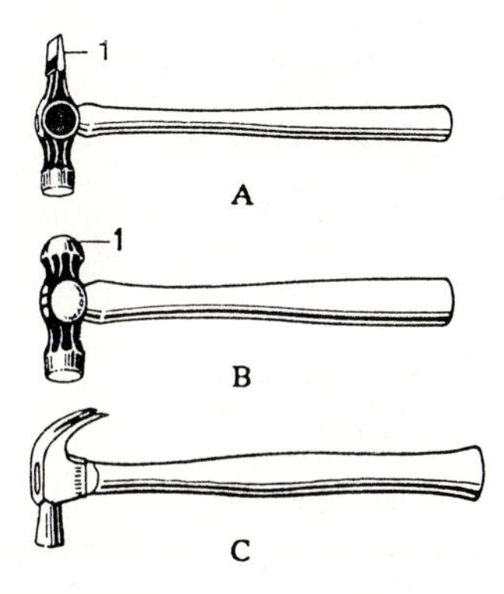

Hammer
Hammer

A tack hammer Keilhammer
1 pane Finne

B ball hammer Rundhammer

C claw hammer Klauenhammer

H

hack-saw Bügelsäge
hair pointer Fadenzeiger
halogen Salzbildner
halve, to halbieren, hälften
hammer Hammer
hand Zeiger, Hand, Arbeiter
hand-feed Handvorschub
handset Handapparat
handed in eingereicht
handicraft Gewerbe, Handwerk
handle, to handhaben, behandeln, abwickeln
handle Griff, Klinke, Henkel
handling Bearbeitung, Handhabung
handy handlich
hangers Gehänge
hang Strang, Strähne
harass, to stören, bedrängen
hard-alloy grinder Horizontalschleifständer
harden, to härten, verfestigen
hardening agent Härtungsmittel
hardness Härte, Festigkeit
hardware Beschlag, Eisenwaren
harm, to schaden
harm Schaden
H armature Doppel-T-Anker
harmonic Oberschwingung, harmonische Oberwelle
harmonics Harmonische
harmonise, to übereinstimmen, abstimmen
harp antenna Fächerantenne
hartley circuit Dreipunktschaltung (Röhre)
harvester Erntemaschine, Selbstbinder, Mäher
hasp Drehriegel, Haspe
hatch, to schraffieren, stricheln
hatch Luke, Klappe, Gatter
hatchet Hacke, Beil
haul, to schleppen, befördern fördern
haulage Förderung, Transport, Treideln, Ziehen
hauler Schlepper
hawse hole Klüse
hawse pipe Ankerklüse
hawser Festmachetau, Tau, Halteleine

hazard, to riskieren, aufs Spiel setzen
hazard Gefahr, Risiko
hazardous riskant, gefährlich, gewagt
haze, to anlaufen
haze Dunst, Nebel, diesiges Wetter
hazy dunstig, unscharf, diesig
H bend H-Bogen
H corner H-Winkel
head Boden, Abschluss, Kopf, Kopfstück
headlamp Scheinwerfer
headlight Fernlicht, Scheinwerfer
headrace Fallwasser, Speisekanal
headstock Reitstock, Spindelbacken, Spindelkasten
headway lichte Höhe, Vortriebsstollen
heap, to häufeln
heap Haufen, Stapel, Stoss
hearing Vernehmung, Gehör, Audienz
hearings mündliches Verfahren
hearth Erhitzungskammer, Feuerraum, Schmelzraum
heat, to anwärmen, heizen, feuern
heat Hitze, Wärme, Charge
heat-resisting hitze-, wärmebeständig
heat-treating Vergüten, Wärmebehandlung
heater Erhitzer, Heizapparat
heater-plug Glühkerze
heating Erwärmung, Erhitzung
heave, to hieven, heben
heave Hebung, Aufwindung
heaving Hebung
Heaviside effect Stromverdrängung
heavy-duty Hochleistungs . . ., dauerhaft
heel, to krängen
heel over, to kentern
heeling Krängung
height Erhebung, Grösse, Höhe
helical schnecken-, spiralschraubenförmig
helically toothed wheel Rad mit Schrägverzahnung
helipot Wendelpotentiometer
helix Kehrwendel, Schnecke
Hell printer Hellschreiber
helm Helm, Ruder(pinne)

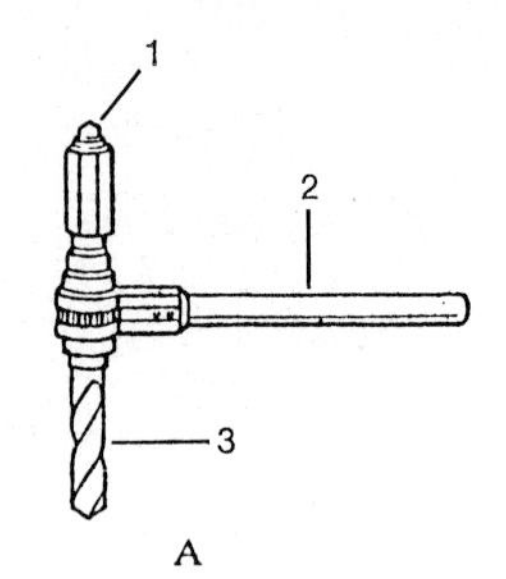

A

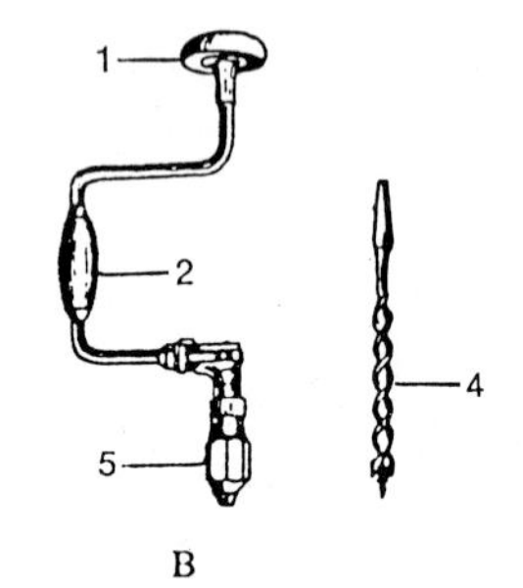

B

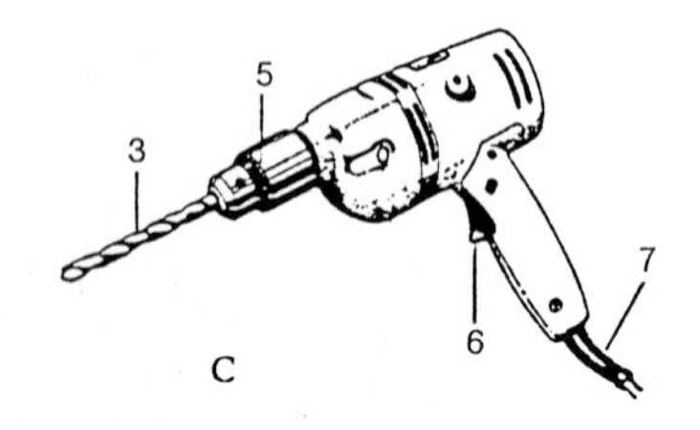

C

Hand drills (drilling machine)
Handbohrer (-bohrmaschine)

A **drill ratchet** Bohrknarre

B **drill brace** Bohrkurbel

C **electric hand drill** elektrische Handbohrmaschine

1 **pressure rest** Druckauflage (-fläche, -point)
2 **handle** Handgriff
3 **twist drill** Spiralbohrer
4 **wood drill (bit)** Holzbohrer
5 **chuck** Bohrfutter
6 **switch** Schalter
7 **supply cable** Zuleitungskabel

A

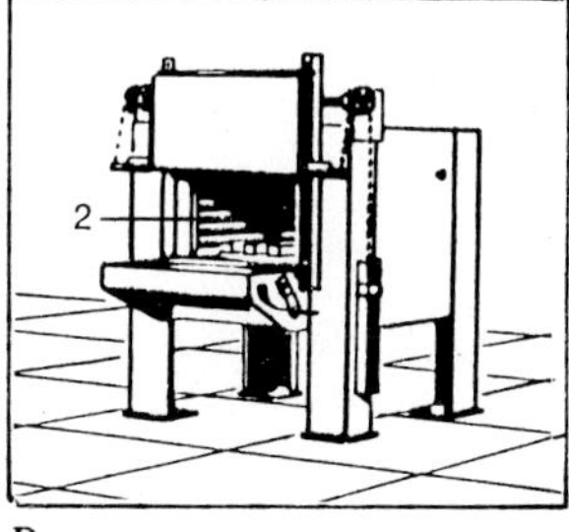

B

1

C

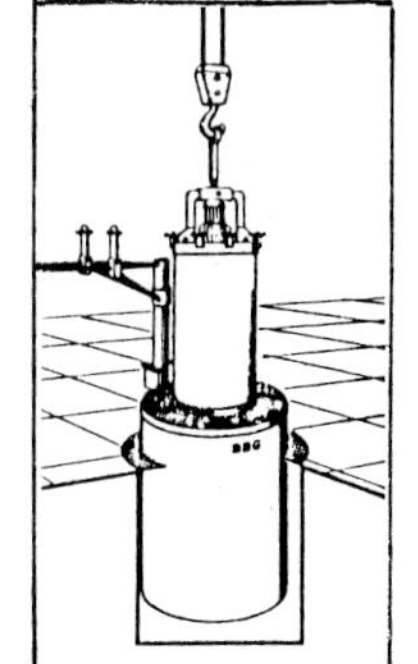

D

Heat-treatment furnaces (resistance-type)
Wärmebehandlungsöfen (widerstandsbeheizt)

A **continuous pulling-chain furnace** Durchzieh-Kettenofen

B **batch furnace** Kammerofen

C **bogie hearth furnace** Herdwagenofen

D **pit furnace** Schachtofen

1 **charge** Einsatzgut

2 **heating rods or heating spirals** Heizstäbe oder Heizwendel

Heat treatments
Wärmebehandlungen

- **annealing** glühen
- **hardening** härten
- **homogenizing** homogenisieren
- **quenching and hardening** vergüten (abschrecken/härten)
- **refining** raffinieren
- **tempering** anlassen (härten)

helpful hint Kniff
hem, to besäumen, einfassen
hem Rand, Saum
hemicycle Halbkreis
hemisphere Erdhalbkugel
hemming machine Saummaschine
hemp Hanf, Hede
heptode Heptode, Siebenpolröhre
hermetic dicht, hermetisch, luftdicht
Hertz effect fotoelektrische Wirkung
heterodyne Überlagerer, Überlagerungs . . .
hew, to abschroten, abschruppen, hacken
hew Abhieb
hex nut Sechskantmutter
high-performance Hochleistungs . . .
high-pressure Hochdruck . . .
hi-lo-check Hoch-Niedrig-Prüfung
hinge, to drehbar anbringen, mit Scharnier versehen
hinged aufklappbar, schwenkbar, mit Scharnieren versehen
hiss, to fauchen, zischen
hit, to (auf)schlagen, treffen, stossen
hit Aufschlag, Treffer, Schlag, Stoss
hitch, to anhaken, einhaken, anspannen
H-network H-Filterglied
hoarse rauh, heiser
hob, to (ab)wälzen, wälzfräsen, verzahnen
hob Abwälz-, Gewindefräser
hobbing Fräsen nach Abwälzverfahren
hoist, to aufziehen, anheben
hoist Aufzug, Flaschenzug, Lift
hold Griff, Laderaum (Schiff)
holder Halter, Besitzer
hole, to aushöhlen, ausschrämmen, ein Loch herstellen
hole Loch, Bohrung, Höhle
hollow, to aushöhlen, vertiefen
hollow Aushöhlung, Aussparung
homogeneous homogen, einheitlich
homopolar Gleichpol . . ., gleichpolig
hone, to honen, ziehschleifen

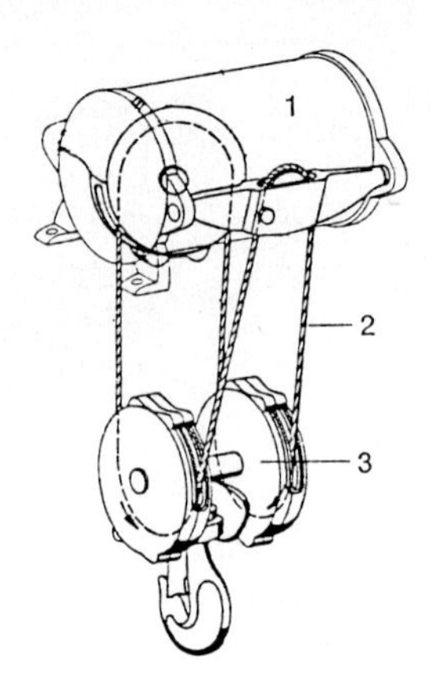

Hoist
Aufzug (Flaschenzug)

1 **drive** Antrieb
2 **rope** Seil
3 **pulley** Seilrolle

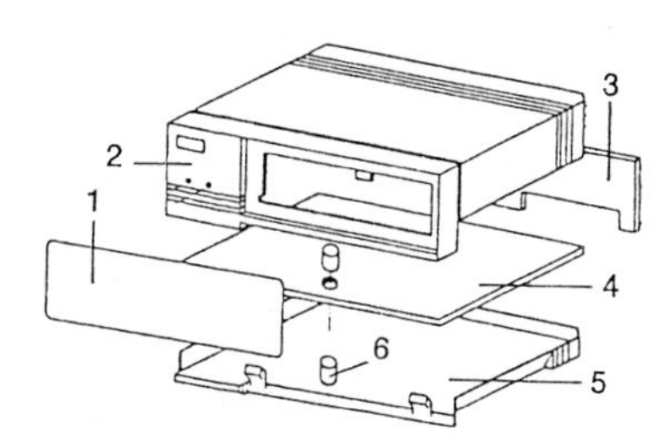

(Instrument) Housing
(Instrumenten-)Gehäuse

1 **cover plate** Abdeckplatte
2 **front plate** Frontplatte
3 **back plate** Rückwand
4 **mounting plate** Bestückungsplatte
5 **base plate** Bodenplatte
6 **spacer** Abstandsbolzen

honing Honen, Ziehschleifen, Abziehen
hood Haube, Kappe, Abzug
hook, to abbiegen, anhaken, einhaken
hook Haken, Henkel, Öse
hopper Behälter, Füllgefäss
horizon Horizont
horizontal horizontal, waagerecht
horn Arm, Horn, Hebel
horn welding Hammerschweissung
horse Anlegetisch, Auflagebock
horse-power Pferdestärke
hose Schlauch
hot-wire ammeter Hitzdraht-ampèremeter
house, to unterbringen
housed eingeschlossen
housing Gehäuse, Unterbringung
hover, to kreisen, schweben
hub Narbe, Lagerbüchse, Knotenpunkt
hue Färbung, Farbton
hueless farblos
hull Bootskörper, -rumpf, Hülle
humid dampfhaltig, feucht
humidifier Luftbefeuchtungsanlage
hunt, to nachlaufen, jagen
hunting Bildverschiebung, Pendelschwingung, Nachlaufen
hurt, to schaden, verletzen
hydraulic hydraulisch
hydrocarbon Kohlenwasserstoff
hydroelectric hydroelektrisch
hydrofoil Tragflügel(boot)
hydrogen Wasserstoff
hydrometer Aerometer, Dichtigkeitsmesser, Dichtemesser
hydrous wasserhaltig, wässerig
hygrometer Feuchtemesser
hyperbola Hyperbel
hyphen Bindestrich
hygroscopic wasseraufnehmend

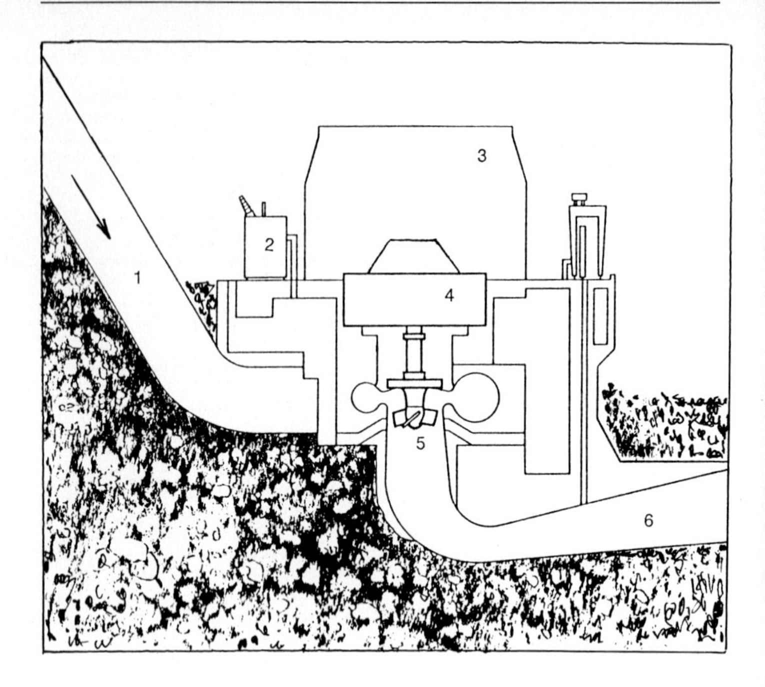

Hydroelelectric power station
Wasserkraftwerk

1 **penstock (water from reservoir)** Druckrohr (Wasser vom Reservoir)
2 **transformer** Transformator
3 **power house** Maschinengebäude
4 **hydrogenerator** Wasserkraftgenerator
5 **hydroturbine** Wasserkraftturbine
6 **draft tube** Abflußrohr

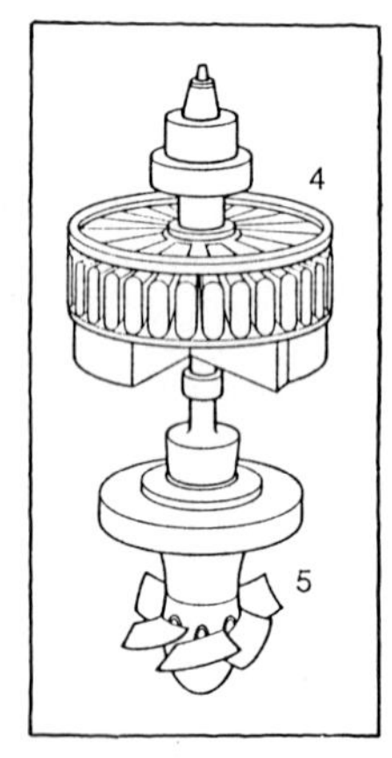

I

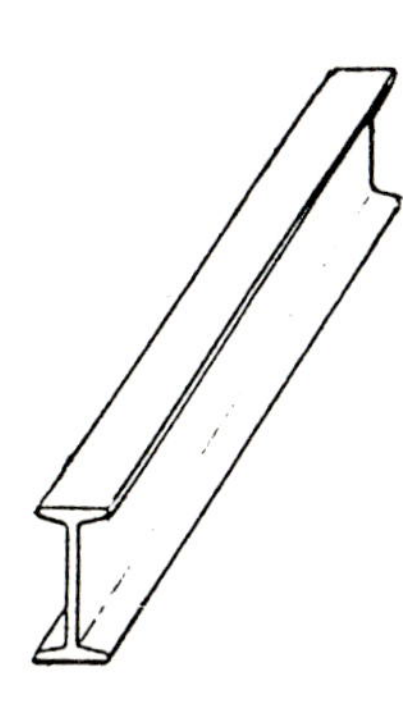

I beam
Doppel-T-Träger

I beam Doppel-T-Träger
identify, to feststellen, identifizieren
identity Identität
idle, to run leerlaufen
idle frei, träge, untätig
idler Führungsrolle (Förderanlage)
idlers Riemenspanner, Tragrollen
I girder Doppel-T-Träger
ignitable zündbar, zündfähig
ignite, to (an)zünden, entzünden
igniter Anzünder
ignition Entzündung
ignore Auslasszeichen, Leerzeichen
ill effect nachteilige Wirkung
illegible undeutlich
illumination Beleuchtung, Ausleuchtung
illustrate, to illustrieren, bebildern, verdeutlichen
image, to abbilden, bildlich darstellen
image Abbild, Bild
imbalance Unwucht, Ungleichgewicht
imbibe, to saugen, tränken
imbue, to tränken
imitate, to imitieren, nachahmen
imitation Nachahmung
immaterial stofflos, unwesentlich
immediate sofortig, unverzüglich
immense unermesslich, ungeheuer gross
immerse, to (ein)tauchen, untertauchen
immersion Eintauchung, Versenkung
immoderate unmässig
immovable unbeweglich
immune unempfindlich
immutable unwandelbar
imp Gerüststange
impact, to aufschlagen, zudrücken
impact Aufschlag, Wucht, Schlag, Stoss
impair, to beeinträchtigen, verschlechtern
impairing Verschlechterung, Beeinträchtigung

impart, to erteilen, mitteilen (Wissen), geben
impeccable einwandfrei
impedance Scheinwiderstand
impede, to anhalten, arretieren, erschweren
impel, to (an)treiben
impeller Kreisel-, Flügelrad
impenetrable undurchdringlich, undurchlässig
imperil, to gefährden
impermeability Undurchdringlichkeit, Undurchlässigkeit
impermeable undurchdringlich, undurchlässig
impervious dicht, undurchlässig
impetus Antrieb
impinge, to anprallen, aufprallen, auftreffen
implantation Verpflanzung, Implantation
implement Gerät, Werkzeug, Vorrichtung
implements (Pl.) Arbeitsgerät, Handwerkszeug
implicate, to mit eingreifen, verflechten
implode, to platzen, zusammenbrechen
implosion Einfallen
imply, to einbeziehen
impose, to auferlegen
impound, to aufspeichern, stauen
impoundage Eindämmung
impracticable unzweckmässig, unausführbar
impregnate, to durchtränken, sättigen
impregnation Imprägnierung, Tränkung
impression Eindruck, Abdruck
improperly unsachgemäss
improve, to verbessern, aufbessern
improvement Verbesserung, Verfeinerung
improvisation Notbehelf
imprudence Unvorsichtigkeit, Leichtsinn
impulse, to einen Impuls erteilen
impulse Anstoss, Impuls
impulsion Antrieb, Schwung
impure unrein
impurity Fremdstoff, Verunreinigung

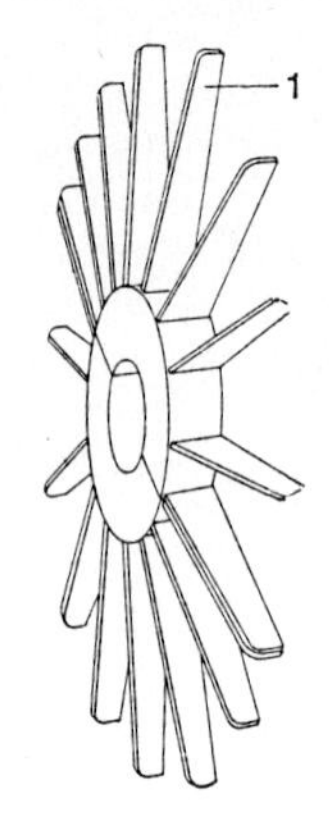

Impeller
Flügelrad
1 blade Flügel

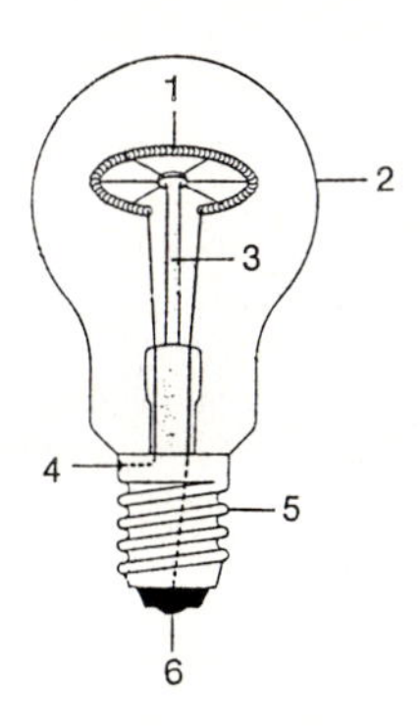

Incandescent lamp (bulb)
Glühlampe (-birne)

1 **filament spiral** Glühfaden-Wendel
2 **glass bulb** Glaskolben
3 **filament holder (glass)** Glühfadenhalter (Glas)
4 **socket contact** Sockelkontakt
5 **threaded socket** Schraubsockel
6 **insulated base contact** isolierter Bodenkontakt

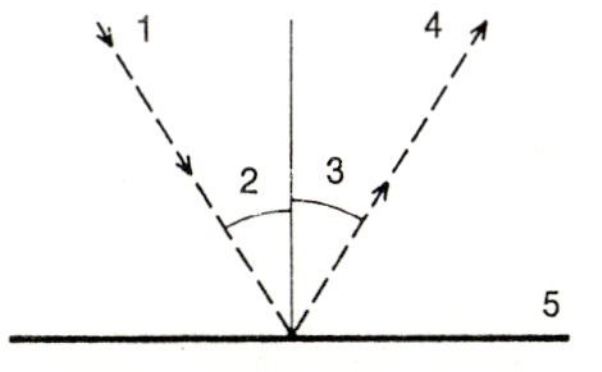

Incident beam
Einfallender Strahl

1 **incident beam** einfallender Strahl
2 **angle of incidence** Einfallwinkel
3 **angle of reflection** Ausfallwinkel
4 **reflected beam** ausfallender Strahl
5 **surface** Fläche

impute, to beimessen, zuschreiben
inability Unfähigkeit
inaccessibility Unzugänglichkeit
inaccessible unzugänglich
inaccuracy Ungenauigkeit
inaccurate ungenau
inactivate, to unwirksam machen
inactive unwirksam, ausser Dienst
inadequacy gap Nachholbedarf
inadequate unzulänglich
inadvertence Versehen, Unachtsamkeit
inadvertent mistake Flüchtigkeitsfehler
inadvertently irrtümlich
inarticulate undeutlich
inattentive unachtsam, unaufmerksam
inaudible unhörbar
inauguration Amtseinführung
incandescent lamp Glühfadenlampe
incapability Unfähigkeit
incapable unfähig
incapacity Unfähigkeit
incase, to (in ein Gehäuse) einschliessen
incentive Anreiz, Ansporn, lohnanreizend
inching operation Tippschaltung
incide, to einfallen
incidence Einfall
incidental zufällig, beiläufig
incinerator Müllverbrennungsanlage
incitation Anregung
incite, to anregen, anreizen, anspornen
incitement Anreizung
inclinable umlegbar, neigbar
inclination Neigung, Schräge
incline, to neigen, kippen abschrägen
include, to einschliessen, beifügen
inconsistency Widerspruch, Unbeständigkeit
inconsistent widersprechend, unverträglich
inconspicuous unauffällig
inconvenience Unbequemlichkeit
inconvenient lästig, ungelegen
incorporate, to eingliedern, einverleiben
incorporation Eingliederung, Einschluss

incorrectly unsachgemäss
increase, to anwachsen, steigen, erhöhen
increase Anwachsen, Steigen, Erhöhung, Zunahme
indenture Lehrbrief
indestructible unzerstörbar
indeterminate unbestimmt, unbekannt
index, to einteilen, einreihen, auf eine Teilmarke stellen
India-rubber Gummi
indicate, to anzeigen, angeben, deuten
indication Angabe, Anzeige, Hinweis, Zeichen
indicator Anzeigegerät
indiscriminate unterschiedslos
indispensible unerlässlich, unabkömmlich
indissoluble unlöslich, unlösbar
indistinct unklar, undeutlich, verschwommen
indistinctness Unklarheit, Undeutlichkeit
individual eigen, individuell
indivisible untrennbar, unzerlegbar
indolent faul, träge
indoor(s) inwendig, Innen ...
indraft Ansaugluftströmung, Einströmung, Saugluft
induce, to induzieren, erzeugen, einleiten, beeinflussen
inducement Anregung, Veranlassung
inducer Vorlaufrad
inductance Induktanz, induktiver Blindwiderstand
induction Induktion, Ansaugvorgang, Folgerung
inductive induktiv, Induktions ...
inductivity Induktivität, Induktanz, Dielektrizitätskonstante
inductor Impedanzspule, Drossel, Induktor
indurable härtbar
indurate, to verhärten
induration Verhärtung, Hartwerden
industries fair Industriemesse
industrious arbeitsam, emsig, geschäftstüchtig, fleissig
ineffective unwirksam, wirkungslos

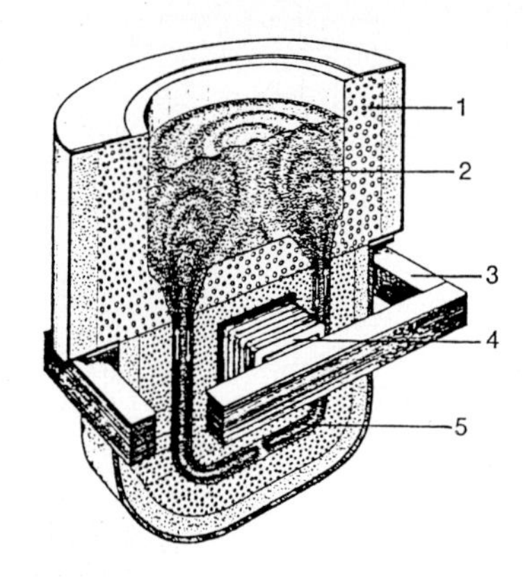

Induction furnace (channel-type) Induktionsrinnenofen

1 **refractory lining** feuerfeste Ausmauerung
2 **liquid iron** Flüssigeisen
3 **magnet core** Magnetkern
4 **inductor coil** Induktorspule
5 **channel** Rinne

Induction melting furnaces Induktionsschmelzöfen ➡

A **crucible-type induction furnace** Induktionstiegelofen
(**melting furnace** Schmelzofen)

B **channel-type induction furnace** Induktionsrinnenofen
(**holding furnace** Warmhalteofen)

1 **furnace coil (water-cooled)** Ofenspule (wassergekühlt)
2 **magnet core** Magnetkern
3 **pouring spout** Abgußschnauze
4 **crucible cover** Tiegeldeckel
5 **supply cables to furnace coil** Zuleitungskabel für Ofenspule
6 **cooling water pipes** Kühlwasserrohre
7 **hydraulic tilting cylinder** Hydraulik-Kippzylinder
8 **tilting pedestal** Kippständer
9 **charging hole** Einfüllöffnung

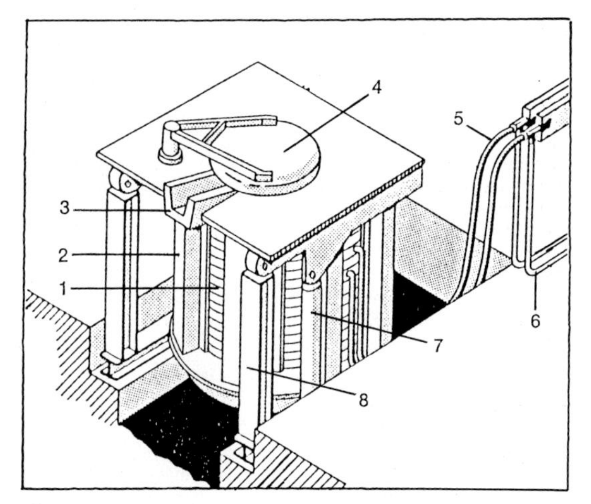

A

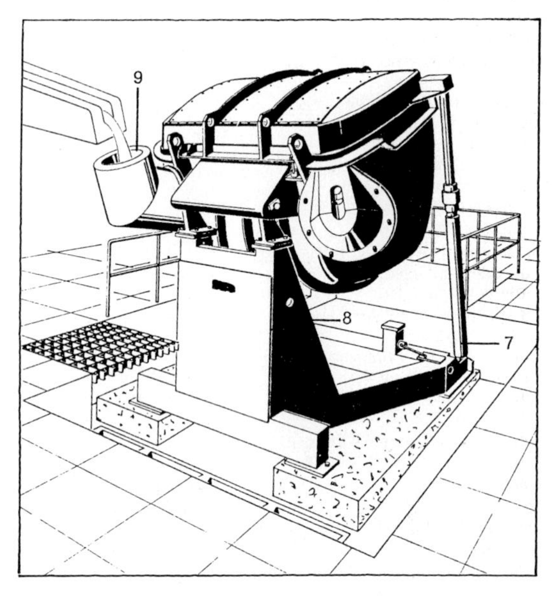

B

ineffectiveness Unwirksamkeit, Wirkungslosigkeit
inefficacy Wirkungslosigkeit
inefficiency schlechte Wirkung, Wirkungslosigkeit
inefficient leistungsunfähig, unwirksam
inelastic starr, unelastisch
inequality Ungleichheit, Ungleichung
inert (Gase) edel, inaktiv, neutral, unentzündbar
inertance Inertanz, Massenwirkung
inertia Trägheit(svermögen)
inertial force Trägheitskraft
inertness Massenwiderstand, Trägheit
inexact ungenau
inexcitability Unanfechtbarkeit
inexhaustible unerschöpflich
inexpansible unausdehnbar
inexpedient unzweckmässig
inexpensive billig
inexperience Unerfahrenheit
inexperienced unerfahren
inexplosive unexplodierbar
inextensible unausdehnbar, undehnbar
inextricable unentwirrbar
infect, to verseuchen, anstecken
infection Verseuchung
in-feed Beschickung, Zustellung
infer, to folgern
inference Folgerung, Hinweis, Rückschluss
inferential meter Durchfluss-messer
inferior geringwertig
inferiority Minderwertigkeit, Unterlegenheit
infiltrate, to eindringen, ein-sickern, durchsetzen
infiltration Einfiltrierung, Einsickerung, Durchsickerung
infinite endlos, unendlich
infinitesimal calculus Infinite-simalrechnung
infinity Unendlichkeit
inflame, to entzünden
inflammability Entflammbarkeit
inflammable brennbar, entzündbar
inflammation Entflammung
inflatable boat Schlauchboot
inflate, to aufblähen, aufblasen
inflation Aufblähung

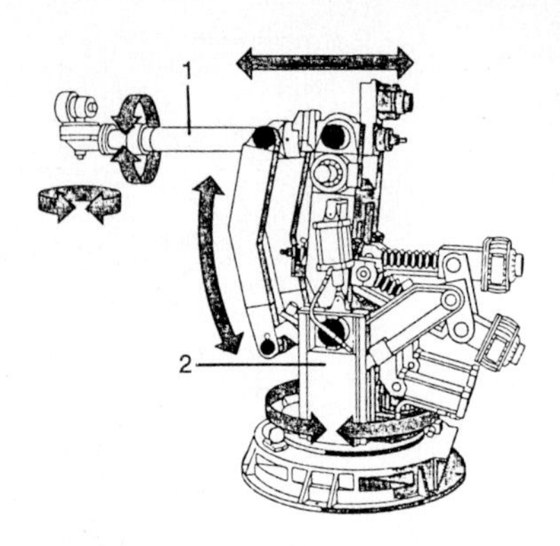

Industrial robot with 5 axes of movement
Industrie-Roboter mit 5 Bewegungsachsen

1 **articulated arm** Gelenkarm
2 **slewing column** Schwenksäule

inflator hose Füllschlauch
inflect, to ablenken, beugen, biegen
inflecting Biegen
inflection Ablenkung, Beugung
inflector Einlenkkondensator
inflexibility Steifheit, Unbiegsamkeit
inflexible starr, steif, unbiegsam
inflexion Knickpunkt
inflict, to beibringen, auferlegen
inflow Ansaugluft, Einfluss
influence, to influenzieren, beeinflussen
influence Beeinflussung, Einfluss, Einwirkung
influx Einfluss, Zufluss, Zustrom
inform, to unterrichten, angeben, avisieren
informal formlos
informality Formlosigkeit, Formfehler
infrared infrarot
infrasonic Infraschall ..., Unterschall ...
infringe, to beeinträchtigen, verletzen (Vorschriften)
infringement Beeinträchtigung, Übertretung
infuse, to aufgiessen, einflössen
infusible unschmelzbar
ingate Trichtereinlauf
ingenious erfinderisch, sinnvoll
ingenuity Findigkeit
ingestion Ansaugen (Triebwerk)
ingot Barren, Block
ingredient Bestandteil, Zugabe
ingress, to eindringen
ingress Eindringen, Eintritt
inherent anhaftend, innewohnend, eigentümlich
inherently von Natur aus
inhibit, to hemmen, entgegenwirken
inhibition Hemmung, Behinderung
initial anfänglich, Anfangs ...
initiate, to ansetzen, beginnen, einleiten
initiating Inbetriebsetzung
initiation Beginn, Einleitung, Einsatz
inject, to injizieren, einblasen, einspritzen
injection Einspritzung, Einblasen

A informative instruction?
Eine sachkundige Anweisung?

injection pump Einspritzpumpe
injector Einspritzpumpe, Dampfstrahlgebläse
injunction Verbot, Untersagung
injure, to beschädigen, verletzen
ink, to (Zeichnungen mit Tusche) ausziehen
ink Tusche, Tinte, Farbe
inked ribbon Farbband
inlay Einlage
inlet Einlass, Eingang, Zuführung, Bucht
inlet valve Einlass-, Ansaugventil
inner innen, inner
innovation Neuerung
inodorous geruchlos
inoffensive harmlos
inoperating contact Ruhekontakt
inoperation Stilliegen, (Anlage) ausser Betrieb sein
inoperative in Ruhestellung befindlich, ruhend, unwirksam
inphase gleichphasig, phasengleich
in-process material Zwischenprodukt
input (Leistungs-)Aufnahme
inquire, to abfragen, sich erkundigen
inquiry Abfragen, Erhebung, Erkundigung
inrush Einströmen, Zustrom
inrush current Einschaltstrom
inscribe, to einschreiben
inscription Aufschrift, Beschriftung
insecure unsicher
insensibility Unempfindlichkeit
insensible unempfindlich
insensitive unempfindlich
inseparability Untrennbarkeit
insert, to einsetzen, einlegen, aufnehmen
insert Einlasstück, Einlage
insertion Einsatz, Einfügen, Ansatz
inset Einsatz, Zwischenlage
inside Innenseite, innerhalb
insight Einblick, Erkenntnis
insignia Abzeichen, Hohheitszeichen
insignificant unbedeutend
insist, to bestehen (auf)
insistant beharrlich
insolubility Unlöslichkeit

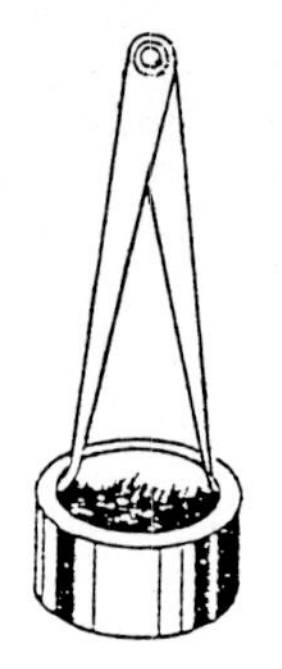

Inside caliper
Innentaster

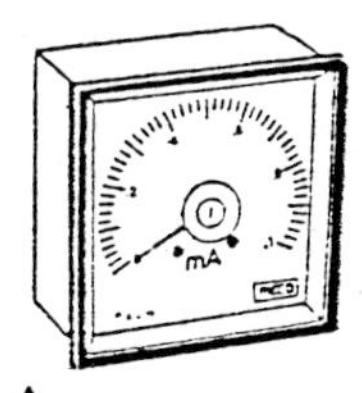

A

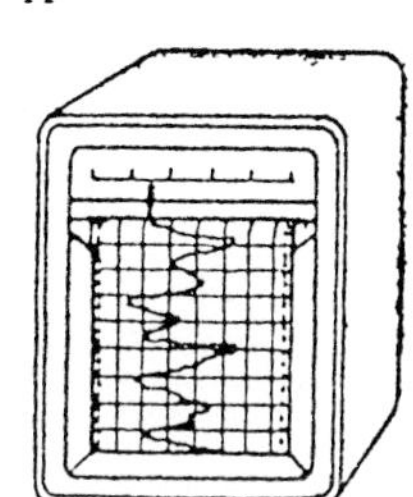

B

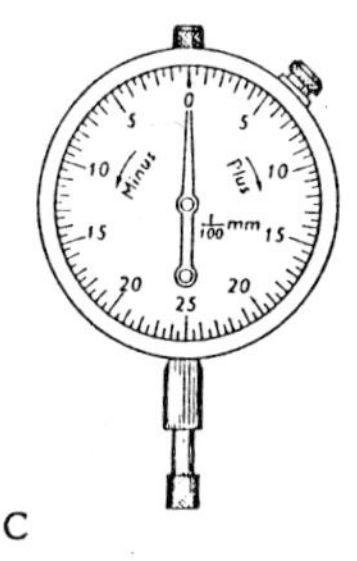

C

Instruments
Instrumente

A **electric measuring instrument (milliammeter)** elektrisches Meßinstrument (Milliampèremeter)

B **electric recording instrument (chart recorder)** elektrisches Schreibgerät (Kurvenschreiber)

C **mecanical measuring instrument (dial gauge)** mechanisches Meßinstrument (Meßuhr)

insolvable nicht mischbar
insolvency Zahlungsunfähigkeit, Konkurs
insolvent zahlungsunfähig
inspect, to beaufsichtigen, inspizieren
inspecting officer Inspekteur
inspection agency Überwachungsstelle
inspector Abnahmebeamter, Aufseher, Inspekteur
inspire, to einatmen, ansaugen, begeistern
instability Instabilität, Labilität
install, to installieren, aufbauen, errichten
installation Installation, Anlage, Anordnung
installer Einrichter
installment Teilzahlung, Rate, Akontozahlung
instantaneous momentan, augenblicklich, unverzögert
instantly sofort, augenblicklich
institute Anstalt
institute of technology Technische Hochschule
instruction Anleitung, Richtlinie, Unterweisung, Instruktion
instructive belehrend, instruktiv, lehrreich
instructor Lehrer, Ausbilder
instrument Gerät, Instrument
instrumentation Instrumentierung
insubmersible unsinkbar
insubordination Achtungsverletzung, Ungehorsam
insufferable unerträglich
insufficient unzureichend, mangelhaft
insulate, to (ab)isolieren
insulation Isolation
insulator Isolator
insurance Versicherung
insure, to versichern
insurer Versicherer
intact intakt, unversehrt
intake Ansaugen (Motor), Ansaugöffnung
integral integral, ein Ganzes bildend
integrate, to integrieren
integrated circuit zu einer Baugruppe zusammengefasste Schaltelemente

integrating amplifier Integrationsverstärker, Summierverstärker
integration Integration, Zusammenfassung
intelligence Nachricht, Nachrichtengehalt
intelligibility Verständlichkeit, Deutlichkeit
intensification Verstärkung, Intensivierung
intensifier Verstärker
intensify, to verstärken, intensivieren
intensity Stärkegrad, Intensität
intensive cooling Tiefkühlung
interact, to in Wechselwirkung stehen, einwirken
interaction Wechselwirkung, Rückwirkung
interblend, to vermischen
intercept, to abfangen, auffangen, begrenzen, abhören
intercept Abschnitt (Math.)
interception activity Abhörtätigkeit
interception service Abhördienst, Horchfunk
interceptor Sammler, Auffänger, Abscheider, Geruchverschluss
interchange, to auswechseln, austauschen
interchange Vertauschung, Austausch
interchangeable austauschbar, auswechselbar
interchanger Austauscher
interchannel modulation Kreuzmodulation
intercom system Gegensprechanlage
intercommunication Wechselverkehr
interconnected verkettet, zwischengeschaltet
interconnect, to zusammenschalten, zwischenschalten, miteinander verbinden
interconnection Verkettung, Zwischenschaltung, Zwischenverbindung
interconversion Umkodierung
interconvertible umsetzbar, umwandelbar
intercooler Zwischenkühler
interdialling Zwischenwahl

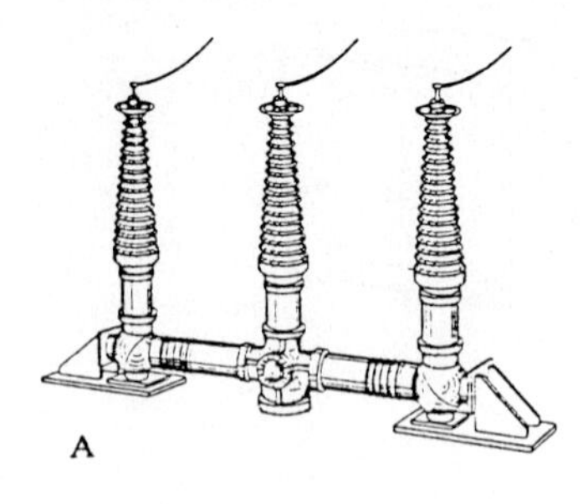

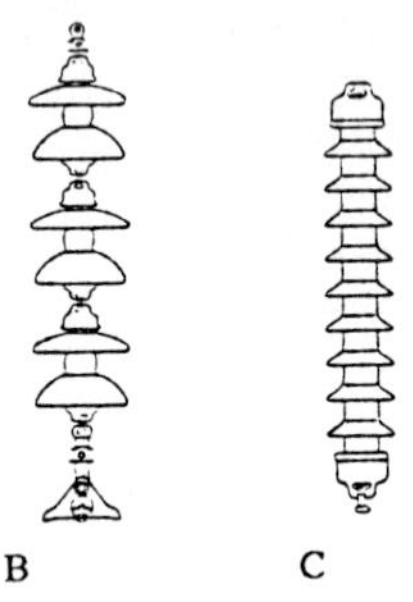

Insulators (high voltage)
Isolatoren (Hochspannung)

A **post-type** Stützerausführung

B **suspension string** Hängekette

C **rod-type** Langstab-Ausführung

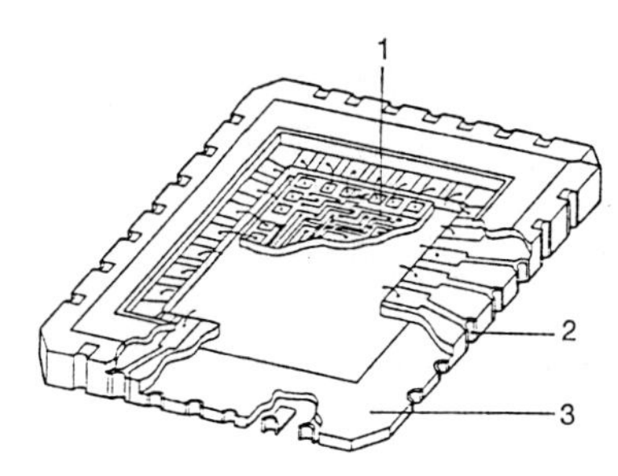

Integrated circuit (chip)
Integrierte Schaltung (Chip)

1 **circuit components** Schaltkreis-Elemente
2 **terminal contacts** Anschlußkontakte
3 **ceramic carrier** Keramik-Träger

interface Grenzfläche, Berührungsfläche, Grenzschicht
interfere, to beeinträchtigen, stören
interference Interferenz, Überlagerung, Beeinflussung
interfusion Verschmelzung
intergalactic travel Flug im intergalaktischen Raum
interior inner, Innen . . . , inwendig
interlace, to verflechten, vernetzen
interlaminar bonding Schichtverband
interlayer Zwischenschicht
interline flicker Zeilenflimmern
interlining Zwischenfutter, Einlagestoff
interlink, to verketten
interlock, to verriegeln, blockieren
interlock Verriegelung, Verblockung
interlocking Sperre, Verriegelung
intermediate Zwischenglied, Zwischen . . .
intermesh, to kämmen (Zahnräder)
intermingled durchwachsen (Min.)
intermittent aussetzend, stossweise, zeitweise
intermix, to vermischen
internal innerlich, Innen . . .
internet traffic Querverkehr
interphase Zwischenphase
interphone Haustelefon
interpolate, to interpolieren
interpole Zwischenpol, Wendepol
interpreter Dolmetscher, Zuordnerkode
interrogate, to abfragen
interrogation Abfragung
interrogator Abfragesender
interrupt, to unterbrechen
interrupter Ausschalter
interruption Unterbrechung
interruptive capacity Abschaltleistung
intersecting angle Schnittwinkel
intersection Durchdringung, Schnittlinie
interspace Zwischenraum
intersperse, to durchsetzen
interstage Zwischenstufe
interstellar kosmisch, interstellar
interstice Zwischenraum, Spalt

interval Zwischenzeit, Zeitspanne
intervalve coupling Röhrenkopplung
intricacy Schwierigkeit, Kompliziertheit
intricate schwierig, kompliziert
intrinsic Eigen . . . , innerlich, zugehörig
intrinsically safe eigensicher
introduce, to einführen, einsetzen
introduction Einführung, Einleitung
intrude, to eindringen, intrudieren
intrusion Durchfressen, Intrusion
inundate, to überfluten, überschwemmen
inundation Überflutung, Überschwemmung
invalid ungültig
invalidity Ungültigkeit
invariable unveränderlich, invariabel
invent, to erfinden
invention Erfindung
inventor Erfinder
inventory Inventar, Inventur
inverse Kehrwert
inverse entgegengesetzt, umgekehrt
inversion Umkehrung, Inversion
invert, to umkehren, umwenden
inverter Wechselrichter
invertible umkehrbar
investigation Erforschung, Untersuchung
investment casting Giessen mit verlorener Giessform
invisibility Unsichtbarkeit
invisible unsichtbar
invoice Rechnung
involute evolventisch
involute Evolvente
inward batter Verjüngung
iodine Jod
ionised atom Atomion
ionosphere Ionosphäre
iris diaphragm Irisblende
iron, to mit Eisen beschlagen, bügeln
iron Eisen
ironmongery Beschläge, Eisenwaren
irradiance Bestrahlungsstärke
irradiate, to bestrahlen
irradiation Bestrahlung
irradiator Strahler

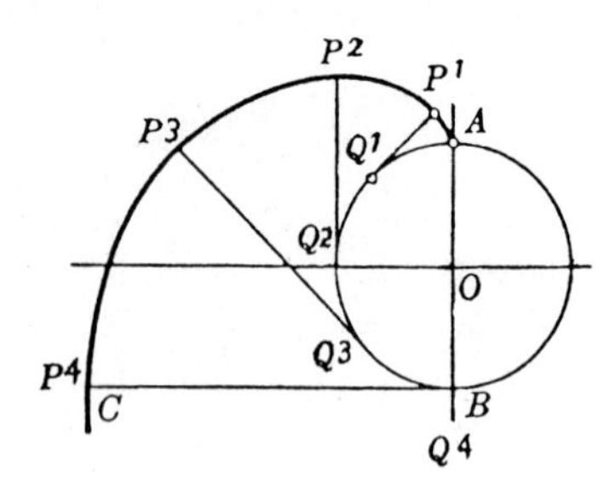

Involute
Evolvente

(involute AC is formed by the locus of point P as line BC unrolls from curve BA Evolvente AC entsteht durch den Lokus von Punkt P beim Abrollen der Geraden BC auf der Kurve BA)

Iris diaphragm
Irisblende

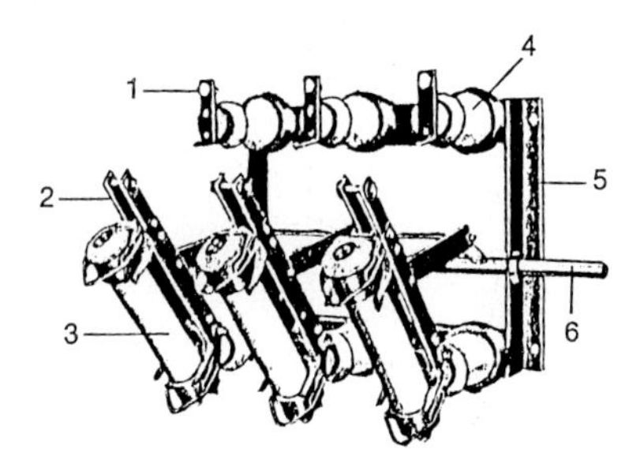

(High-voltage) Isolator, Disconnect switch
(Hochspannungs-) Trenner, Trennschalter

1 **terminal** Anschlußklemme
2 **making contact** Schaltkontakt
3 **fuse** Sicherung
4 **insulator** Isolator
5 **wall-mounting frame** Wandbefestigungsrahmen
6 **operating shaft** Betätigungswelle

irregular ungleichmässig
irregularity Ungleichmässigkeit
irreparable nicht reparierbar
irreplaceable unersetzbar
irresolvable nicht zerlegbar, nicht auflösbar
irreversible nicht umkehrbar, selbstsperrend
irrigate, to bewässern, berieseln
irrigation Bewässerung, Berieselung
irritant Reizmittel
island effect Inselbildung
isolate, to trennen, abschalten, unterbrechen
isolated vereinzelt, einzeln dastehend
isolating capacitor Sperrkondensator, Trennkondensator
isolation Trennung, Abschaltung
isolator Trenner, Trennschalter
isotherm Isotherme, Wärmegleiche
isothermal isotherm
isotope Isotop
issue, to in Umlauf setzen, in Kurs setzen
issue Austritt, Ausgang, Ausgabe
item Gegenstand, Einzelheit, Punkt, Posten
itemise, to spezifizieren
itemised costs Einzelkosten
iteration Wiederholung

J

jack, to anheben, hochwinden, verrücken
jack Hebebock, Heber, Presse, Winde
jacket, to verkleiden, umhüllen, ummanteln
jacket Mantel, Umhüllung, Verkleidung
jacketed ummantelt
jackhammer leichter Bohrhammer
Jacquard loom Jacquardstuhl
jag, to einkerben, verstufen
jag Zacken, Kerbe, Einschnitt
jam, to blockieren, festfressen, klemmen, steckenbleiben, sich stauen
jam Blockierung, Stockung
jammer Störsender, Störungsfunkstelle
jamming Blockieren, Stocken, Festfressen, Störung
jar, to rattern, rütteln, vibrieren
jar Gefäss, Flasche, Krug, Tiegel, Glaskolben
jarring Rütteln, Erschütterung
jaw Klaue, Backe
jemmy kleine Brechstange
jenny Laufkatze
jerk, to stossen, rütteln, rucken
jet, to heraussprudeln, -spritzen
jet Strahl, Düse, Strahltriebwerk
jetcrete Spritzbeton
jib, to scheuen, ausweichen, ablehnen
jib Ausleger, Krankbalken
jig, to setzen, durchsetzen
jig Vorrichtung Spannvorrichtung, Bohrschablone
jig builder Vorrichtungsbauer
jigger Kopplungstransformator, Selbstinduktionsspule
jigging Vorrichtungsbau
job Anstellung, Arbeit, Arbeitsstück
jobbing Einzelteilfertigung (auftragsweise)
jobsite Baustelle
jog, to langsam bewegen, stossen
jog Absatz, Sprung
jogger Schüttelmaschine
jogging Feineinstellung, Rucken
joggle Nut und Feder

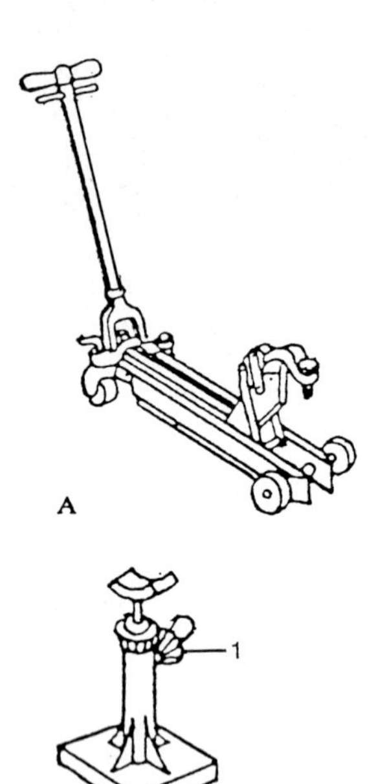

Jack
Heber, Hebebock

A **wheeled car jack** fahrbarer Wagenheber

B **pedestal jack** Hebebock
1 **gear** Übersetzung

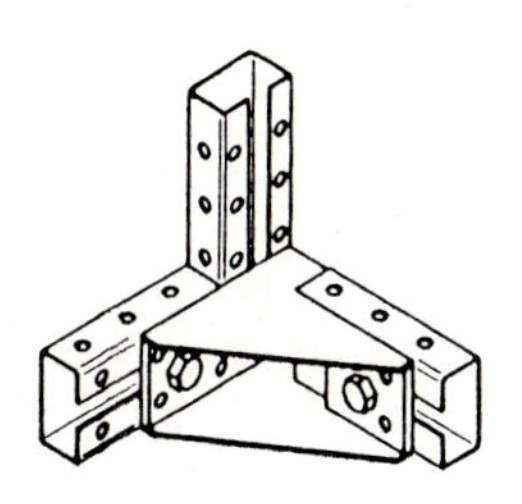

Joint (corner joint of a steel frame)
Verbindung (Eckverbindung eines Stahlrahmens)

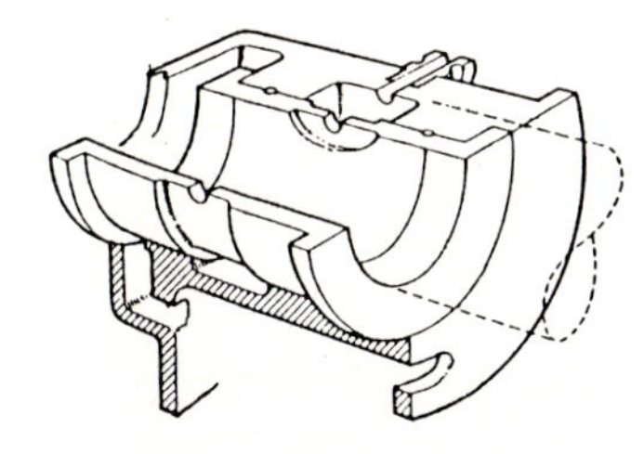

Journal bearing
Gleitlager, Zapfenlager

join, to verbinden, fügen, zusammenschalten
joint Fuge, Stoss, Gelenkstück, Verbindungsstelle
jointer Kabellöter
jointing Verbinden, Vermuffen
jolt, to stauchen, schleudern, rütteln
jolt Stoss, Rütteln
journal, to lagern (masch.)
journal Zapfen, Wellenzapfen, Fachzeitschrift
journeyman Facharbeiter, Geselle
jump, to (über)springen, überschlagen
jump Sprung, Überschlag
jumper Kurzschlussbrücke, Überbrückungskabel
junction Abzweigung, Übergangszone, Halbleiterübergang, Knotenpunkt
juncture Verbindungspunkt
junk Schrott, Altmaterial

K

K degrees Kelvingrade
Kaplan turbine Kaplan-Turbine
Kapp-type phase advancer Kapp-Phasenschieber
keel Kiel
keep, to (er)halten, festhalten, lagern, aufbewahren
keeper Halteeinrichtung, Halter, Wächter
keeping Verwahrung
kerosene Kerosin, Paraffinöl
kettle Wasserkessel
key, to verdübeln, verkeilen, verzahnen
key Längskeil, Feder, Schlüssel
key and slot Feder und Nut
key industry Grundindustrie, Schlüsselindustrie
keyboard Tastatur, Einstellwerk
keygroove, to Keilnuten ziehen
key circuit Tastkreis
keyseat Keilnut
keyway Keilnut
kick, to stossen, antreten
kick Stoss, Schlag
kickout Auslöseeinrichtung
kill, to unwirksam machen, beruhigen, niederschlagen
killed acid Lötwasser
kiln Brennofen, Trockenraum
kind Art, Sorte, Gattung, Güte
kindle, to anzünden, entzünden
kink, to sich verknoten, in Schleifen legen (Leitungen)
kink Schleife
kirner Handbohrer
kit Werkzeugkasten, Ausrüstung
klaxon elektrische Hupe
knee Knie(stück)
knife Klinge, Messer, Schneide
knife switch Messerschalter
knob Knopf, Drehknopf
knock, to aufschlagen, prallen, zusammenstossen, klopfen (Motor)
knot Knoten, Knorren
knockproof klopffest
knuckle joint Kardan-, Kreuzgelenk
knurl Rändel(rad)
knurled screw Rändelkopf-Schraube

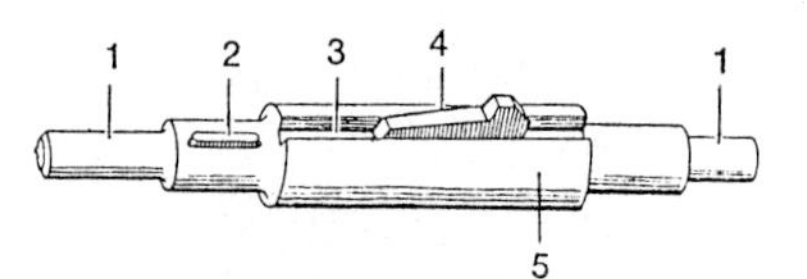

(Shaft) Key
(Wellen-) Keil

1 **journal** Zapfen
2 **fitted key** Einlegekeil
3 **key slot (key way)** Keilnut
4 **headed key** Nasenkeil
5 **shaft** Welle

Keyboard
Tastatur, Tastenfeld

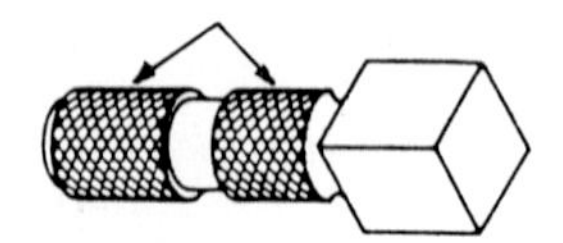

Knurled handle
Rändelgriff

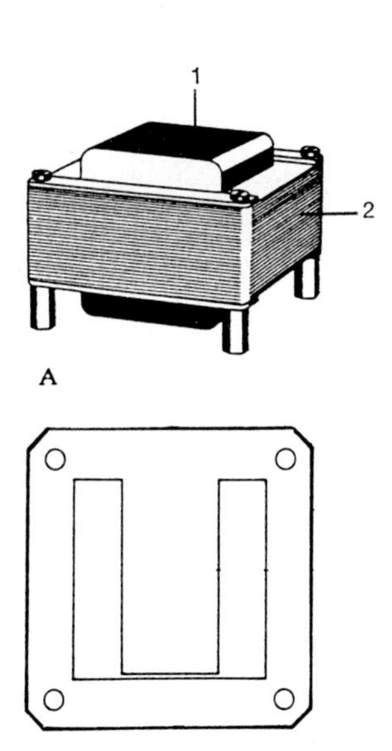

Lamination
Lamenierung

A **Low-voltage transformer** Niederspannungstransformator
1 **winding** Wicklung
2 **laminated iron core** laminierter (geschichteter) Eisenkern

B **Lamination sheet** Lamenierungsblech

L

label, to etikettieren, markieren beschriften, bezeichnen
label Etikett, Kennzeichen, Markierung
labelling Markierung
laboratory Laboratorium
labour, to bearbeiten, bebauen, arbeiten
labour Arbeit
labourer (ungelernter) Arbeiter
lace, to (ver)binden, schnüren
lace Litze, Schnur, Spitze
lack, to ermangeln, nicht haben, nicht besitzen
lack Mangel, Fehlen, Nichtvorhandensein
lacquer, to lackieren, lacken
lacquer (Lack-)Farbe
lacquerer Lackierer
ladder Leiter, Treppe
lade, to beladen, belasten
lading Ladung, Beladen
ladle, to ausschöpfen
ladle Kelle, Pfanne
ladleman Giesser, Pfannenführer
lag, to nacheilen, zurückbleiben, sich verzögern
lag Nacheilung, Verzögerung, Zurückbleiben
lamella Lamelle, Plättchen
lamellar lamellar, geschichtet
lamina Folie, Schicht, Lamelle
laminable streckbar, walzbar
laminate, to schichten, beschichten
lamination Dopplung, Schichtung, lamelliertes Blech
lamp Lampe, Leuchte
landmark Ortungspunkt
lane Fahrspur, Fahrbahn, Gasse
lantern Laterne
lap, to läppen, überlappen
lap Überlappung, Läppen
large-scale production Massenproduktion
laser (light amplification by stimulated emission of radiation) Laser (Lichtverstärkung durch induzierte Emission von Strahlung)
lash, to festbinden, befestigen, zurren
last, to aushalten, dauern, währen

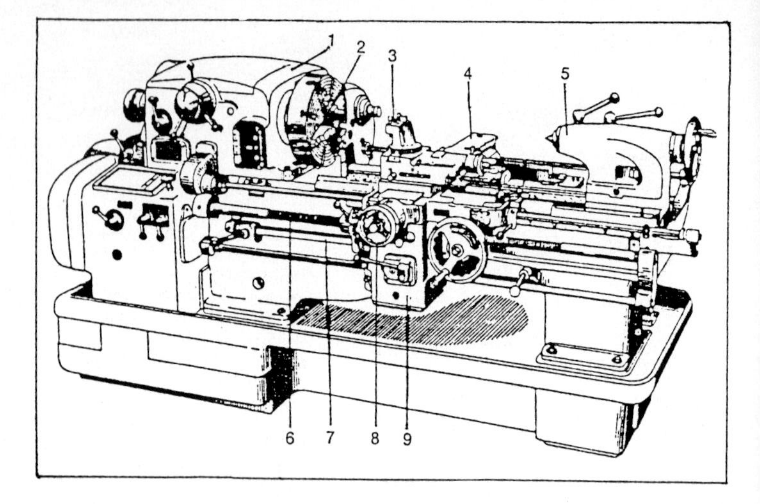

Lathe (conventional, manually operated type)
Drehbank (konventionelle, handbetätigte Ausführung)

1 **headstock** Spindelstock
2 **face plate with dogs (chuck)** Planscheibe mit Spannfutter
3 **tool-post Werkzeughalter**
4 **cross-slide** Querschlitten
5 **tailstock** Reitstock
6 **lead-screw** Leitspindel
7 **feed-shaft** Zugspindel
8 **feed control** Vorschubsteuerung
9 **apron** Support-Schloßplatte (Hauptschlitten

Lattice steel tower of an overhead power line
Gitterstahlmast einer Freileitung

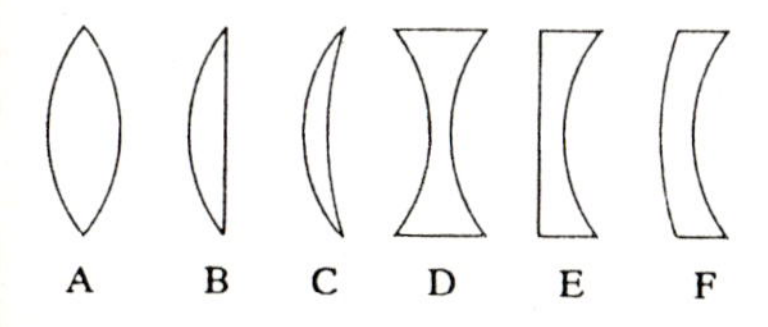

Lens shapes
Linsen-Formen

Converging lenses Sammellinsen

A **Convexo-convex** bikonvex
B **Plano-convex** plankonvex
C **Convexo-concave** konvexkonkav

Dispersing lenses Streulinsen

D **Concavo-concave** bikonkav
E **Plano-concave** plankonkav
F **Concavo-convex** konkavkonvex

latch Klinke, Schnappschloss
lathe Drehbank
lattice Gitter(werk)
launch Lancierung, Stapellauf, Start
law Gesetz
layer Schicht, Lage
laying Verlegung
layout Lageplan, Anordnung, Grundriss
layshaft Vorgelegewelle
leach, to (aus)laufen, herauslösen
leaching agent Laugemittel
lead, to leiten, führen, voreilen
lead Leitung, Leiter (Strom), Steigung
lead, to [led] verbleien, mit Blei auskleiden
lead [led] Blei
leaded fuel verbleiter Kraftstoff
leaf Blatt, Folie, Torflügel
leak Undichtheit, Leckstelle, Ableitung
leakage Undichtigkeit, Leckstelle
leakless lecksicher
leakproof lecksicher, dicht
leaky undicht
lean mager
learner-driver Fahrschüler
leather Leder
lecture, to vortragen, Vorträge halten
lecture Vorlesung
left and right reversed seitenverkehrt
leg Schenkel, Teilstrecke, Bein, Kathete
legend Legende, Zeichenerklärung
legible lesbar
lending library Leihbibliothek
length Länge, Strecke, Dauer
lengthen, to verlängern, strecken
lengthening Verlängerung
lens Linse, Objektiv, Lupe
Lenz's law Lenzsche Regel
Leonard control Leonard-Schaltung
lessen, to vermindern, verringern
letter, to mit Buchstaben bezeichnen, beschriften, stempeln
letter Zeichen, Buchstabe, Brief
letting down Anlassen (Stahl)
level, to abziehen, ebnen, glätten, ausrichten
level Pegel, Niveau, Höhe, Richtgerät

levelling Nivellieren, Abziehen, Ausgleichen, Ebnen, Planieren
lever Hebel, Arm, Schwengel
leverage Gestänge, Hebelwirkung, Hebelwerk
lewis bolt Ankerschraube, Steinschraube
Leyden jar Leydener Flasche
liable for damages schadenersatzpflichtig
liberate, to loslösen, befreien, entwickeln
librarian Bibliothekar(in)
library Bibliothek
licence Lizenz, Zulassung, Bewilligung
licence, to zulassen, Lizenz erteilen
licensee Lizenzinhaber
lid Deckel, Klappe
Lieben-Reiss relay Lieben-Röhre
life Lebensdauer, Betriebsdauer
lift, to heben, anheben, abnehmen
lift Aufzug, Fahrstuhl, Abhebung
lifter Heber
light, to beleuchten, anzünden
light leicht, hell
lighter Anzünder, Feuerzeug, Leuchter
lighting Beleuchtung, Lichttechnik
lightness Leichtheit, Helligkeit
lignite Leichtbraunkohle
limb Schenkel, Gradbogen
limit, to begrenzen, beschränken
limit Grenze, Toleranz
limitation Begrenzung, Einschränkung
limiter Begrenzer
limitless unbegrenzt
limits and fits Toleranzen und Passungen
limp schlaff
line, to ausfüttern, ausgiessen, ausschlagen, auslegen
line Linie, Reihe, Strich, Leitung, Strang, Bahnlinie
linearity Linearität, Geradlinigkeit
lineman Freileitungsmonteur
liner Zylinderlaufbuchse, Unterlegscheibe, Auskleidung, Deckmantel, Ausguss
lineshaft drive Transmissionsantrieb

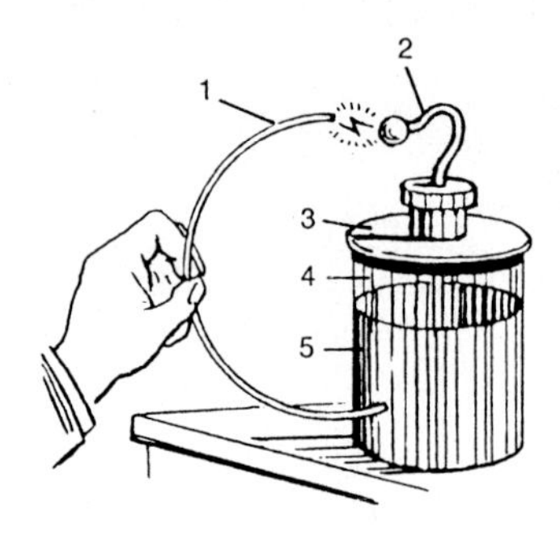

Leyden jar (from the early days of electrical technology)
Leydener Flasche (aus den Anfängen der Elektrotechnik)

1 **discharging wire** Entladungsdraht
2 **brass electrode** Messing-Elektrode
3 **ebonite stopper** Ebonit-Stopfen
4 **glass jar** Glasflasche
5 **tin foil** (Weiß-) Blechfolie

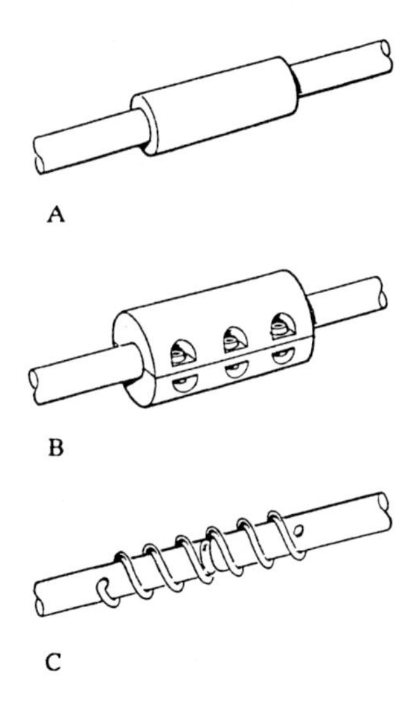

Links
Verbindungen

Examples: means of linking two shaft ends
Beispiele: Mittel zum Verbinden zweier Wellenenden

A **Metal or plastic sleeve** Metall- oder Kunststoffhülse

B **Slit metal muff (bolted)** geteilte Metallmuffe (geschraubt)

C **Coiled steel spring (ends fixed)** gewundene Stahlfeder (Enden fixiert)

lining Belag, Auskleidung, Bremsbelag, Futter
link, to verketten
link Glied, Gelenk, Verbindungsstück, Bindung
linkable verkettbar
linkage Gelenkgetriebe, Verbindung, Aufhängung
linking Verkettung
lip Rand, Kante, Schneidlippe
liquefy, to verflüssigen
liquid Flüssigkeit
list, to eintragen (in eine Liste)
list Liste, Verzeichnis
listen, to anhören, horchen
litter Abfallstoffe, Kehricht
live stromführend, spannungsführend, mitlaufend
load, to belasten, beanspruchen, beschicken, zuführen
load Belastung, Beanspruchung, Last, Ladung, Beschickung
loadability Belastbarkeit
loader Ladegerät, Lader
loading Laden, Aufladen, Belastung, Beanspruchung
lobe Flügel, Lappen, Keule, Zipfel
local örtlich
localiser Landekurssender
locality Ort, Örtlichkeit
locate, to plazieren, anordnen, in Stellung bringen
location Stellung, Lage, Standort
locator Anflugfeuer
lock, to verriegeln, sperren, sichern
lock Riegel, Schloss, Sperre
locker Schrank, Spind
lockhandle Spannhebel
locking Festklemmen, Verriegeln, Arretierung
lockplate Sicherungsplatte
locksmith Schlosser
lockwasher Unterlegscheibe
loctal tube Loktalröhre, Allgasröhre
lodestone Magneteisenstein
lodge, to umlegen
loft Boden, Speicher
log, to aufzeichnen, in Tabelle eintragen, Bäume fällen
log Logarithmus, Logbuch
logger Datenschreiber, Fischereifahrzeug

logging Registrieren, Ausdrucken, Holzfällen
logic Logik
longitude Länge, Längengrad
longitudinal Längsspant, Seiten-
loom Isolierschlauch, Webstuhl
loop, to in Schleife legen, zusammenbinden
loop Masche, Schleife, Rahmen-
looping mill Umsteckwalzwerk
loose locker, lose, abnehmbar, schlaff
loosen, to lösen, lockern, abschrauben
lop, to abschneiden, abrunden
lorry Last(kraft)wagen
loss Abfall, Verlust, Dämpfung
lossless verlustlos, verlustarm
loudspeaker Lautsprecher
louvre Luftschlitz, Jalousie
loupe Lupe
low niedrig, tief, schwach
lower, to senken, herablassen
lowering Herablassen, Senken
lowermost unterst
L.P.G. tanker (liquified petroleum gas tanker), Flüssiggastanker
lubricant Schmiermittel
lubricate, to schmieren, fetten, ölen
lubrication Schmierung
lubricator Oeler, Schmiereinrichtung
luffing crane Hebekran, Einziehkran
lug Henkel, Öse, Klemme
luke-warm handwarm, lauwarm
lumber Nutzholz, Gerümpel
lumeter Lumenmesser
luminaire Lampe, Leuchte, Leuchtkörper
luminance Leuchtdichte
luminesce, to lumineszieren
luminescence Lumineszenz
luminous leuchtend, Licht . . . , Leucht . . .
lump, to zusammenballen
lump Klumpen, Stück
lumpy großstückig
lustre Schimmer
lute, to verschmieren, verkitten
lute Dichtungsmasse, Kitt
luting agent Dichtungskitt
lye Lauge

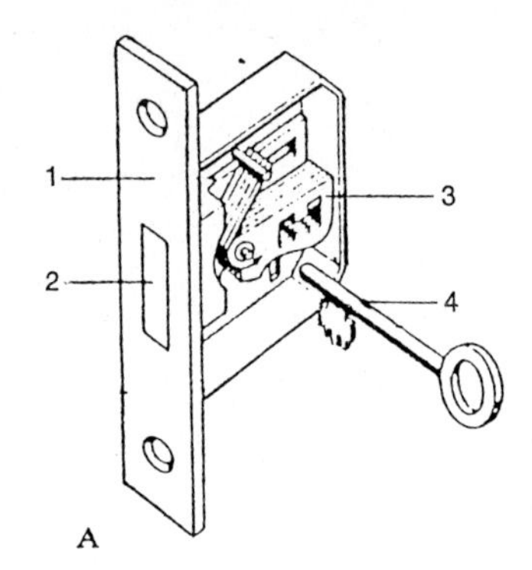

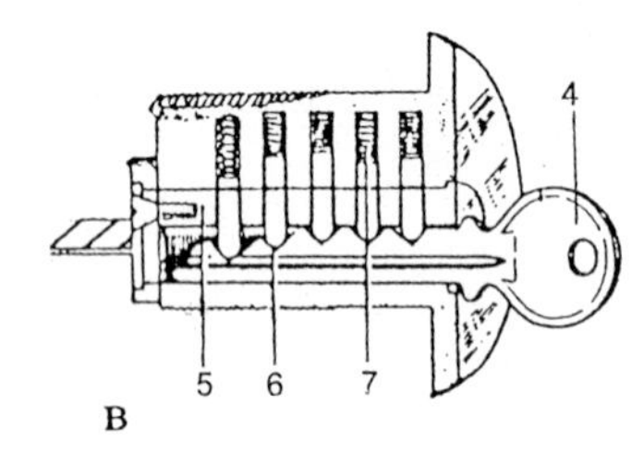

(Door) Lock
(Tür-) Schloß

A **Mortise lock** Kasten(Steck)schloß
1 **face plate** (Abdeck-) Stirnplatte
2 **bolt** Riegel
3 **tumbler** Zuhaltung
4 **key** Schlüssel

B **Cylinder lock** Zylinderschloß
5 **cylinder** Zylinder
6 **key notch and bit** Schlüsselkerbe und -zacken
7 **tumbler pin and spring** Zuhaltungsstift und Feder

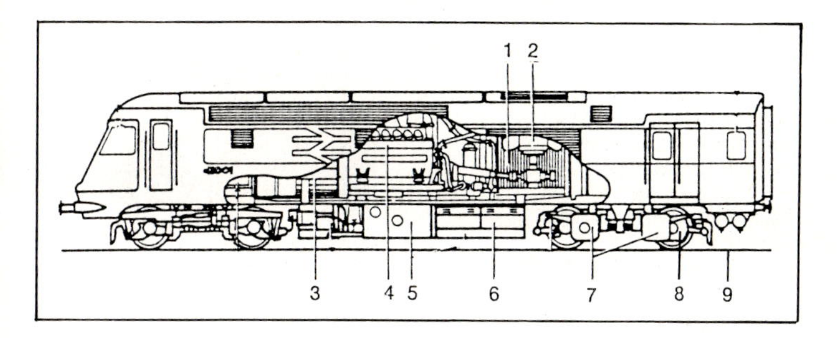

Locomotive (diesel-electric)
Lokomotive (diesel-elektrisch)

1 **radiators** Radiatoren
2 **cooling fan** Lüfter
3 **generator** Generator
4 **diesel engine** Dieselmotor
5 **fuel tank** Treibstofftank
6 **batteries** Batterien
7 **traction motors** Bahnmotore
8 **driving wheel** Antriebsrad
9 **track** Gleis, Schienenstrang

M

Mach number Mach-Zahl
machinability Spanbarkeit, Bearbeitbarkeit
machinable bearbeitbar, spanbar
machine, to bearbeiten, spanen, fertigen
machine tool Werkzeugmaschine
machinery Maschinenpark, Maschinenwesen
machining Fertigung, Maschinenarbeit
made to scale massgerecht hergestellt
magamp (magnetic amplifier) Magnetverstärker
magazine Warenlager, Speicher, Magazin
magnetisable magnetisierbar
magnetise, to magnetisieren
magnification Vergrösserung
magnifier Vergrösserungsglas, Lupe
magnify, to vergrössern
magnifying Vergrössern
magnitude Grösse, Wert, Betrag, Menge
magslip Drehmelder
main Haupt . . .
mains Netz, Speiseleitung
maintain, to aufrechterhalten, instandhalten, pflegen, warten, unterhalten
maintaining Instandhaltung
maintenance Wartung, Instandhaltung, Pflege, Unterhaltung
major grössere(r), Haupt . . .
majority Majorität
make, to herstellen, erzeugen, fertigen, produzieren, aufbauen
make Fabrikat, Bauart
maker Hersteller, Herstellerfirma
makeshift Notbehelf
maladjustment Justierfehler
malfunction Störung, Versagen, Nichtfunktionieren
mall Schlegel
malleable unter Druck verformbar
malleablise, to glühfrischen, tempern
malleablising Tempern
mallet Fäustel, Schlegel

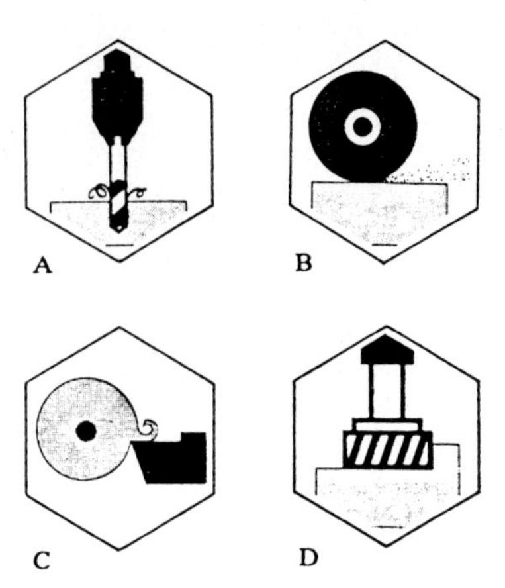

Machining
Maschinenbearbeitung
A **Drilling** Bohren
B **Grinding** Schleifen
C **Turning** Drehen
D **Milling** Fräsen

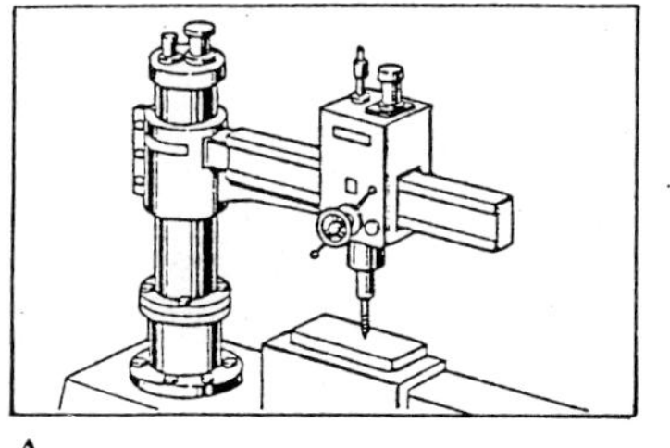

A

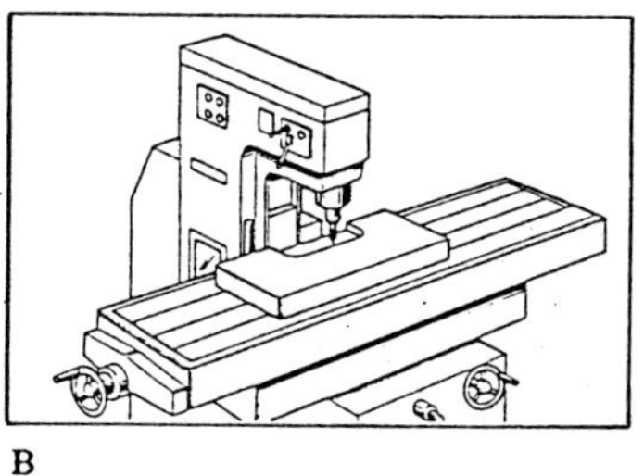

B

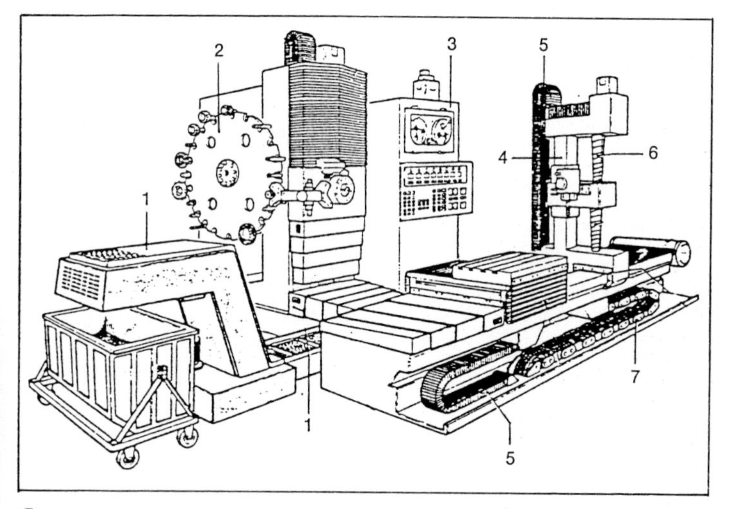

C

Machine tools
Werkzeugmaschinen

A **Drilling automatic** Bohrautomat

B **Milling automatic** Fräsautomat

C **Drilling center** Bohrsystem
1 **swarf transport** Späneförderer
2 **tool changer** Werkzeugwechsler
3 **control station** Steuerschrank
4 **tool column** Werkzeugsäule
5 **cable duct** Leitungsführung
6 **telescopic spring** Teleskop-Feder
7 **positioning chain** Führungskette

maltese cross Malteserkreuz
man, to bemannen, besetzen
manageable manövrierfähig, handlich
management Betriebsleitung, Direktion, Handhabung, Leitung
manganese Mangan
manhole Mannloch, Einsteigöffnung
manifold Rohrverzweigung, Sammelrohr, Krümmer
manipulate, to bedienen, betätigen, handhaben
manipulator Ferngreifer, Manipulator, Kantvorrichtung
manless unbemannt
manned bemannt
manner Art, Weise, Methode
Mannesmann process Schrägwalzverfahren
manning Bemannung, Besetzung
manpower Arbeitskraft
mantle Überform, Glühkörper
manual Handbuch, Handakte
manual von Hand, manuell
manually operated handbetätigt
manufacture, to fertigen, herstellen, produzieren
manufacture Fertigung, Herstellung, Produktion
manysided vielseitig
manystage mehrstufig, Mehrstufen . . .
map, to abbilden, aufzeichnen, aufnehmen
map Karte, Landkarte
mar, to beschädigen, zerkratzen
mar resistance Kratzfestigkeit
margin, to (auf Ansprechwert) einstellen
margin Grenze, Toleranz, Rand, Fasenbreite
marginal am Rande befindlich, Rand . . . , Grenz . . .
marine engineering Schiffsmaschinenbau
marine fouling Schiffsbewuchs
mark, to bezeichnen, markieren, kennzeichnen, ankörnen
mark Marke, Zeichen, Impuls
mastic Mastix, Kitt, Harz
mat matt, glanzlos, mattiert
match, to aufeinanderpassen, anpassen, angleichen

A

B

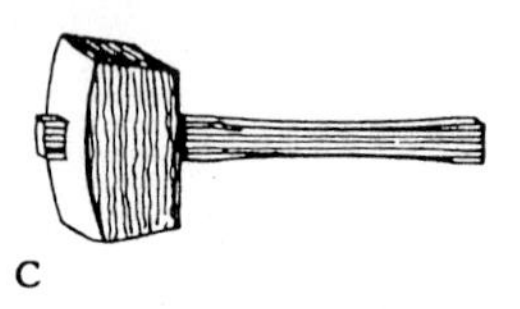

C

Mallets
Fäustel, Schlegel

A **mason's mallet** Steinhauer-Fäustel

B **tinsmith's mallet** Blechschmiede-Schlegel

C **carpenter's mallet** Schreiner-Schlegel

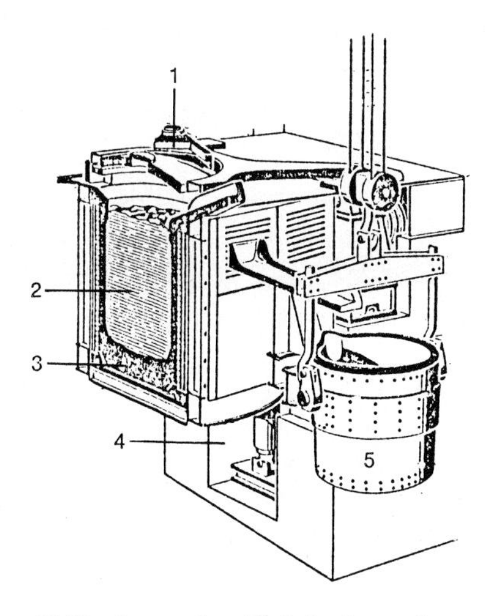

Melting furnace (crucible induction type)
Schmelzofen (Induktionstiegelofen)

1 **swing lid** Schwenkdeckel
2 **melt (liquid iron)** Schmelze (Flüssigeisen)
3 **refractory lining** feuerfeste Auskleidung
4 **furnace pit** Ofengrube
5 **pouring bucket** Abgußkübel

match Zündholz, Zündlunte
matching Anpassung, Zusammensetzen
mate, to ineinandergreifen, kämmen, fügen
materials Baustoffe, Werkstoffe
materiology (zerstörungsfreie) Werkstoffprüfung, Materialprüfung
mathematician Mathematiker
mathematics Mathematik
matter Materie, Stoff, Masse
mature, to reifen, ablagern, altern
maul Holzhammer
mechanical engineering Maschinenbautechnik
mechanics Mechanik
megger Megohmmeter
melt, to schmelzen
melt Schmelze, Charge
member Bauglied, Bauteil, Organ, Element
memory Speicher, Datenspeicher, Gedächtnis
mend, to flicken, reparieren
mental arithmetic Kopfrechnen
merchant marine Handelsmarine
merchant mill Feineisenstrasse
mercury Quecksilber
merit Übertragungsgüte (Funksprechverkehr)
Mesa configuration Mesastruktur
mesh, to eingreifen, kämmen, ineinandergreifen
mesh Geflecht, Masche, Raster
meshing Zahneingriff
message Information, Nachricht, Bescheid
metadyne Metadyne (Elektromaschinenverstärker)
metalliferous metallhaltig, metallführend
metallising Metallauftragung, Metallisierung
metallurgical metallurgisch
metallurgist Metallurge
metallurgy Metallurgie, Hüttenwesen
meteorological chart Wetterkarte
meteorology Meteorologie
meter, to messen
meter Messgerät, Zählwerk
metering Messen, Dosieren, Zählung
methane Methan
metre Meter

Melting shop (case in point: melting cast-iron in induction furnaces)
Schmelzanlage (Fallbeispiel: Schmelzen von Gußeisen in Induktionstiegelöfen)

1 **electrical equipment** Elektro-Ausrüstung
2 **unloading of charging material** Entladen von Einsatzmaterial
3 **conveyor** Transportanlage
4 **charging bin** Beschickungsbehälter
5 **preheating stationn** Vorwärmstation
6 **charging gantry crane** Beschickungs-Kranbrücke
7 **melting floor gantry crane** Schmelzhallen-Kranbrücke
8 **crucible induction furnace** Induktionstiegelofen
(cut-off view showing melt aufgeschnittene Ansicht mit Schmelze)
9 **pouring bucket** Abgußkübel

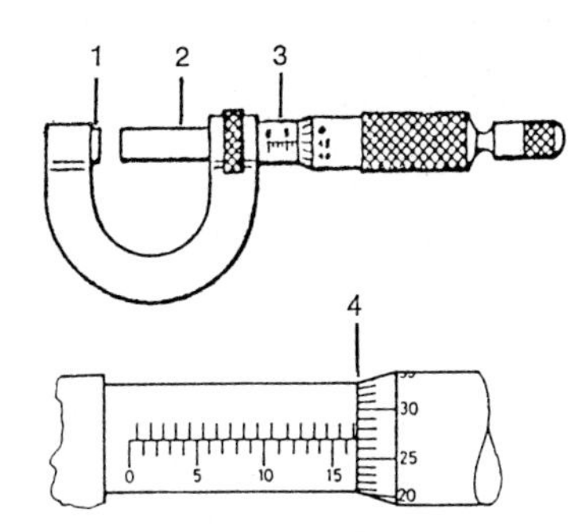

Micrometer
Mikrometer

1 **anvil** Anschlag (Amboß)
2 **spindle** Spindel
3 **scale** Skala
4 **reading example: 16.77 mm** Ablesebeispiel: 16,77 mm

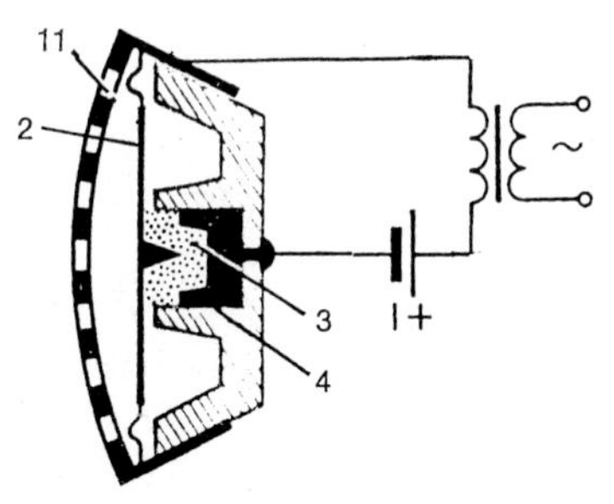

Microphone (carbon type)
Mikrophon (Kohle-Ausführung)

1 **protective cover of perforated material** Schutzkappe aus Maschenmaterial
2 **carbon diaphragm** Kohlemembrane
3 **carbon granule** Kohlegrieß
2 **carbon electrode** Kohleelektrode

metric metrisch
microcrystalline feinkörnig
micron Mikron (1/1000 mm)
micrometer Mikrometer
microniser Feinstmahlvorrichtung
microphone Mikrophon
microstrip Streifenleitung (Wellenleiter)
microstructure Feingefüge
microswitch Mikroschalter
middle Mitte
midget Kleinst . . ., Miniatur . . .
midget base Mignonsockel
migrate, to wandern (Ionen)
migration Wanderung (Ionen)
mil (Längeneinheit) (1 mil = 25,4 Mikrometer = 0,001 Zoll)
mild-carbon steel kohlenstoffarmer Stahl
mile Meile (1 statute mile = 1,609 km)
milage zurückgelegte Strecke (in Meilen)
mill, to fräsen, walzen
mill Mühle, Walzwerk, Fräsmaschine, Fabrik
miller Fräsmaschine
milling Fräsen, Mahlen
mine, to Bergbau betreiben, abbauen, schürfen, gewinnen
mine Bergwerk, Grube, Zeche
miner Bergmann
minimise, to vermindern, auf das kleinste Mass herabsetzen
minority carrier Minoritätsträger
mint Münze
minute klein, sehr klein
mirror, to spiegeln, widerspiegeln
mirror Spiegel
misadjusted falsch eingestellt
misfire Fehlzündung, Versager
misnomer Fehlbezeichnung
misphased phasenvertauscht, phasenfalsch
misprint Druckfehler, Fehldruck
miss, to versagen, verfehlen, fehlschlagen
missile Rakete, Geschoss
mist, to beschlagen, schwitzen
mist Sprühnebel, feiner Regen
mistake Fehler, Irrtum
mistune, to verstimmen
mix, to (ver)mischen, vermengen, anrühren

mixing Mischen, Vermischen, Vermengen
mixture Gemisch, Mischung, Gemenge
mobile beweglich, transportabel, fahrbar
mobility Beweglichkeit, Wendigkeit
mock Schein . . . , unecht
mode Art, Weise, Methode, Form
model, to gestalten, modellieren, formen
model Modell, Konstruktion
moderate, to mässigen, mildern, abbremsen, moderieren
moderator Brennstoff, Moderator
modification Abänderung, Abwandlung
modify, to abändern, modifizieren
modular Modul . . .
modulate, to modulieren, modeln (modulgerecht bauen)
modulating Modulations . . .
modulator Modulator, Zerhacker, Kanalumsetzer
module Baustein, Element, Modul, Grundmass
moist feucht, nass
moisten, to anfeuchten, benetzen, befeuchten
moisture Feuchtigkeit, Nässe
molten geschmolzen
moly steel Molybdänstahl
momentary momentan, kurzzeitig, Momentan . . .
momentum Bewegungsgrösse, Impuls
monitor, to kontrollieren, überwachen, vermitteln, abhören
monitoring Überwachung, Kontrolle, Mithören
monkey wrench Schraubenschlüssel, Universalschraubenschlüssel
monobloc aus einem Stück gegossener Zylinderblock
monolithic monolithisch, aus einem Stück gearbeitet
monostable monostabil
monotooth alternator Wechselstromgenerator mit konzentrierter Wicklung

Mobile transformer
Fahrbarer Transformator

1 **roller wheel** Laufrolle
(**castor wheel** schwenkbare Laufrolle)

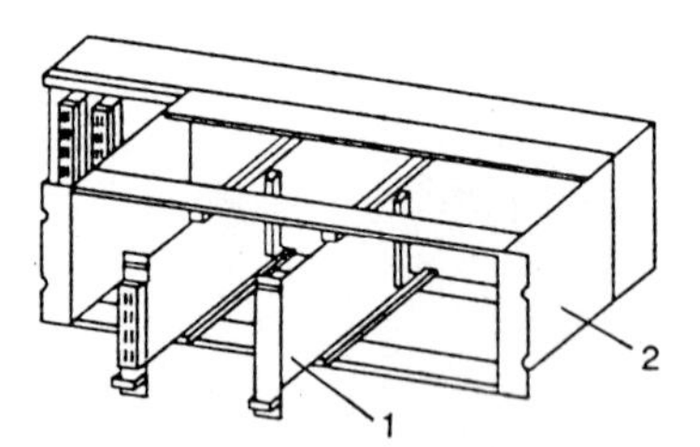

Modular design
Modulbautechnik

1 **module consisting of several electronic components** Baugruppe aus mehreren elektronischen Bausteinen bestehend
2 **rack** Baugruppenträger

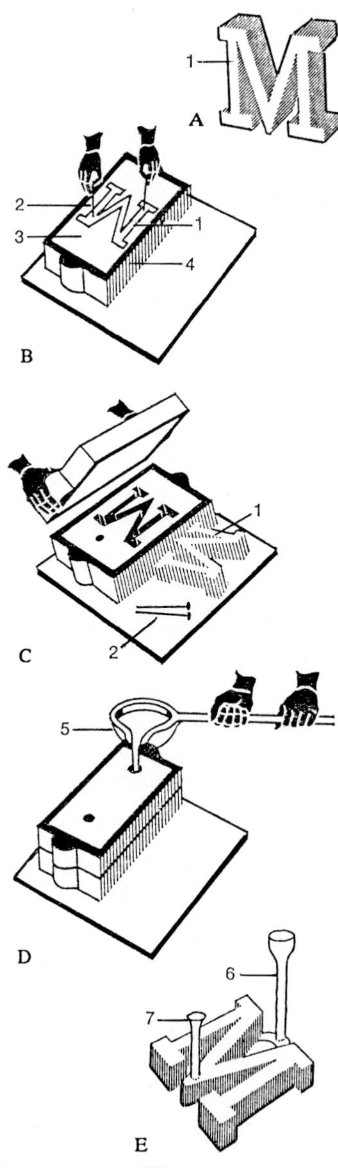

(Legende s.Seite 138)

moor, to anlegen, festmachen
mordant Beize, Ätzung
mordant beizend, ätzend
mortise, to ausstemmen, einschlitzen
mortise Stemmloch, Schlitz
mossy lead Bleischwamm
motion Bewegung, Lauf, Gang
motionless bewegungslos, stillstehend
mould, to formen, gestalten, giessen, pressen
mould Kokille, Form, Modell, Presswerkzeug
mouldable formbar, verpressbar
mount, to aufstellen, montieren, zusammenbauen, anbringen
mount Montierung, Halterung, Fassung
mouth Öffnung, Mündung, Maul, Abstichloch
movable beweglich, verstellbar
move, to bewegen, verschieben, laufen, verholen
movement Bewegung, Lauf, Verschiebung, Uhrwerk
mud Schlick, Schlamm, Moder
muffle, to dämpfen (Schall)
muffle Schmelztiegel, Löschrohr
muffler Schalldämpfer, Auspufftopf
muller mixer Kollergang
multi Viel..., Mehr...
multigrade oil Mehrbereichsöl, Multigradöl, Öl für alle Jahreszeiten
multiple mehrfach, vielfach
multiplexer Mehrfachkoppler
multiplicable vervielfachbar, multiplizierbar
multiplier Vervielfacher
multiply, to multiplizieren
multipoint recorder Punktschreiber
musa aerial Mehrfachrautenantenne
mush Interferenz, Störung (Radar)
mute, to abdämpfen
mute Dämpfer, Schalldämpfer
muting Dämpfung, Abschwächung
mutual gegenseitig, Wechsel...
muzzle Mündung

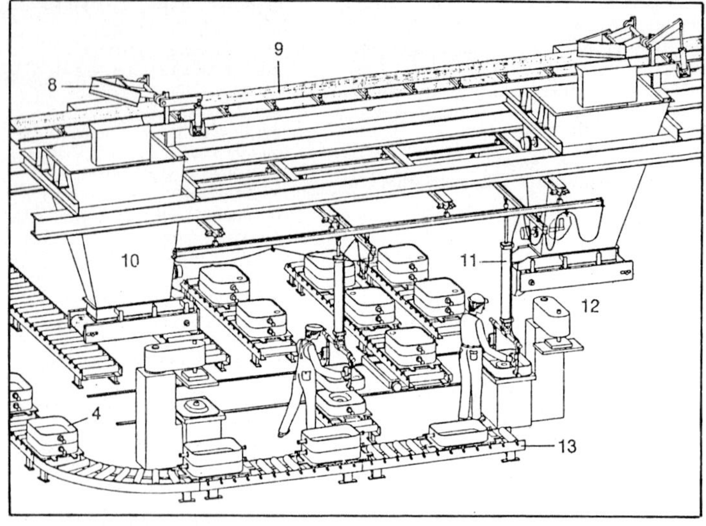

F

Moulding
Formen

A **Model** Modell

B **Mould with solid model inserted in moulding sand** Form mit in Formsand eingebettetem Modell

C **Mould with model removed leaving "M"-shaped cavity in moulding sand** Form mit entferntem Modell und Hohlraum in "M"-Form im Sand hinterlassend

D **Pouring** Gießen

E **Casting** Gußstück

F **Moulding shop** Formerei

1 **model** Modell
2 **moulding pin** Drahtnagel
3 **moulding sand** Formsand
4 **moulding box** Formkasten
5 **pouring ladle** Gießpfanne
6 **runner** Läufer
7 **riser** Steiger
8 **plough** Abstreifer
9 **sand conveyor belt** Sand-Transportband
10 **sand bunker** Sandbunker
11 **pneumatically operated hoist** pneumatisch betätigtes Hebezeug
12 **sand filling station** Sandeinfüllstation
13 **roller conveyor** Rollenbahn

(Siehe auch vorhergehende Seite)

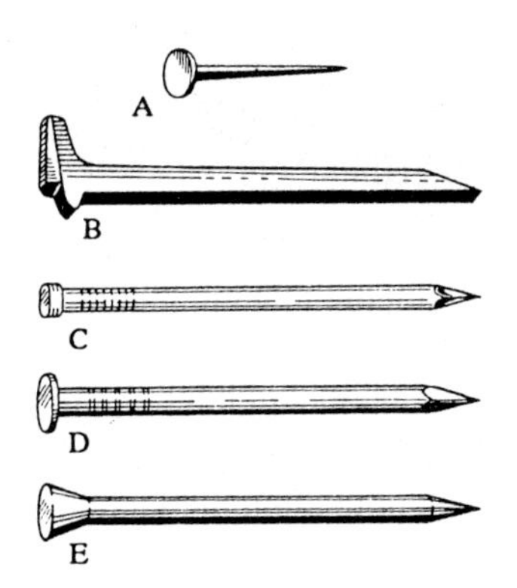

Nails
Nägel

A **Tack** Stift, Teks

B **Rail dog** Schienennagel

C **Lost-head oval wire brad** ovaler, kopfloser Drahtnagel

D **Common round nail** normaler Rundnagel

E **Masonry hard-steel nail** Mauer-Stahlnagel

N

n-type n-leitend, überschussleitend
nacelle Gondel, Rumpf
nail, to nageln
nail Nagel
naked blank, offen, bloss
narrow, to sich verjüngen, enger werden
narrow eng
natural natürlich, Natur . . ., Eigen . . .
nautical nautisch, See . . .
nautics Nautik
naval Marine . . .
navigable befahrbar (Gewässer)
navigate, to befahren (Schiff)
navigating bridge Kommandobrücke
navigation Schiffahrt, Navigation
navigator Nautiker, Seefahrer
navvy Erdarbeiter
navy Kriegsmarine
nc (numerically controlled) numerisch gesteuert
neat rein, sauber, unvermischt
neatly grouped übersichtlich angeordnet
necessary notwendig, zwangsläufig
neck, to einstechen, einschnüren
neck Hals (Welle), Ansatz, Zapfen
need Bedarf, Mangel
needle Nadel
negligible vernachlässigbar
NEITHER-NOR circuit WEDER-NOCH-Schaltung
neodymium Neodym
neon signal lamp Glimmlampe
neoprene Neopren
net Netz
network Netz, Netzwerk
neutral Nulleiter, Mittelleiter, Sternpunkt(leiter)
neutralise, to aufheben, neutralisieren, kompensieren
newly discovered neuentdeckt
nib Ende, Spitze, Stahlfeder
nibble, to ausschneiden, aushauen (Blech)
nibbler Nibbelschere
niccolite Kupfernickel
niche Nische
nichrome Chromnickelstahl

nick, to einkerben, schnitzen
nick Kerbe, Einschnitt, Schlitz
nickelise, to vernickeln
nigging chisel Scharriereisen
niobium Niob
nip off, to abkappen, abkneifen, abzwicken
Nipkow disk Nipkow-Scheibe
nipper Kneifzange
nippers Ziehzange, Beisszange, Kneifzange
nipple Nippel, Ansatz
nitride, to nitrieren
nitralloy Nitrierstahl
nitratine Chilesalpeter
n-n junction nn-Übergang
no-clearance spielfrei, ohne Spiel
no-delay call Schnellgespräch
no-load Leerlauf . . ., unbelasteter Zustand
node Knotenpunkt, Schwingungsknoten, Wellenknoten
nodular kugelig
nog Holzdübel, Holzstift
noise Lärm, Geräusch, Rauschen
noiseless geräuschlos
noisy geräuschvoll
nomenclature Nomenklatur, Terminologie, Benennung
nominal Nominal . . ., nominell, Nenn . . .
non nicht . . ., un . . ., . . . frei
Norton type gear Nortongetriebe
nose Nase, Ansatz, Kopf, Spitze, Schnauze
nosing Ausladung, Kante, Schutzleiste
NOT-AND circuit NICHT-UND-Schaltung
not to scale nicht maßstäblich
not-under-control manövrierunfähig
notable wahrnehmbar, feststellbar
notation Darstellung, Bezeichnung, Schreibweise
notch, to ausklinken, schlitzen, einschneiden, kerben
notch Kerbe, Raste, Schnitt
note Notiz, Merkmal, Note
notice Bericht, Nachricht, Gedanke, Begriff
notion Begriff
nowel Unterkasten (Formerei)
nozzle Düse, Öffnung, Mundstück
n-p junction np-Übergang

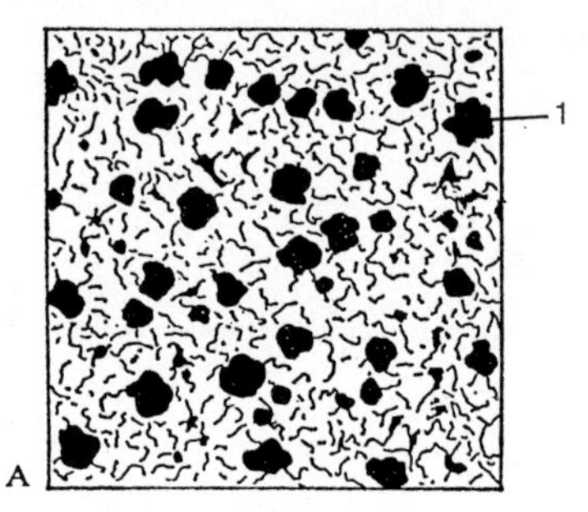

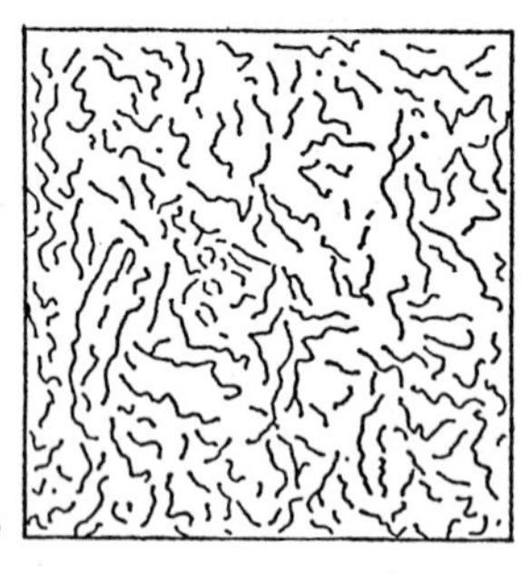

Nodular-graphite (spheroidal) grey iron
Kugelgraphit-Graueisen (Sphäroguß)

A **Microstructure with nodular graphite** Mikrostruktur mit Kugelgraphit
1 **graphite nodules** Graphitkügelchen
B **Microstructure of normal grey iron** Mikrostruktur von normalem Graueisen

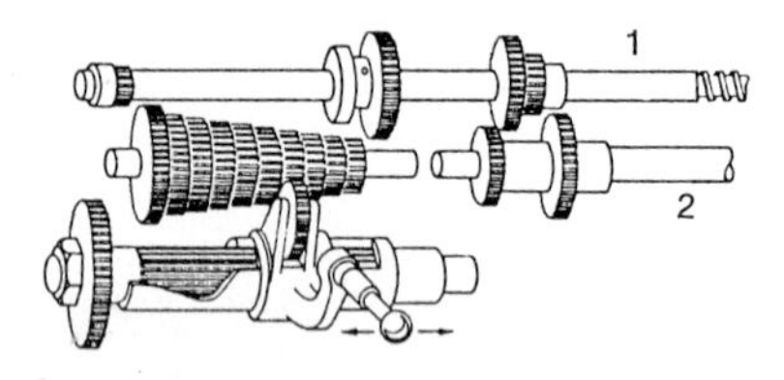

Norton gear
Nortongetriebe (Vorschubgetriebe)
1 **leading shaft** Führungswelle
2 **feed shaft** Vorschubwelle

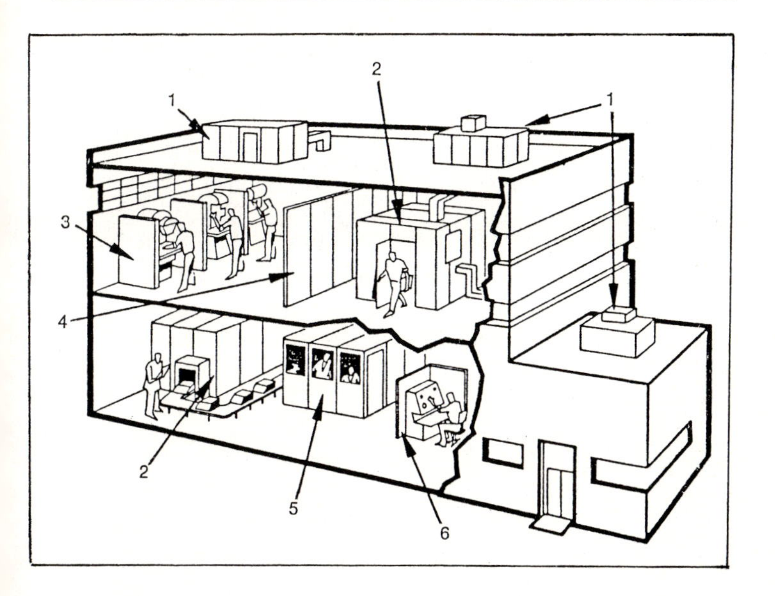

Noise abatement
Lärmbekämpfung
1 **noise-containing air exhaust and vent enclosures** lärmabschirmende Luftaustritts- und Ventilationsgehäuse
2 **noisy machinery enclosure** Verkapselung lärmstarker Maschinen
3 **noise partition** Schallscheidewand
4 **noise absorption wall** Schallschluckwand
5 **sound-proof personnel cabin** schalldichte Personalkabine
6 **sound barrier** Schallbarriere

nuclear Kern . . . , nuklear
nuclear fission Kernspaltung
nuclear power station Kernkraftwerk
nucleus Kern
nugget Nugget, Klumpen
nuisance area Störgebiet, Interferenzzone
nullify, to annullieren, nullen
number, to zählen, beziffern
number Nummer, Zahl, Anzahl, Ziffer
numbering Numerierung, Bezifferung
numerator Zähler (Bruch)
numerical control numerische Steuerung
nut (Schrauben-)Mutter
nutating-piston meter Taumelscheibenzähler
nuts Nusskohlen
nylon reinforced nylonverstärkt

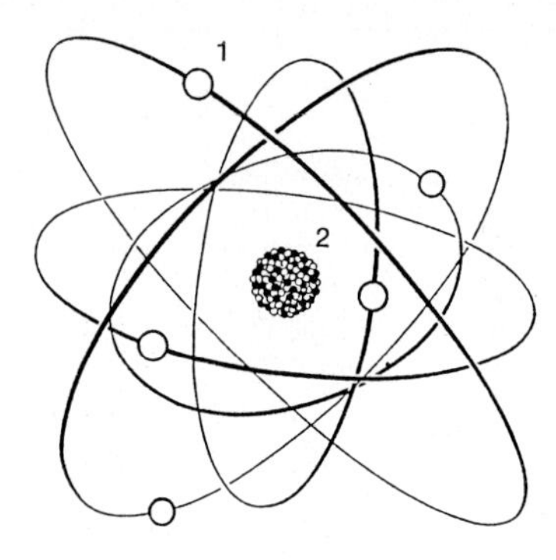

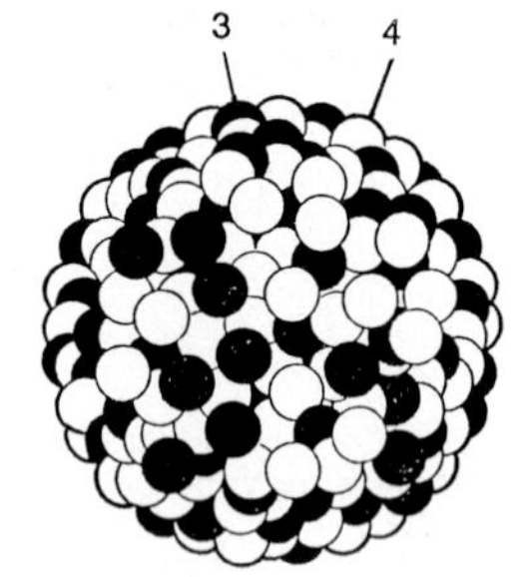

Nuclear structure
Atomstruktur

1 **electron** Elektron
2 **nucleus** Atomkern
3 **proton** Proton
4 **neutron** Neutron

O

Off-shore drilling platform
Der Küste vorgelagerte Bohrinsel

1 **shore** Küste
2 **main platform** Hauptplattform
3 **working platform** Arbeitsplattform
4 **steel pillar** Stahltragsäule

O-ring Dichtungsring mit rundem Querschnitt
oakum Kalfaterwerg
object Ding, Gegenstand
oblate abgeplattet, abgeflacht
oblateness Abplattung
oblique schief, schräg
obliqueness Schräge, Schrägstellung
obliterated verwaschen (Bild)
obliteration Auslöschung, Tilgung
oblong Rechteck, länglich, verlängert
obscuration Verdunkelung
obscure, to verdunkeln, verdecken
obscured glass Milchglas, Mattglas
observation Überwachung, Beobachtung
observatory Sternwarte
observe, to beobachten, messen, beachten
observer Beobachter
obsolete veraltet
obstacle Hindernis, Widerstand
obstruct, to verstopfen, behindern, hemmen
obstructive hinderlich
obtain, to empfangen, erhalten, gewinnen, erreichen
occupation Beruf, Besetzung
occupational beruflich
occupy, to beschäftigen, besetzen, belegen, einnehmen
occurence Ereignis, Vorkommen, Verbreitung
oceanography Meereskunde
octagon Achteck, Achtkant
octagonal achteckig, achtkantig
octane Oktan
octet shell Achtpolröhre
octode Oktode, Achtpolröhre
octuple, to verachtfachen
odd ungerade, unpaarig
oddments Einzelstück (Ladung), Ladereste
odour control Geruchsbeseitigung
off aus, geschlossen
off-shore drilling platform Bohrinsel
off-the-shelf ab Lager

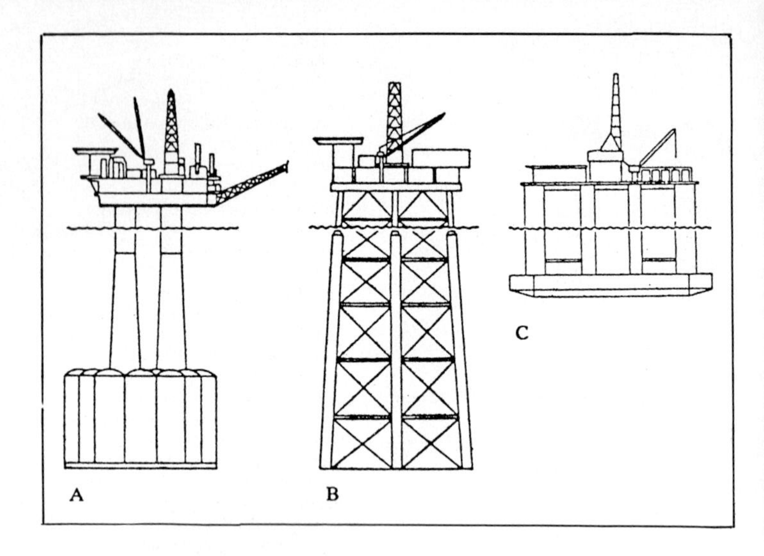

Off-shore drilling platforms (means of anchorage)
Bohrinseln (Verankerungsmethoden)

A **Gravity platform** Schwerkraft-Plattform

B **Steel structure platform** Stahlgerüst-Plattform

C **Semi-submersible (floating) platform** Schwimm-Plattform

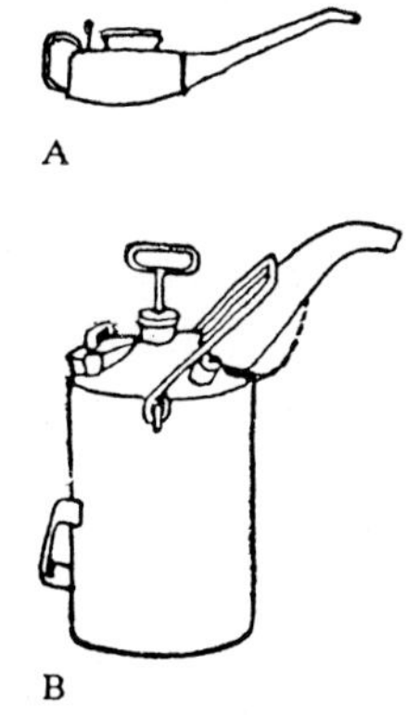

Oil can
Ölkanne

A **Hand can** Handkanne

B **Pressure mixing can** Druck-Mischkanne

offal Abfall
offcut Verschnitt, Abfallpapier
offer, to anbieten, offerieren
office Amt, Büro
offset, to versetzen, abbiegen, absetzen
offset Absatz, Kröpfung, Abbiegung
offtake Abzugskanal, Ableitungsrohr
oil, to ölen, schmieren
oil Öl
oiled geölt, geschmiert
oiler Öler, Schmierkanne
oilskin Ölzeug
oleic acid Ölsäure
omission Wegfall, Auslassung
omit, to weglassen, auslassen
on-line eingegliedert, mitlaufend
on-line computer Prozessrechner
ondometer Frequenzmesser, Wellenmesser
ooze, to durchsickern, langsam austreten
oozing basin Sickerbecken
opaque trüb, milchig
open, to öffnen, unterbrechen, trennen, einmünden
open-air Freiluft . . .
opencast Tagebaubetrieb
opening Öffnung, Loch, Kanal, Durchführung, Mündung
operable betriebsbereit
operand Rechengrösse
operate, to bedienen, betätigen, handhaben, arbeiten
operating Betrieb, Betätigung, Betriebs . . .
operation Bedienung, Betätigung, Betrieb, Verfahren
operational betrieblich, betriebsmässig
operations area Bedienungsbühne
operator Bedienungsmann, Arbeiter, Maschinenarbeiter
oppose, to entgegenwirken
opposed entgegengesetzt, gegenläufig, Gegen . . .
opposite Gegenteil, gegenüber
opposite number Gegenspieler
opposition Gegenläufigkeit
optic optisch, Seh . . .
optics Optik
optimal Best . . . , optimal

Opencast browncoal mine
Braunkohlen-Tagebau

1 **bucket-chain dredger** Eimerkettenbagger
2 **conveying bridge** Förderbrücke
3 **overburden** Abraum
4 **coal seam (upper seam)** Kohlenflöz (Oberflöz)
5 **bucket-wheel dredger** Schaufelradbagger
6 **spoil** (Abraum-) Zwischenmittel
7 **lower coal seam** Unterflöz

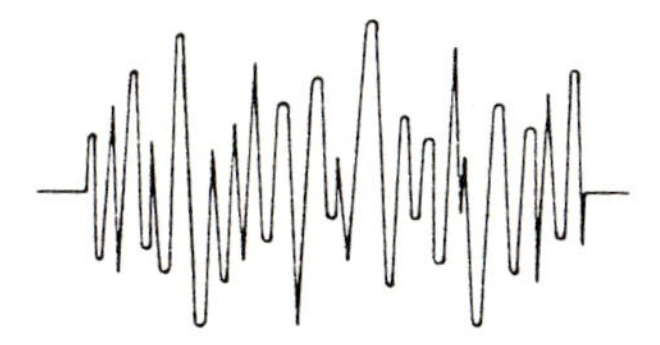

(High-freqency) Oscillations
(hochfrequente) Schwingungen

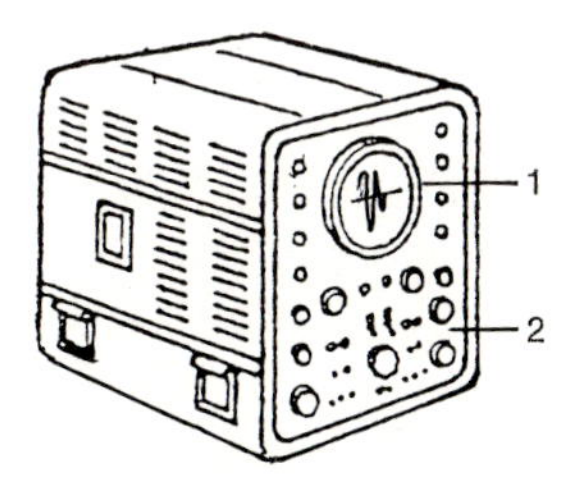

Oscilloscope
Oszilloskop

1 **screen with graph** Bildschirm mit Kurvenbild
2 **controls** Steuerknöpfe

optimisation Optimierung
optimise, to optimieren
optimiser Optimisator, Optimierungseinrichtung
optimising control Bestwertregelung
optional wahlweise
OR-buffer ODER-Schaltung
OR-circuit ODER-Schaltung
OR-element ODER-Glied
OR-ELSE ENTWEDER ODER
OR-gate ODER-Glied, ODER-Schaltung
orbit, to in Umlauf bringen, auf eine Umlaufbahn bringen
orbit Bahn, Umlaufbahn, Kreisbahn
orbital Bahn . . . , Hüllen . . .
order, to bestellen, befehlen, vorschreiben
order Auftrag, Befehl, Ordnung, Grössenordnung
ordinary gewöhnlich, ordentlich, normal
ore Erz
orifice, to drosseln
orifice Düse, Öffnung, Mündung
origin Ursprung, Ausgangspunkt
original Original . . . , original, ursprünglich
orphan (alte) Maschine, für die keine Ersatzteile erhältlich sind
orphitron Rückwärtswellenoszillatorröhre
orthojector circuit-breaker Ölstrahlschalter
oscillate, to schwingen, oszillieren
oscillating Schwingung(s . . .)
oscillation Schwingen, Schwingung, Oszillation
oscillator Oszillator, Schwingungserzeuger, Schwinger
oscillatory oszillierend, schwingend
oscillograph loop Oszillographenschleife
oscillograph valve Oszillographenröhre
oscillometer Dekameter, DK-Meter
oscilloscope Oszillograph, Oszilloskop
osculating Krümmungs . . .
osculation Berührung

osmosis Osmose
Otto cycle Otto-Verpuffungs-verfahren
Otto engine Ottomotor
ounce Unze (28,34953 g)
out-of-balance Unwucht
outboard bearing Nebenlager, Gegenlager, Aussenlager
outburst Ausbruch
outdated veraltet
outdoor Freiluft . . . , Frei . . . , aussen, draussen
outer Aussenleiter
outer aussenbefindlich, Aussen . . . ,
outermost am weitesten aussen, äusserst
outfall Abflussleitung
outfit Ausrüstung, Ausstattung, Geräte, Apparatur
outflow, to abfliessen, ausfliessen, ausströmen
outflow Ausfluss, Abflussmenge, Auslass
outgas, to entgasen
outgoing abgehend, austretend, abgegeben
outgrowth of crystals Kristall-wachstum
outlast, to überdauern, länger halten als
outlet Austritt, Auslass, Steckdose
outline, to umreissen, aufzeichnen
outline Umriss, Begrenzung, Zeichnung
outperform, to in der Leistung übertreffen
outphase, to phasenverschieben, aus der Phase bringen
outphased phasenverschoben
output Leistung, Ausgang, Förderung, Produktion, Ausstoss
outreach Ausladung
outrigger Ausleger
outside Aussenseite, Aussenfläche, Stirnfläche
outside aussen, ausserhalb, Aussen . . .
outspeed, to an Schnelligkeit übertreffen
outturn sheet Ausfallmuster
outward Aussen . . . , äusserer, aussen befindlich
outwash, to auswaschen, ausschlämmen

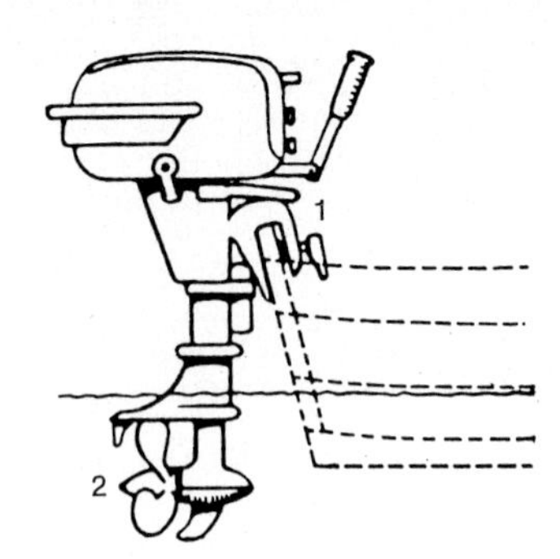

Outboard motor
Außenbordmotor

1 **attachment to boat** Bootsbefestigung
2 **propeller** Motorschraube

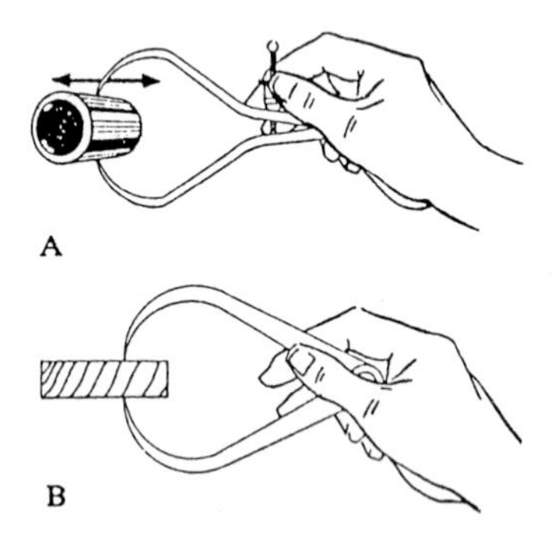

Outside caliper
Außentaster

A **measuring outer diameter of a pipe** beim Messen des Außendurchmessers eines Rohres

B **measuring thickness of a workpiece** beim Messen der Dicke eines Werkstückes

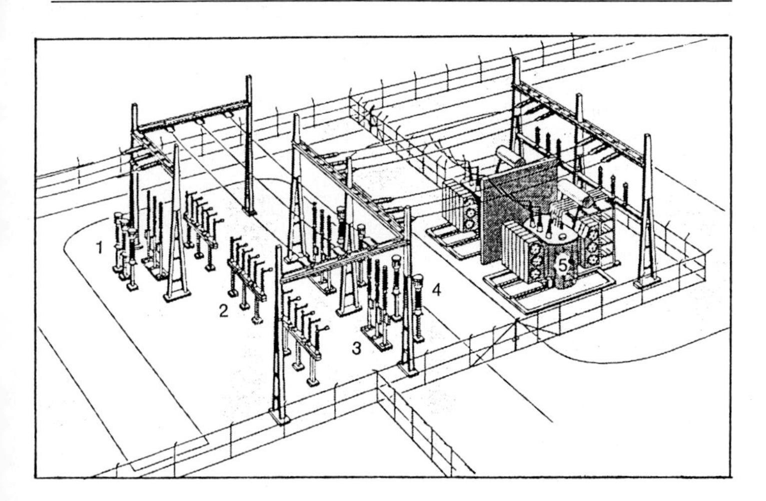

Outdoor switchyard (substation)
Freiluftschaltanlage (-Unterstation)

1 **surge arrester** Überspannungsableiter
2 **disconnect (isolator)** Trenner
3 **circuit breaker** Leistungsschalter
4 **instrument transformer** Meßwandler
5 **power transformer** Leistungstransformator

outwork Heimarbeit
oven Ofen
overal Total . . . , Gesamt . . .
overamplification Übersteuerung
overarch, to überwölben
overcharge, to überladen
overcharge Überladung, Überlast, Überdruck
overcurrent Überstrom
overdrive, to übersteuern
overdrive Schnellgang, Schongang
overrich mixture zu fettes Gemisch
overrunning Überlaufen, Weiterlaufen
overshoot, to überschwingen
overshoot Durchschwingen, Überschwingen
oversize Übergrösse, Übermass
overspeed Schleuderzahl
overspeed, to überdrehen
overstress, to überbeanspruchen, überlasten
overtime Überstunden, Mehrarbeit
overtone harmonische Oberschwingung
overtravel, to überregeln, überschreiten, überschwingen
overtravel Überlauf
overwind, to überdrehen (Uhrfeder)
oxidate, to oxydieren
oxidation Oxidation, Sauerstoffaufnahme
oxldise, to oxydieren, Sauerstoff anlagern
oxidulated copper ore Rotkupfererz
oxygen Sauerstoff
oxygenated mit Sauerstoff gesättigt
oxyhydrogen Knallgas
ozone Ozon
ozoniferous ozonreich

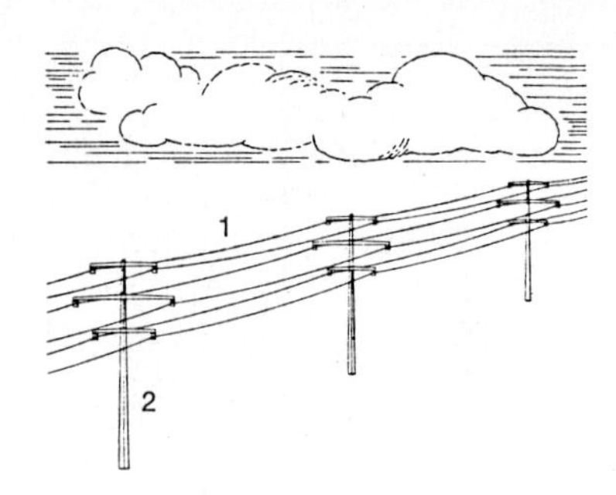

Overhead line
Freileitung

1 **conductor cable** Leiterkabel
2 **pole** Mast

P

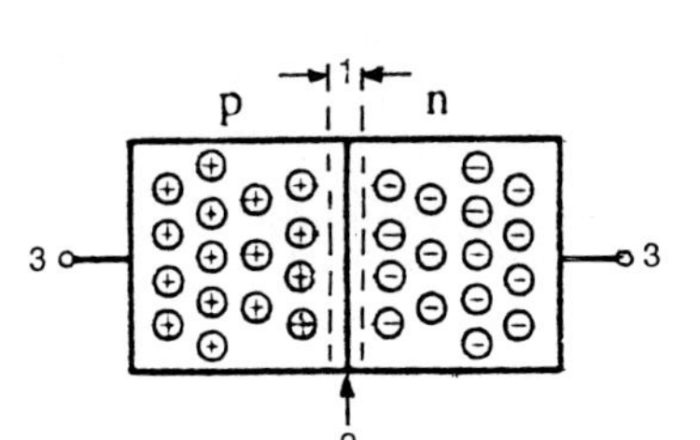

p-n semiconductor element (diode)
pn-Halbleiterelement (Diode)

1 **p-n boundary** pn-Grenzschicht
2 **p-n junction** pn-Übergang
3 **terminal** Anschluß
⊕ **donator** Donator
⊖ **acceptor** Akzeptor

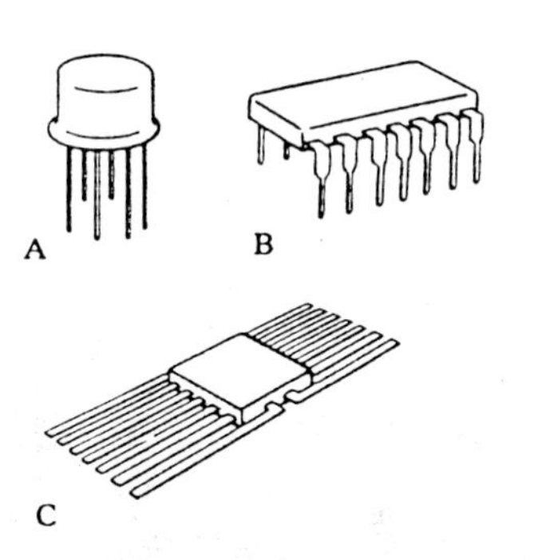

Semiconductor **Package**
Halbleiter-Gehäuse, -Verkapselung

A **Cap-type package** Haubengehäuse

B **Dual-in-line package** Gehäuse mit Anschlüssen in Doppelreihe

C **Flatpack** Flachgehäuse

p-conducting p-leitend
p-donor p-Donator
p-hole Defektelektron
p-material p-leitendes Material
p-n boundary pn-Grenzschicht
p-n junction pn-Übergang
p-type p-leitend, defektleitend
p-type conduction p-Leitung Löcherleitung, Mangelleitung
p-type conductivity p-Leitfähigkeit
pace Schritt, Gangart
pack, to abdichten, stopfen, verpacken
pack Bündel, Paket
package Paket, kompakte Baugruppe, Verpackungsbehälter
packaging Verpackung
packing Packen, Abdichtung, Dichtungsmittel, Futter
pad, to polstern, wattieren
pad Kissen, Polster, Puffer, Stossdämpfer, Dämpfungsglied, Schreibblock
paddle Schaufel, Rührarm
padlock Vorhängeschloss
page, to paginieren, umbrechen
page Seite
pail Eimer
paint, to malen, anstreichen, tünchen
paint Farbe, Anstrich, Überzug
pallet Werkstückträger, Ladeplatte, Palette
pan Schale, Pfanne, Napf, Schüssel, Trog
pancake coil Scheibenspule, Flachspule
pane, to aushämmern
pane Scheibe, Tafel, Pinne
panel Tafel, Platte, Feld, Beplankung
panelling Täfelung, Wandbekleidung, Seitenverkleidung
panscales Kesselstein
paper Papier, Zeitung, wissenschaftliche Abhandlung, Beitrag zu wissenschaftlicher Tagung
parabola Parabel
parabolic parabelförmig, Parabel ...

parachute Fallschirm
parallel, to parallelschalten
parallel Parallele, Richtschiene
paralleling Parallelschaltung
paralleling switch Synchronisierschalter
parameter Parameter (Hilfsvariable), Kenngrösse, Kennwort
parasite Parasit, Schmarotzer
parasitic unerwünscht, parasitär, Wirbel . . . , Kriech . . .
particular besonders, partikulär
particulate aus Einzelheiten bestehend
parting Lösen, Trennung, Scheidung
partition leichte Trennwand, Zwischenwand
pass, to durchlaufen, durchfliessen, durchgehen, passieren
pass Durchgang, Arbeitsgang
p-band Durchlassbereich, Durchlassband
passage Durchgang, Durchlauf, Übergang
passed for press für den Druck freigegeben
passenger Fahrgast, Passagier
passing Durchgang, Durchfluss, Durchfahrt
passivate, to passivieren
passivating Passivierung
past gegen, vorbei
paste, to bekleben
paste Paste, Kleister, Klebstoff
patch, to ausbessern, flicken, zusammenschalten
patch Flickstelle, Fleck, Korrekturbefehl
patchcord Verbindungskabel, Steckerschnur
patent, to patentieren lassen
patentable patentierbar
patentee Patentanmelder, Patentinhaber
path Bahn, Weg, Verlauf, (Strom-)Pfad
pattern Muster, Zeichnung, Schablone, Ausführung, Schema
patternmaker Modelltischler
pave, to pflastern
pavement Pflasterung, Gehweg, Strassendecke, Fahrbahn
pawl, to einrasten, einklinken

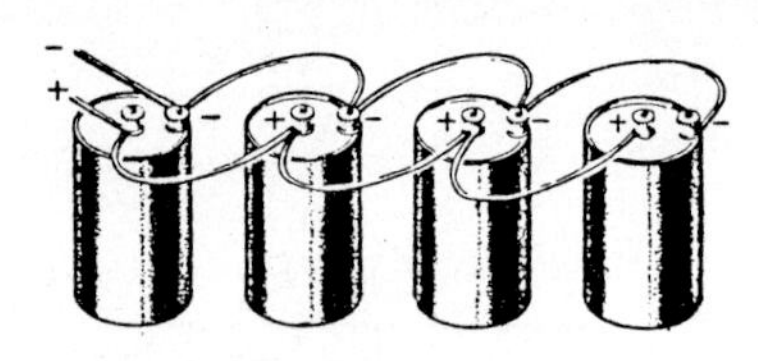

Parallel connected dry-cells
Parallelgeschaltete Trockenbatterien

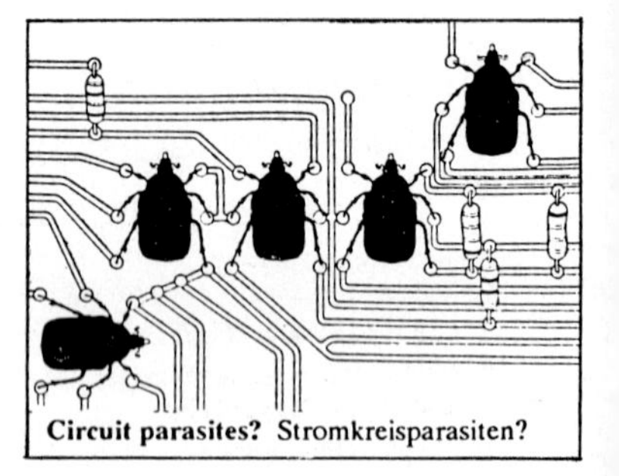

Circuit parasites? Stromkreisparasiten?

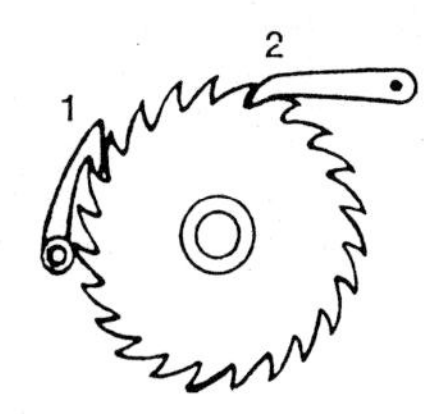

Pawl
Sperrklinke

1 **hooked pawl** Hakenklinke
2 **straight pawl** gerade Klinke

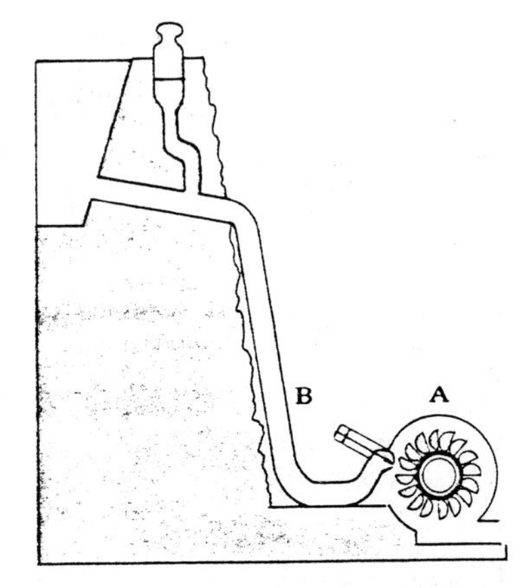

A **Pelton turbine** Peltonturbine
B **Penstock** Druckleitung
(**hydro-power plant** Wasserkraftwerk)

pawl Schaltklinke, Sperrklinke, Sperrhebel, Mitnehmerklinke
pay, to bezahlen, entlohnen
pay Bezahlung, Gehalt, Lohn, Heuer
payable zahlbar
payload Nutzlast
payment Bezahlung, Entlohnung
peacock ore Buntkupferkies, Bornit
peak Spitze, Gipfel, Höhepunkt, Scheitelwert
peakedness Kurvensteilheit
peaking Spitzenwertbildung
pebble Kieselstein
pebbles körnige Trägersubstanz, Körner
pedal Pedal, Fusshebel, Tritt
pedestal Sockel, Bock, Auflager, Ständer
peel, to abschälen, abblättern, entrinden
peen, to aushämmern, ausrichten
peep hole Schauloch
peg, to verstiften, verdübeln
peg Stöpsel, Dübel, Drehstift
pellet Kügelchen, Pille, Perle, Tablette
Peltier cell Peltier-Zelle
Peltier-effect Peltier-Effekt
Pelton turbine Peltonturbine, Freistrahlturbine
pen Feder, Einsatz
pencilled drawing Bleistiftzeichnung
pendant schwebend
pendant Schwenkarm, Hängelampe
pending angemeldet, beantragt
pendulous Pendel . . . , pendelartig
pendulum Pendel
penetrate, to durchdringen, eindringen, einsickern
penetration Durchdringung, Eindringen
penetrative durchdringend, Durchdringungs . . ., Durchschlags . . .
penstock Rohrzuleitung, Druckleitung, Düsenstock
pentagrid Heptode, Siebenpolröhre
penthouse Schutzdach, Anbau
pentode Pentode, Fünfpolröhre
perceivable wahrnehmbar
perceptible erkennbar

perception Wahrnehmung, Empfindung
perch, to beschauen, absuchen
perch Stange, Längenmass ($5^1/_2$ yard)
perchlorate Perchlorat, Salz der Perchlorsäure
percolation Sickerung, Durchsickerung
percolator Perkolator, Filtrierapparat
percuss, to abklopfen
percussion Schlag, Stoss, Erschütterung
percussion drilling Schlagbohrverfahren
percussion drilling machine Schlagbohrmaschine
perfect, to vervollkommnen, vollenden
perfect vollkommen, einwandfrei
perfection Vervollkommnung, Vollendung
perforate, to durchbohren, perforieren, lochen
perforating Perforieren, Lochen
perforation Perforation, Lochung
perforator Locher, Stanzer
perform, to ausführen, leisten
performance Kenndaten, Leistungsfähigkeit, Betriebsverhalten
performance check Funktionsprüfung
perigon Vollwinkel (360°)
perimeter Umfang, Umriss
period Dauer, Periode, Frist, Schwingungsdauer
periodic periodisch, Perioden . . .
peripheral peripher, Umfangs . . . , Rand . . .
periphery Umfang, Peripherie
periscope Periskop, Sehrohr
perishables leicht verderbliche Güter
permanence Permanenz, Dauer, Beständigkeit
permanent permanent, dauernd, bleibend, ständig, Dauer . . .
permeability Durchlässigkeit, (magnetische) Permeabilität
permeable durchlässig, durchdringbar, permeabel
permeance Permeanz, magnetischer Leitwert

Perforated (punched) tape
Lochstreifen

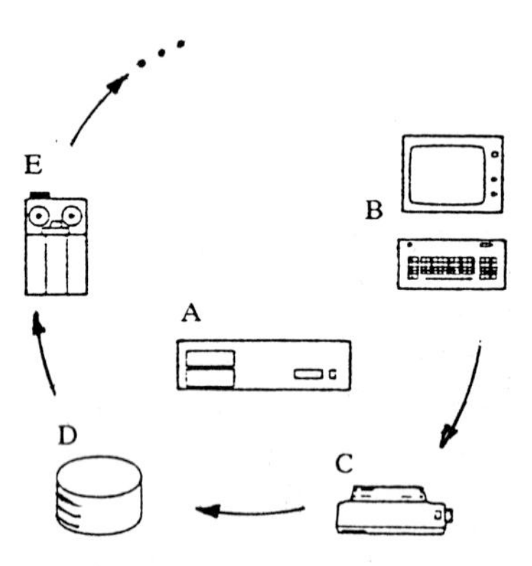

Peripheral devices of a computer system
Peripheriegeräte eines Rechnersystems

A **Central processing unit** (CPU) Zentraleinheit
B **Terminal** Bedienungsgeräte
C **Printer** Drucker
D **Disks** Platten
E **Tape** Band
B, C, D **Peripheral devices** Peripheriegeräte

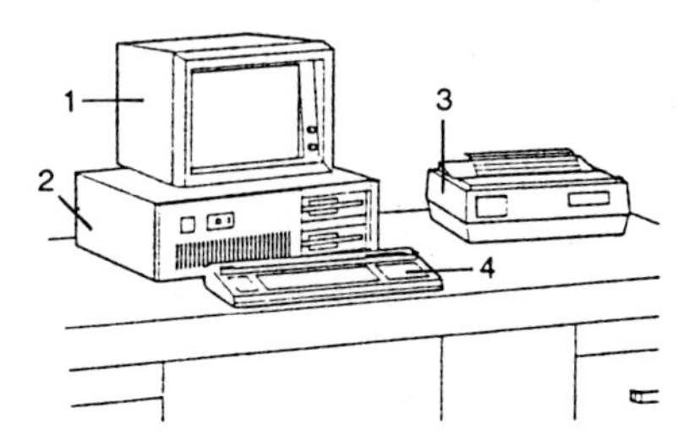

Personal computer (PC) system
Persönlicher (Arbeitsplatz-) Rechner

1 **visual display unit** Bildschirm
2 **computer** Rechnereinheit
3 **printer** Drucker
4 **keyboard** Tastatur

permeate, to durchdringen, durchfluten
permeation Durchdringung, Sättigung
permissibility Zulässigkeit
permissible zulässig
permit, to erlauben, zulassen, aufnehmen
permittivity Dielektrizitätskonstante
permutation Vertauschung, Permutation
permutator Kommutatorgleichrichter
perpendicular senkrecht, rechtwinklig
perpetual beständig, ewig
Persian leather Indisches Leder
persist, to beharren, bestehen bleiben, nachleuchten (Kathodenstrahlröhre)
persistence Beständigkeit, Persistenz, Nachleuchtdauer (Kathodenstrahlröhre)
personal computer (PC) persönlicher (Arbeitsplatz-) Rechner
personnel Belegschaft, Mannschaft
perspex Plexiglas (Handelsname)
perturbation Störung, Störeffekt
pervious durchlässig, permeabel
perviousity Durchlässigkeit
pet cock Probierhahn, Kondenswasserhahn, Kompressionshahn
petrification Verhärtung, Versteinerung
petrified versteinert
petrify, to versteinern
petrochemical Erdölchemie . . . , petrochemisch
petrol Benzin
petticoat Isolatorglocke
pewter Weissmetall
pewtery Zinngiesserei
pH-acidometer pH-Messer
pH meter pH-Messgerät
phanotron Phanotron, ungesteuerte Gleichrichterröhre
phantom Phantom, Phantomschaltung, Viererschaltung
phase, to in Phase bringen, gleichphasig machen
phase Phase
phase advancer Phasenschieber
phase balance relay Phasenunterbrechungsrelais

phasing Phaseneinstellung
phenomenon Phänomen, Effekt, Erscheinung
phial Phiole, Flakon, gläsernes Fläschchen
phone, to telefonieren
phone Telefon
phone Hörer, Kopfhörer
phonic phonisch
phonogram Tonaufzeichnung, zugesprochenes Telegramm
phonograph Plattenspieler
phonometer Phonometer, Lautstärkemesser
phosphate, to phosphatieren
phosphate Phosphat, Salz der Phosphorsäure
phosphoresce, to phosphoreszieren
phosphorus Phosphor
photoactive fotoaktiv, lichtelektrisch
photocell Fotozelle, lichtelektrische Zelle
photoconductive fotoleitend, lichtelektrisch
photoelectric fotoelektrisch, lichtelektrisch
photoemitter Fotoemitter, Fotokathode
photograph Fotografie
photon Photon, Lichtquant
photophone Lichttelefoniegerät
photoradio transmission Bildfunk
photoresistor Fotowiderstand
photosensitive lichtempfindlich
photosensitise, to lichtempfindlich machen
photostat Fotokopie
photostimulated durch Licht angeregt
physical physikalisch
physicist Physiker
physics Physik
pick, to sammeln, auslesen, sortieren, auswählen
pick Hacke, Haue, Schrämmeissel
picker Anfangsbohrer
pickle, to pickeln, ätzen, beizen
pickle Beize, Pökellauge
picklock Dietrich
pictorial bildlich, illustriert
picture Bild
piece Stück, Werkstück, Bauteil
piece by piece stückweise
piece of work Werkstück

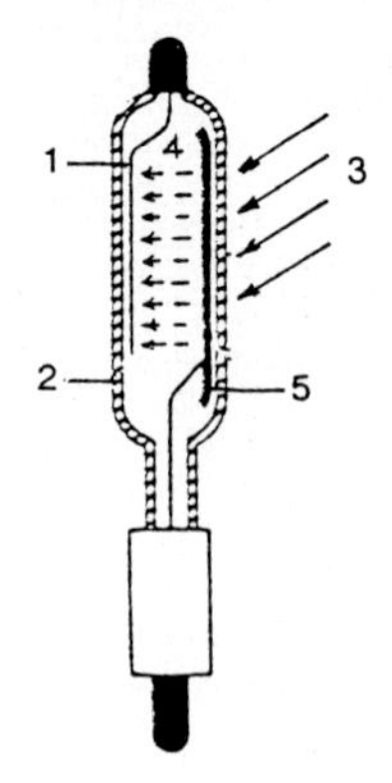

Photocell
Fotozelle

1 **anode** Anode
2 **glass bulb** Glaskolben
3 **light** Licht
4 **emission of electrons** Elektronenemission
5 **cathode** Kathode

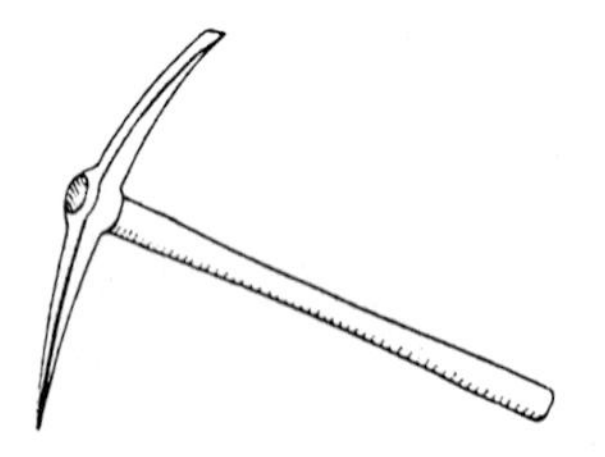

Pickaxe
Spitzhacke (Pickel)

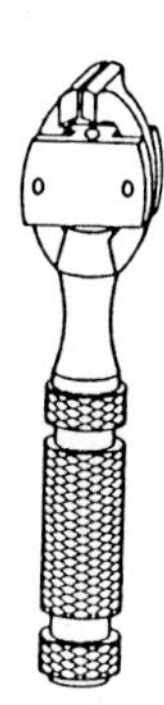

Pin vice
Stiftkloben

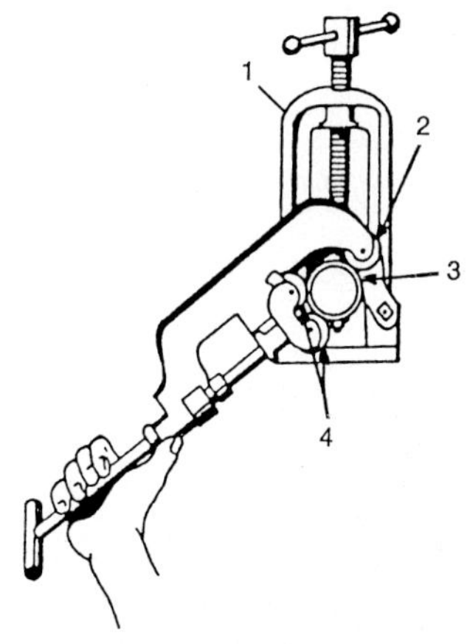

Pipe cutter
Rohrschneider

1 **screw clamp** Schraubzwinge
2 **cutting wheel** Schneidscheibe
3 **pipe** Rohr
4 **guide roller** Führungsrolle

pierce, to lochen, durchbohren, perforieren, stanzen
pierced solder tag Lötöse
piercer Lochstanze
piercing Lochen, Perforieren
piezoelectric piezoelektrisch, druckelektrisch
pig Barren, Block
pig iron Roheisen
pigtaily Lötfahne, Lötöse, geflochtene Litze
pile, to schichten, stapeln, rammen
pile Haufen, Stapel, Pfahl
pilfer-proof abschraubsicher
pilger mandrel Pilgerdorn, Pilgermandrill
pillar Pfeiler, Pfosten, Säule, Ständer
pillow Zapfenlager
pilot, to führen, steuern, lenken
pilot Pilot, Lotse, Führungszapfen, Kraftschalter, Schnellverstärker
pilot plant Versuchsanlage
pin, to verstiften, verbinden, verbohren
pinch, to zusammendrücken, kneifen, klemmen, quetschen
pinch Quetschung
pinched base Quetschfuss (Röhre)
pinching Blockierung, (ungewollte) Sperrung von Röhren
ping, to klingeln (Motor)
pinhole feines Loch
pinion Ritzel, Kammwalze, Nuss
pinking (leichtes) Klopfen des Motors
pinned verstiftet
pinpoint, to genau festlegen
pinpoint accuracy höchste Genauigkeit
pint Pint (0,568 l in GB; 0,473 l in den USA)
pintle Gelenkstift, Achse, Zapfen, Düsennadel
pip kurzer Impuls
pipage Rohrnetz, Rohrlegung
pipe, to (in Rohrleitung) leiten
pipe Rohr, Leitungsrohr
piping Rohranlage, Rohrnetz, Verlegung von Rohren, Leitungssystem
piston Kolben, Stempel

pitch, to errichten, aufstellen, werfen, schleudern, neigen, teeren
pitch Pech, Ganghöhe, Steigung Abstand, Neigung
pitchblende Pechblende
pitchy pechartig
Pitot tube Pitotrohr
pitted narbig (Oberfläche)
pitting Lochfrass, Auskolkung
pittings Lochfrasskorrosion
pivot, to gelenkig verbinden, (sich) um einen Zapfen drehen
pivot Zapfen, Tragzapfen, Wellenzapfen, Drehpunkt
pivotally mounted gelenkig angebracht
pivoted drehbar gelagert, drehbar eingesetzt
pivoting Drehung, Zapfenlagerung
pivoting drehbar, schwenkbar
place, to anordnen, plazieren, anlegen
placing Aufsetzen, Aufstellen, Anordnen
plain eben, flach, blank, unlegiert, einfarbig
plaited filter Faltenfilter
plan, to planen, entwerfen
plan Plan, Entwurf, Zeichnung
planmilling Fräsen mit Planeten-spindel-Fräsmaschine
planar plan, eben
planchet Platine, Rohling
plane, to glätten, hobeln
plane plan, eben
planer Hobelmaschine
planetary planetarisch, Planeten . .
planish, to glattwalzen, polierschlagen
planished sheet poliertes Blech
planisher Richtmaschine
plant Anlage, Betrieb, Werk
plasma physics Plasmaphysik
plastainer Kunststoffbehälter
plaster Putz, Putzmörtel
plate, to plattieren, überziehen, mit Schutzschicht versehen
plate Platte, Blech
plated plattiert, metallüberzogen, plattenförmig
plateworking machine Blechbearbeitungsmaschine
platform Bühne, Bedienungsstand, Plattform

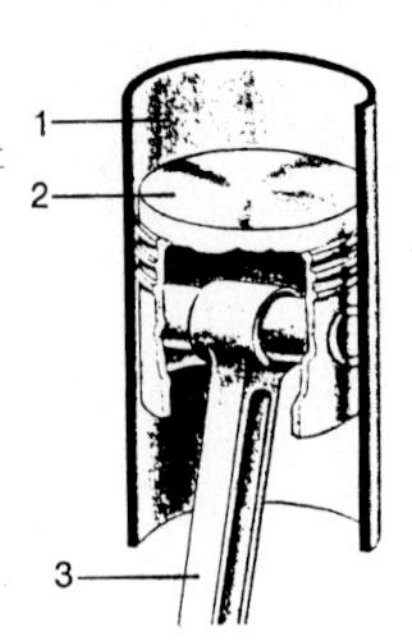

Piston (in a combustion engine)
Kolben (in einem Verbrennungsmotor)

1 **zylinder** Zylinder
2 **piston** Kolben
3 **connecting rod** Pleuelstange

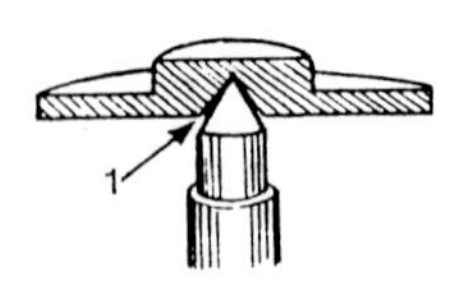

Pivot
Zapfenlager

1 **pivoting point** Drehpunkt

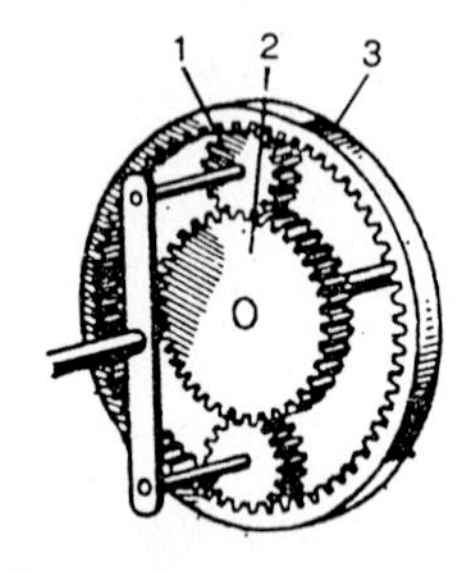

Planetary gear
Planetengetriebe

1 **planet pinion** Planeten-Zahnrad
2 **centre (sun) wheel** Sonnen-Zahnrad
3 **toothed circumferential wheel** Zahnkranz

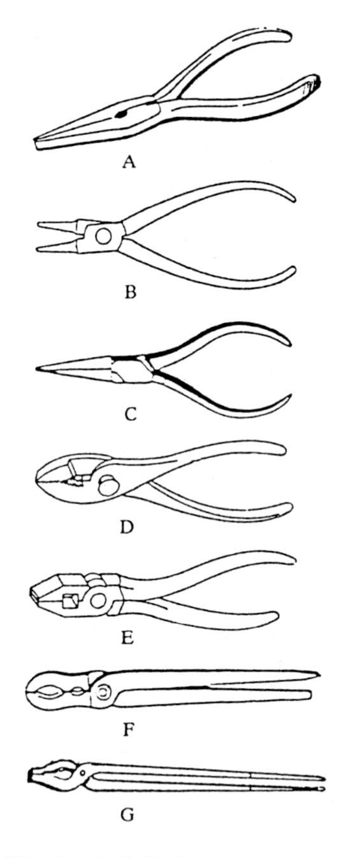

Pliers (a pair of pliers)
Zangen (Zange)

A **Flat-nosed pliers** Flachzange
B **Round-nosed pliers** Rundzange
C **Needle-nosed pliers** Spitzzange
D **Plumber's pliers** Klempnerzange
E **Combination pliers** Kombinationszange
F **Pipe-fitter's pliers** Rohrzange
G **Blacksmith's pliers** Schmiedezange

plating Galvanisierung, galvanischer Überzug
platinum Platin
play Spiel(raum), Luft (Lager)
playback Abspielen, Wiedergabe
pleat, to falten
pleat Falte
pleating Faltung
pliability Biegsamkeit, Geschmeidigkeit
pliable biegsam, geschmeidig
pliant biegsam, geschmeidig
pliers Zange, Kombizange, Kneifzange
plot, to aufzeichnen, auftragen, eintragen, diagrammatisch darstellen
plot Diagramm
plotter Kurvenschreiber, Zeicheneinrichtung
plug, to verstopfen, stöpseln
plug into connection, to an die Steckdose anschliessen
plug Pfropfen, Stöpsel, Stecker, Zündkerze
plugger Handbohrhammer
plugging Verstopfen, Verstopfung
plumb, to bleien, löten, verlöten
plumb Lot, Senkblei
plumb lotrecht, im Lot
plumbago Graphit
plumbe künstliches Echo (Radar)
plumber Klempner, Installateur
plumbing Klempnerarbeit, Loten
plumbism Bleivergiftung
plummer block Lauflager, Bocklager
plummet Lot, Senkblei
plunge, to eintauchen, versenken
plunger Druckkolben, Stössel, Plunger, Kolben
pluviometer Regenmesser
ply Schicht, Furnierplatte
plywood Sperrholz
pneumatic pneumatisch, Druckluft . . .
pneumatic tube conveyor Rohrpost
pneumatics Pneumatik, Mechanik der Gase
pocket Tasche, Sack
pocketing Gesenkfräsen
pockwood pillow Pockholzlager
pod Sockel, Halter
point, to spitzen, schärfen
point Punkt, Spitze, Stelle, Nadel

pointed spitz, zugespitzt, Spitz . . .
pointer Zeiger, Zunge
points Weiche
pointsman Weichensteller
poise, to balancieren, konstant halten
poise Gleichgewicht, Gewicht
poison Gift
poisoning Vergiften, Einbrennen (Röhre)
poke, to stochern, schüren
poker Schürer, Schürstange
polar polar, gerichtet, polig
polariscope Polarisationsapparat
polarity Polarität
polarisable polarisierbar
polarise, to polarisieren
pole, to polen
pole Mast, Stange
polish, to polieren, glätten, schlichten
polish Politur, Glanz, Glätte
polishing Polieren, Glanzschleifen
poll, to kappen, beschneiden
pollute, to verunreinigen, verschmutzen
pollution Verunreinigung, Verschmutzung
polygon Vieleck, Polygon
polygonal vieleckig, vielkantig, polygonal
polymer Polymer, Polymerisat
polymeric polymer
polynuclear mehrkernig
polyphase mehrphasig
ponderable wägbar
pony motor Anwurfmotor
pool, to ausstemmen, ausmeisseln
pool Schmelzbad (Schweissen)
poor schlecht, geringwertig, mager
poorly conducting schlecht leitend
poorness Minderwertigkeit
pop mark Messmarke
poppet Spannbock, Spannschraube, Ventilkegel
populated besetzt, bevölkert
population Gesamtheit, Bevölkerung
porcelain Porzellan
porcupine Nadelwalze
pork pie furnace Maerz-Boelens-Ofen
porosity Porosität
porous porös, porig

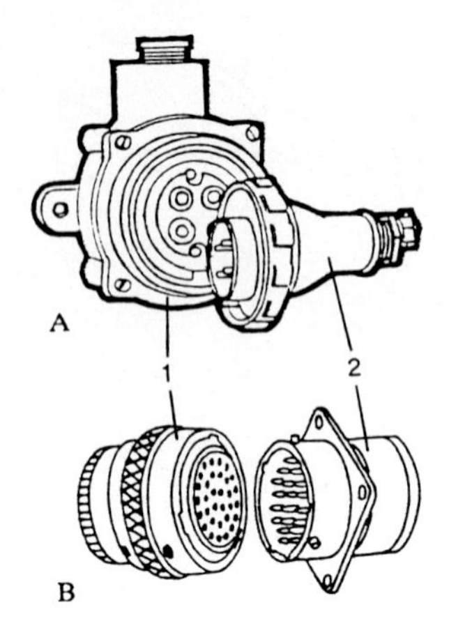

Plug-and-socket connector Steckverbinder

A **Heavy-current type** Starkstromausführung

B **Low-current-type** Schwachstromausführung

1 **socket** Steckdose
2 **plug** Stecker

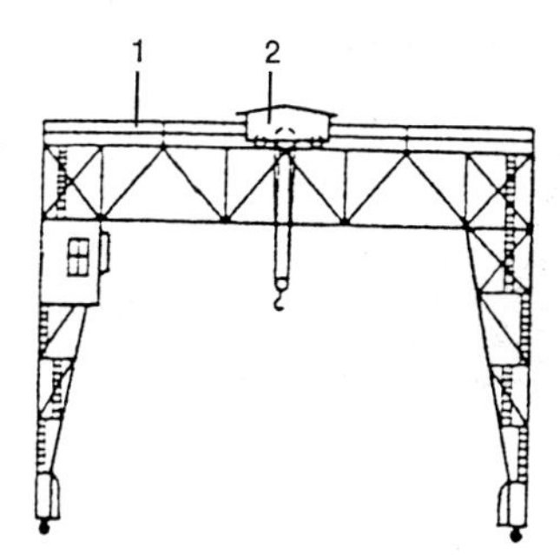

Portal crane (gantry crane)
Portalkran

1 **gantry** Kranbrücke
2 **travelling crab** Laufkatze

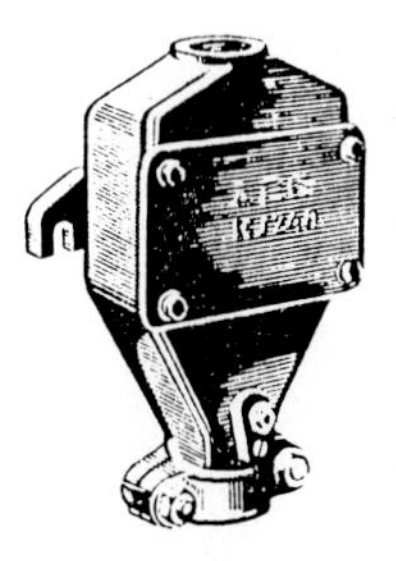

(Cable) Pothead
Kabelendverschluß

portable transportabel, tragbar, ortsbeweglich
portal crane Portalkran
porthole Bullauge
portion Anteil, Menge
portrait size Hochformat
position, to positionieren, in Stellung bringen, orten
position Lage, Stellung, Position
positioning Positionierung, Punktsteuerung (Werkzeugmaschine)
positive positiv, fest
positively actuated zwangsläufig betätigt
post, to absenden
post Pfosten, Säule, Pfeiler, Stütze
postal service Postdienst
pot, to giessen, ausgiessen
pot Topf, Tiegel, Gefäss, Kübel
potable trinkbar
potassium Kalium
potential Potential, Spannung
pothead Kabelendmuffe
potted capacitor Becherkondensator
pound, to zerstossen, schlagen, klopfen
pound Pfund; USA: 0,3732 kg; GB: 0,45359 kg
pounder Stössel
pour, to giessen, schütten
pour point Stockpunkt
pourable vergiessbar
pouring Vergiessen, Abguss
powder, to pulverisieren, zerreiben, mahlen
powder Pulver, Puder
powdered pulverisiert
power Leistung, Stärke, Potenz, Kraft, Kapazität
powerful leistungsfähig
power station Kraftwerk
practicable durchführbar
practical praktisch, verwendbar
practice Praxis, Übung, Ausführung
practise, to ausführen, durchführen, praktisch anwenden
pre-... Vor...
precaution Vorsichtsmassnahme
precious edel, kostbar
precipitate, to (sich) niederschlagen, absetzen

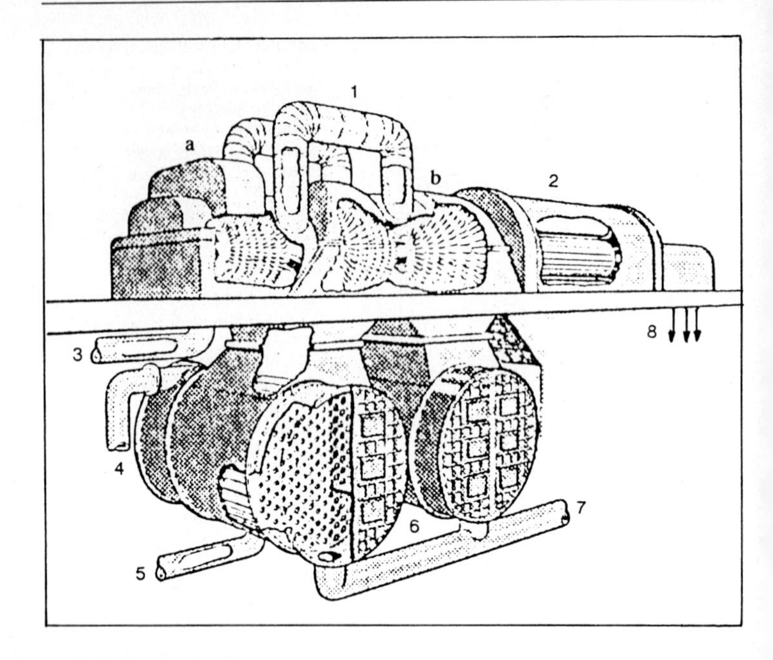

Power generating plant
Krafterzeugungsanlage

1 **turbine** Turbine
 a **high-pressure part** Hochdruckteil
 b **low- pressure part** Niederdruckteil

2 **generator** Generator
3 **steam from boiler** Dampf vom Kessel
4 **cooling water outlet** Kühlwasserabgang
5 **condensed steam** Kondensat
6 **condensers** Kondensatoren
7 **cooling water inlet** Kühlwassereinlaß
8 **power supply** Stromabgang

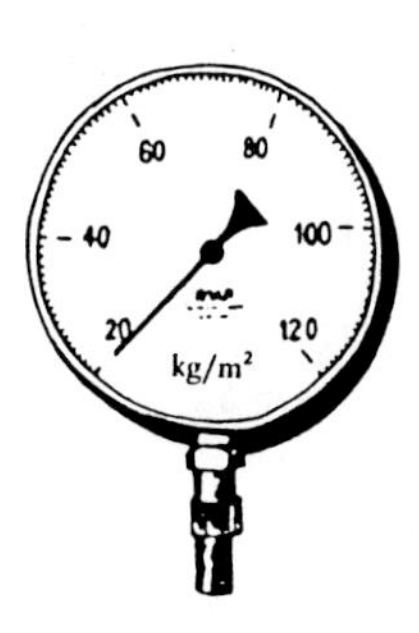

Pressure gauge (am. **gage**)
Manometer

Prick punch
Schlagdorn

precipitate Niederschlag, Abscheidung
precipitation Ausfällung, Ausscheidung
precise genau, präzis
precision Präzision, Genauigkeit
precoat, to grundieren, vorstreichen
precoat Grundierung, Grundanstrich
predecessor Vorgänger
predetermination Vorausberechnung
predetermine, to vorausberechnen
prediction Vorhersage
prefabricate, to vorfertigen
prefabricated vorgefertigt
prefabrication Vorfertigung
prefix (signal) Vorbereitungszeichen, Vorimpuls
preheater Vorwärmer, Vorerhitzer
preliminary Vor . . . , einleitend, vorbereitend
premature vorzeitig
preparation Vorbereitung, Aufbereitung
preparatory vorbereitend
prepare, to vorbereiten, vorrichten
preselection Vorwahl
preselector Vorwähler, Vorwahlschalter
presentation Darstellung
preservation Konservierung, Erhaltung
preserve, to konservieren, erhalten
preset, to voreinstellen, vorgeben
presetting Voreinstellung
press, to pressen, drücken, verdichten, quetschen
press Presse
pressboard Presspappe, Preßspan
pressure Druck
pressurise, to unter Druck setzen
pressuriser Druckregler, Druckbehälter
prestore, to vorspeichern
prestress, to vorspannen
preventive Schutzmittel
prick, to stechen (mit Nadel)
prickle Dorn, Stachel
prick punch Schlagdorn
prillion Schlackenzinn
primary primär
prime, to grundieren
prime Prim . . .

primer Grundierung, Anlasskraftstoff
principal Haupt ...
principle Prinzip
print, to drucken, schreiben
print Druck, Kopie, Lichtpause
printed gedruckt
printer (Messwert-)Drucker
printing Drucken, Kopieren, Lichtpausen
prism Prisma
pritchel Lochdorn
probability Wahrscheinlichkeit
probable wahrscheinlich
probe Sonde, Messfühler
procedure Verfahren, Prozedur
proceed, to vorgehen
proceed-to-dial (Wähl-)Bereitschaftszeichen
proceeding Vorgehen, Arbeitsgang
proceeds Reinertrag, Gewinn
process, to bearbeiten, fertigen, verarbeiten
process Prozess, Verfahren, Methode
processible verarbeitbar
produce, to herstellen, erzeugen, produzieren
producer Hersteller
product Produkt, Erzeugnis
productive produktiv, leistungsfähig
productivity Produktivität
profession Beruf
professorship Professur, Lehrstuhl
profile, to fassondrehen, profilieren
profile milling machine Kopierfräsmaschine, Nachformfräsmaschine
profiled profiliert, geformt, Form ...
profiler Kopierfräsmaschine
profitability Rentabilität
profitable rentabel
programme, to programmieren
programme control Programmsteuerung
programmer Programmierer
progress Fortschritt, Vorwärtsbewegung
progression Folge, Reihe
progressive fortschreitend
project, to projizieren, projektieren

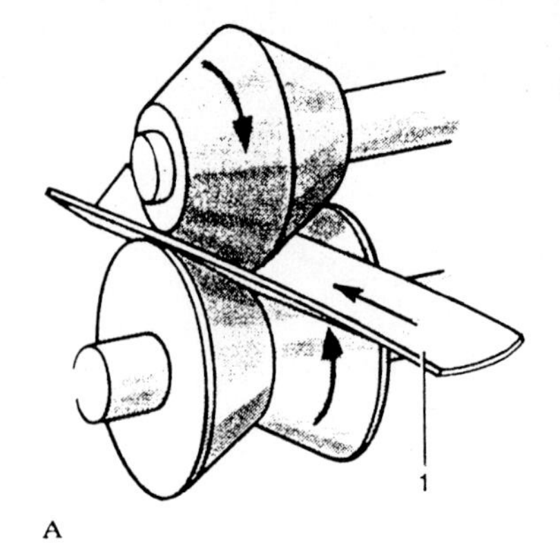

A

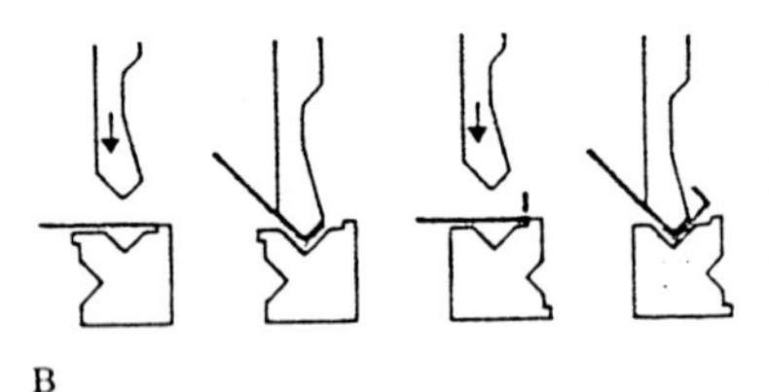

B

Profiling
Profilieren

A **Profiling tape steel by rolling** Profilieren von Bandstahl durch Preßwalzen
1 **tape steel** Bandstahl
B **Profiling by step-by-step punching** Profilieren durch stufenweises Abkantpressen

projecting vorspringend, vorstehend, ausladend
prolate verlängert, ausgedehnt
prolong, to verlängern, ausdehnen
prolongation Verlängerung, Ausdehnung
prong Gabel, Zacken, Kontaktstift
proof, to dicht machen, abziehen
proof Abzug, Beweis, Dichtigkeit
proof beständig, dicht
proofing Dichtprüfung
prop, to absteifen, abstützen
prop Stütze, Steife, Pfahl
propagate, to sich fortpflanzen, vordringen, sich ausbreiten
propagation Ausbreitung, Fortpflanzung
propane Propane
propel, to vorwärtstreiben
propellant Treibstoff, Treibladung
propeller Propeller, Schraube (Schiff, Flugzeug usw.)
proper sachgemäss, angemessen
property Eigenschaft
proportion, to bemessen
proportion Verhältnis, Proportion
proportioned dimensioniert
proportioning Dimensionierung, Zumessung
proposed scheme Anlageprojekt
proposition Satz, Lehrsatz
propulsion Antrieb
propulsiv Antriebs . . .
prospective current theoretisch zu erwartender Strom
protect, to schützen
protected geschützt, Schutz . . .
protecting schützend, Schutz . . .
protection Schutz
prototype Ausgangsbautyp, Muster, Prototyp
protract, to verlängern, zeichnen
protractor Winkelmesser, Transporteur
protrude, to herausragen, vorstehen
provable beweisbar
prove, to beweisen, sich erweisen
provide, to vorsehen, liefern, sorgen für
provisional vorläufig, provisorisch
proximity Nähe, Nachbarschaft
proximity fuse Näherungszünder
Prussian blue Berliner Blau, Preussischblau

A prospective customer?
Ein potentieller Kunde?

pry Brecheisen, Hebezeug
public address system Lautsprecheranlage
publication Veröffentlichung
publish, to veröffentlichen
publisher Verleger, Verlag
puck Scheibe
puddle, to puddeln, stochern
pull, to ziehen, zerren
pull in, to einziehen (Kabel)
pull Zug, Zugkraft
pull switch Zugschalter
pulley Flaschenzug, Rolle
pulp, to zu einem Brei anrühren, einstampfen
pulp Brei, Masse, Pulpe
pulsate, to pulsieren, schwingen, schütteln
pulsation Schwingung, Schwankung
pulse Impuls, Stoss
pulse code modulation Pulskodemodulation
pulsed impulsgesteuert
pulsing Impulsgabe
pulverisation Feinmahlung, Pulverisierung
pulverise, to feinmahlen, pulverisieren
pummel Stampfer, Ramme
pump, to pumpen
pump Pumpe
pumpage Pumpwirkung
pun, to rammen, stampfen
punch, to lochen, ankörnen, stanzen
punch Stempel, Locheisen
punch and form shaper Form- und Stempelhobelmaschine
punch card Lochkarte
puncheon Stempel, Pfosten, Ständer
puncher Lochstanze
punctiform punktförmig
puncture, to punktieren, durchstechen
puncture Einstich, Reifenpanne
pungent scharf (schmeckend)
pupin coil Pupinspule
pupinised cable pupinisiertes Kabel
purchase Kauf
purchasing department Bestellungsabteilung, Einkaufsabteilung

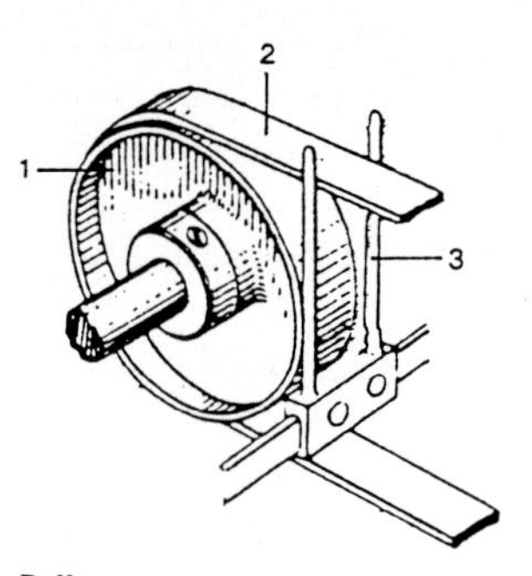

Pulley
(Riemen-) Scheibe, Rolle

1 **pulley** Riemenscheibe
2 **belt** Riemen
3 **guide** Führung

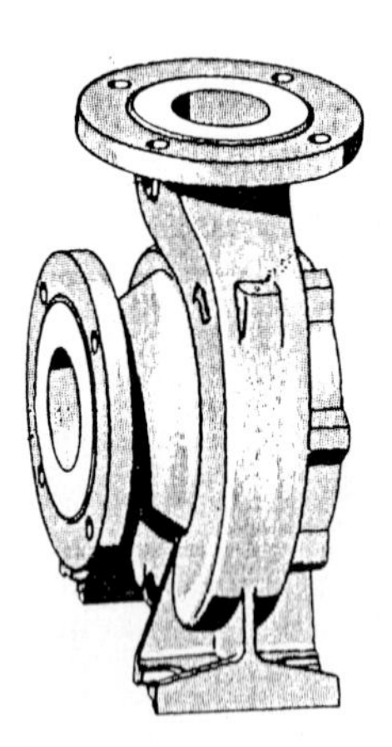

Pump casing
Pumpengehäuse

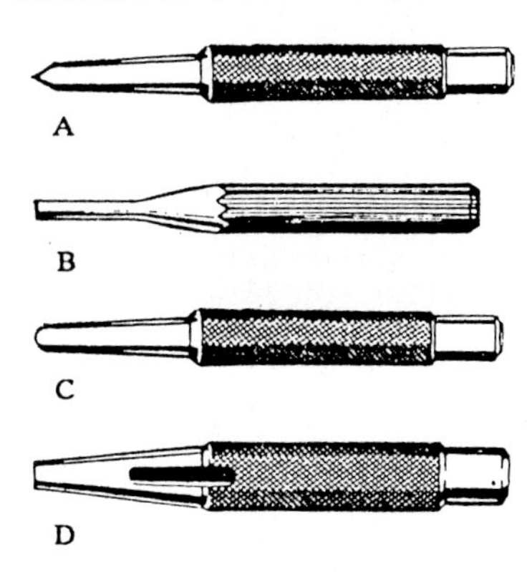

Punch
Durchschlag, Schlagstempel, Körner

A **Center punch** Körner

B **Pin punch** Stiftdurchschlag

C **domed punch** gewölbter Schlagstempel

D **Hole punch (with hollow shank)** Lochdurchschlag (mit hohlem Schaft)

pure rein, gediegen
purification Reinigung, Aufbereitung
purifier Reinigungsgerät
purify, to reinigen
purity Reinheit, Feingehalt
purl, to wirbeln, rieseln
purple Purpur
purpose Zweck, Aufgabe
push, to stossen, drücken, schieben
push Stoss, Schub
push-button Druckknopf, Drucktaster
push-pull cascade Gegentaktstufe
put, to setzen, stellen, legen
put a voltage across, to Spannung an . . . legen
put in circuit, to einschalten
put into operation, to in Betrieb setzen
put through, to durchstellen (Telefon)
putting out of action Ausserbetriebsetzen
putty, to spachteln, auskitten
putty Spachtelmasse, Kitt
pylon Mast, Pylone

Q

quad, to zum Vierer verseilen
quad Viererkabel
quadding machine Verseilmaschine
quadrant Quadrant
quadrate, to quadrieren
quadrature-axis component Querfeldkomponente
quadruple vierfach
qualify, to sich qualifizieren
quality Qualität, Güte
quality of finish Oberflächengüte
quantity Menge, Masse, Betrag, Quantität
quantity of electricity Elektrizitätsmenge
quantity production Massenherstellung, Serienfertigung
quart Volumeneinheit; 1,136 l (GB), 0,946 l (USA)
quarter Viertel, Quartier
quartz Quarz
quench, to löschen (Lichtbogen), abschrecken (Stahl)
quench Abschreckmittel
quenching (Lichtbogen-)Löschen, Abschrecken, Härten
quenching chamber Löschkammer
quick access store Schnellzugriffspeicher
quicksilver Quecksilber
quiescent ruhend, ruhig
quiet ruhig, geräuscharm
quietness Geräuschlosigkeit
quiet in operation ruhiger Lauf
quill, to falten, knittern
quill Hülse, Pinole
quinary quinär, zur Basis 5
quintuple fünffach
quirk Hohlkehle, Nut

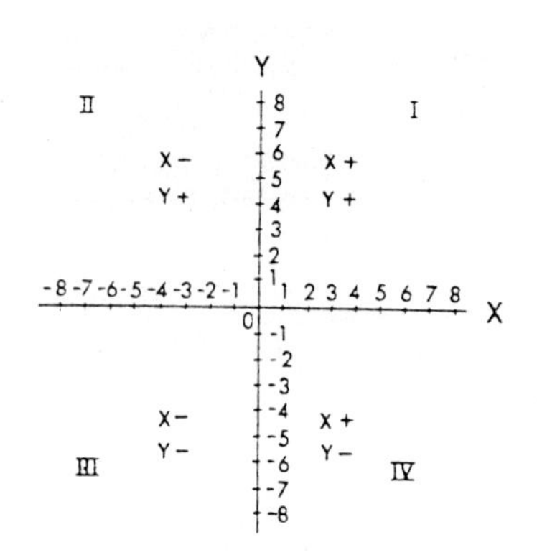

Quadrants I, II, III, IV in a coordination system
Quadranten I, II, III, IV in einem Koordinatensystem

R

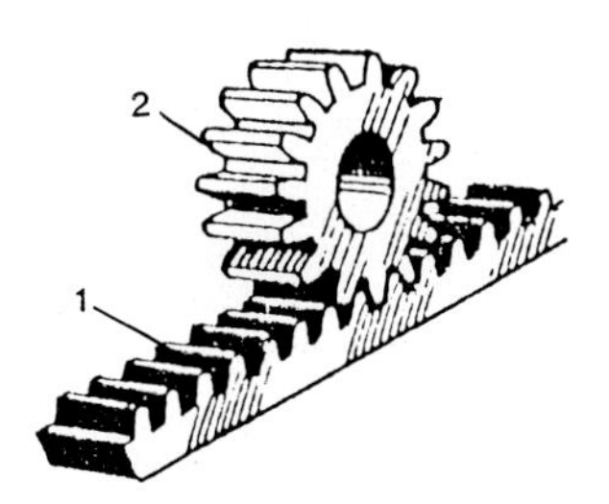

Rack-and-pinion gear
Zahnstangengetriebe

1 **toothed rack** Zahnstange
2 **pinion** Ritzel

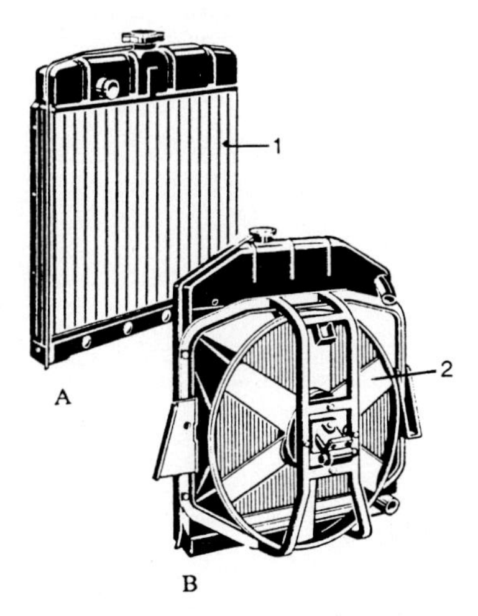

Radiator
(Wasser-) Kühler

A **Front view** Frontansicht

B **Rear view** Rückansicht

1 **grill** Grill
2 **fan** Lüfter

rabbet, to fugen, falzen
rabbet Fuge, Falz
rabble, to rühren
race, to durchgehen (Motor)
race Wettfahrt, Strom, Strömung, Laufbahn
racer Laufring (Lager)
raceway Kabelkanal, Leitungsrohr, Laufbahn (Lager)
rack, to stapeln, einlegen
rack Regal, Gestell, Zahnstange
radar (radio detection and ranging) Funkmesstechnik
radial radial, sternförmig
radiant Radiant (Bogenmass)
radiant strahlend
radiate, to strahlen, (aus)senden
radiation Ausstrahlung
radiative strahlend
radiator Radiator, Heizkörper, Kühler
radio Radio, Rundfunk, Funk . . .
radio direction finder Funkpeiler
radioactive (radio)aktiv
radioactivity Radioaktivität
radiobiology Strahlenbiologie
radiogram Funktelegramm, Röntgenaufnahme
radiograph, to durchstrahlen, röntgen
radiograph Röntgenaufnahme
radiography Radiographie
radiologist Radiologe, Röntgenfacharzt
radiology Radiologie
radionavigation Funknavigation
radiotron Summerröhre
radix Wurzel, Basis
radome Radarkuppel
raft Floss
rag Lappen, Lumpen
ragged zackig
rail Schiene, Führung, Geländer
railing Reling, Geländer
rainproof regendicht
raise, to (hoch)heben, erhöhen, aufstellen
raise Erhöhung, Steigerung
rake, to rechen, schüren
ramification Verzweigung, Gabelung, Aufspaltung

ramify, to verzweigen, gabeln, (sich) aufspalten
rammer Ramme, Stampfer
ramp, to steigen, fallen
ramp Rampe
random Zufall
random zufällig, regellos, wahllos
random access wahlfreier Zugriff (Rechenspeicher)
random access storage Speicher mit beliebigem Zugriff
range, to sich erstrecken, reichen, einordnen
range Bereich, Strecke, Reichweite
rank Rang, Reihe
rap, to abklopfen, losklopfen
rapid schnell, Schnell . . . , Eil . . .
rapidity Schnelligkeit, Geschwindigkeit
rarefied evakuiert, verdünnt
ratch Sperrstange
ratchet Sperre, Ratsche, Klinkenrad
rate, to bemessen, taxieren, auslegen
rate Rate, Grösse, Menge, Wert
rate of air flow Luftdurchflussmenge
rated bemessen, Nenn . . . , Nominal . . .
rated voltage Nennspannung
rating Auslegung, (Nenn-)Leistung, Nennbereich
ratio Verhältnis, Übersetzung
rationalise, to rationalisieren, rational machen
rattles Rattern, Rattergeräusch
raw roh, unbearbeitet, unverarbeitet
rawness Rohzustand
RC network RC-Schaltung, RC-Glied
reach, to erreichen
reach Reichweite, Strecke, Bereich
react, to reagieren, einwirken, entgegenwirken
reactance Reaktanz, Blindwiderstand
reaction Reaktion, Gegenwirkung, Rückwirkung
reactive reaktiv, reagierend, reaktionsfähig
reactive power Blindleistung
reactor Reaktor, Drossel(spule)
read, to (ab)lesen, abtasten

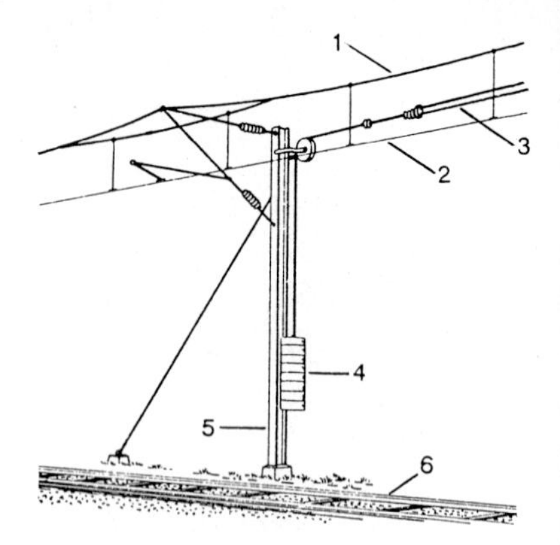

Railway overhead power line
Fahr-Oberleitung

1 **carrier rope** Tragseil
2 **contact wire** Fahrdraht
3 **tension rope** Abspannseil
4 **weight providing tension** Gewicht, um Seil unter Spannung zu halten
5 **pole** Mast
6 **track** Gleis

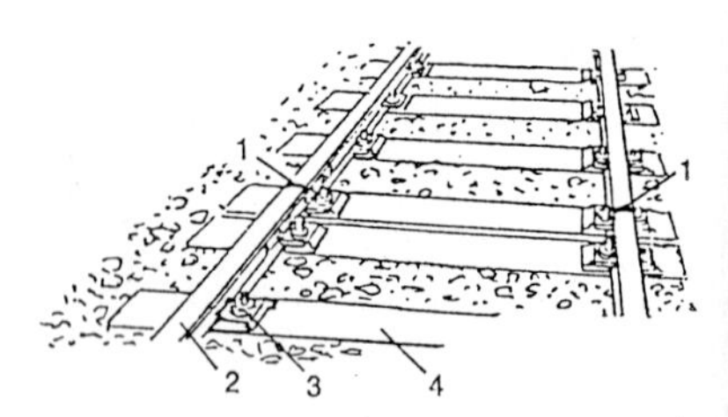

Railway track
Gleisanlage

1 **butt joint** Stoß
2 **rail** Schiene
3 **rail dog and clamping plate** Schienennagel und Griffplatte
4 **sleeper** Schwelle

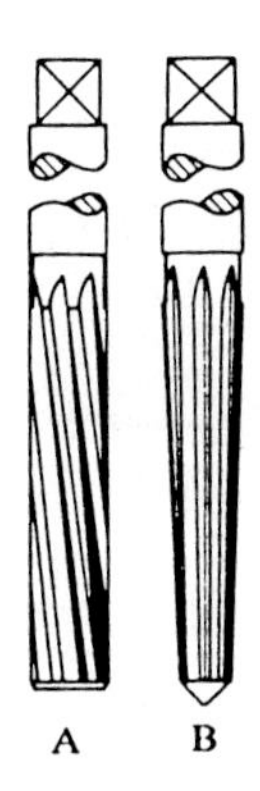

Reamer
Reibahle

A **Parallel reamer** zylindrische Ahle

B **Tapered reamer** konische Ahle

read proofs, to Korrektur lesen
read-only memory Auslese-speicher
readable (ab)lesbar
readjust, to verstellen, nachstellen
readjustable nachstellbar
readjustment Nachstellung, Verstellung, Nachregelung
ready for operation betriebsbereit
real reell (Zahl)
real measure Sollmass
real time Echtheit, Realzeit
realign, to nachrichten
ream, to reiben, aufreiben
reamed bolt Paßschraube, eingepasste Schraube
reamer Reibahle, Räumer
reaming Reiben, Ausreiben
reanneal, to nachglühen
rear Rückseite, Hintergrund
rear Hinter . . . , rückwärtig, hintere
rearrange, to neu ordnen, umordnen
rearrangement Umordnung, Neuordnung
reassemble, to wieder zusammenbauen
rebalance, to nachwuchten
rebalancing Nachwuchten
rebate Falz, Vergütung, Rabatt
rebated gefalzt, genutet
rebating cutter Falzfräser
rebore, to nachbohren
rebound, to abspringen, zurückprallen
rebound Rückprall, Rückschlag, Rückstoss
rebuild, to wiederaufbauen, umbauen
rebuilding Wiederaufbau, Umbau
recalibrate, to nacheichen
recalling key Rufschalter
receipt of goods Wareneingang
receive, to erhalten, empfangen, aufnehmen
received band Empfangsband
receiver Empfänger, Behälter, Auffänger
Behälter, Steckdose
receptacle Sammelbecken, Steckdose
reception Empfang, Annahme, Aufnahme
recess, to einstechen, aussparen

recess Vertiefung, Einstich, Aussparung, Nische
recharge, to nachladen, nachfüllen
recharging Aufladung, Auffüllung
rechuck, to umspannen, wieder einspannen (Werkstück)
recipe Rezept, Vorschrift
reciprocal Reziprokwert
reciprocal reziprok, invers, umgekehrt
reciprocate, to hin- und hergehen
reciprocating engine Kolbenmaschine, Kolbenmotor, Kolbentriebwerk
reciprocation Hin- und Herbewegung
recirculation Umlauf, Wiederaufnahme in den Kreislauf
reckon, to rechnen, errechnen
reclaim, to zurückgewinnen, wiedergewinnen
reclamation Rückgewinnung
reclamp, to umspannen (Werkstück)
reclose, to wiedereinschalten
reclosing fuse Sicherung mit selbsttätiger Wiedereinschaltung
reclosing relay Wiedereinschaltrelais
recoil, to zurückprallen, zurückstossen
recoil Rückprall, Rückstoss
recoiler Aufwickeleinrichtung
recombination Rekombination, Wiederverbindung, Wiedervereinigung
recommend, to empfehlen
recommended value Richtwert
recondition, to instandsetzen, ausbessern, überholen
reconditioning Instandsetzung, Überholung
reconnection Wiedereinschaltung, Umschaltung
reconstruct, to umbauen, umgestalten, umarbeiten
reconstruction Wiederherstellung, Rekonstruktion, Umbau
recool, to rückkühlen
recooling Rückkühlung
recooling tower Rückkühlturm, Rückkühler
record, to aufzeichnen, eintragen, registrieren, schreiben, aufnehmen

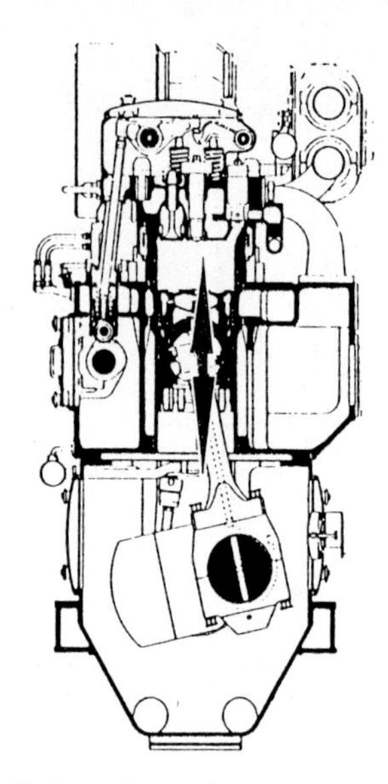

Reciprocating engine
Kolbenmaschine (Verbrennungsmotor)

↕ **reciprocating motion of piston**
Hin- und Her- Bewegung des Kolbens

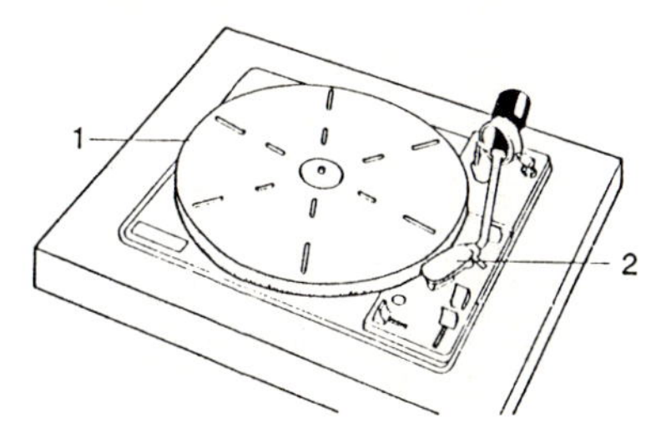

Record-player
Plattenspieler

1 **turntable** Plattenteller
2 **pick-up head** Abnahmekopf

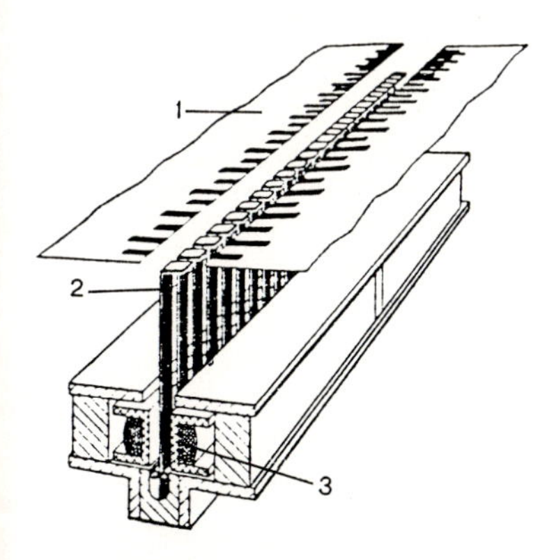

Reed (vibration) frequency measuring unit
Zungen- (Vibrations-) Frequenzmeßwerk

1 **scale** Skala
2 **reed** Zunge
3 **electro-magnet** Elektromagnet

record Aufzeichnung, Eintragung, Registrierung
recover, to wiedergewinnen, sich erholen, bergen
recovery Rückgewinnung, Wiedergewinnung, Bergung
recreation period Erholungspause
rectangle Rechteck
rectangular rechteckig
rectification Gleichrichtung, Demodulation, Richten
rectifier Gleichrichter, Entzerrer
rectify, to gleichrichten, entzerren
recuperate, to zurückgewinnen, wiedergewinnen
recurrent circuit Kettenleiter
recycle, to zurückführen, umpumpen, wieder in Umlauf bringen
recycling Wiederverwendung, Rückführung, Wiedereinsetzen
red brass Rotguss, Rotmessing
red lead Bleimennige
reddle Mennige, Roteisenerz
redesign, to umkonstruieren, umarbeiten, umgestalten
reduce, to reduzieren, verkleinern, herabsetzen, vermindern
reducer Reduzierstück, Übergangsstück
reduction Reduktion, Verkleinerung
reduction of cross section Querschnittsverengung
redundancy Überzähligkeit, Überflüssigkeit, Redundanz
redundant überzählig, redundant
reed Zunge, Blatt, Ried
reed frequency meter Zungenfrequenzmesser
reed relay Schutzgasrelais, Reedrelais, Zungenrelais
reefer (ship) Kühlschiff
reel, to spulen, aufwickeln, abhaspeln
reel Rolle, Spule, Haspel, Aufspuler
reference Verweis, Quellennachweis, Literaturverzeichnis
reference Vergleichs . . . , Bezugs . . .
refill, to betanken, nachfüllen
refine, to raffinieren, reinigen, verfeinern
refinery Raffinerie

reflect, to reflektieren, zurückstrahlen, spiegeln
reflection Reflexion, Spiegelung
reflex Reflex, Spiegelbild
reforge, to umschmieden
refract, to brechen (Licht, Schall)
refraction Brechung, Refraktion
refractories feuerfeste Steine, Schamottsteine
refractoriness Feuerfestigkeit, Feuerbeständigkeit
refractory feuerfester Baustoff
refractory feuerfest, wärmebeständig
refrigerant Kältemittel, Kühlmittel, Kälteträger
refrigerate, to kühlen
refrigerating Kühl . . . , Kälte . . .
refrigeration Kälteerzeugung, Kältetechnik, Kühlung
refrigerator Kälteanlage, Kühlschrank
refuel, to nachbunkern, tanken
refuse Müll, Abfall, Kehricht
refusion Schlacke umschmelzen
regain Zunahme, Wiedergewinnung
regenerate, to regenerieren, auffrischen, wiedergewinnen, wiederaufbereiten
regenerative regenerierend, wiedererzeugend, rückkoppelnd, selbstansaugend (Pumpe)
regenerator rückgekoppelter Verstärker, Regenerator, Wärmespeicher
region Bereich, Bezirk, Zone, Fläche
register, to aufzeichnen, eintragen, aufnehmen, zählen
register Register, Verzeichnis, Tabelle, Zählwerk, Speicherzelle
regrind, to nachschleifen, nachschärfen
regular regulär, regelmässig
regularity Regelmässigkeit, Stetigkeit, Gleichmässigkeit
regulate, to regeln, stellen, stabilisieren
regulating regelnd, Regulier . . .
regulation Regelung, Vorschrift
regulator Regler, Regelschalter
reheat, to anlassen, nachhitzen, tempern

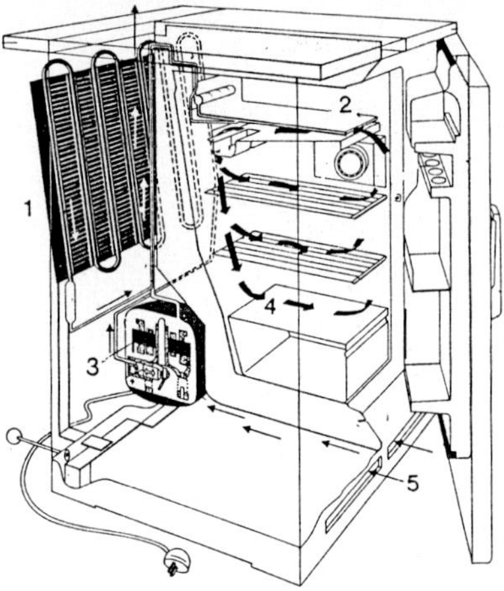

Refrigerator
Kühlschrank

1 **liquidizer** Verflüssiger
2 **evaporator** Verdampfer
3 **motor/compressor** Motor/Verdichter
4 **air circulation** Luftumlauf
5 **ventilation** Belüftung

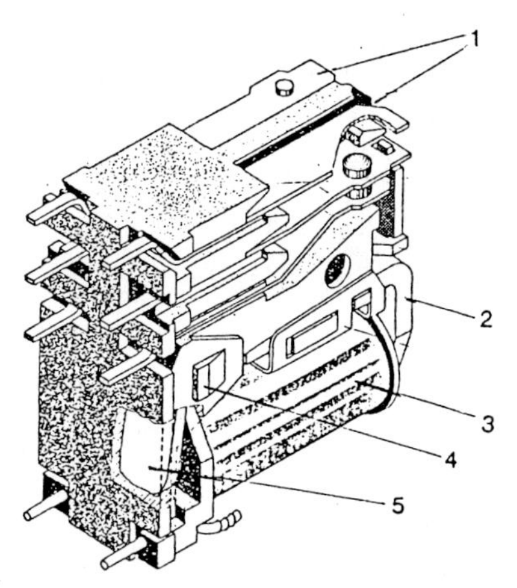

(Control) Relay with cover removed
(Steuer-) Relais ohne Schutzkappe

1 **spring contacts** Federkontakte
2 **magnet yoke** Magnetjoch
3 **magnet coil** Magnetspule
4 **clip for cover** Einrastnase für Schutzkappe
5 **armature** Anker

reheat Zwischenüberhitzung, Nachverbrennung, Wiedererwärmung
reheating furnace Nachwärmofen
reinforce, to verstärken, versteifen, abstützen, armieren, bewehren
reinforced concrete Stahlbeton
reinforcement Verstärkung, Bewehrung, Armierung
reject, to zurückweisen
reject Ausschussteil
rejection Beanstandung, Zurückweisung
rejector Ausstosser, Sperre, Stromresonanzkreis
rejects Ausschuss, Abfall
relation Beziehung, Verhältnis
relative relativ, verhältnismässig, Relativ . . .
relaxation Kippen, Erschlaffung, Entspannung, Relaxation
relaxed gelockert, entspannt, entlastet
relay Relais
release, to freigeben, auslösen, loslassen, entriegeln, ausspannen
release Freigabe, Auslösung, Auslöser, Entlastung, Abfall
reliability Betriebssicherheit, Zuverlässigkeit
relief erhöhte Form, Freiwinkel, Abhebung, Entlastung, Freiwinkel
relieve, to hinterdrehen, hinterschleifen, entlasten
reload, to umladen, neu einlegen
relocate, to verlagern, verrücken, umstellen, umspannen, verstellen, umsetzen
reluctance Reluktanz, magnetischer Widerstand
reluctivity spezifische Reluktanz, spezifischer magnetischer Widerstand
remainder Rest, Rückstand
remanence Remanenz, remanenter Magnetismus
remanent zurückbleibend, remanent
remedy Gegenmittel, Heilmittel
remote entfernt, abseits, Fern . . .
remote control Fernbedienung, Fernbetätigung
remotely controlled ferngesteuert, fernbetätigt

removable abnehmbar, auswechselbar
removal Beseitigung, Entfernung, Abnahme
remove, to entfernen, beseitigen, ausbauen, demontieren, abführen
render, to machen, reproduzieren, ausschmelzen
renew, to erneuern, auswechseln
renewable erneuerbar, auswechselbar
renewal Erneuerung
repair Reparatur, Instandsetzung
repeat, to wiederholen, weitergeben
repeater Verstärker, Übertrager, Folge . . .
repel, to abstossen, zurückstossen
reperforation Empfangslochung
reperforator Empfangslocher
repetition Wiederholung
repetitive wiederkehrend, sich wiederholend
replace, to ersetzen, auswechseln, erneuern
replacement Ersatz, Auswechselung
replenish, to auffüllen, nachfüllen,
repoussé punch Schlagstempel
represent, to darstellen, verkörpern
representation Darstellung, Darstellungsweise
reprint Nachdruck, Sonderdruck
reprocess, to wiederaufarbeiten, wieder verarbeiten
reprocessing Aufarbeitung
reproduce, to reproduzieren, wiedergeben, kopieren
reproduction Wiedergabe, Reproduktion
repulse, to abstossen
repulse Rückstoss, Abstossung, Repulsion
repulsion Abstossung, Repulsion
repulsion motor Repulsionsmotor
repulsive abstossend
repulsion Abstossung, Repulsion
require, to erfordern, verlangen
reroute, to umleiten
rerun, to wiederholen (Dat.)
rescue Rettung
research Forschung
reseat, to einschleifen, nachschleifen (Ventile)

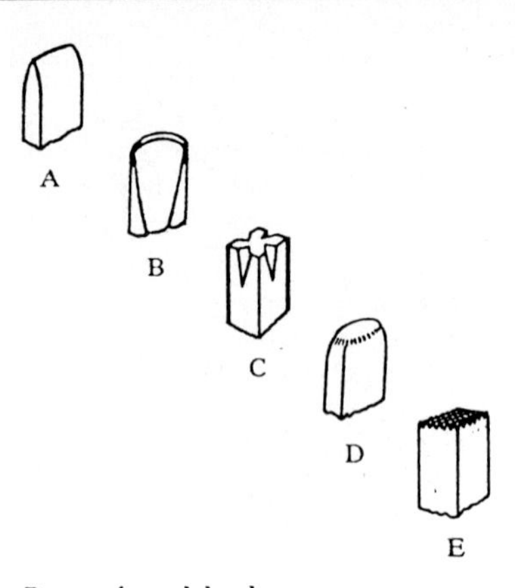

Repoussé punch head
Schlagstempelkopf

for für/zum

A **straight tracing** gerade Linien
B **curved tracing** Kurven
C **decorating** Verzieren
D **chasing** Treiben
E **matting** Mattieren

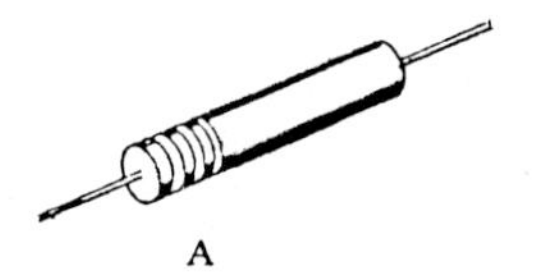

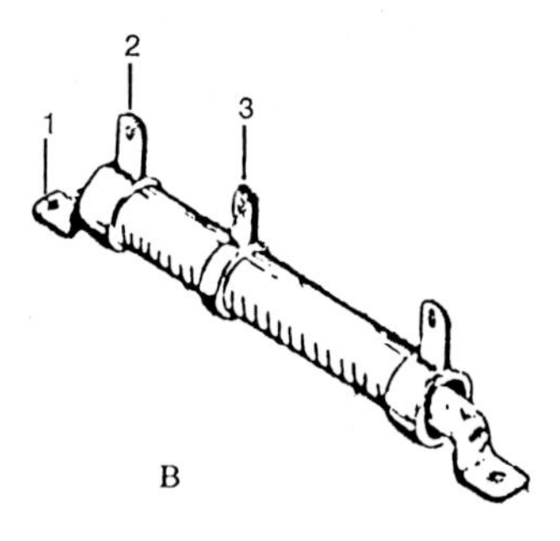

(electric) Resistor
(elektrischer) Widerstand

A **carbon resistor** Kohlewiderstand (**low wattage** niedrige Leistung)

B **wire-wound resistor** drahtgewickelter Widerstand (**high wattage** hohe Leistung)

1 **fixing lug** Befestigungslasche
2 **terminal** Anschluß
3 **tapping** Anzapfung

reserve Reserve
reservice, to instandsetzen, reparieren, aufarbeiten
reservicing Reparatur, Wiederinstandsetzung, Aufarbeitung
reset, to zurückstellen, neu einstellen, zurückgehen, umspannen, umrüsten
reset Rückstellung
residual Rest . . ., Eigen . . ., übrigbleibend, remanent
residuary resistance Formwiderstand
residue Rückstand, Rest
resile, to zurückfedern, elastisch sein
resilience Zurückspringen, Rückprall, Federwirkung
resiliency Elastizität, Steifigkeit
resilient elastisch, federnd
resin, to harzen, mit Harz tränken
resin Harz
resist, to widerstehen, aufnehmen, widerstandsfähig sein gegen,
resistance Widerstand, Festigkeit, Beständigkeit
resistant widerstandsfähig, beständig, fest
resistive widerstandsfähig, mit Widerstand
resistivity spezifischer Widerstand, Widerstandsfähigkeit
resistor Widerstand, Widerstandselement
resistron Resistron
resnatron Resnatronröhre
resoluble lösbar
resolution Auflösung, Trennung, Trennschärfe
resolve, to auflösen, zerlegen (Kräfte)
resolver Koordinatenwandler, Vektorzerleger
resonance Resonanz, Mitschwingen
resonant resonant, mitschwingend
resonate, to auf Resonanz bringen, mitschwingen
resonator Resonator, Schwinger, Schwingkreis
resorb, to resorbieren, aufsaugen, einsaugen
resound, to widerhallen, schallen
respond, to ansprechen, reagieren auf

responder Antwortsender, Empfänger (Radar)
response Ansprechen, Anziehen (Relais), Anzeige (Messinstrument), Verhalten
rest, to ruhen, in Ruhelage sein
rest Auflage, Stütze, Schlitten, Support, Stillstand, Ruhe
restoration Wiederherstellung, Instandsetzung
restore, to wiederherstellen, instandsetzen
restraint Beschränkung, Zwang, Hinderung
restrict, to beschränken, drosseln, verengen
restriction Beschränkung, Einschränkung, Begrenzung
retard, to verzögern, verlangsamen
retardation Verzögerung, Hemmung, Bremsung
retarder Hemmwerk, Verzögerer
retention Beibehaltung, Zurückhaltung
retentive zurückhaltend, festhaltend
retort Retorte, Kolben, Destillierblase
retool, to umrichten, umrüsten
retorting Destillieren
retouch, to retuschieren
retrace Rücklaufspur (Kathodenstrahl)
retract, to zurückziehen einziehen
retractable einziehbar
retransmit, to weitersenden
retransmitter Zwischensender, Ballsender
retreat, to nachbehandeln
retreatment Nachbehandlung
retrieval of digital information Wiederauffinden digitaler Information
retro rückwärts, Rück . . .
retroact, to rückwirken
retroactive rückwirkend, rückgekoppelt
return, to zurückkehren, zurückführen
return Wiederkehr, Rückkehr
reusable wiederverwendbar
reverberate, to zurückwerfen, reflektieren, nachhallen

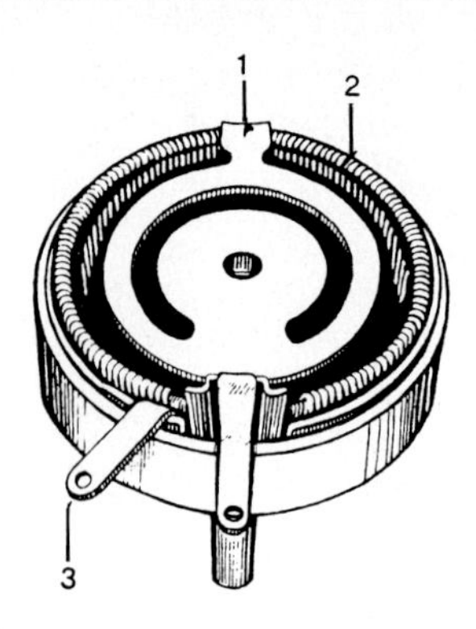

Rheostat
Drehwiderstand (Rheostat)

1 **sliding contact** Schleifkontakt
2 **resistance coil** Widerstandsspule
3 **connecting lug** Anschlußfahne

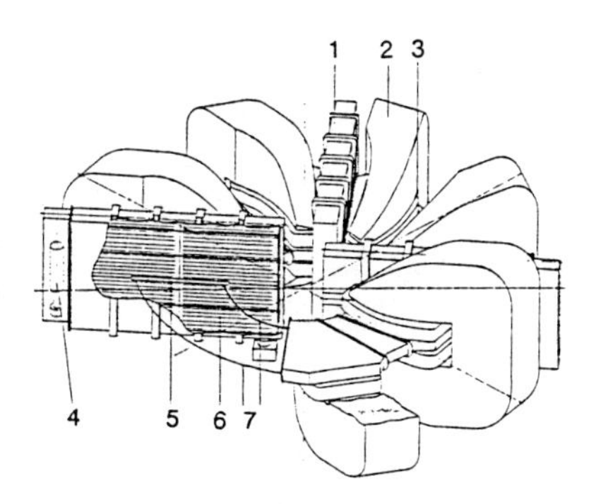

Ring magnet accelerator (cyclotron)
Ringmagnetbeschleuniger (Zyklotron)

1 **vacuum chamber** Vakuumkammer
2 **sector magnet** Sektormagnet
3 **main excitation coil** Haupterregerspule
4 **50-MHz end stage** 50 MHz-Endstufe
5 **acceleration gap** Beschleunigungsspalt
6 **50-MHz resonance cavity** 50 MHz-Resonanzkavität
7 **proton paths** Protonenbahnen

reversal Umkehr, Umsteuerung, Umkehrung
reverse, to umdrehen, umsteuern, umschalten
reverse Umkehr, Umsteuerung, Rücklauf, Gegenstück
reverse umgekehrt, rückwärts, Umkehr . . .
reversible umsteuerbar, umkehrbar, umschaltbar
reversion Umkehr, Rückkehr
revive, to auffrischen
revolution Umdrehung, Umlauf
revolutions per minute Umdrehungen pro Minute
revolve, to (sich) drehen, umlaufen, umdrehen
revolving rotierend, drehbar, Dreh . . .
rewind, to zurückspulen, umspulen
rewind Rücklauf, Rückwicklung, Umwicklung
rewire, to Leitungen neu verlegen, neu installieren
rework, to aufarbeiten, umarbeiten
r. f. radio-frequency (Hochfrequenz)
rheostat Rheostat, Regelwiderstand
rhomb Rhombus
rhombic rhombisch
rib, to rippen, mit Rippen versehen
rib Rippe, Steife, Spant
ribbed verrippt, durch Rippen versteift
ribbon Band, Streifen
rich reich, reichlich, fündig
ricochet, to abprallen
ricochet Abprall
riddle, to (ab)sieben, durchlöchern
riddle Schüttler
ridge, to riefen, furchen
ridge Gang (Gewinde), Riefe Kamm, Rücken
riffle, to furchen, riffeln
riffle Riffelung
riffler Lochfeile, Riffelblech
rift, to aufreissen, springen
rift Schlucht, Kluft, Riss Spalte
rig, to (Schiff) auftakeln, ausrüsten

rig Vorrichtung, Anlage, Aufbau, Ausrüstung, Gerät, Bohranlage
rigging Verspannung, Takelage (Schiff)
righting Aufrichten
rigid starr, steif, unbiegsam
rigidity Steifigkeit, Starrheit
rimmed unberuhigt vergossen
rimming unberuhigtes Vergiessen (Stahl)
rimose rissig
ring, to klingen, klingeln, läuten, (an)rufen
ring Ring, Öse, Anruf
ringer Rufstromgeber
rinse, to (ab)schwemmen, (ab)spülen, auswaschen
rinsing agent Spülmittel
rip, to reissen, auftrennen
rip Riss, Schlitz
ripple, to brummen, riffeln, (sich) kräuseln
ripple Brummen, Störgeräusch, Welligkeit
rise, to (an)steigen, zunehmen, sich steigern, erhöhen
rise Steigerung, Anstieg, Erhöhung, Zunahme
riser Steigleitung
rivet, to nieten
rivet Niet(e)
riveted genietet
riveting Nieten, Nietung
rms root mean square (Effektivwert, quadratischer Mittelwert)
roast, to rösten, abschwelen
robot Automat, Roboter
robotize, to (am.) automatisieren
rock, to schaukeln, wiegen, erschüttern
rock Fels, Gestein
rocker Schwinge, Kipphebel, Wippe
rocket Rakete
rocketeer Raketenfachmann
rocking schwingend, oszillierend
Rockwell hardness Rockwellhärte
rod Stab, Stange, Rundstahl, Draht
rods Gestänge, Stabstahl
Roentgen apparatus Röntgenapparat
roll, to rollen, walzen
roll Rolle, Walze, Haspel
rolled gewalzt, Walz . . .

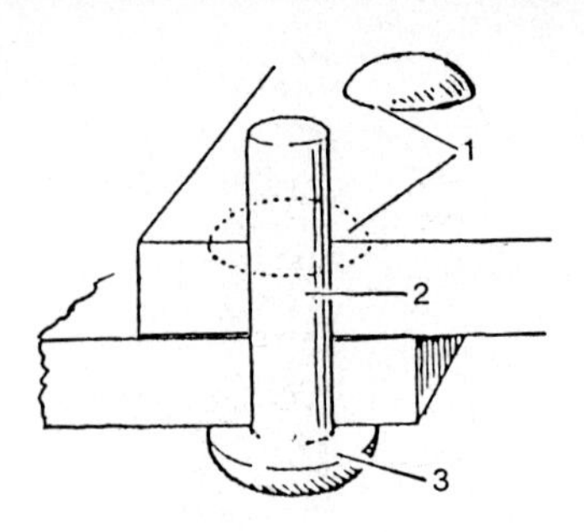

Riveting
Vernieten

1 **snap head** Schließkopf
2 **rivet shaft** Nietbolzen
3 **rivet head** Setzkopf

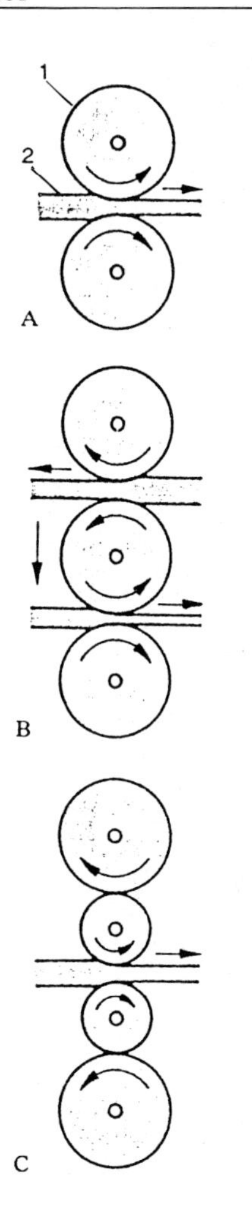

Rolling stand
Walzgerüst

A **Two-high rolling stand** Duo-Walzgerüst

B **Three-high rolling stand** Trio-Walzgerüst

C **Four-high rolling stand** Quarto-Walzgerüst

1 **roll** Walze

2 **material to be rolled** Walzgut

roller Läufer, Rolle, Walze, Laufrolle
rolling Rollen, Walzen
roof Dach, First, Deckel
root Wurzel
rope, to anseilen, seilförmig verdrallen
rope Seil, Tau
ropebelt conveyor Bandförderer
ropy fadenziehend, zäh(flüssig)
rot, to vermodern, verrotten
rot Fäulnis
rotable drehbar
rotary rotierend, drehend, drehbar, Dreh ...
rotate, to (sich) drehen, umlaufen
rotation Drehung, Rotation, Umlaufbewegung
rotor Läufer, Rotor
rough, to rauh werden, aufrauhen, schruppen
rough rauh, grob
roughen, to aufrauhen
roughing Schruppen, Vorbearbeitung, Grobwalzen
roughness Rauheit
roundness Rundheit
rout, to ausschneiden, ausarbeiten, nachfräsen
route Linienführung, Trasse, Leitungsführung
router Langlochfräser, Nachformfräsmaschine, Grundhobel
routing Ausfräsen, Umrissfräsen, Aushobeln
row Reihe, Schicht, Zeile
rub, to (ab)reiben, abziehen, abwischen
rubber Gummi, Kautschuk
rubbish Müll, Abfall
ruby Rubin
rudder Ruder
rude rauh, unpoliert
rugged kräftig, stabil, nicht störanfällig
ruggedness Unempfindlichkeit, geringe Störanfälligkeit
ruin, to beschädigen, zerstören
ruin Verfall, Baufälligkeit
rule, to anordnen
rule Regel, Anordnung, Schiene
ruler Lineal
run, to laufen (lassen), fahren, betreiben, Kabel verlegen
run Lauf, Verlauf, Gang, Weg

runway Start- und Landebahn
runner Läufer, Laufrolle, Laufrad
rupture, to (zer)reissen, zerbrechen
rupture Bruch, Reissen, Durchschlag (Isolation)
rupturing capacity Ausschaltleistung, Schaltleistung
rush Ausbruch, Andrang, Ansturm
rust, to rosten
rust Rost
rustless rostfrei, rostbeständig
rusty rostig, verrostet

A

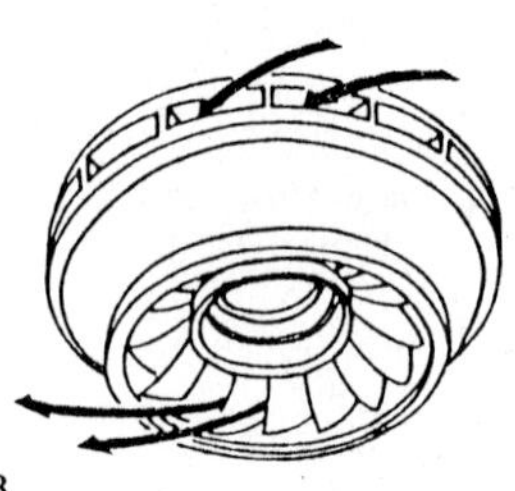

B

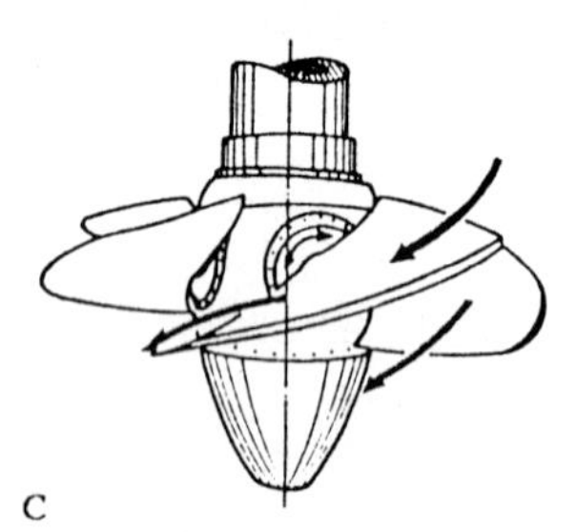

C

Runner
Laufrad

A **Runner of a Pelton hydro-turbine**
Pelton-Turbinenlaufrad

B **Runner of a Francis hydro-turbine**
Francis-Turbinenlaufrad

C **Runner of a Kaplan hydro-turbine**
Kaplan-Turbinenlaufrad

A

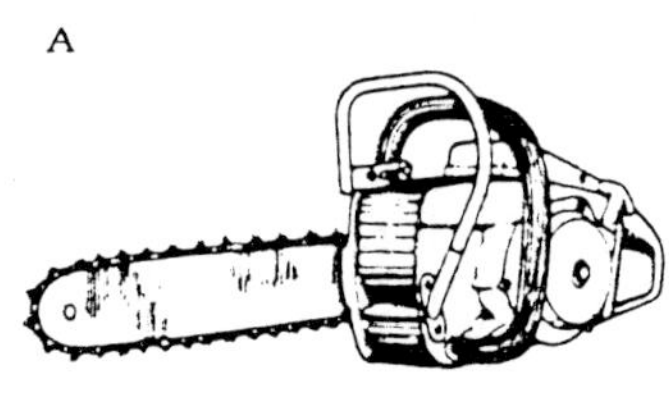

B

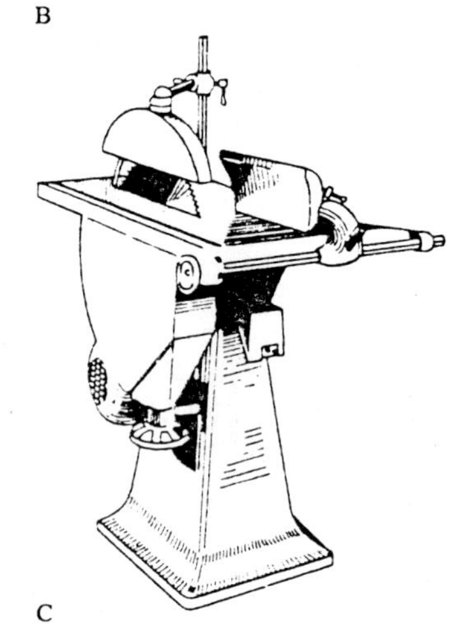

C

Saw
Säge

A **Hand saw** Handsäge (Fuchsschwanz)

B **Motor-driven chain saw** Motor-angetriebene Kettensäge

C **Machine circular saw** Maschinen-Kreissäge

S

safeguard Schutz, Sicherung
safety Sicherheit
safety Sicherheits . . ., Schutz .
sag, to durchhängen, (Leitung), absinken (Kurve)
sag Durchhang, Durchbiegung, Senkung
saggar Brennkapsel, Kerntrockenkasten
salary Gehalt, Salär
sale Verkauf
sales analysis Umsatzstatistik
salient vorspringend
salient pole ausgeprägter Pol
saline salzig, salzartig
salinity Salzhaltigkeit
saltpetre Salpeter
salvage, to bergen, retten, wiedergewinnen
salvage Bergung, Wiedergewinnung
salve, to bergen, retten
satisfy, to zufriedenstellen, genügen
saturable reactor Sättigungsdrossel
saturated gesättigt
saturation Sättigung
sausage aerial Reusenantenne
save, to sparen, erhalten, retten
save-all Siebtisch, Rückgewinnungsanlage
saw, to sägen
saw Säge
scaffold, to Gerüst bauen, einrüsten
scaffold Baugerüst, Gerüst
scale, to abblättern, messen, wiegen, verzundern, abschuppen
scale Schuppe, Skala, Masseinteilung, Zunder
scalp, to vorsieben, grobsieben
scan, to abtasten, zerlegen
scan Abtastung, Auflösung
scanner Abtastgerät, Drehantenne
scarf, to anschärfen, abschrägen, mit schrägem Stoss verbinden
scarf Laschenverbindung
scatter, to zerstreuen
scatter Streuung, Streuecho
scavenge, to durchspülen, reinigen
scavenge Spülluft

sceleton Schiffsgerippe
scend Tauchschwingungs-amplitude
schedule, to planen, aufstellen, ansetzen
schedule Liste, Plan, Programm
schematic schematisch
schematic diagram Schaltbild, Schaltschema, Schemazeichnung
scheme Schema, Plan, Anordnung
schist Schiefer
scrape, to schaben
scraper Spachtel
scratch, to zerkratzen, ritzen
scratch Kratzer, Schramme
screech, to flattern
screen, to aussieben, filtern, abschirmen
screen Sieb, Filter, Schirm, Schutzwand
screenings Ausgesiebtes
screw, to festschrauben, schneiden (Aussengewinde)
screw Schraube, Bolzen, Schnecke
screwdriver Schraubenzieher
screwed angeschraubt, verschraubt
scribe, to anreissen
scribe awl Reissnadel
scroll Schnecke, Spirale, Rolle
scrub, to scheuern, schrubben
scrutinise, to genau untersuchen, prüfen
scud Schleim, Schmutz
scuff, to abnutzen, verschleissen, fressen
scuffing Verschleiss, Fressen
scuttle, to anbohren, Seeventil öffnen
scuttle Springluke
seal, to abdichten, absperren, verschliessen
seal Dichtung, Absperrung, Plombe
sealant Abdichtmittel, Isolier-mittel
sealing Dichtung, Absperrung, Plombe
seam, to säumen, falzen
seam Saum, Naht, Lötstelle
seamless nahtlos
sear, to welken, sengen, brennen
search, to suchen
searchlight Scheinwerfer
season, to ablagern, altern, aus-trocknen

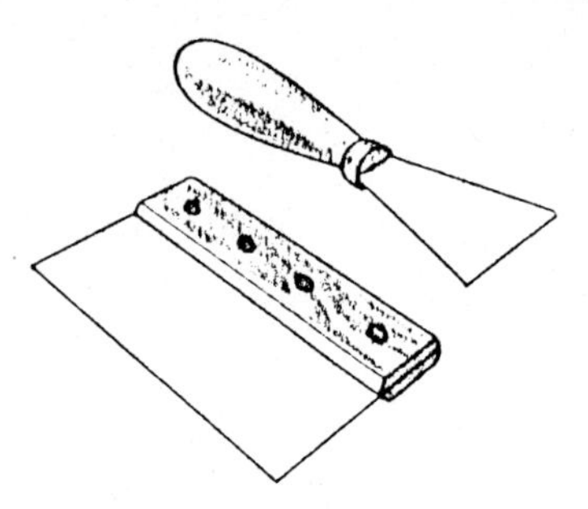

Scraper
Spachtel

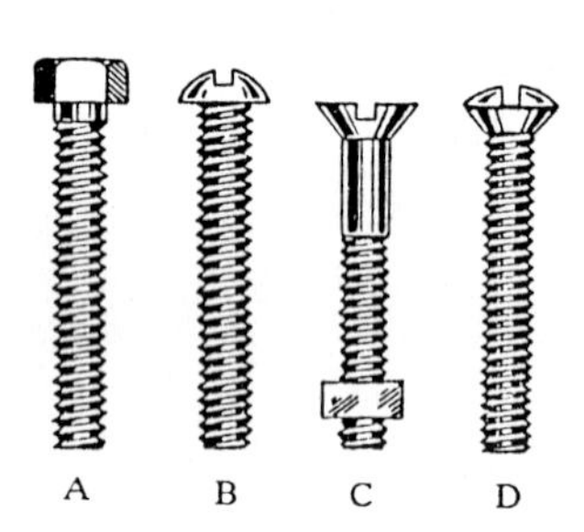

Screw
Schraube

with mit

A **hexagon head** Sechskantkopf

B **round head (slotted)** Rundkopf (mit Schlitz)

C **countersunk head** Senkkopf

D **raised countersunk head** überhöhter Senkkopf

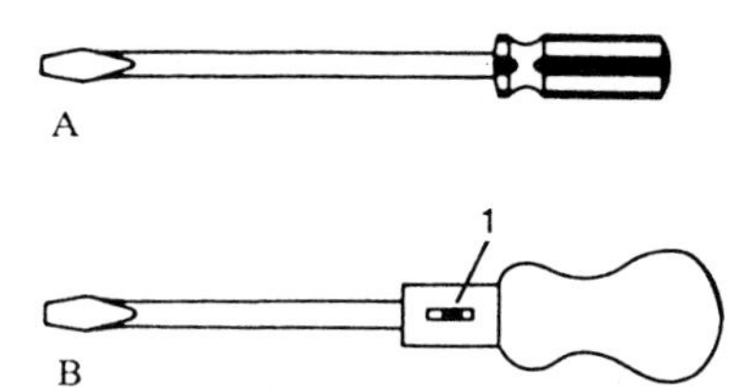

Screwdriver
Schraubenzieher

A **Ordinary type** gewöhnliche Ausführung

B **Ratchet type** Ausführung mit Ratsche
1 **positioning "screw/unscrew"** Einstellung "an-/abschrauben"

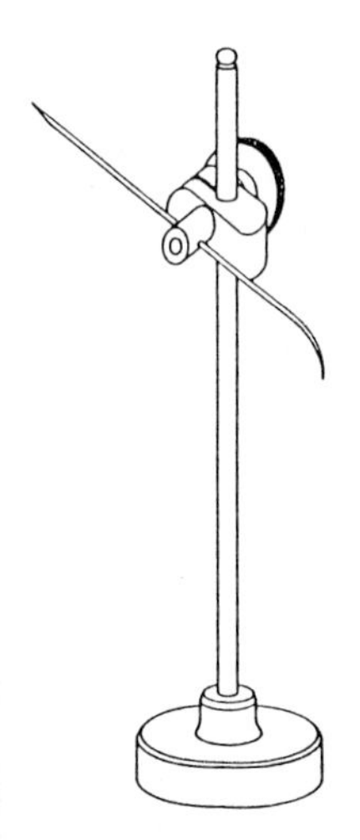

Scribing block
Anreißer (Parallelreißer)

seat, to einpassen, einsetzen, aufnehmen
seat Sitz, Auflagefläche, Anlagefläche
seating Einpassen, Einsetzen
secondary sekundär
sectile teilbar, abtrennbar
section Abschnitt, Abteilung, Profil
sectionalise, to in Abschnitte zerlegen
sectioning Schnittdarstellung, Schraffieren von Schnittflächen
secure, to sichern, befestigen
security Sicherheit
sedimentate, to absetzen, niederschlagen
seep, to versickern
segment Segment, Abschnitt, Teilstück
segregate, to absondern, trennen
segregation Entmischung, Seigerung
seize, to (sich) festklemmen, hängenbleiben
seizing Festklemmen, Hängenbleiben
seizure Fressen (Lager)
select, to wählen
selection Auswahl
selectivity Selektivität, Trennvermögen
selector Wähler, Wahlschalter
selectron Selektron(röhre)
selenium Selen
self Selbst . . .
selling expenses Vertriebskosten
selsyn Drehmelder, Selsyn, Gleichlaufanlage
semi Halb . . .
semicircle Halbkreis
semiconductor Halbleiter
send, to senden, verschicken
sensation of heat Wärmeempfindung
sense, to abtasten, fühlen, erfassen
sensibility Empfindlichkeit
sensing element Messfühler
sensitive empfindlich, feinstufig
sensor Messfühler, Fühler
separable trennbar
separate, to trennen, schneiden, separieren
separate getrennt, isoliert
separation Trennung, Scheiden

separator Trennanlage, Abscheider
sequence, to anreihen, einreihen
sequence Reihenfolge, Arbeitsablauf
sequential aufeinanderfolgend
serial Reihen . . . , Serien . . . , laufend, aufeinanderfolgend
serially connected in Reihe geschaltet
series Reihe, Serie
serrate, to riefen, riffeln
serve, to (be)dienen, versorgen
service, to bedienen, warten
servicing Wartung
set, to (ein)stellen, abbinden, einrichten, aushärten
set Satz, Gerät, Lage, Haltung
settle, to absetzen, lagern
settling Senkung, Bodensatz
sever, to trennen, brechen
sewage Abwässer
shade, to schraffieren, abstufen, abtönen
shade Schattierung, Farbton
shaded schraffiert, dunkel getönt
shaft Welle, Schacht
shake, to schütteln, rütteln
shank Schaft, Griff
shape, to verformen, bilden, gestalten, kurzhobeln
shape Gestalt, Form, Umriss
shapeless formlos
shaper Waagerechtsstossmaschine
share Anteil
sharp scharf, spitz
sharpen, to schärfen
shatter, to zerschmettern, zertrümmern
shatter Scherben, Splitter
shear, to scheren, schneiden
shear Schere, Wange
sheath, to armieren
sheathing Ummantelung Umhüllung
sheave Scheibe, Rolle
shed, to verschütten, fallen lassen
shed Schuppen, Fach
sheer rein, einfach, unvermischt
sheet, to auswalzen
sheet Blech, Tafel, Blatt, Platte
shell Schale, Aussenhaut
shelter, to schützen
shelter Schutzraum, Schutz
shelve, to mit Fächern versehen

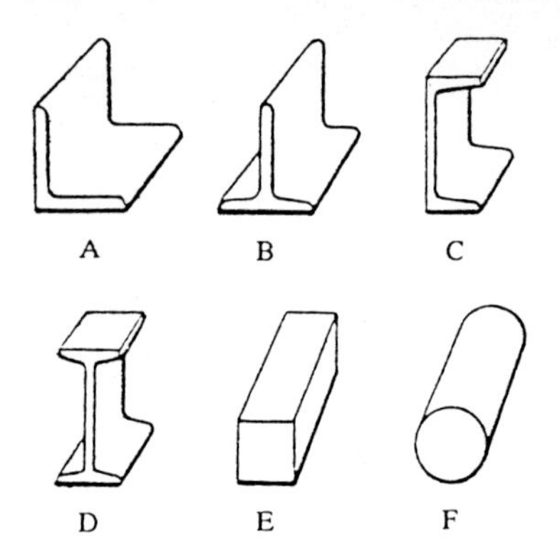

Sections (rolled steel)
Profile (Walzstahl)

A **Angle section** Winkelprofil
B **Tee-section** T-Profil
C **Channel** U-Profil
D **H-section** Doppel-T-Profil
E **Square section** Vierkantprofil
F **Round section** Rundprofil

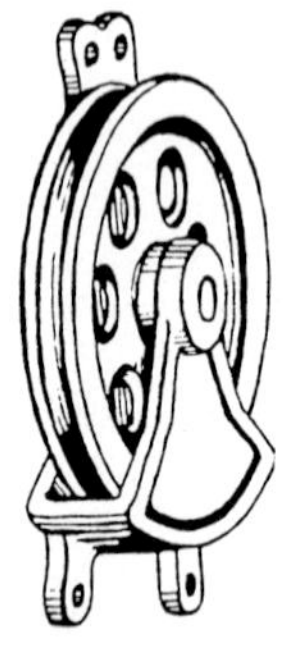

Sheave
Seilrolle

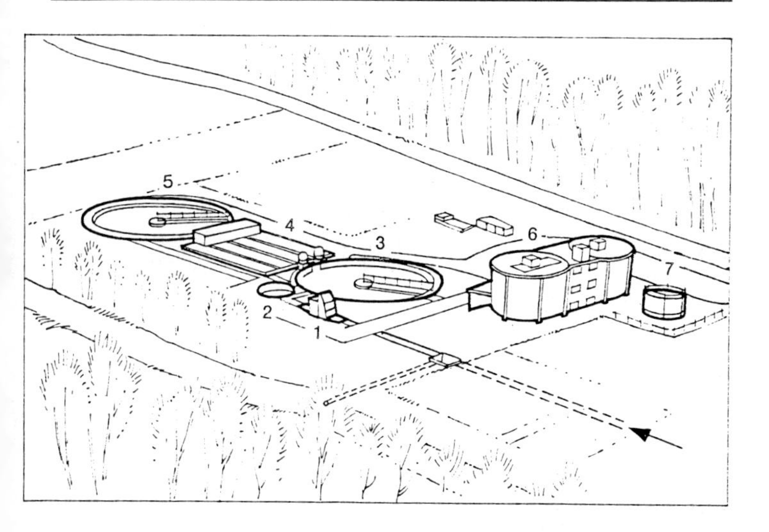

Sewage treatment plant
Abwasser-Kläranlage

1 **grid/crusher** Rechen/Zerkleinerer
2 **sand trap** Sandfang
3 **settling basin (separation of solids by mechanical means**
Vorklärbecken (Abscheidung von festen Stoffen durch mechanische Behandlung)
4 **airing basin** Belüftungsbecken
5 **biological-treatment basin** Nachklärbecken (biologische Behandlung)
6 **putrefaction rooms** Faulräume
7 **gas tank** Gasbehälter

shield, to abschirmen, schützen
shield Abschirmung, Schild, Schutz
shielding Abschirmung
shift, to verschieben, verstellen, schalten, verrücken
shift Schicht (Arbeit), Verschiebung, Verstellung
shim, to unterlegen, unterbauen
shim Beilage, Zwischenlage, Beilegscheibe
shimmy, to flattern, vibrieren
shine, to scheinen, glänzen
shipbuilding Schiffbau
shipment Verladung, Verschiffung
shipping Versand, Schiffahrt
shock Schlag (el.), Stoss, Erschütterung
shoe Schuh, Schleifstück, Gleitschuh
shore, to stützen, absteifen
short-circuit, to kurzschliessen
short-circuit Kurzschluss
shortage Knappheit, Mangel
shorted kurzgeschlossen
shotting Granulieren
shoulder Absatz, Schulter, Kröpfung
shovel, to schaufeln, schippen
shovel Schaufel, Löffelbagger
show, to zeigen, darstellen, sichtbar machen
show-case Schaukasten
shred, to zerreissen, zerkleinern
shredder Reisswolf, Schnitzelmühle
shrink, to schrumpfen, eingehen
shrinkage Schrumpfung, Schwund
shroud, to einhüllen
shunt, to parallelschalten, rangieren
shunt Nebenschlusswiderstand, Nebengleis
shutter Verschluss, Klappe, Riegel
shuttle, to pendeln
sidewise seitlich, seitwärts
sieve, to sieben
sign, to anzeichnen, markieren
sign Zeichen, Marke
signal, to signalisieren, melden
signal Signal, Meldung
signaller Signalgeber
significance Bedeutung
significant bedeutsam, wichtig
silence Ruhe, Stille

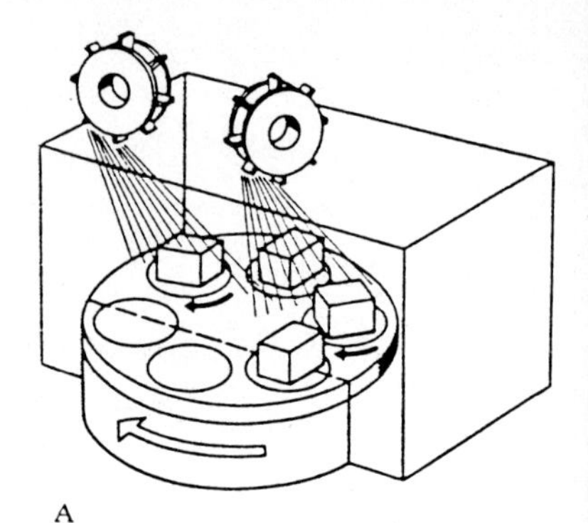

A

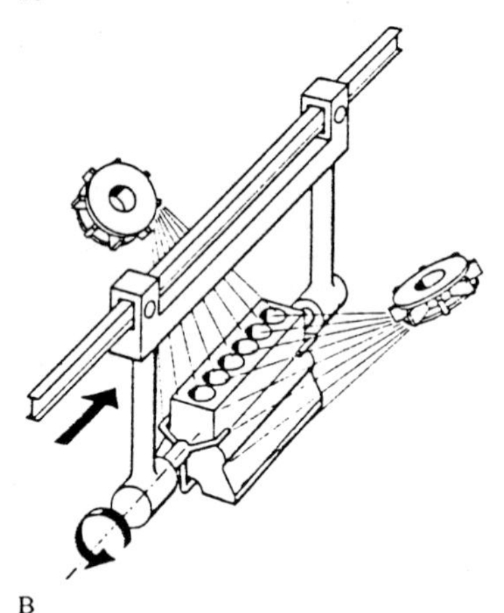

B

Shot-blasting of castings
Strahlputzen von Gußstücken

A **Rotating-table blaster** Drehtisch-Strahlanlage

B **Rail-roll blaster** Fahrjoch-Strahlanlage

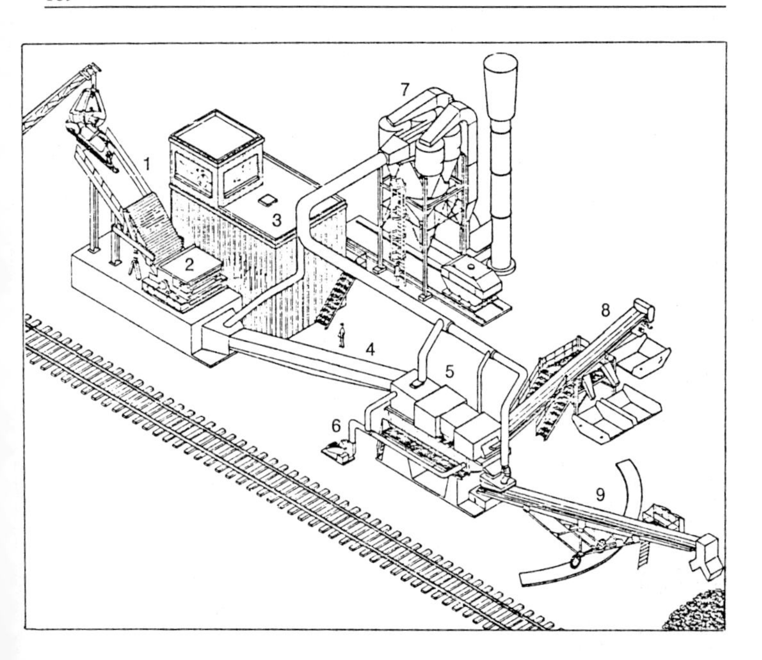

Shredder plant
Verschrottungsanlage
1 **scrap feeder (chain conveyor)** Kettenförderer zum Einziehen des Verschrottungsgutes
2 **hammer mill shredder** Hammermühlen-Zertrümmerungsanlage
3 **machine house** Maschinenhaus
4 **enclosed conveyor** geschlossenes Förderband
5 **separation station (magnetic-sorting)** Separationsanlage (Magnetsortierung)
6 **cooling ventilator** Kühlluftventilator
7 **cyclone filter plant** Zyklonfilteranlage
8 **dump discharge** Schuttaustrag
9 **mobile conveyor for shredded scrap** fahrbares Verladeband für End-Schrott

silencer Schalldämpfer
silent ruhig, still
silicon Silizium
silk Seide
sill Grundschwelle, Süll, Unterzug
silver-plate, to versilbern
similar ähnlich, gleichartig
simmer, to sieden, wallen
simplified analysis Näherungsverfahren
simplify, to vereinfachen
simulate, to nachbilden, nachahmen
simultaneity Gleichzeitigkeit
simultaneous gleichzeitig, simultan
sine Sinus
singe, to brennen, flammen, sengen
single-break switch Schalter mit Einfachunterbrechung
sink Ausguss, Spülbecken
sinter, to sintern, fritten
sinusoidal sinusförmig, Sinus . . .
siphon Heber, Siphon
siren Sirene
site, to aufstellen, anbringen, unterbringen
site Platz, Baustelle, Aufstellungsort, Lage
size, to auf Endmass bringen, kalibrieren
size Grösse, Abmessung
sizer Kalibrierwerkzeug, Abrichtmaschine
sizing Dimensionieren, Kalibrieren, Klassieren
sizzle, to zischen, knistern
skate Gleitkontakt
skeleton Gerippe
skelp rolling mill Rohrstreifenwalzwerk
sketch, to skizzieren
sketch Skizze
skew Schrägverzerrung (Bildfunk)
skew schräg, verzerrt
skewness Schräge, Asymmetrie
skid, to gleiten, rutschen
skid Gleitbahn, Gleitschiene
skids Unterleghölzer
skill Geschick, Handfertigkeit
skillet Gußstahltiegel
skim, to abheben, abschöpfen
skimmings Schlacke, Schaum
skin, to abisolieren, abziehen
skin Haut, Hülse
skinner isoliertes Drahtende

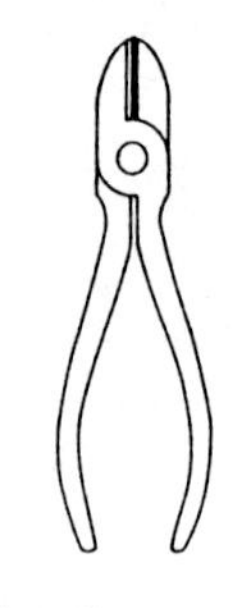

Side-cutter
Seitenschneider

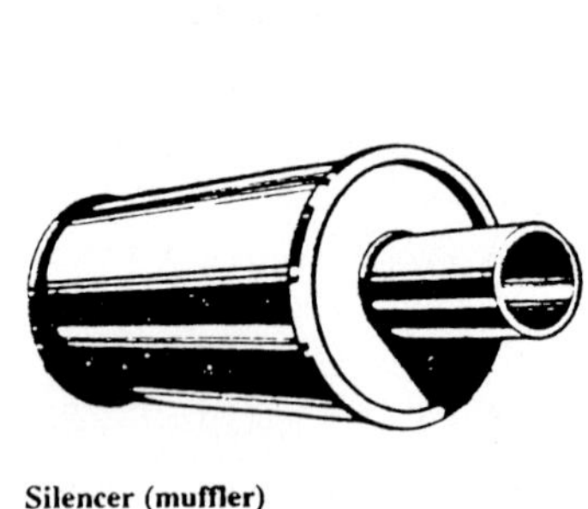

Silencer (muffler)
Schalldämpfer (Auspufftopf)

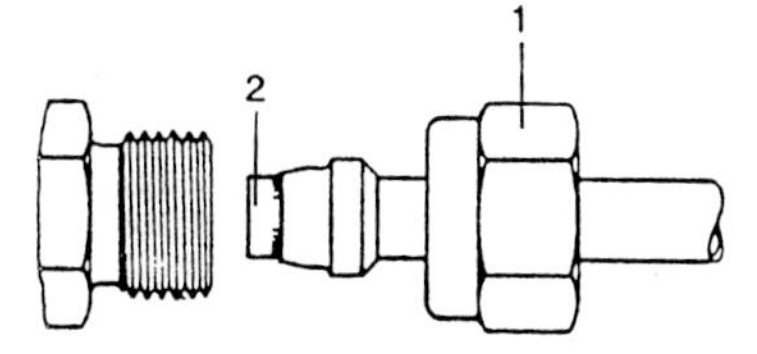

Sleeve and nozzle of a tubular connection
Hülse und Düse einer Rohrverbindung
1 **sleeve** Hülse (Überwurf)
2 **nozzle** Düse

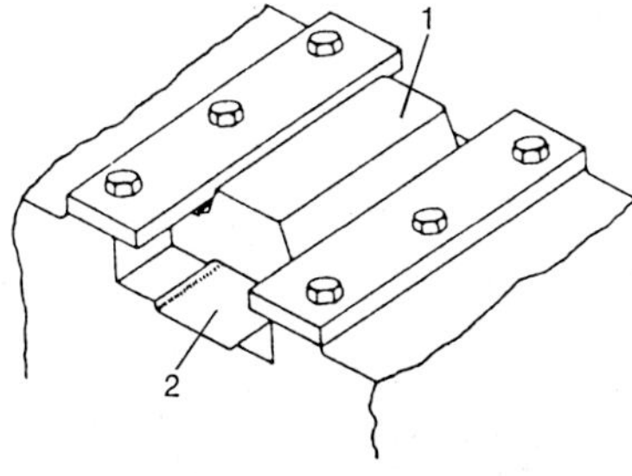

Sliding block
Führungsschlitten, Kulissenstein
1 **sliding block** Führungsschlitten (-stein)
2 **sliding path** Gleitbahn

skip, to überspringen, auslassen
skip Fördergefäss, Skip
skirt, to ausspritzen
skirt Rand, Saum, Grenze
skull Pfannenbär (Met.)
skylight Oberlicht
slab, to abblättern, ablösen, flachwalzen, Flächen bearbeiten
slab Platte, Tafel, Bramme
slack, to lockerlassen, nachlassen
slack Durchhang einer Leitung, Zuschlag zur Leitungsdrahtlänge, Gruskohle
slack locker, lose, schlaff
slacken, to sich lockern, lösen (Schraube)
slackening off Lockern, Lösen, Zurückdrehen
slackness Spiel, Schlaffheit, Durchhängen
slag, to schlacken, ausschlacken, sintern
slag Schlacke, Asche
slagging Schlackenbildung, Entschlackung
slaggy schlackig, schlackenartig
slake, to (Kalk) löschen
slant, to schräg liegen (sitzen), (sich) neigen
slant Neigung, Schrägung, Gefälle
slanting Schrägung
slash, to schlitzen
slat Latte, Steg
slate Schiefer
slavearm Arbeitsarm (Fernbedienungsgerät)
sleek, to glätten
sleeve Buchse, Hülse, Muffe
slender schlank
slew, to schwenken, drehen
slice, to in Platten (Scheiben) schneiden
slice Scheibe
slicer Schneidmaschine
slidable verschiebbar
slide, to gleiten, rutschen
slide Gleitfläche, Support, Führungsschlitten, Diapositiv
slider Gleitkontakt, Lauffläche, Schieber
slideway Führung, Gleitbahn
slight gering, schwach, leicht
sling, to schleudern, werfen
sling Schlinge, Stropp
slip, to rutschen, schlüpfen
slip Rutschen, Schlupf, Zettel

slipper Gleitstück, Gleitschuh
slipway Helling
slit, to schlitzen, spalten
slog schwere Arbeit, grosser Span
slope, to abfallen, ansteigen (lassen)
slope Gefälle, Neigung, Schrägfläche, Anstieg
slot Kerbe, Nut, Schlitz
slotter Senkrechtstossmaschine
slotting machine Senkrechtstossmaschine
slow, to verlangsamen, verzögern
slow down, to verlangsamen, abbremsen
slub Wulst
sludge Schlamm
sludging Entschlammung
slug Block, Rohling, Metallklumpen
sluggish träge, zähflüssig
sluggishness Trägheit
sluice, to spülen
sluice Schleuse, Schütz (Schleusenventil)
slurry Brei, Schlamm
slush Schmiere, Schmutz
smallwares Kurzwaren
smash, to zerschmettern, zerbrechen, zertrümmern
smear, to schmieren, streichen
smell Geruch
smelt, to schmelzen
smelter Schmelztiegel, Schmelzer
smith forging Freiformschmieden, Freiformschmiedeteil
smithy Schmiede
smoke Rauch
smooth, to glätten, abschleifen, abziehen
smooth glatt, ruhig, stossfrei
smoothness Glattheit
smoulder, to schwelen, glimmen
smudge, to beschmutzen, verschmieren
smudge Schmutz, Schmutzfleck
snag, to abgraten, putzen
snagging Abgraten, Putzen
snaked wire verdrillter Draht
snap, to schnappen, abreissen
snap-action contact Sprungkontakt
snarl, to verwickeln, überdrehen
snarl Schleife, Kräuselung
sneak current Kriechstrom, Fremdstrom
snick Einschnitt

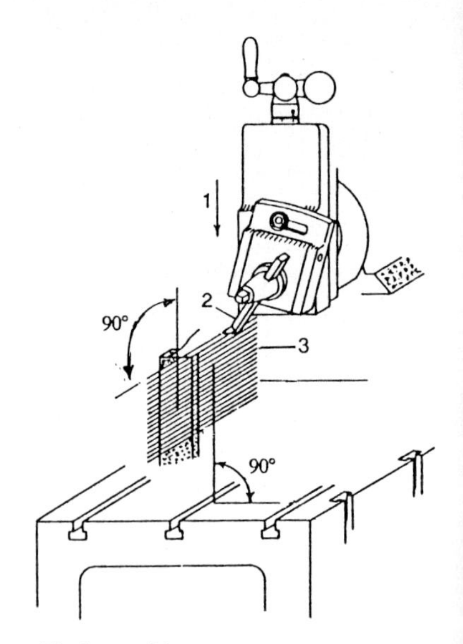

Slotting machine
Vertikal-Stoßmaschine
1 **vertical feed** Senkrecht-Vorschub
2 **tool** Werkzeug
3 **plane swept out by feed** vom Vorschub bestrichene Fläche

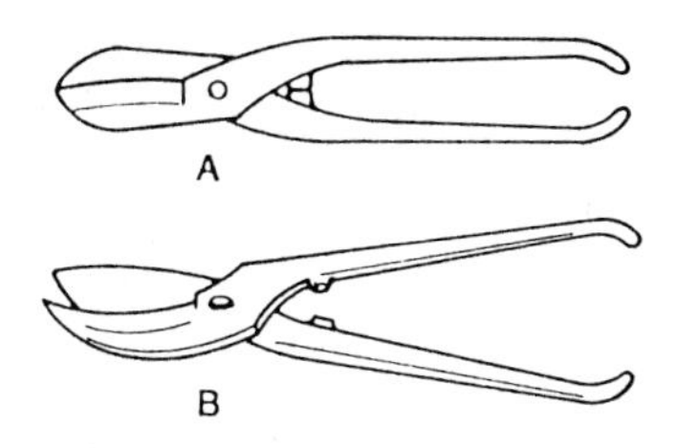

Snips
Blechschere

A **with straight cuts** mit geraden Schneiden

B **with curved cuts** mit gebogenen Schneiden

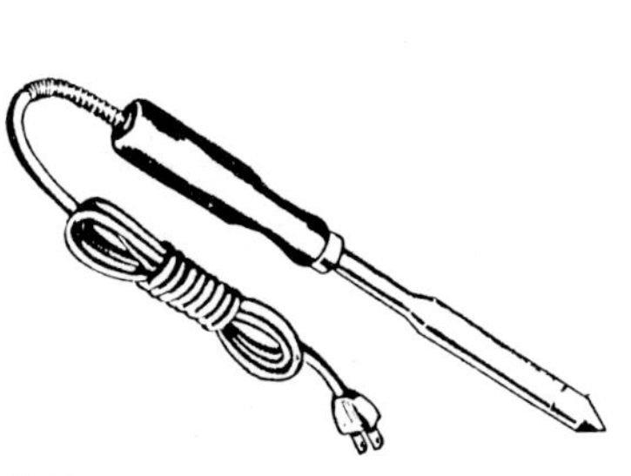

Soldering iron
Lötkolben

snifting valve Schnüffelventil
snips Handschere, Blechschere
snout Düse, Mundstück
snug fit enge Passung
soak, to aufsaugen, tränken, einweichen
soar, to in die Höhe schnellen
socket Steckdose, Hülse, Muffe, Sockel
sodium Natrium
soft weich, mild
soften, to erweichen, aufweichen
soil, to verschmutzen
soil Boden, Grund
solar Sonnen . . .
solder, to löten
solder Lot, Weichlot
soldering Löten, Weichlöten
sole Sohle, Unterseite
sole alleinig, Allein . . .
solenoid Zylinderspule, Solenoid
solenoid operation magnetische Betätigung
solenoid switch Magnetschalter
solid fester Körper, Festkörper
solid fest, massiv, einteilig
solidify, to erstarren, fest werden
solidly earthed system starr geerdetes Netz
soluble löslich
solution Lösung, Auflösung
solve, to lösen
solvent lösend, löslich
sonic Schall . . ., Ton . . .
soot, to berussen, verrussen, verschmutzen
soot Russ
sophisticated verfeinert, hochgezüchtet (Gerät), kompliziert
sorbent adsorbierender Stoff, Adsorbens
sordine Dämpfer
sorption Sorption, Aufnahme
sort, to sortieren, auslesen, aussondern
sorting Sortieren
sough Abflussgraben
sound, to tönen, schallen, loten
sound fehlerfrei, lunkerfrei
sound Klang, Schall, Ton, Sund
sounder Lotgerät
sour sauer
source, Quelle, Spannungsquelle
sourdene Schwingungsdämpfer (Freileitungen)
souring bath Säurebad

space, to in Abstand bringen, teilen
space Raum, Abstand, Platz
space flight Raumfahrt
spaced unterteilt
spacer Abstandshalter, Distanzstück
spacing Abstand, Zwischenraum, Teilung
spacious geräumig, weiträumig
spade drill Spitzbohrer
span, to überbrücken, überspannen
span Spannweite, Stützbereich
spangle Metallfolie
spanner Schraubenschlüssel
spare, to einsparen, sparen
spare Ersatzteil
spare Ersatz . . . , Reserve . . .
spark, to funken, feuern
spark Funke
sparkle, to funkeln, glitzern, glänzen
sparry iron ore Spateisenstein
spatial räumlich, Raum . . .
spatter, to herausschleudern, verspritzen
spear Stange
special Spezial . . . , Sonder . . . , speziell, besonders
specific spezifisch
specification Bauvorschrift, Pflichtenheft, Beschreibung, Vorschrift
specifications Hauptabmessungen, technische Daten
specify, to spezifizieren, detailliert angeben, vorschreiben
specimen Exemplar, Muster, Probestück, Prüfling
speck Flecken
speckle, to tüpfeln, sprenkeln,
spectrum Spektrum, Frequenzband
spherically seated kugelig gelagert
spider Drehkreuz, Spinne, Radstern
spigot Zapfen, Drucklager
spike Impulszacke
spill, to vergiessen, verschütten
spin Spin, innerer Drehimpuls
spindle Spindel
spine Rücken, Gitterstab
spiral Spirale
spiral Spiral . . . , spiralförmig
spire spitzer Körper, Turmspitze
splash, to spritzen

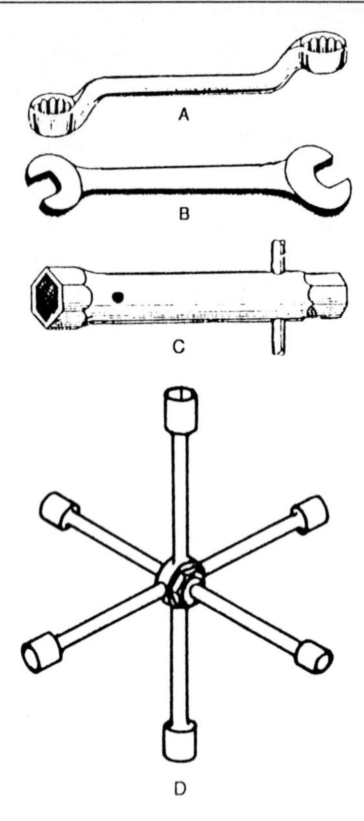

Spanner
Schraubenschlüssel

A **Ring spanner** Ringschlüssel

B **Open spanner** Maulschlüssel

C **Box spanner** Hohlschlüssel

D **Spider** Kreuzschlüssel

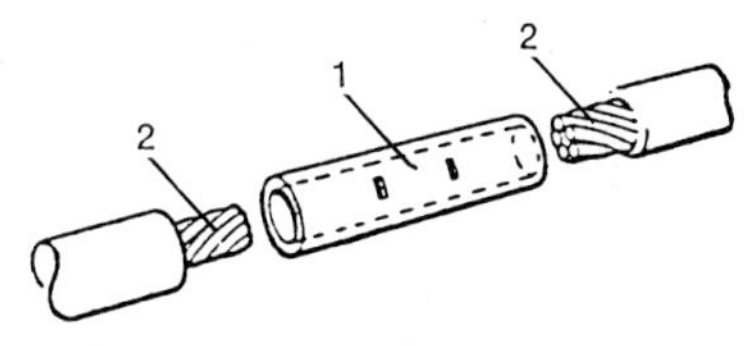

Splicing connection
Spleißverbindung

1 **splicer** Verbindungstülse
2 **tinned conductor end** verlötetes Leiterende

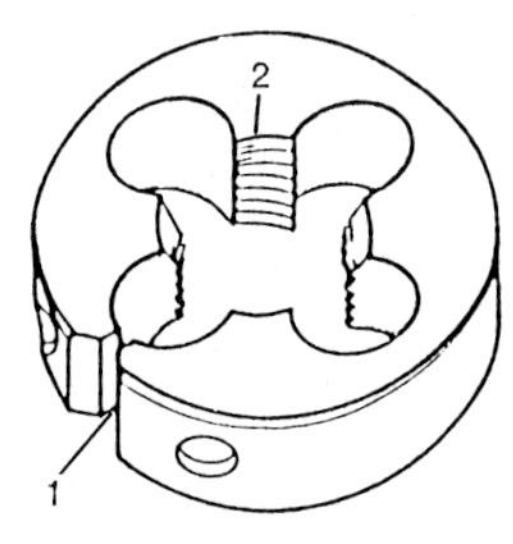

Split threading die
Geschlitztes Gewindeschneideisen

1 **slit** Schlitz
2 **cutting thread** Schneidgewinde

splasher Spritzschutz, Spritzblech
splatter Störung durch einen Nachbarkanal
splice, to spleissen
splice Spleissung, Stoss
spline, to längsnuten
spline Passfeder, Gleitfeder, Schiebekeil
splining Keilnutenfräsen
splint Splint
split, to spalten, aufreissen, schlitzen
split Schlitz
split geschlitzt, geteilt, gespalten
spoil, to verderben
spoil Aushub
spoilage Ausschuss
spoke Speiche, Sprosse
spoked wheel Speichenrad
sponge Schwamm
sponson Ausleger, ausladende Plattform
sponsored research Auftragsforschung
spontaneous spontan
spool, to spulen, aufwickeln
spool Spule, Haspel
spoon Löffel, Kelle
sporadic sporadisch, vereinzelt
spot, to Position feststellen, orten
spot Fleck, Stelle, Lichtmarke
spotfacer Stirnsenker, Plansenker
spotting Fleckigwerden
spout Auslauf, Schnauze, Ausgussrinne, Wellenleiteröffnung
sprag, to abstemplen, verspreizen
sprag Spreizstempel, Strebe, Bolzen
spray, to spritzen, besprühen
spray Zerstäubung, Spritzung, Sprühnebel
spread, to auftragen, ausbreiten, verteilen, streuen
spread Auftrag, Aufstrich, Streuung, Ausbreitung
sprig Drahtstift, Kernnagel
spring, to springen, splittern, zuschnappen
spring Feder, Quelle, Becken
springback Rückfederung
springy rückfedernd
sprinkle, to berieseln, anfeuchten, bespritzen
sprinkler Berieselungsapparat, Brause
sprocket Kettenrad, Zahntrommel

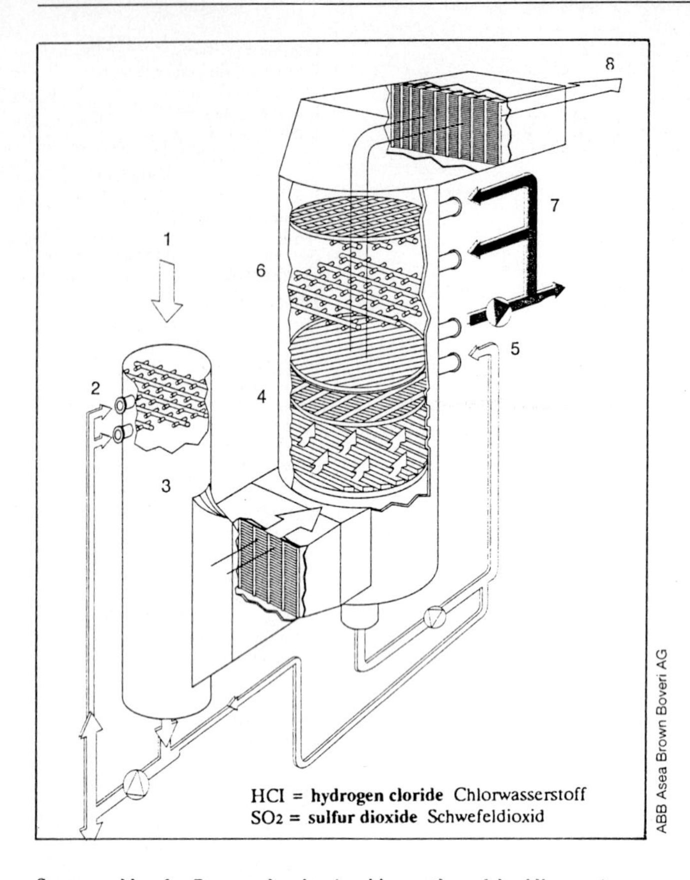

Spray scrubber for flue-gas cleaning (washing out harmful acidic gases)
Sprühwaschanlage für Rauchgasreinigung (Auswaschen von sauren Schadgasen)

1 **flue-gas entry** Eintritt der Rauchgase
2 **liquid sprayer** Einsprühen von Flüssigkeit
3 **quencher (flue-gas is cooled to its wet-bulb temperature)** Löschanlage (Rauchgase werden bis zur Kühlgrenztemperatur gekühlt)
4 **separation of HCl (first scrubbing stage)** Abscheidung von HCl (erste Waschstufe)
5 **scrubbing agent sprayed into gas flow** Einsprühung der Waschflüssigkeit
6 **separation of SO_2 (second scrubbing stage)** Abscheidung von SO_2 (zweite Waschstufe)
7 **scrubbing agent with addition of calcium hydroxide** Waschflüssigkeit mit Hinzugabe von Kalziumhydroxid
8 **exit of cleaned flue-gas** Austritt der gereinigten Rauchgase

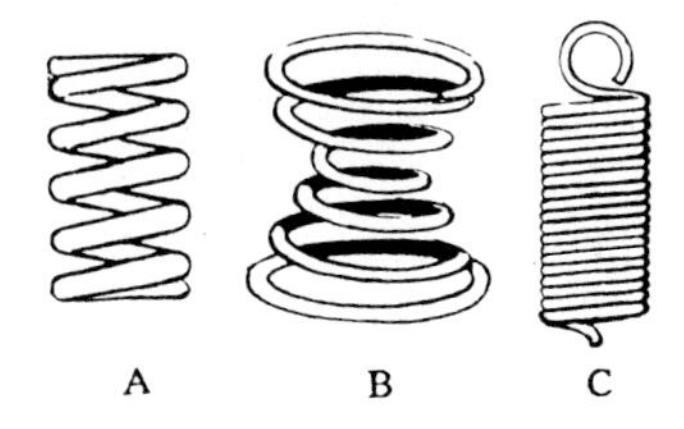

Spiral spring
Spiralfeder

A, B **Compression spring** Druckfeder

C **Extension spring** Zugfeder

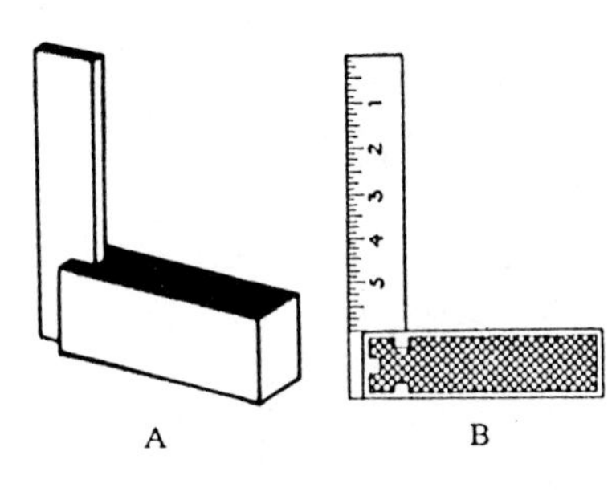

Square
Anschlagwinkel

A **plain square** einfacher Winkel

B **with scale** mit Meßskala

sprue Giesstrichter, Anschnitt
spuding bit Flachmeissel
spur Sporn, Strebe
spurious falsch, künstlich, unerwünscht
spurious frequency Störfrequenz
sput concrete Schleuderbeton
sputter, to sprühen, zerstäuben
spy hole Schauloch
square, to ins Quadrat erheben, rechtwinklig schneiden
square Quadrat, rechteckiger Platz
square quadratisch, kantig, rechtwinklig
squared hoch zwei, Quadrat
squaring Bearbeiten auf rechten Winkel
squash, to quetschen, zerdrücken
squeal Heulen, Pfeifen, Quieken
squeeze, to drücken, pressen, quetschen
squeeze Druck, Quetschung
squeezer Quetschwalzwerk, Pressformmaschine
squegger Sperrschwinger, Pendeloszillator
squirrel cage Käfiganker
squirrel cage induction motor Motor mit Kurzschlussläufer
squirt, to spritzen
squirt-gun welding haulbautomatisches Unterpulverschweissen
stab, to durchstechen, steppen, abteufen
stability Stabilität, Beständigkeit
stabilisation network Stabilisierungsnetz
stabilise, to stabilisieren
stabiliser Stabilisator
stable stabil
stack, to stapeln, lagern, aufschichten
stack Schacht, Schornstein, Stapel, Stoss
staff Personal, Belegschaft, Stab, Stock
stage Stufe, Stadium, Bühne
stagger, to versetzen, schwanken, taumeln
stagger versetzte Anordnung
staging Stellage, Gerüst
stagnant unbewegt, stillstehend
stain, to beflecken, anlaufen, korrodieren, verfärben

stain Fleck, Verfärbung, Rost
stainless rostfrei, nichtrostend, fleckenfrei
stair Treppe
staircase Treppe, Treppenhaus
stalk Stengel, Einspannzapfen
stall, to stillstehen, zum Stillstand bringen, abwürgen
stall Stillstand, Stand
stamina Widerstandskraft, Ausdauer
stamp, to stampfen, stanzen, prägen, aufdrucken
stamp Stempel, Marke
stamped poles lamellierte Pole
stanchion Stütze, Pfosten, Strebe
stand Gestell, Gerüst
standby Bereitschaft, Reserve ...
standard Standard, Norm, normal, geeicht
standardisation Standardisierung
standardize, to standardisieren
standpipe Standrohr, Hydrant
standstill Stillstand
stanniferous zinnhaltig
stannum Zinn
staple, to stapeln, sortieren
staple Stapel, Rohstoff, Klammer
star-connected sterngeschaltet
star-delta connection Sterndreieckschaltung
start, to starten, beginnen, anfangen
starter Anlasser
starve, to ungenügend Füllung haben
state Zustand, Beschaffenheit
statement Angabe, Darlegung, Feststellung, Information
static atmosphärische Störungen, statisch, ruhend
statics Statik, Mechanik ruhender Körper, (atmosphärische) Störungen
stationary ortsfest, ruhend, feststehend
statistics Statistik
stator Stator, Ständer
statute mile = 1,609 km (am.)
staunch wasserdicht, luftdicht
stay, to bleiben, stützen
stay Stütze, Anker, Ständer, Strebe
staying Verspannung, Abspannung
steadiness Stetigkeit, Standfestigkeit

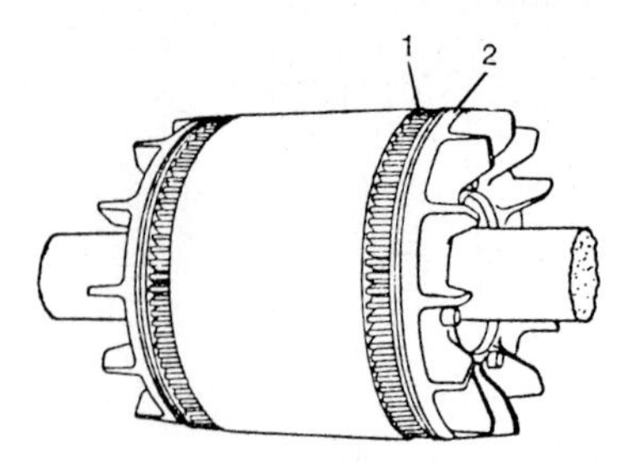

Squirrel-cage rotor (induction motor)
Käfigläufer (Kurzschlußmotor)

1 **conductor bar** Leiterstab
2 **short-circuit ring with cooling fins** Kurzschlußring mit Kühlrippen

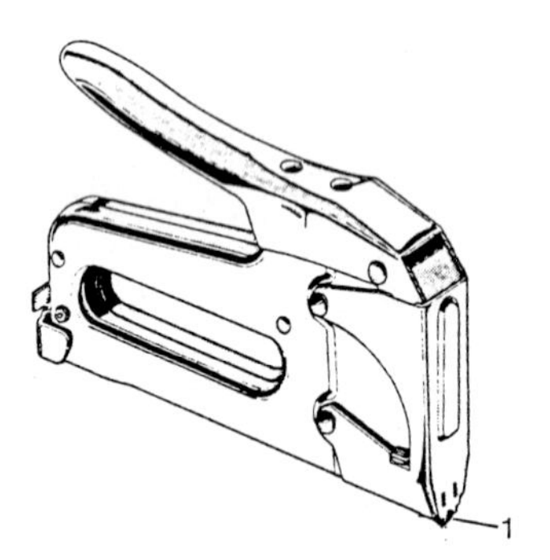

Staple gun (tacker)
Heftpistole

1 **staple ejection** Heftklammer-Ausstoß

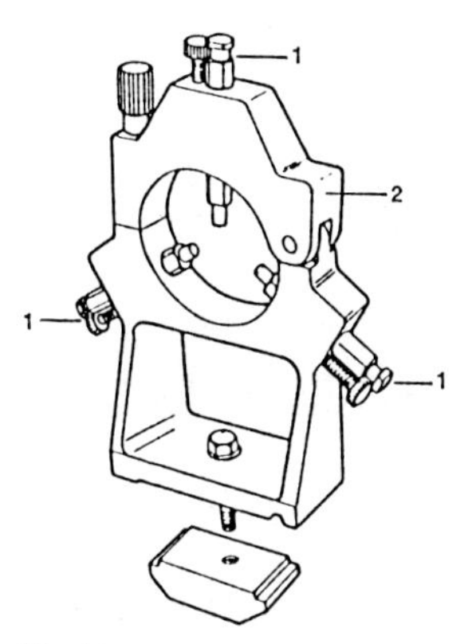

Steady
Setzstock

1 **three-point adjustment** Dreipunkt-Setzeinstellung
2 **hinged upper part** aufklappbares Oberteil

steady, to stabilisieren
steady stetig, gleichmässig, standfest, Setzstock
steam, to dampfen
steam Dampf
steel, to verstählen
steel Stahl
steep steil, schroff
steeping Imprägnierung
steepness Steilheit
steer, to steuern, lenken
steering Lenkung
stem, to dichten, verstopfen, anstauen
stem Stamm, Stiel, Stengel, Rippe
stench trap Geruchverschluss
stencil, to schablonieren
stencil Schablone, Matrize
step, to stufen, absetzen
step Schritt, Stufe, Schwelle
stepless stufenlos
stepped treppenförmig, abgestuft
stern Heck, Gatt
sternpost Achtersteven
stibium Antimon
stick, to stecken, schmieren haften, verklemmen
stick Stock, Stange, Stiel
sticky klebrig
stiff steif, starr, stramm
stiffen, to versteifen
stiffener Versteifung, Steife
stiffness Steifigkeit
stifle, to ersticken (Brand)
stile Senkrechtstreifen, Durchgang
still ruhend, unbewegt
stillage Pritsche, Ladeplatte, Plattform
stillpot Absatzbecken, Klärbecken
stillson wrench Rohrzange
stimulus Wirkungsgrösse, Anreiz
stipple, to tupfen
stir, to rühren
stirrup Bügel, Steigeisen
stitch, to heften, steppen
stitch Stich, Masche
stock, to aufbewahren, versorgen
stock Lager, Vorrat, Halde
stocking Lagerung, Lagern
stockpiling Stapeln, Lagern, Aufspeichern
stoke, to beschicken, schüren, heizen
stokehold Heizraum, Kesselraum

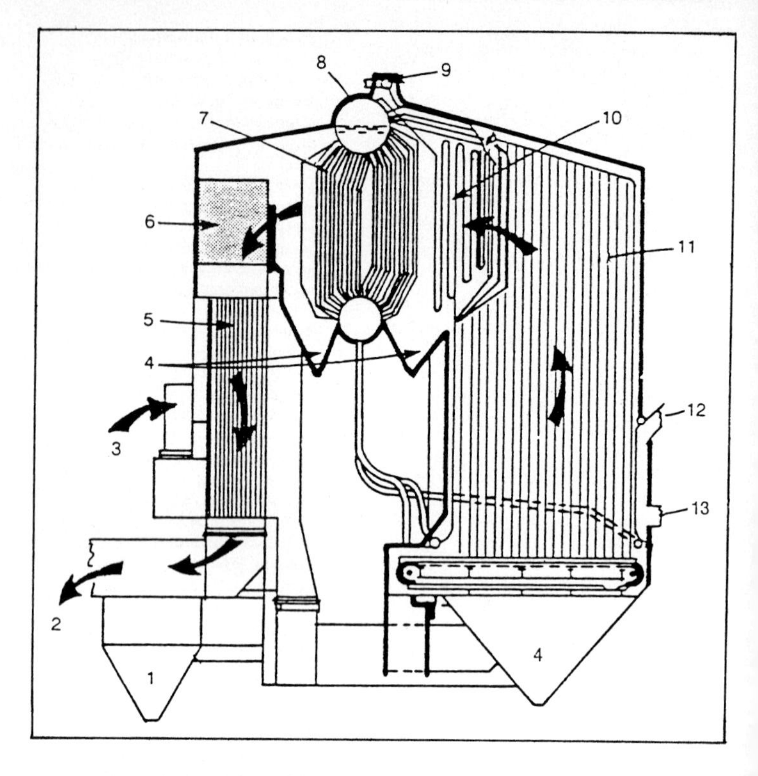

Steam generator (boiler)
Dampferzeuger (Dampfkessel)

1 **dust collector** Staubabscheider
2 **air to induced-draft fan** Luft zum Abzugsventilator
3 **air from forced-draft fan** Luft vom Frischluftventilator
4 **ash hoppers** Aschentrichter
5 **air heater** Lufterhitzer
6 **economizer** Vorwärmer
7 **water tubes (heat transfer by convection)** Wasserrohre (Konvektions-Wärmeübertragung)
8 **steam drum** Dampfsammelbehälter
9 **steam outlet** Dampfaustritt
10 **superheater** Überhitzer
11 **water walls (radiant heat transfer)** Wasserwände (Strahlungs-Wärmeübertragung)
12 **fuel distributor** Brennstoffverteiler
13 **stoker** Brennstoffzuführung

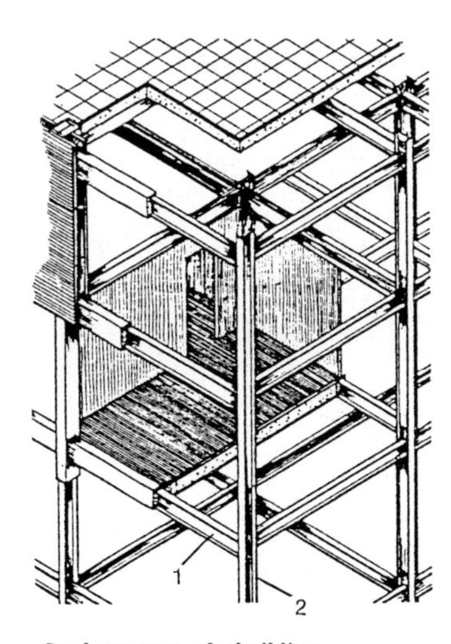

Steel structure of a building
Stahlkonstruktion eines Gebäudes

1 **girder** Träger
2 **stanchion** (Stütz-)Pfeiler

strengthen, to verstärken, versteifen
strengthening Verstärkung (mech.)
stress, to spannen, beanspruchen, belasten
stress Spannung (mech.), Beanspruchung
stretch, to strecken, dehnen, straffen
stretch Strecke, Länge, Ausdehnung
stretchforming Streckformen
stria Schliere, Riefe, Furche
strickle Schablone, Lehre
striction Einengung, Verengung
strident schrill, schneidend (Ton)
strike, to schlagen, prägen, stossen auf, prallen auf
striking Auftreffen, Aufprallen, Zündung (Lichtbogen)
string, to binden, schnüren, knüpfen
string Schnur, Bindfaden
stringer Längsbalken, Längsträger, Stützbalken, Holm, Stringer
strip, to abisolieren, abstreifen, ablösen, abkratzen
strip Band, Streifen, Leiste
stripe Streifen
stripped nackt (Draht)
strobing Signalauswertung
strobeglow Stroboskop mit Neonröhre
stroboscope Stroboskop
stroke Hub, Takt, Schlag, Strich
strong fest, stark
struck abgebaut (Gerüst), abmontiert
structural baulich, konstruktiv, Gefüge . . . , Struktur . . .
structure Struktur, Gefüge, Konstruktion, Bauwerk
strut, to versteifen, verstreben, abstützen
strut Strebe, Verstrebung, Stützsäule
strutted pole verstrebter Mast
stubby gedrungen
stud Stehbolzen, Stiftschraube, Anschlag, Steg, Runge
study, to untersuchen, studieren
stuff, to vollstopfen, polstern
stuff Grundstoff, Zeug, Stoff, Materie

stoker Heizer
stone Stein; 1 stone = 6,350 kg
stool Bock, Schemel
stop, to anhalten, stillsetzen, arretieren
stop Halt, Stillstand, Anschlag, Sperre, Hubbegrenzer
stopcock Absperrhahn
stopper, to verstöpseln
stopper Stopfen, Pfropfen, Stöpsel
store, to speichern
store Speicher, Lager, Abstellraum
storehouse Lagerhaus, Magazin, Speicher
storey Stockwerk, Etage
storm guyed pole Abspannmast
stove, to einbrennen, härten
stove Ofen
stow, to stauen
stowage Stauraum, Stauen
stowaway Abstellraum, abgestellte Güter, blinder Passagier
straggling Streuung, Schwankung, Straggling
straight gerade, geradlinig
straighten, to begradigen, ausrichten
straightener Blechrichtmaschine
strain, to anspannen, beanspruchen, dehnen, durchpressen
strain Dehnung, Deformation, Beanspruchung
strainer Filtereinsatz
strainfree spannungsfrei (mech.)
strand, to verlitzen, verseilen
strand Litze, Faserbündel
stranded wire Litzendraht
strap, to festbinden, anschnallen, verlaschen
strap Spannband, Riemen, Gurt
strapped joint Laschenstoss
stratum Schicht, Ablagerung, Lage, Flözschicht
straw Stroh
strawboard Strohpappe, Strohplatte
stray, to streuen, abweichen, vagabundieren
straying Streuung, Streuverlust
streaking Nachziehen
stream, to strömen
stream Strömung, Strom
streamline, to stromlinienförmig gestalten
strength Festigkeit, Stärke, Kraft

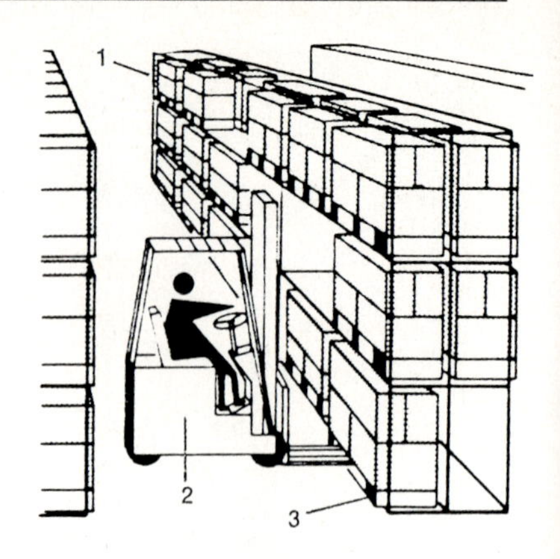

Storage room (storehouse)
Lagerraum (Lagerhaus)

1 **rack** Gestell
2 **fork-lift truck** Gabelstapler
3 **palette** Palette

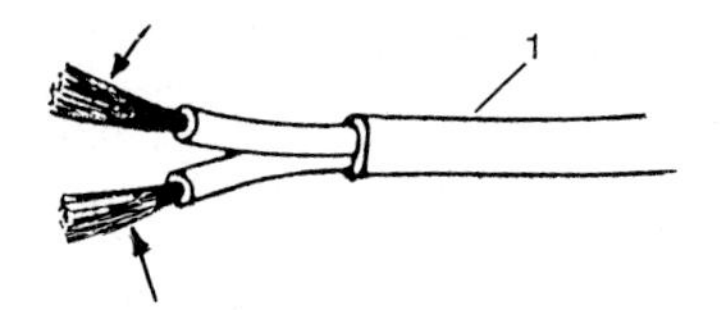

Stranded wire
Litzendraht

1 **rubber/plastic insulation** Gummi-/Plastik-isolation

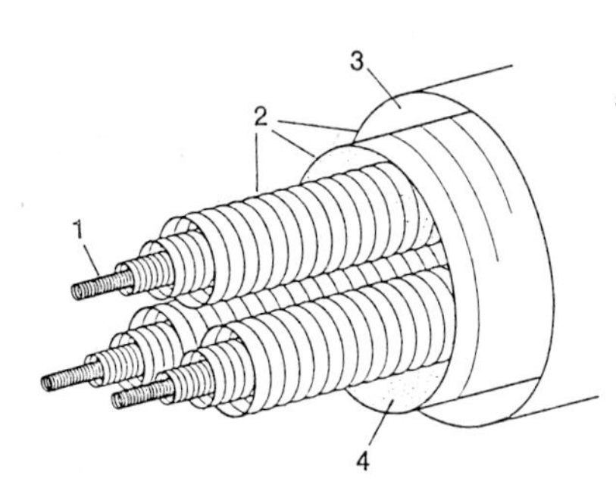

Superconducting three-phase cable
Supraleitendes Drehstromkabel

1 **conductor** Leiter
2 **cold-protective tubes** Kälteschutzrohre
3 **vacuum** Vakuum
4 **liquid nitrogen** flüssiger Stickstoff

sturdiness Stabilität, Festigkeit, Widerstandsfähigkeit
sturdy stabil, robust, kräftig
stylus Fühlerstift, Taster, Nadel
subassembly Bauteil, Baueinheit, Teilmontage
subcarrier Zwischenträger
subcircuit Abzweigung, Abzweig-stromkreis
subcontractor Unterlieferant
subdivide, to unterteilen
subdue, to dämpfen, unterdrücken
subjected to beansprucht auf
subject Fachgebiet, Thema
submarine Unterseeboot, Unter-wasser . . . , unterseeisch
submerge, to überfluten, eintauchen
submersible tauchfähig, wasser-dicht (Maschine)
subrepeater Hilfsverstärker
subscriber Teilnehmer
subside, to (sich) senken, setzen
subsidiary Hilfs . . . , Neben . . . , Tochtergesellschaft
subsonic Unterschall . . .
substance Stoff, Substanz
substation Unterwerk, Unterstation
substitute, to austauschen, er-setzen, einsetzen
substitute Austauschstoff, Ersatz(stoff)
substitutional resistance Ersatz-widerstand
substrate Schichtträger, Substrat
subtract, to abziehen, subtrahieren
succession Folge, Reihenfolge
successive aufeinanderfolgend
suck, to saugen
sucking coil Tauchkernspule
suction Saugen, Ansaugung
sue for damages, to auf Schaden-ersatz verklagen
suggestion box Einwurfkasten für Verbesserungsvorschläge
suggestions for improvements Verbesserungsvorschläge
suit, to passen, sich eignen, taugen
sulphur Schwefel
sum, to addieren, summieren
sum Summe
summarise, to zusammenfassen
summary Zusammenfassung

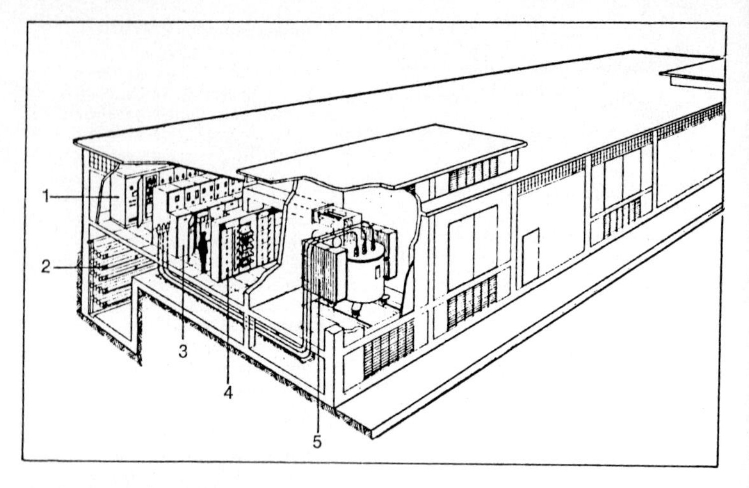

(Power distribution) Substation
(Stromverteilungs-) Unterstation

1 **low-voltage switchboard** Niederspannungs-Schalttafel
2 **cable run** Kabelgang
3 **high-voltage switchgear** Hochspannungs-Schaltanlage
4 **control and relay board** Steuer- und Relaytafel
5 **transformer** Transformator

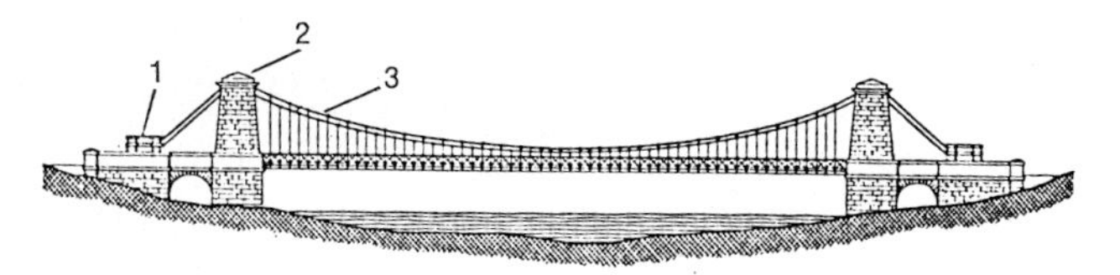

Suspension bridge
Hängebrücke

1 **anchorage** Verankerung
2 **tower** Pylon
3 **carrier cable** Tragkabel

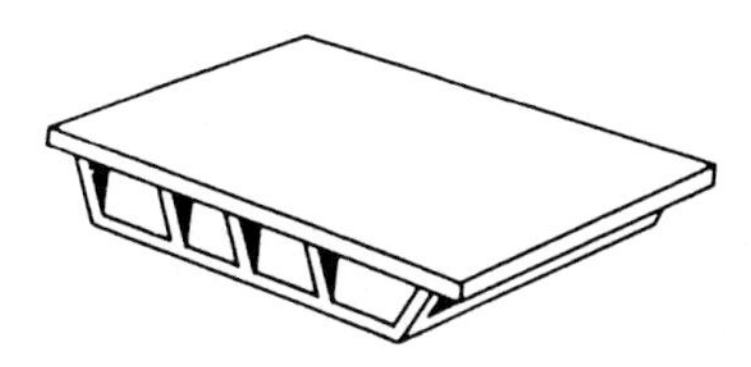

Surface plate
Anreißplatte

summit Spitze, Gipfel
sump (Oel-)Sumpf
supercharge, to aufladen (Verbrennungsmotor)
supercharge Aufladung
supercharger Aufladegebläse
superconductive supraleitend
superconductivity Supraleitfähigkeit
superconductor Supraleiter
superfinish, to feinstbearbeiten
supergrid Hochspannungsnetz
superheater Überhitzer
superimpose, to überlagern
superpose, to überlagern
supersonic Ultraschall . . . , ultraschallfrequent
supervise, to überwachen
supply, to versorgen, liefern, zuführen
supply Versorgung, Lieferung, Zuführung, Speisung
support, to stützen, tragen, aufliegen
support Auflage, Stütze, Lagerung, Halter
suppress, to unterdrücken, dämpfen, sperren
suppression Unterdrückung
suppressor Begrenzerschalter, Bremsgitter (Röhre)
surface, to plandrehen, abrichten
surface Fläche, Oberfläche, Aussenseite
surfacing Flächenbearbeiten, Planen
surge Spannungsstoss, Überstrom
surmount, to bedecken, überragen
surplus Überschuss
surveillance Überwachung
survey, to vermessen, überwachen
survey Überwachung, Überblick, Vermessung

susceptance Blindleitwert, Suszeptanz
susceptibility Empfindlichkeit, Anfälligkeit
susceptible empfindlich, anfällig
suspend, to aufhängen, frei tragen, schweben
suspension Aufhängung, Aufschlämmung
sustained Dauer . . . , ununterbrochen, ungedämpft
swage, to tiefziehen, fassonschmieden, gesenkdrücken
swaging Gesenkschmieden, Gesenkdrücken, Tiefziehen
sway, to schwingen
sweep, to fegen, kehren, ablenken
sweep Schwung, Abtastung, Bereich
sweep deflection Kippablenkung
sweeping Wobbeln, Durchlauf
sweeps Ablenkspannungen
swell, to schwellen, aufquellen
swell Schwellung, Ausbauchung
swill, to spülen, waschen
swing, to schwingen, schwenken
switch, to schalten, rangieren
switch Schalter, Weiche
switchboard Schalttafel, Vermittlungsschrank
switchgear Schaltgerät(e), Schaltanlagen
switching Schalten, Schaltung, Rangieren
switchyard Hochspannungsschaltanlage (Freiluft)
swivel, to schwenken, drehen, schrägstellen
swivel Drehteil, Drehscheibe, Spannschloss
symmetrical symmetrisch
synchro Drehmelder
synchromesh gear Synchrongetriebe
synchronisation Synchronisierung
synchronise, to synchronisieren
synchroniser Synchronisiervorrichtung
synchronous synchron, Synchron . . .
synthetic künstlich, synthetisch
syntonising coil Abstimmspule
syntony Abstimmung, Resonanz
system Anlage, System

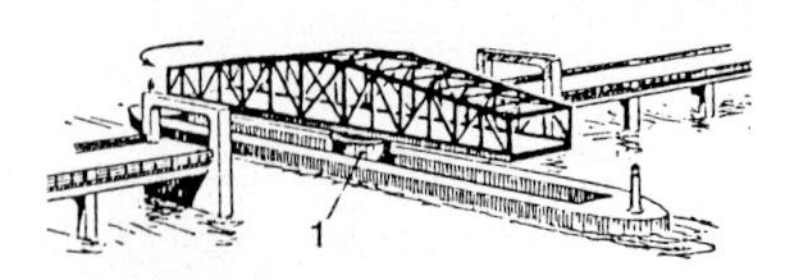

Swing bridge
Drehbrücke

1 **turn-table** Drehscheibe

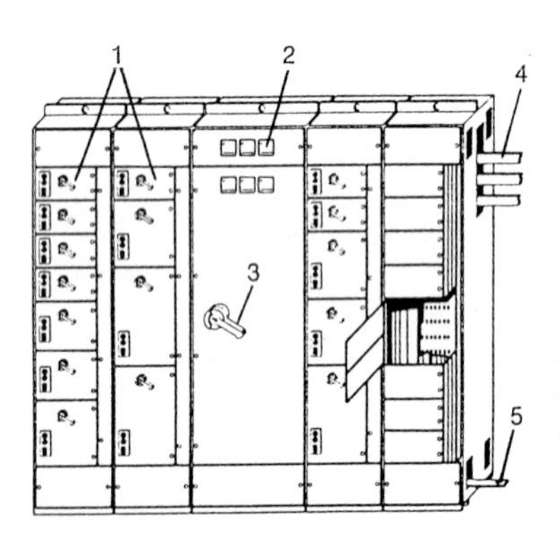

Switchboard
Schaltschrank (-tafel)

1 **controls** Steuereinrichtungen
2 **measuring instruments** Meßinstrumente
3 **switch lever** Schalthebel
4 **busbar** Sammelschiene
5 **earthing bar** Erdungsschiene

T

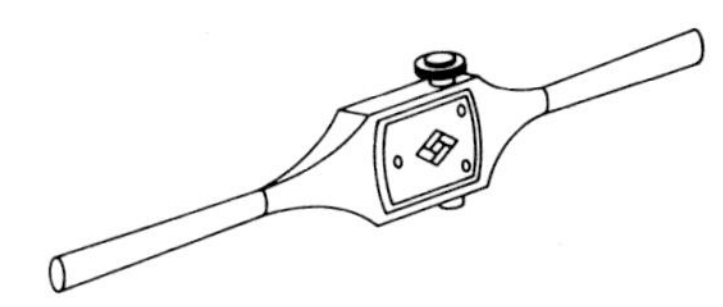

Tap wrench
Windeisen

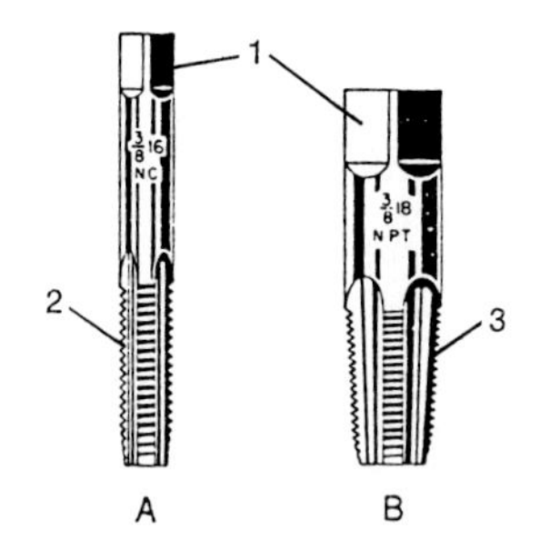

Tap
Gewindebohrer

A **Machine tap** Schraubengewinde-Bohrer

B **Pipe tap** Rohrgewinde-Bohrer

1 **square head for tap wrench** Vierkant-Kopf für Windeisen
2 **cylindrical cutting thread** zylindrisches Schneidgewinde
3 **tapered cutting thread** sich verjüngendes Schneidgewinde

tab Beschriftungsschild, Aufreissband, Drucktaste
table Tabelle, Tafel, Planscheibe, Tisch
tabular tabellarisch, tafelförmig
tack, to (leicht) befestigen, heften, nageln
tack Zwecke, Stift, Nagel
tackle Spannzeug (Freileitungsbau), Flaschenzug, Gerät, Ausrüstung
tag Anhänger, Aufkleber, Lötfahne
tagged mit Lötösen versehen
tagger dünnes Feinblech
tail off, to allmählich abklingen
tail Schwanz, Heck
tailstock Reitstock (Masch.)
take, to nehmen, aufnehmen, fassen, annehmen
take-apart model zerlegbares Modell
taken from abgeleitet von
talk-back circuit Gegensprechschaltung
tally, to nachprüfen, stimmen, registrieren
tally Probe, Kontrollrechnung, Kontrollmarke
tangle, to verwirren, verfitzen
tantalum Tantal
tap, to anzapfen, abgreifen, abhören
tap Anzapfung, Abgriff, Gewindebohrer
tape, to umwickeln
tape Band, Streifen, Isolierband
taper, to sich verjüngen, kegelig machen
taper Kegel, Verjüngung, Keil
taping Steuerung durch Lochstreifen, Bandwicklung
tapper Gewindebohrmaschine, Abgreifer
tappet Greifer, Knagge, Mitnehmer
tar Teer
target Ziel, Auffangplatte, Prallplatte, Fangelektrode
tarnish, to anlaufen, glanzlos werden
task Aufgabe

taut straff, gespannt
taxi, to rollen (Flugzeug)
teaching aid Lehrmittel
tear, to reissen, zerreissen, verschleissen
technician Techniker
technology Technik, Technologie
ted, to wenden, ausbreiten
tee, to abzweigen (el.)
teem, to in Kokillen abgiessen
teeth Verzahnung, Zähne
telecast, to durch Fernsehen übertragen
telecast Fernsehsendung
telecommunication(s) Fernmeldewesen, Nachrichtentechnik
telecontrol Fernsteuerung
telephone Telefon (Fernsprech-)
teleprinter Fernschreibmaschine
telex Fernschreiber
telltale Anzeiger
telltale board Anzeigetafel
temper, to anlassen (Metall), stimmen
temper Härtegrad
template Lehre, Schablone
temporary zeitweise, temporär
tenacious zäh, widerstandsfähig, beharrlich
tend, to warten, bedienen
tendency Neigung, Tendenz
tender, to Kostenanschlag einreichen
tender Kostenanschlag
tending Wartung, Pflege
tenon Zapfen, Vorsprung
tensible zugbelastbar, streckbar
tension, to auf Zug beanspruchen, strecken
tension Zug, Spannung
tepid lauwarm
term, to benennen
term Begriff, Fachausdruck
terminal Klemme, Anschluss, Endamt, Endstation
terminate, to abschliessen, begrenzen
termination Abschluss, Beendigung, Begrenzung, Endverschluss
terms of delivery Lieferbedingungen
ternary aus drei Einheiten bestehend
test, to prüfen, untersuchen

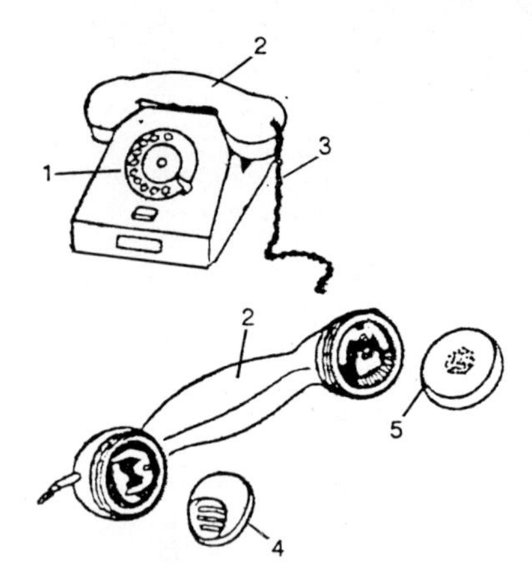

Telephone set
(Fernsprech-) Telefonapparat

1 **dial** Wählscheibe
2 **receiver** Hörer
3 **cable cord** Kabelschnur
4 **mouthpiece** Sprechkapsel
5 **earpiece** Hörmuschel

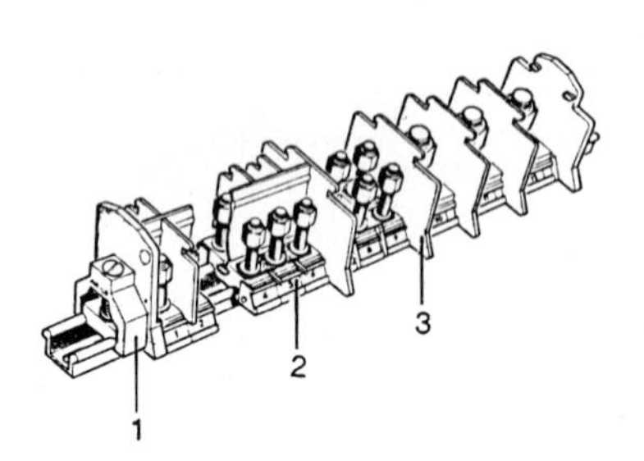

Terminal strip
Klemmenleiste

1 **clamp** Klammer
2 **terminal** Klemme
3 **insulation barrier** Isolationsabtrennung

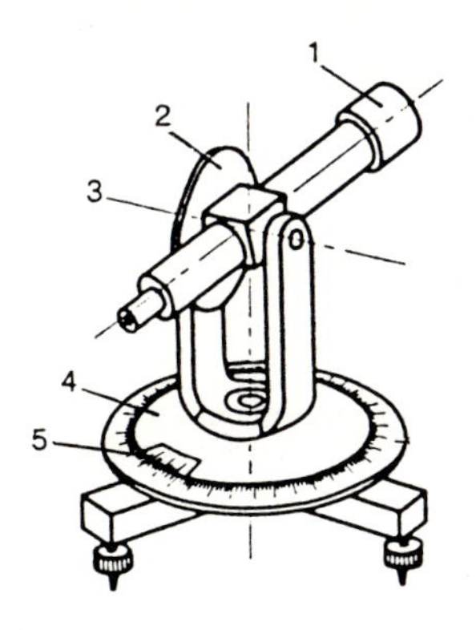

Theodolite (angle measuring instrument)
Theodolit (Winkelmeßgerät

1 **telescope** Zielfernrohr
2 **vertical vernier** Höhengradkreis
3 **tilt axis** Kippachse
4 **horizontal vernier** horizontaler Gradkreis (Limbus)
5 **reading** Ablesung

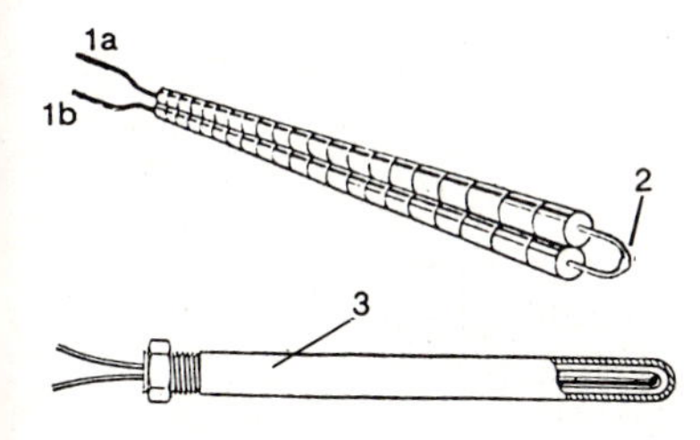

Thermocouple
Thermoelement

1a, 1b **conductors of different metals** Leiter aus verschiedenen Metallen
2 **contact point where Neebeck effect occurs** Kontaktstelle an der Neebeckeffekt auftritt
3 **thermowell** Tauchbehälter

test Prüfung, Versuch
texture Gewebe, Schicht, Gefüge
thaw point Taupunkt
theodolite Theodolit
theorem Lehrsatz
theory Theorie, Lehre
thermal thermisch, Wärme
thermionic glühkathodisch
thermionics Theorie der Elektronenröhrentechnik
thermistor Thermistor, Heissleiter
thermocouple Thermoelement
thesis These, Dissertation, Diplomarbeit
thick dick, stark
thickness Dicke
thimble Hülse, Muffe, Fingerhut
thin, to verdünnen
thin dünn
thin-film semiconductor Dünnschichthalbleiter
thread, to Gewinde schneiden, durchführen
thread Gewinde, Faden
threaded mit Gewinde
threshold Schwelle
thrive, to gedeihen, gut wachsen
throat Gicht (Hochofen), Rachen, Kehle
throttle, to drosseln
throttle Drossel(klappe)
through-and-through bore Durchgangsbohrung
throw, to werfen, schleudern, kippen
thrust Druck, Schub
thumb nut Flügelmutter
tick, to ticken
ticker Zerhacker, Schnellunterbrecher
tickler Rückkopplungsspule
tide Gezeiten (Pl.)
tie, to verbinden, befestigen
tie Zuganker, Verbindungsstelle, Strebe (Mast)
tier, to stapeln
tier Etage, Stapel
tight dicht, fest straff, undurchlässig
tighten, to anziehen (Mutter), spannen, festklemmen
tile, to kacheln, fliesen, decken
tile Kachel, Fliese, Platte
tilt, to kippen, schrägstellen
timber Bauholz, Nutzholz

time, to zeitlich bemessen
timed control Programmregler
timer Schaltuhr, Stoppuhr, Zeitrelais
timing Einstellung des Zeitpunktes
tin, to verzinnen
tin Zinn, Blechbüchse (verzinnt)
tinman Zinngiesser, Klempner
tinman's snips Blechschere
tinned verzinnt
tint, to aufhellen, abtönen
tint Farbaufhellung
tip, to Plättchen auflöten, abkippen, bestücken
tip Spitze, Ende, Schneide, Abladeplatz
tipper Kipper
tissue Gewebe, Stoff
titanium Titan
title block Schriftfeld (Zeichnung)
toggle Hebel, Gelenk
toggle switch Kippschalter
tolerable zulässig
toll Läuten
tone Ton, Klang
tongs Zange
tool, to mit Werkzeug bearbeiten, einrichten, aufspannen (Werkzeuge)
tool Werkzeug, Meissel
torch Brenner, Flamme, Fackel
torque Drehmoment
torsion Verdrehung, Torsion
tortuous (mehrfach) gewunden
total Gesamt(betrag), Gesamt ..., gesamt, total
totally enclosed geschlossen, gekapselt
tote box Transportbehälter
touch, to berühren
touch Berührung, Tast ...
tough zäh
toughness Zähigkeit
tow, to schleppen
tow Schleppzug
tower Turm, Gittermast
toxic giftig, toxisch
trace, to zeichnen, nachziehen, abtasten, suchen
trace Spur
tracer Fühler, Taster
track, to schleppen, spuren
track Spur, Bahn
trackless vehicle schienenloses Fahrzeug

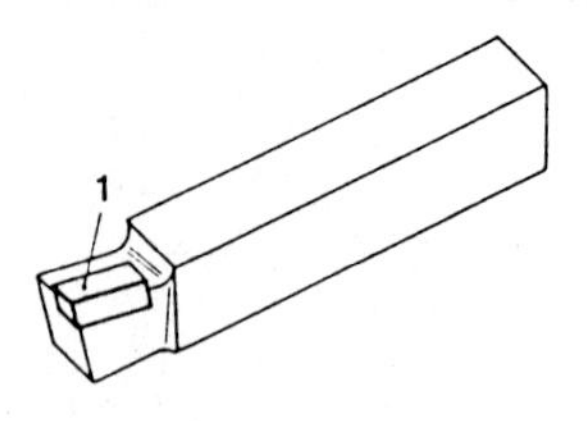

Tipped lathe tool
Bestückter Drehmeißel

1 **brazed tip** hart-eingelötete Meißelspitze

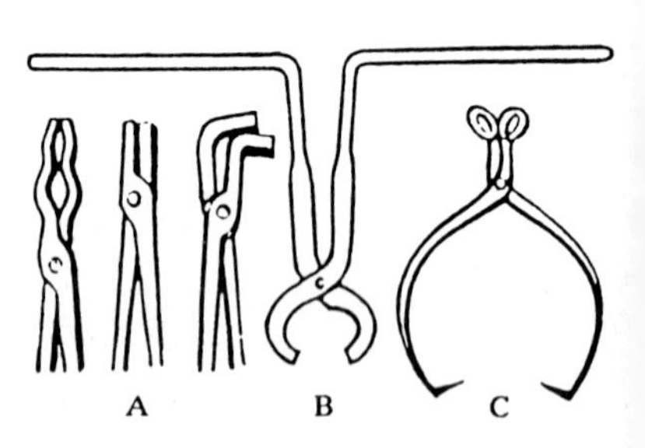

Tongs
Greifzange

A **Blacksmith's tongs** Schmiedezangen
B **Rail tongs** Schienenzange
C **Ice tongs** Eiszange

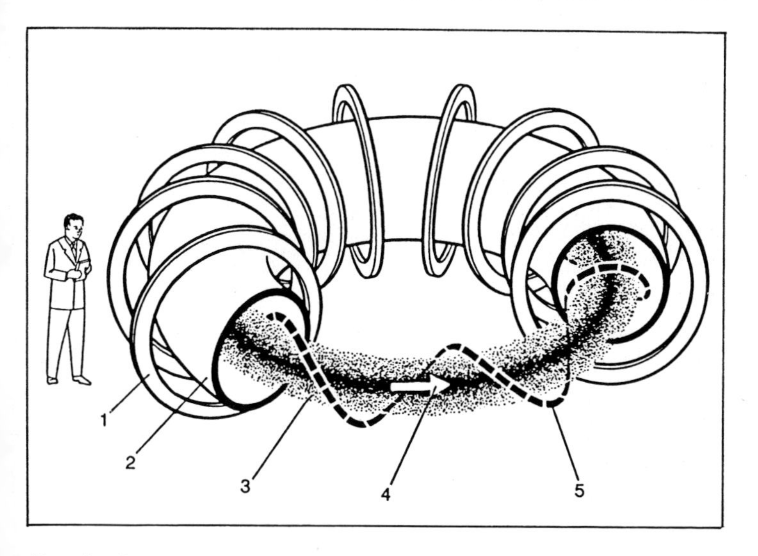

Torus for plasma-physical research
Torus für plasma-physikalische Forschung

1 **magnetic field coil** Magnetfeldspule
2 **vacuum tube (torus)** Vakuumröhre (Torus)
3 **plasma** Plasma
4 **axial current flow** axialer Stromfluß
5 **magnetic field** Magnetfeld

traction Zug
trade Handel, Gewerbe, Beruf
traffic Verkehr
trail, to schleppen
trail Nachlauf
trailer Anhänger
trailing nacheilend
train, to ausbilden, richten
train Reihe, Zug
trainee Anlernling, Praktikant
training Ausbilden
transaction Abhandlung
transatlantic cable Überseekabel
transceiver Sende-Empfangsgerät
transcribe, to umschreiben, abgreifen
transcriber Übersetzer (Rechner)
transducer (Messgrössen-)Wandler
transfer, to übertragen, transportieren
transfer Übertragung, Transport
transform, to umwandeln, umsetzen, umspannen
transformer Transformator
transient Einschwingvorgang, Ausgleichsvorgang, Einschaltstoss
transient einschwingend, kurzzeitig, momentan
transit Durchgang, Durchlauf
transition Übergang
translatory fortschreitend, translatorisch
translucent transparent
transmission Übertragung, Sendung
transmit, to übertragen, senden, leiten (Strom)
transmitter Sender, Geber
transparency Transparenz
transponder Antwortsender
transpose, to versetzen, vertauschen
transversal quer, Quer . . . , transversal
transverse quer, transversal
transverter Umrichter
trap, to einfangen, einschliessen
trap Falle, Abschneider, Auffangvorrichtung
trapped mit Geruchverschluss
trash Abfall, Ausschuss
travel, to (sich) bewegen, verschieben
travel Weg, Bewegung

Trammel
Stangenzirkel

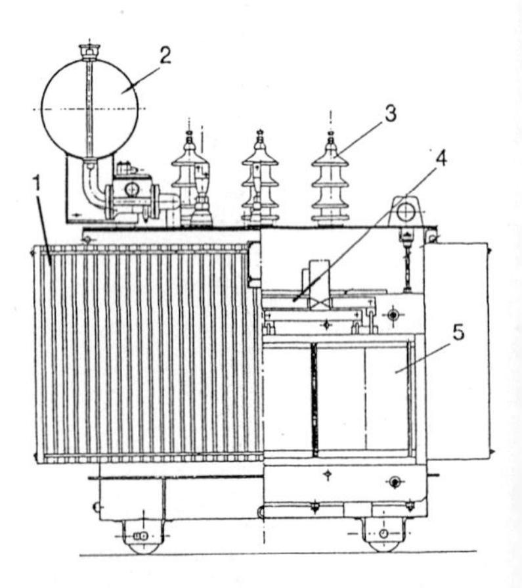

Transformer
Transformator

1 **transformer tank with cooling fins** Transformatortank mit Kühlrippen
2 **oil expansion tank** Ölexpansionsgefäß
3 **insulator bushing** Isolatordurchführung
4 **laminated iron core** Eisenkern aus Magnetblechen
5 **windings** Wicklungen

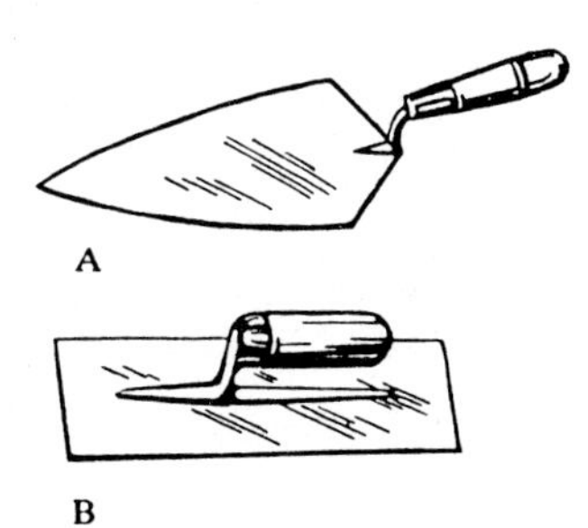

Trowel
Kelle

A **Brick-layer's trowel** Maurerkelle
B **Plastering trowel** Verputzkelle

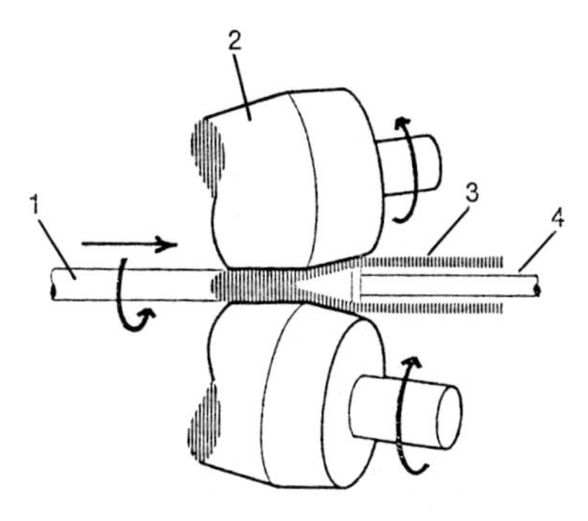

Seamless tube making (Mannesmann principle)
Nahtlose Rohrherstellung (Mannesmann-Prinzip)

1 **hot steel rod** erhitzte Stahlstange
2 **roll** Walze
3 **seamless tube** nahtloses Rohr
4 **nose-piece** Treibdorn

traverse, to überqueren, fahren, verschieben
traverse Bewegung (Masch.), Verschiebung
tray Mulde, Trog, Schale
tread Treppenstufe, Laufkranz
treat, to behandeln, verarbeiten, vergüten
trellised mast Gittermast
trench Graben
trepan, to hohlbohren, ringbohren
trestle Bock, Gerüst
trial Versuch, (Abnahme-)Prüfung
triblet Dorn
trickle, to tropfen
trickle charge Pufferladung
trigger, to auslösen, einleiten, triggern
trigger Auslöser, Trigger(schaltung)
trim, to abgraten, putzen, zurichten, trimmen
trio mill Drillingswalzwerk
trip, to auslösen (Relais)
trip Auslösung
triple, to verdreifachen, dreifach sein
tripod Dreibein, Dreifuss, Stativ
trolley Laufkatze, Förderwagen
trough Trog, Schale, Wanne
truck Lastkraftwagen (Am.)
true richtig, genau, masshaltig
truncate, to abflachen
trundle, to rollen
trunk Stamm, Schaft, Sammelschiene, Koffer
trunnel Dübel
trunnion Achse, Auflager, Drehzapfen
truss, to unterstützen, halten, versteifen
try, to versuchen
try Versuch
tub Bottich, Wanne
tube Rohr, Röhre, Schlauch, Untergrundbahn
tubular Rohr . . . , rohrförmig
tuck, to falten
tug, to schleppen
tug Schlepper
tune, to (ab)stimmen
tuned abgestimmt
tungsten Wolfram
tuning Abstimmen

turbine Turbine
turbine-generator set Turbogeneratorsatz
turbocharge, to mit Abgasturbolader aufladen
turbocharger Turbolader
turbulence Turbulenz, Wirbelung
turn, to drehen, wenden, kreisen
turn Drehung, Wendung, Windung
turner Dreher
turning Drehen
turnover Umsatz
turnpike Autobahn (Am.)
turnstile Drehkreuz
turret Revolver(kopf)
turret lathe Revolverdrehmaschine
twist, to verdrillen, verwinden
type Bauart, Modell, Typ, Ausführung
typewriter Schreibmaschine
tyre Reifen
tyred luftbereift

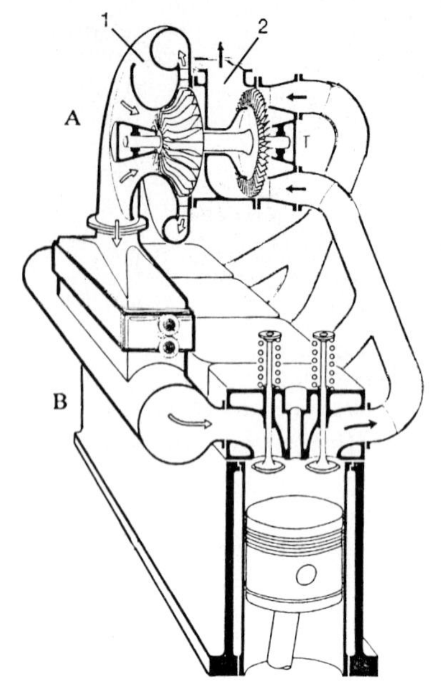

Turbocharger (working principle)
Turbolader (Arbeitsprinzip)

A **turbocharger** Turbolader
B **diesel engine** Dieselmotor
1 **compressor part (supplies compressed air to the combustion chamber)** Kompressorteil (führt komprimierte Luft der Verbrennungskammer zu)
2 **turbine part (drives the compressor by means of the exhaust gases)** Turbinenteil (treibt mit Hilfe der Abgase den Kompressor an)

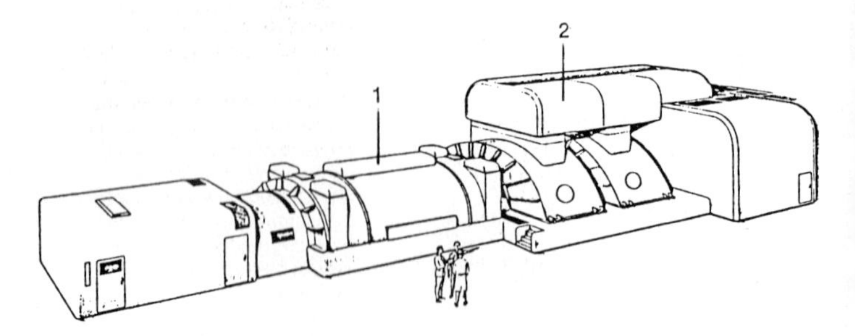

Turbine-generator set
Turbogeneratorsatz

1 **generator** Generator
2 **steam turbine** Dampfturbine

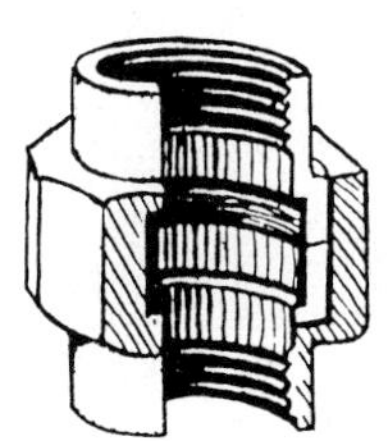

Union socket (pipe union)
Verbindungs(Rohr)muffe

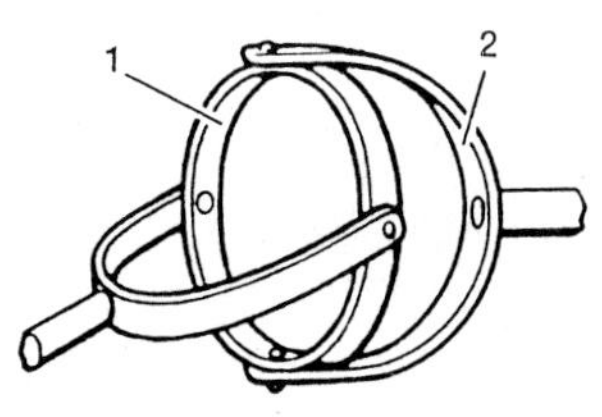

Universal joint
Universalgelenk

1 **ring** Ring
2 **fork** Gabel

U

U-bend Doppelkrümmer, U-Rohr
u.h.f. (ultra high frequency) ultrahohe Frequenz
U-section U-Profil
ultimate äusserst, höchstmöglich
ultrared infrarot, ultrarot
ultrasonic Ultraschall . . .
umbrella Schirm
unattended wartungsfrei, unbewacht
unbalance Unwucht
unclutch, to entkuppeln
undercut, to unterschneiden
undercut Unterschnitt, Untergriff
undercut angle Spanwinkel
underrate, to zu niedrig auslegen
undersize Untermass
undertighten, to zu locker anziehen
undo, to aufmachen, lösen, losschrauben
undue unzulässig, unangemessen
unequal spacing ungleiches Teilen
unfinished unbearbeitet, nicht fertig bearbeitet, nicht geschlichtet
uniform indexing Durchführung gleicher Teilungen
unilateral einseitig
union Anschlussstück
union nut Überwurfmutter
union socket Verbindungsmuffe
unique einzig
unit Einheit, Aggregat
universal joint Kardangelenk
universal shaft Gelenkwelle
unlatch, to ausklinken, lösen
unload, to entladen, löschen
unlock, to lösen, Festklemmung lösen, entriegeln, ausspannen
unmachinable unzerspanbar
unmachined unbearbeitet
unrelieved nicht hinterschliffen, ohne Freiwinkel
unscrew, to abschrauben, lösen
unskilled ungelernt
unstable unbeständig
usable nutzbar
use, to verwenden, benutzen
use Gebrauch, Verwendung, Nützlichkeit
utilise, to ausnutzen, nutzbar machen
utility Versorgungseinrichtung

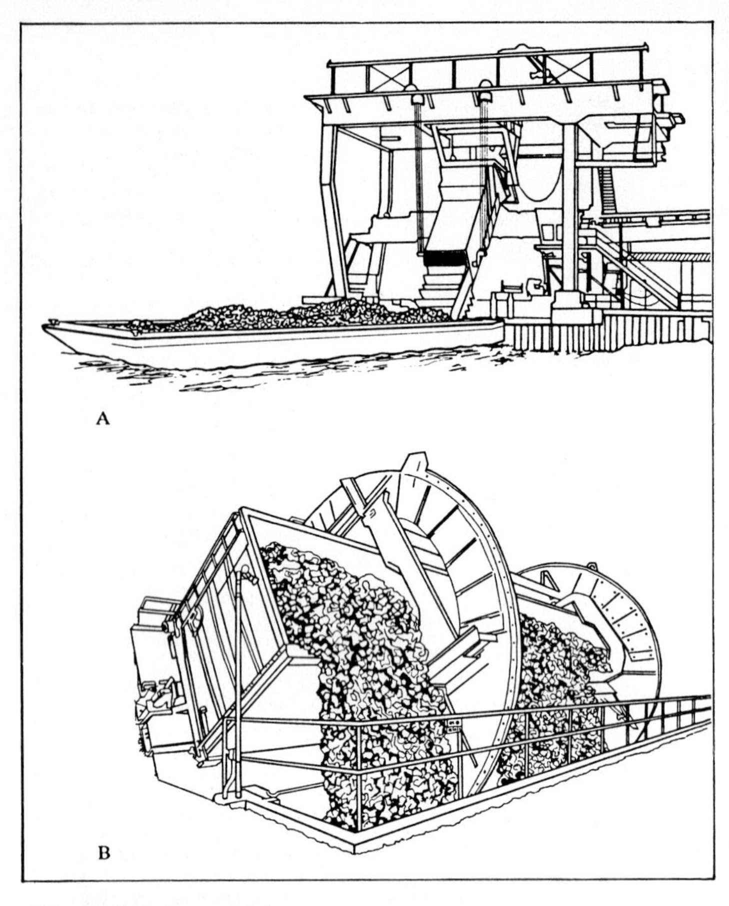

Unloaders for bulk material
Entlade-Einrichtungen für Schüttgüter

A **Bucket-chain elevator unloading a barge** Eimerketten Förderanlage zum Entladen eines Lastkahns

B **Tilting-car dumper** Kippwagen-Entlader

V

V-belt Keilriemen
V-shaped V-förmig
vacancy Leerstelle
vacant unbesetzt, leer, frei
vacuum breaker Vakuumschalter
value, to bewerten, abschätzen
value Wert
valve Ventil, Klappe, Schieber, Röhre
vane Blatt, Flügel, Schaufel
vanish, to verschwinden
vapour Dampf, Dunst
variability Veränderlichkeit
variable veränderlich, variabel
variation Änderung, Abweichung, Ablenkung
variety Vielzahl, Abart, Sorte
varnish, to lackieren, tränken
varnish Lack, Firnis
varnished cambric Isolierband
vary, to variieren, abändern, schwanken
varying component Wechselstromkomponente

Vee-type diesel engine for marine propulsion
V-Dieselmotor für Schiffsantrieb

1 **pistons arranged in vee-form contrary to in-line arrangement** in V-Form angeordnete Zylinder im Gegensatz zur Reihenanordnung

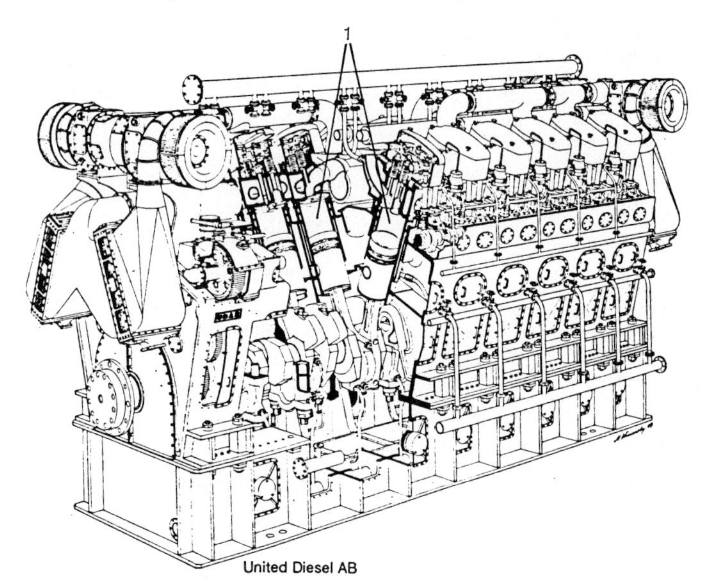

United Diesel AB

vee engine V-Motor
vee V-förmig
vehicle Fahrzeug
veil Schleier
velocity Geschwindigkeit
vent, to entlüften
vent Entlüftungsöffnung, Abzug
ventilate, to ventilieren
verge Rand, Aussenkante, Welle
verification Bestätigung, Beweis, Kontrolle
verify, to nachprüfen
versatile anpassungsfähig, vielseitig einsetzbar
versability Vielseitigkeit, vielseitige Verwendbarkeit
vertex Scheitelpunkt, Spitze, Gipfelpunkt
vertical boring and turning mill Karusselldrehmaschine
vertical milling machine Senkrechtfräsmaschine
vessel Schiff, Behälter, Hohlkörper
vial Ampulle, Fläschchen
vibrate, to vibrieren, schwingen
vibrationless schwingungsfrei
vice Schraubstock, Spanner
vibratory schwingend, Schwing . . .
view Bild, Ansicht (Zeichnen)
violent heftig, stark
viscosity Viskosität
viscous viskos, dickflüssig
visibility Sicht
visible sichtbar
vitreous gläsern, glashart
voice Stimme
void Leere, Hohlraum, Pore
volatile flüchtig
voltage Spannung
voltaic galvanisch
volume Volumen, Lautstärke
vortex Wirbel, Strudel
voucher copy Belegexemplar
vulcanise, to vulkanisieren
vulgar fraction gemeiner Bruch

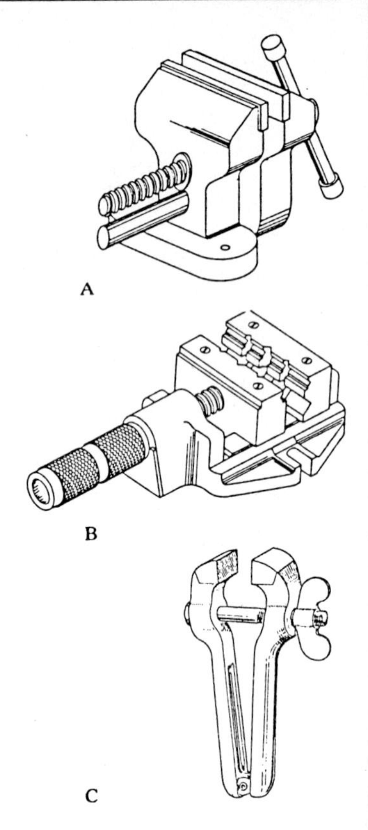

Vice
Schraubstock

A **bench vice** Werkbank-Schraubstock

B **machine vice** Maschinen-Schraubstock

C **hand vice** Handkloben

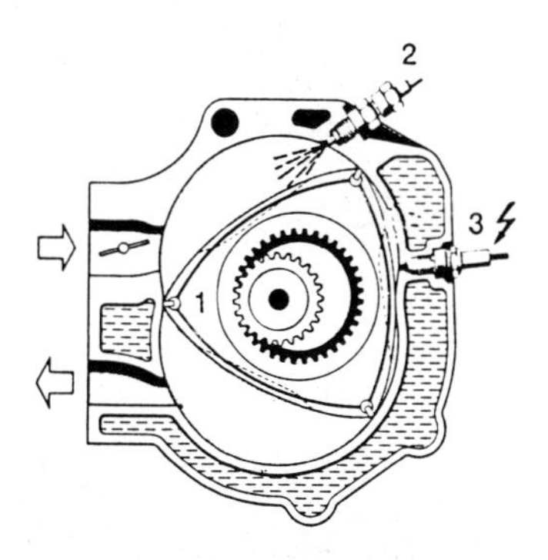

Wankel engine combustion chamber
Verbrennungsraum eines Wankel-Motors

1 **wobble plate** Taumelscheibe
2 **fuel injection nozzle** Kraftstoffeinspritz-düse
3 **ignition plug** Zündkerze

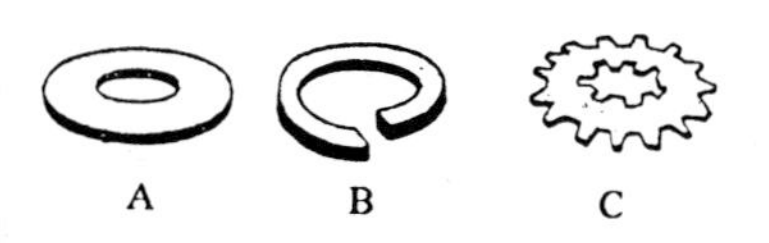

Washer
Unterlegscheibe

A **flat washer** flache Unterlegscheibe
B **lock (spring) washer** Federring
C **toothed wrinkle washer** gezackte und gewellte Unterlegscheibe

wabble, to schwanken, taumeln
wafer Scheibe
wages Arbeitslohn, Heuer
wagon Waggon, Lore
wall Wand(ung), Mauer
Wankel engine Wankel-Motor
warble, to wobbeln
Ward-Leonard set Leonhardsatz
warmed-up betriebswarm, warm-gelaufen
warp, to sich krümmen, verziehen
warning Warnung, Alarm
washer Unterlegscheibe
waste Abfall, Vergeudung
waterproof wasserdicht
watertight wasserdicht
watts input Leistungsaufnahme
wave Welle
waveguide Wellenreiter
weak schwach
wear, to tragen, abnutzen, verschleissesn
wear Abnutzung, Verschleiss
wearless verschleissfrei
web Steg, Schenkel, Versteifung
wedge Keil
weigh, to wiegen, wägen
weight Gewicht
weld, to schweissen
weld Schweissung
welder Schweisser
welding Schweissen
weldless tube nahtloses Rohr
well Bohrung, Brunnen, Schacht
wet nass, feucht
wheel Rad, Ruder
whet, to abziehen, wetzen
whip, to peitschen, schlagen
whirpool combustion chamber Wirbelkammer
whistle tone Pfeifton
whizzer Trockner, Zentrifuge
widen, to verbreitern, aufweiten
width Breite, Dicke, Weite
winch Winde, Kurbel
winding Wicklung, Umwickeln
windlass Ankerspill, Ankerwinde
windscreen Windschutzscheibe
wing Flügel, Tragfläche, Kotflügel
wipe, to (ab)wischen
wire, to verdrahten, Leitung verlegen

Waste incineration plant
Müllverbrennungsanlage
1 **refuse (waste) delivery and storage** Müll-Anlieferung und -Lagerung
2 **boiler house** Kesselhaus
3 **dust collection bags** Staubabscheidesäcke

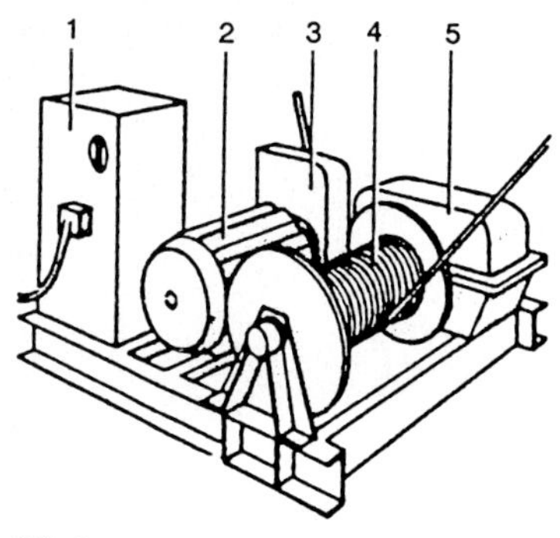
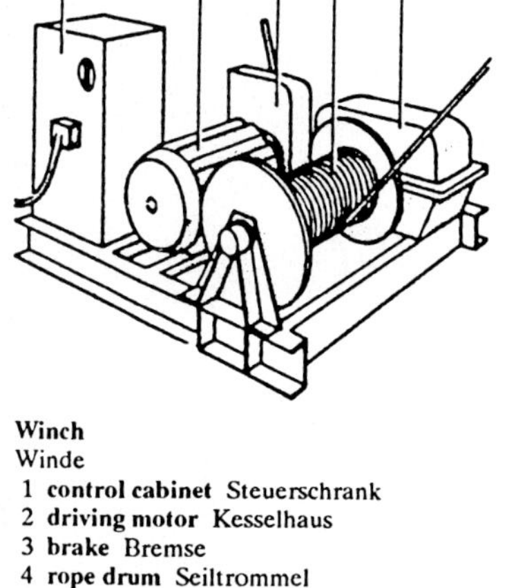

Winch
Winde
1 **control cabinet** Steuerschrank
2 **driving motor** Kesselhaus
3 **brake** Bremse
4 **rope drum** Seiltrommel
5 **gear** Getriebe

wire Draht, Leiter, Ader
wireless drahtlos, Rundfunk . . .
wiring Verdrahtung
withdraw, to zurückziehen
withstand, to widerstehen, standhalten
withstand voltage Stehspannung
witness line Bezugslinie, Masshilfslinie
wobble, to flattern, taumeln
wobble engine Taumelscheibenmotor
work Arbeit, Werkstück
workmanlike fachkundig
workmanship Fachkönnen, Werkstattarbeit
workpiece Werkstück
workshop Werkstatt
worm Schnecke
worn abgenutzt, abgetragen
wrap, to einwickeln, einpacken
wrench Schraubenschlüssel
wrinkle washer federnde, gewellte Unterlegscheibe
wrist Kurbelzapfen
wrought bearbeitet, zugerichtet, geschmiedet
wrought iron Schmiedeeisen

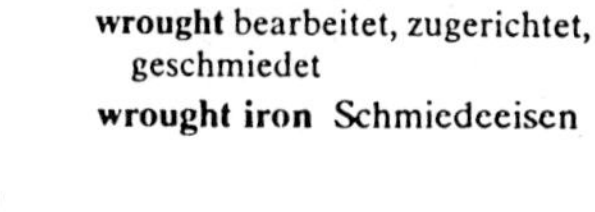
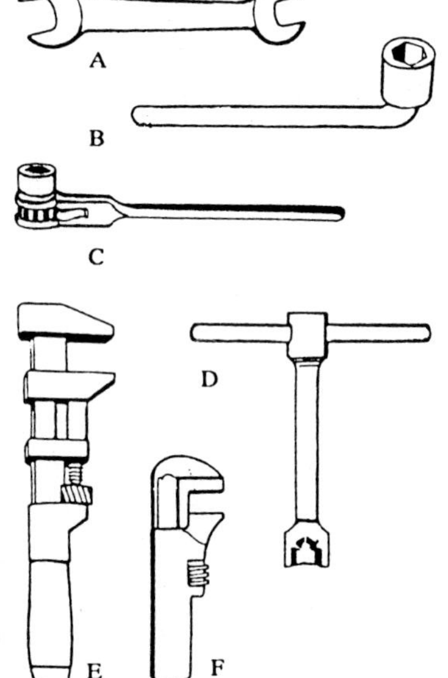

Wrench
Schraubenschlüssel

A **open-mouth wrench** Maulschlüssel
B **offset wrench** abgewinkelter Hohlschlüssel
C **ratchet wrench** Ratschenschlüssel
D **socket wrench** Steckschlüssel
E **monkey wrench** Engländer
F **screw wrench** Franzose

X

x deflection X-Ablenkung
xenon lamp Xenonlampe
X-ray, to röntgen
X-rays Röntgenstrahlen

Y

yard Hof, Lagerplatz
yardstick (Vergleichs-)Maßstab, Yardstock
yield, to erbeuten, ergeben, nachgeben
yield Ausbeute, Ergiebigkeit, Ertrag
Y-joint Gabelmuffe (Kabel)
yoke Joch, Waagebalken
Y-pipe Hosenrohr, Rohrverzweigung
yttrium Yttrium

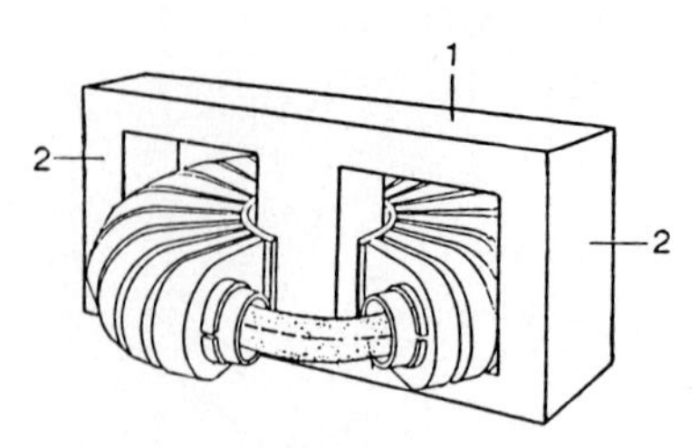

Yoke of a magnet core
Joch eines Magnetkerns

1 **yoke** Joch
2 **limb** Schenkel

Z

Zener diode Zener-Diode
zero Null, Nullpunkt
zero level Nullpegel
zigzag connection Zickzackschaltung
zinc Zink
zincify, to verzinken
zonal zonal, Zonen . . .
zone Bereich, Zone
zooming stetiges Vergrößern/Verkleinern

Technisches Vokabular

Deutsch - Englisch

A

abändern to modify, to alter, to change
Abänderung modification, alteration, change
abätzen to etch, to corrode off
Abbau mining, exploitation, demounting, dismantling
abbauen to mine, to quarry, to exploit, to demount, to remove, to decompose
Abbe-Längenmessgerät Abbe vertical metroscope
abbeizen to pickle
Abbeizmittel pickling (agent), paint (varnish) remover
abbiegen to bend (off), to deflect
Abbiegung bend
Abbild image, picture
abbilden to map, to form an image, to depict
Abbildung mapping, image, depiction, projection
Abbildungsmagnet focusing magnet
abbinden (Kabel) to tie up, to tie off, to lace
abbinden to harden, to set, to cure
abblasen to blow off, to blow down, to quench, to blast
Abblaseventil exhaust valve, blow-out valve, bleeder
Abblasöffnung vent hole
abblättern to flake off, to peel off, to scale off
Abblendeinrichtung anti-dazzle device
abblenden to dim, to stop down, to dip
Abblenden dimming, stopping down, dipping
Abblendfaden dimming filament
Abblendlicht passing light, dimmed light, anti-dazzle light
Abblendschalter dimming switch, dimmer, dipper, anti-dazzle switch
Abbrand melting loss by oxydation, burn-up
abbrechen to break off, to discontinue, to demolish, to pull down
abbremsen to brake, to slow down
Abbremsung braking, slowing-down
abbrennen to burn up, to burn off
Abbrennkontakt arcing tip
Abbrennschweissen flash welding
abbröckeln to crumble, to spall
Abdampf exhaust steam, waste steam
abdampfen to vapourise, to evaporate
Abdampfleitung vent pipe
Abdampfturbine exhaust steam turbine
Abdeckblech cover plate
abdichten to seal off, to pack, to stuff, to make tight
Abdichtmittel sealant
abdrehen to turn off
abdrosseln to load with chokes, to throttle, to stall
Abdrückschraube forcing screw
abfahren to leave, to start, to depart, to set out
Abfahrt departure, leaving, starting, setting-out
Abfall waste, refuse, scrap, garbage; drop, fall, decay
Abfallbehälter waste bin
Abfallbeseitigung waste disposal, refuse disposal
Abfallbeseitigungsanlage disposal plant
abfallen to drop, to fall, to decrease; to slant, to rake
Abfallverwertung waste utilisation
Abfallverzögerung releasing delay
abfangen to support (building)
abfärben to stain, to bleed
abfasen to chamfer, to bevel, to cant
Abfasung chamfer
abfertigen to register, to dispatch
abfliessen to flow off, to leak off
Abfluss drain, discharge, leakage, flowing off
Abflussgeschwindigkeit velocity of flow, leakage rate
Abfragebetrieb direct trunking (tel.)
Abfrageimpuls interrogation pulse (memory)
abfragen to interrogate (memory), to accept a call, to answer

Abfragesender interrogator
abfräsen to mill off, to cut off
abfühlen to scan
Abfuhr removal, delivery, dissipation (heat)
abführen (Wärme) to dissipate (heat)
abfüllen to fill, to bottle
Abgabe delivery, output (power), dissipation (heat))
Abgangsschalter outgoing switch, feeder (switch)
Abgas exhaust gas
Abgasheizung exhaust-operated air heating
Abgasturbine exhaust-gas turbine
Abgasturbogebläse exhaust-turbo-driven-blower, exhaust-type turbo-blower (or: supercharger)
Abgasvorwärmer economiser, waste gas feed heater
abgeben to deliver, to yield, to dispense, to release, to emit, to radiate, to give off
Abgleich balance, balancing, gauging, setting up
abgleichen to adjust, to balance, to set up, to compensate, to level, to make even
Abgleichfehler unbalance, alignment error
Abgleichkondensator balance capacitor, trimmer
abgleiten to glide off, to slip off
abgraten to deburr, to burr, to flash
Abgratmaschine deburring machine, burr removing machine, flash trimmer
abheben to rise, to relieve, to raise, to lift off (tel.)
abheften to file (away)
abhören to intercept, to tap the wires
abisolieren to strip, to skin, to bare (a wire)
abkanten to bevel, to bend
Abkantpresse bending-off press
abklemmen to disconnect, to pinch off
abklingen to fade, to abate, to die out, to be damped out
Abklingkurve decay curve, decay characteristic
abkneifen to pinch off, to nip off
abkratzen to scrape off
abkühlen to cool, to refrigerate, to chill, to quench (steel)
Abkühlmittel cooling agent
Abkürzung abbreviation
abladen to unload, to discharge
ablagern to deposit, to settle, to sedimentate
Ablagerung deposit, sediment, deposition, settlings
Ablauf drainage, run-off, discharge
ablaufen to run down
ableiten to drain off, to bleed off, to conduct, to deduct (math.), to derive
Ableiter arrestor, charge suppressor, charge eliminator
Ablenkbarkeit deviability
Ablenkblech deflector plate
ablenken to deflect, to deviate
Ablenkschaltung electronic sweep circuit
Ablenkspule deflector, sweeping coil, scanning yoke
Ablenkung deflection, deviation, scanning
ablesbar readable
ablesen to read off, to take readings
Ablesung reading
ablösen to relieve the watch (ship), to loosen (mech.)
ablöten to unsolder
Abmessung dimension, measuring size
Abmessungen dimensions, data,
abmontieren to detach, to remove, to demount, to dismantle, to take off
Abnahme acceptance
Abnahme der Spannung fall in voltage, voltage drop
abnahmebereit ready for delivery
Abnahmeprüfung acceptance test
abnehmbar detachable, demountable, removable
abnehmen to detach, to demount, to remove, to take off, to decrease, to drop, to accept (test)
abnutzen (sich) to wear out
Abnutzung wear (and tear), abrasion

Abonnement subscription
Abonnent subscriber
abplatzen to flake, to slab, to peel off, to spall off
abprallen to rebound
Abraum rubbish, trash, waste
abrichten to level, to plane, to dress (grind.)
Abrichter dresser (grind.)
Abrichtmaschine smooth planer
Abrieb abrasion, wear
Abriss summary, extract demolition, pulling down, break (building)
absägen to saw off
absaugen to exhaust, to draw off, to evaporate
Absaugpumpe vacuum pump
abschälen to peel off, to strip off
abschalten to switch off, to cut off, to deenergise, to break
Abschaltleistung breaking capacity
Abschaltstrom breaking current
Abschaltung disconnection, switching-off
abschätzen to evaluate, to estimate
Abschätzung estimation, estimate, evaluation
abscheiden to precipitate, to separate, to exude
Abscheider precipitator, separator, trap, interceptor
Abscheidung, precipitation, deposit, removal
abscheren to shear off
Abscherfestigkeit shearing strength, shearing resistance
abschirmen to shield, to screen off
Abschirmkabel screened cable
Abschluss seal, filling, termination
Abschlusskabel termination cable
Abschlussprüfung final test
Abschmelzdauer fusing time
abschmelzen to fuse, to melt off, to seal (valve)
Abschmelzstrom fuse current, blowing current
abschmieren to lubricate
abschmirgeln to abrade with emery
abschneiden to cut off, to chop (waves)
Abschnitt section, segment, sector, cut-off (piece), intercept (math.)
abschnüren to pinch off
abschrauben to screw off, to unscrew, to loosen
abschrecken to quench (hardening), to chill (casting)
abschreiben to depreciate
Abschreibung depreciation
Abschrift, beglaubigte certified copy
abschroten to cut off with a chisel, to chisel off
abschwächen to attenuate, to reduce, to diminish, to soften
Abschwächer reductor, reducer, attenuator
Abschwächung attenuation, damping, fading, fall, reduction
Abschwächungsglied attenuator
abseihen to strain off, to filter
absenken to sink, to lower, to dip, to bring down
Absenkungskurve depression curve
absetzen to settle, to deposit, to set up
Absetzen settling, settlement, sedimentation
Absetztank settling tank
absichern to protect by fuse, to fuse
absondern to segregate, to isolate, to separate
Absorber absorber
absorbieren to absorb, to attenuate
Absorption absorption, attenuation
Absorptionskältemaschine absorption-type refrigeration machine (or: refrigerator)
abspalten to split off, to crack
Abspannisolator terminal insulator
Abspannkette insulator chain
Abspannklemme anchor clamp
Abspannmast straining tower, straining pole
Abspannseil span rope, aerial support, guy rope
Abspannung rigging, guying, support

absperren to shut off (steam), to seal, to close, to throttle, to isolate
Absperrhahn stopcock, cut-off cock
Absperrkreis antiresonant circuit
Absperrschieber shut-off valve, gate valve
Absperrventil stop valve, blocking valve, shut-off valve
abspulen to reel off, to unspool, to wind off, to uncoil
abspülen to rinse, to cleanse
Abspulmaschine reeler, reeling frame
Abstand distance, spacing, space, clearance
Abstand von Mitte zu Mitte centre-to-centre spacing
Abstandhalter spacer, spacing stay, spacing block, distance piece
Abstandshülse distance tube
Abstandsmesser telemeter, distance meter
Abstandsschelle spacing clamp, spacing clip, spacer
abstechen to cut off, to part off (metal cutting), to tap (furnace)
Abstechmaschine cutting-off machine
Abstechmeissel parting-off tool
abstehend projecting
absteifen to prop up, to brace, to strut, to shore, to stiffen
Absteifung propping, stiffening, bracing, strutting
abstellen to stop, to disconnect, to switch off, to shut down
Abstellgleis railway siding, siding track, shunt line
Abstellhebel stop motion lever
Abstellraum store, storage room
abstemmen to chisel off
Abstich tapping (furnace)
Abstichbühne tapping platform
Abstichofen drossing oven
Abstichöffnung tapping door, tap-hole, metal notch
Abstichpfanne tap ladle
Abstichrinne tapping spout
Abstieg descent
Abstimmanzeige indication of tuning
Abstimmauge magic eye, tuning eye
abstimmbar tunable
Abstimmeinheit tuner
Abstimmempfindlichkeit tuning
abstimmen to tune
Abstimmkreis tuned circuit
Abstimmschärfe selectivity, clearness of tuning
Abstimmspule tuning coil, inductance
Abstimmung tuning
abstossen to repel, to repulse, to push off
Abstossung repulsion
abstrahlen to emit, to radiate
Abstrahlung emission, radiation
abstreifen to wipe off, to sweep off, to strip off
Abtaster scanner, scanning aerial, sampler
Abtastfolge scanning cycle, scanning sequence
Abtastfrequenz sampling rate
Abtastimpuls sampling pulse
Abtastkreis sweep circuit (radar)
Abtastung scanning, sensing, sampling, tracing
abteilen to divide, to partition, to classify
Abteilung department, compartment, division
Abteilungsleiter head of department
abteufen to sink (shaft)
Abteufung sinking
abtrennen to separate, to isolate, to detach, to part off
Abtrennen separation, isolation, parting off (metal working)
Abtrieb driven side, output
Abtriebsrad follower, driven gear
abtropfen to trickle down, to drop off
Abwälzbewegung relative rolling motion between the work and the cutter (hobbing)
Abwälzschleifen generation of grinding
Abwärme waste heat, distilled heat
Abwärmeverwertung waste heat utilisation
Abwärtshub downstroke

Abwärtstransformator step-down transformer
Abwasser waste water, sewage
Abwässer sewage
abwechseln to alternate
abwechselnd alternate
abweichen to deviate, to vary
Abweichung deviation, variation, error, variance
abwerfen to disconnect (load)
Abwerfen der Last disconnection of load
abwickeln to unwind, to uncoil
Abwicklung eines Schaufelrades diagram of a blade wheel
Abwicklungskurve evolvent, involute
abwischen to wipe off, to dust
abwracken to scrap, to break up, to dismantle
abwürgen to stall (engine)
abzählen to denumerate
Abzählung denumeration
abzapfen to draw off, to bleed off, to drain, to tap
Abzapfhahn drawing-off cock
Abzapfventil bleeder
abzeichnen to copy
abziehen to take off (material), to hone (tool), to sharpen, to level, to screed, to slag off
Abziehfeder pull-off spring
Abziehstein oilstone, whetstone, hone
Abziehvorrichtung puller, extractor
Abzug proof (printing), copy, print
Abzweig branch, tap, subcircuit
Abzweigdose junction box, distribution box
abzweigen to branch, to shunt, to tap
Abzweigkasten distribution box, joint box
Abzweigklemme tee-joint, branch terminal
Abzweigleitung branch line, branch conductor, shunt line (hydraulics)
Abzweigmuffe tee-joint, joint box
Abzweigstromkreis branch circuit, subcircuit
Achsantrieb axle drive
Achsbolzen front-axle pivot pin
Achsbolzenbuchse king-pin bush
Achsdruck axle load
Achse axle, axis (math.)
Achsengelenk pin joint
Achsenkreuz system of coordinates
Achsstand axle base
Achssturz axle dip
Achteck octagon
achteckig octagonal
achtfach eightfold, octuple
AD-Umsetzer analog-to-digital converter
Adapter adapter
Adapterkabel extension cable, extension lead
Addierimpuls add pulse
Addierschaltung adding circuit
Addierstufe counter stage
Addierwerk adder
Additionszeichen plus sign
Ader core, wire, conductor
Aderbruch cable fault
Adhäsion adhesion
Admittanz admittance
Adresse address
Adressenwerk address unit
Adsorptionsmittel adsorbent
Aerosol aerosol
Aggregat set (engineering), compound machinery
Aggregatzustand state of aggregation
Akkumulator accumulator, storage battery
Akrylfaser acrylic fibre
Akte file, record
Aktiengesellschaft joint stock company
Aktionsrad impulse wheel (turbine)
Aktionsturbine impulse turbine, action turbine
aktivieren to activate, to sensitise
Aktivruder active rudder
Akustik acoustics
akustisch acoustic(al)
Akzeptor acceptor
Alarmglocke alarm bell
Alkohol alcohol
Alleinerfinder sole inventor
Allesförderer general-purpose elevator
Alleskleber all-purpose adhesive
Allpassfilter all-pass filter
allseitig polydimensional
Allstromgerät all-mains set, ac-dc set

Allwellenantenne multi-band aerial
Allzweckgerät general-purpose tool
Alphabetlochprüfer alphabetical verifier
Alteisen iron scrap, scrap iron
altern to age, to season
Alterung ag(e)ing, seasoning
Altöl used oil, waste oil
Aluminium aluminium, aluminum (USA)
Aluminiumwalzwerk aluminium rolling mill
Amateurfunker amateur operator
Amboss anvil
Ammoniak ammonia
amorph amorphous
Amperemeter ammeter, ampere meter
Amperewindung ampere-turn
Amplidyne amplidyne (generator)
Amplitude amplitude
Amplitudenanstieg amplitude increase
Amplitudenbegrenzer peak limiter, clipper
Amplitudenentzerrung correction of amplitudes
Amplitudenhub amplitude swing
Ampulle ampoule, vial
Amt office, station, exchange
Amtswähler junction selector
Amtszeichen dialling tone
analog analogous, analog
Analog-Digital-Konverter (oder: Wandler) analog-digital converter
Analogie analogy
Analysator analyser
Analyse analysis
Analysenwaage chemical balance
analysieren to analyse
Analytiker analyst, analytical chemist
anbauen to attach, to extend, to fit
Anbohren spot-drilling, starting the hole
Anbohrer spotting drill
Anbohrkopf trepanning head
Anbohrmaschine centring lathe
anbringen to fit, to fix, to attach, to mount
andauern to continue
ändern to change, to alter to modify, to transform
Änderung change, alteration, variation, modification, transformation
andrehen to crank, to start an engine, to switch on
Andrehkurbel crank, crank handle, starting crank
Andrehmotor pony motor
andrücken an to press on, to force against
Andruckmagnet pressure solenoid
aneinanderhaften to cling together
Anerkennung recognition, certification acknowledgement, acceptance
anfahren to start up (engine, plant)
Anfahrleistung starting power
Anfall yield, output
Anfang start, beginning, initiation
anfangen to start, to begin, to initiate
anfasen to chamfer, to debur
anfetten to grease
anfeuchten to moisten, to damp
Anfeuern eines Kessels firing-up of a boiler
anflanschen to flange
anfordern to require, to claim, to ask for, to order
anfressen to pit, to corrode
Anfressung corrosion, erosion
Anfuhr delivery, supply
Angabe information, indication, specification
Angaben data, specification
angeben to state, to indicate, to specify, to rate (capacity)
Angebot und Nachfrage supply and demand
angegossen cast-on, cast integral
angemeldet filed, pending (Patent)
Angestellter employee
angreifen to attack, to affect, to pit, to corrode
Angriffspunkt einer Kraft point of application of a force
Anguss runner
anhaken to hook on, to mark off, to check
anhalten to stop, to stay, to bring, to rest

anhaltend continuous
Anhang appendix
anhängen to affix, to annex, to join, to trail (Fahrzeug)
Anhänger trailer, tag (Schild, Zettel)
anhäufen, sich to pile up, to accumulate
Anhäufung accumulation, cluster, conglomeration
anheben to elevate, to lift, to raise, to hoist, to move out
Anhebeschlitten elevating slide (Räumwerkzeug)
anheften to pin to
anheizen to warm up, to heat up, to set fire
Anheizen warm-up, heat-up
Anker anchor (Schiff), armature (el. Maschine)
Ankaufspreis prime cost
Ankerblech armature lamination
Ankerbolzen anchor bolt, fishtail bolt
Ankerrückwirkung armature reaction
Ankerwicklung armature winding
Ankerwinde windlass, gipsy
anklammern to fasten with cramps
ankleben to glue to, to paste on, to agglutinate
anklemmen to clamp, to secure under terminals
anknipsen to switch on,to snap on
ankörnen to punch, to centre punch
Ankörnen punching
ankreuzen to check, to mark
Anlage plant, installation, equipment
anlassen to start, to start up, to crank, to temper (Stahl)
Anlasser starter
Anlassermotor pony motor
Anlassfarbe temper colour
Anlasshärtung artificial aging
anlegen (Spannung) to apply (a voltage)
Anlegewandler split-core type current transformer
anleimen to glue on
anleiten to instruct, to guide
Anleitung instruction
anliegend adjacent
anlöten to solder to
anmelden (Patent) to apply for (a patent)
Annäherung approximation
Annäherungsverfahren approximation method
Annahme supposition, assumption, acceptance
annehmen to assume, to accept
annieten to fix by riveting
annullieren to cancel
Anodenanschluss anode lead
anordnen to arrange, to lay out, to specify, to dispose
Anordnung arrangement, lay-out, array, configuration, assembly
anpassen, to fit, to adjust, to match, to adapt
Anpassung adaption, matching
anpassungsfähig flexible, adaptable
Anpassungskreis matching circuit
anpeilen to bear, to take a bearing
Anpeilung bearing, direction finding
anprallen to collide, to impinge
anregen to excite, to energise, to agitate
Anregung excitation, agitation, stimulation
anreichern to enrich, to concentrate
Anreicherung enrichment, concentration
Anreicherungsanlage enrichment plant
Anreissen marking, layout, scribing
Anreissplatte marking-out table, marking-off table, playing-out plate
Anreisswerkzeug marking-off tool, layout tool
Anruf call, telephone call
Anruf im Fernverkehr trunk call
anrufen to call, to call up, to ring
Anrufer calling party
ansammeln to accumulate
Ansammlung accumulation
ansaugen to draw in, to prime
Ansaugen suction, drawing-in
Ansaugventil intake valve
Anschaffungskosten prime cost, first cost, initial cost
Anschlag stop, trip dog, dog
anschlagbetätigt dog-actuated

anschliessen to join, to connect
Anschluss connection, contact
anschneiden to enter the cut, to begin the cut
anschnüren to tie up
anschrauben to screw on, to bolt to
anschweissen to weld to
anseilen to rope
ansenken to spot-face
Ansenken spot-facing
Ansicht view, elevation
anspannen to tighten, to stress
Ansprechbereich range of sensitivity, sensitive region
Ansprechempfindlichkeit response sensitivity
ansprechen to react, to respond, to pick up
Ansprechen response, actuation, picking-up
anstechen to open the furnace, to tap
ansteigen to rise, to rake, to slant, to ascend
Anstieg rise, slope
anstossen to butt, to abut
anstrahlen to floodlight
anstreichen to paint, to coat with, to brush
Anstrich coating of paint, paint coat
Anteil portion, share
Antenne aerial, antenna, array
Anthrazit anthracite, hard coal
Antiklopfmittel antiknock component
Antrag application
antreiben to drive
Antrieb drive, propulsion
Antriebsanlage propelling plant
Antwortbake responder beacon, transponder
Antwortsender responder
anwachsen to increase, to rise, to grow to
anwärmen to heat up, to preheat
anwenden to apply
Anwendung application
Anwendungsgebiet field of application, field of use
anwerfen to crank
Anzapfdampf bled steam
Anzapfdampfmaschine bleeder-tape steam engine
anzapfen to tap
Anzapfung tap, tapping
Anzapfungsumschalter tap changer
anzeichnen to mark
Anzeige display, indication, reading
Anzeigegerät indicator, display unit
Anzeigelampe indicating light, tell-tale light, pilot bulb
anzeigen to indicate, to display, to read
Anzeiger indicator
Anzeigetafel indicator panel
anziehen to attract, to draw in, to tighten, to draw home
Anziehung attraction
Anziehungskraft force of attraction
anzünden to light, to ignite, to kindle
Apparat apparatus, device, appliance, unit, equipment
äquivalent equivalent
Arbeit work, labour, job, task
arbeiten to work, to operate
Arbeitsablauf cycle, machining cycle, sequence of operation
Arbeitsaufwand expenditure of labour, amount of work, energy expended
Arbeitsbedarf amount of work required
Arbeitsersparnis labour saving
Arbeitsfolge operation sequence, sequence of operations
Arbeitsgang operation, cycle, procedure, process of manufacture
Archiv record office
Arithmetik arithmetic
Arm arm, spider, lever, support, bearer
Armatur instrument, element, armature, fitting
armieren to reinforce, to armour, to sheathe
Armierung armouring, sheathing, reinforcement, braid
arretieren to arrest, to block
Arretierung detent, catch
Art kind, species, sort, variety, class, pattern
Artikel article, item, material
Asbest asbestos

Asche ash, cinders
Assistent assistant
Astronautik astronautics
Asymmetrie asymmetry
asynchron asynchronous
Asynchronmotor asynchronous motor
Atemschutzgerät respiratory protection apparatus
Atemschutzmaske respirator
Äther ether
Äthyl ethyl
Atmosphäre atmosphere
Atomkraftwerk nuclear power station
Atomphysik atomic physics
Ätzkali caustic alkali
ätzen to etch
Ätzen etching
Ätzmittel etchant, etch, corroding agent, caustic
ätzpolieren to attack-polish
Ätzung engraving, etching, corrosion
Audionverstärker amplifying detector
aufarbeiten to repair, to rework, to rebuild, to use up
Aufarbeitung reprocessing, regeneration, treatment, treating, reconditioning, rebuilding
Aufbau erection, building, assembly, arrangement, construction, structure, establishment
aufbauen to build up, to erect, to construct, to assemble, to rig up, to compose
Aufbauinstrument surface mounted instrument
aufbereiten to rework, to reclaim, to recover, to treat, to prepare, to refine, to upgrade
Aufbereitungsanlage preparation plant, cleaning plant
Aufbereitungsgut product of dressing
aufbewahren to stock, to store, to keep, to preserve
aufblasbar inflatable
aufblasen to inflate, to blow up
aufblitzen to flash up
aufbocken to jack up
aufbohren to bore, to enlarge a hole
aufbrechen to force open, to break open
aufbuchsen to bush on
aufdecken to uncover
aufdornen to put on a mandrel, to enlarge, to expand, to pierce
aufdrehen to screw open
aufdrücken (Spannung) to impress a voltage
Aufeinanderfolge succession
aufeinanderfolgend consecutive, successive
Auffangelektrode collecting electrode, collecting plate
auffangen to trap, to collect
Auffänger target, trap, catcher
auffrischen to regenerate, to revivify
auffüllen to refill, to replenish, to fill, to recharge
Aufgabe task, function, purpose
aufgebaut built-on
aufgeben to charge, to forward, to feed, to discontinue, to give up, to abandon
aufgehängt suspended
aufgekeilt auf keyed to, feathered to
aufgliedern to classify, to break down, to subdivide
Aufgliederung classification, breakdown, subdivision
aufhängen to suspend
Aufhängung suspension, linkage
aufheizen to heat up
aufhellen to lighten, to tin, to brighten, to clarify
Aufhellung lightening, tinting, brightening, clarification
aufhören to stop, to end
aufkeilen to key to
aufklappbar folding, hinged
aufklappen to fold upwards, to turn up
aufkleben to glue on, to paste on
Aufladedruck supercharge pressure
Aufladegebläse supercharger, booster
aufladen to load, to charge, to supercharge, to boost
Aufladung charging, supercharging, boosting, loading

Auflage support, rest, bearing surface, seating
Auflager support, bearing
Auflagerbock bearing pad
auflagern auf to bear on, to be supported by
Auflaufbremse loading brake, overrunning brake
aufleuchten to light up, to flash, to be illuminated
auflösbar soluble, resolvable
auflösen to resolve, to solve, to dissolve, to decompose
Auflösung resolution, solution, dissolution, decomposition
Auflösungsmittel solvent
Auflösungsvermögen resolution power, resolving power
auflöten to solder on, to brace on
Aufnahme absorption, taking-up, admission, input
aufnehmen to absorp, to take up, to admit, to put in
Aufprall impingement, impact, bound
aufpumpen to inflate
Aufputzmontage surface installation, surface mounting
Aufputzschalter switch for surface mounting, surface switch
Aufputzsteckdose surface socket
aufrauhen to rough up, to roughen
Aufräumarbeit clearance operation
aufräumen to clear up, to tidy up
Aufräumung clearance, clearing
aufrecht upright
aufrechterhalten to maintain
aufreiben to ream
aufreissen to scribe, to plot, to lay out
aufrichten to erect, to right
Aufriss front view
aufsammeln to pick up
aufsaugen to soak, to absorb
Aufsaugen absorption
aufschaukeln to amplify, to build up
aufschichten to stack, to pile up
aufschieben to delay, to postpone, to push open
aufschliessen to unlock
aufschrauben to bolt on, to screw on
aufschreiben to write down, to record
aufschrumpfen to shrink on
Aufschrumpfen shrinking-on
aufschweissen to weld on, to open with a welding torch
aufschwimmen to float
Aufseher supervisor
aufsetzen to put on, to attach
Aufsicht supervision, control
aufsintern to sinter-fuse
aufspalten to split (up), to break up, to crack, to cleave
Aufspaltung splitting (up), fission, cleaving, dissociation
aufspannen to chuck, to clamp, to mount
Aufspannfutter chuck
Aufspanntisch clamping table, worktable, work-holding table
Aufspannung clamping, setup
Aufspannvorrichtung clamping fixture, clamping device, work holding fixture
Aufsteckdorn stub arbor
aufstecken to mount on, to slip over
aufsteigen to rise, to ascend, to take to the air
aufstellen to erect, to assemble, to allocate, to mount, to stack
Aufstellung assembly, installation erection, mounting position
Aufstieg rise, ascent, climb
Auftankung refuelling
Auftastschaltung gate circuit
Auftrag order
Auftragnehmer contractor
Auftragsschweissen deposit welding, hard surfacing
Auftragsforschung sponsored research
Auftreffelektrode target electrode
auftreffen to strike, to impinge, to encounter
Auftrieb buoyancy, lifting power, upward pressure
Aufwand expenditure, expense, power input
aufwärtsbewegen to ascend, to move upwards
Aufweitung bell mouthing
aufwickeln to wind up, to reel up

Aufzeichnung record, recording, plot, plotting
Aufzug elevator, lift, hoist
Auge eye, lug
ausarbeiten to elaborate, to prepare, to work out
ausbessern to repair, to mend
ausbeulen to bulge, to buckle
Ausbeute yield, output, produce, efficiency
ausbeuten to exploit, to mine
Ausbeutung exploitation, mining
ausbilden to train, to instruct
Ausbildung training, education
Ausblasventil blow-off valve
Ausbohrmaschine boring machine
ausbrechen to break out, to chip, to pinch
Ausbreitung propagation, expansion
ausbrennen to burn out
ausbuchten to bulge, to buckle
ausdehnen to expand, to stretch, to extend, to dilate
Ausdehnung expansion, extension, stretch, dilatation
auseinandernehmen to dismantle, to take apart
ausfällen to precipitate
Ausfallrate failure rate
Ausfällung precipitation, deposit
ausfiltern (Frequenz) to reject (trap out) a frequency
ausfluchten to align
Ausfluchten alignment
Ausfluss discharge, outflow, outlet
ausführen to perform, to accomplish, to construct
Ausführung construction, design
ausgleichen to compensate, to equalise
to level, to even out, to smooth, to balance
Ausgleichskreis balancing circuit
ausglühen to anneal
aushalten to withstand, to resist, to stand
Ausheber lifter
aushöhlen to hollow out
auskehlen to fillet, to groove
auskuppeln to declutch, to disengage the clutch
auslaugen to leach, to extract, to macerate
Ausleger boom, cantilever, jib, outrigger, arm
auslenken to sweep, to deflect, to displace
auslesen to read-out, to classify, to sort out, to separate, to grade
Auslesespeicher read-out memory, read-only memory
ausliefern to deliver, to supply
Auslöser trigger, detent, tripping device
Auslösung trip-out, release
ausloten to fathom, to sound
Ausmass extent
ausmauern to line
ausnutzen to utilise, to exploit
auspacken to unpack
Auspuff exhaust, escape, eduction
auspuffen to exhaust, to escape
Auspuffkrümmer exhaust manifold, exhaust stack
Auspuffrohr exhaust pipe
Auspufftopf exhaust silencer, exhaust muffler
auspumpen to pump out, to suck out, to evacuate
ausrechnen to calculate, to compute, to figure out
ausrichten to align, to line up, to straighten
Ausrichten alignment, lining-up, straightening
ausrücken to disengage, to disconnect, to declutch
ausrunden to radius, to fillet
Ausrunden radiusing, filleting
ausrüsten to equip, to fit, to furnish
Ausrüstung equipment, outfit
ausschaben to scrape
ausschachten to excavate
Ausschachter navvy
Ausschachtung excavation, pit
Ausschaltdauer break-time, breaking period
ausschalten to switch off, to cut off, to disconnect, to de-energise
Ausschalter cut-out, disconnecting switch
Ausschaltleistung breaking capacity
Ausschaltlichtbogen break arc

Ausschaltspitzenstrom cut-off current (fuse)
Ausschaltstrom breaking current
Ausschaltung circuit-breaking, disconnection
Ausschaltvermögen breaking capacity
ausscheiden to precipitate
Ausscheidung precipitation, deposit
ausschieben to exhaust (combustion engine)
ausschlachten to cannibalise
ausschlacken to draw the slags, to clear from dross
ausschlagen to deviate, to deflect
ausschleifen to grind internally
ausschmieden to draw out, to hammer
ausschneiden to cut out
ausschreiben to invite tenders
Ausschuss scrap, scrap parts, rejects
ausschussfrei scrapless
ausschütten to pour out, to dump (excavator)
ausschwenkbar swing-out
Ausschwingen decay, dying out, dying away
Ausschwingstrom decay current
Aussehen appearance
ausseigern to liquate, to segregate
Aussenabmessung external dimension
Aussenbackenbremse outside shoe brake
Aussenbelag outer coating
Aussenbeleuchtung outdoor lighting
aussenden to emit, to transmit
Aussendurchmesser outer diameter
Aussengewinde external thread
Aussenhaut hull, outside planking
Aussenkabel outside cable, external cable
Aussenkante outside edge
aussenken to counterbore
Aussenken counterboring
Aussenlinie contour
Aussenmantel outer surface, covering (cable)
Aussenpolgenerator overhung-type alternator (or: generator)
Aussenrundschleifen externally grinding cylindrical parts
Aussenschale outer shell
Aussentaster outside caliper(s)
Aussentemperatur outside temperature, outdoor temperature, ambient temperature
Aussentransformator free-air transformer, outdoor transformer
Aussenwand exterior wall, external wall
Aussenwiderstand load resistance
Ausserbetriebsetzen putting out of operation
aussermittig off-centre, eccentric
Aussertrittfallmoment pull-out torque
Aussetzbetrieb intermittent operation, periodic duty
aussetzen to interrupt, to fail
Aussetzen interruption, failure
aussetzend intermittent, interrupted
aussieben to screen, to filter out
aussondern to segregate, to separate
aussortieren to sort (out), to eliminate
ausspannen to unclamp (work)
Ausspannen unclamping
aussparen to block out, to leave a recess, to recess
Aussparung recess, pocket, block-out
ausspülen to rinse, to wash out
ausstatten to furnish, to equip, to fit out
Ausstattung equipment, outfit
«Aus»-Stellung off-position
Ausstellung exhibition, exposition
ausstemmen to chisel out
aussteuern to control, to modulate
Aussteuerung drive, driving, control
ausstopfen to stuff
ausstossen to turn out, to produce, to manufacture, to expel, to eject
ausströmen to flow out, to escape
Ausströmung outflow, effusion, exhaust
austasten to blank out, to gate
Austastung blanking, gating
Austastzeit gate time
Austausch exchange

austauschbar interchangeable, exchangeable
austauschen to exchange, to substitute
Austauscher exchanger
Austritt exit, outcome, outflow, leakage
Austrittskammer exhaust chamber
austrocknen to dry out, to dehumidify
Austrocknung drying out, dehumidifying
ausüben to exercise, to practise
Auswahl selection, choice, sampling
auswählen to select, to choose, to sample
auswechselbar interchangeable, exchangeable, replaceable
auswechseln to interchange, to replace, to substitute
ausweiten to expand, to extend, to widen
auswerfen to eject
Auswerfer ejector
auswerten to evaluate, to analyse
Auswerter analyst
Auswertung evaluation
auswuchten to balance
Auswuchtung balancing
auszeichnen to trace, to mark out, to display
ausziehbar extractable, telescopic
ausziehen to pull out, to draw out, to extract
autogen autogenous
Automatenstahl free cutting steel
automatisieren to automate
Automatisierung automation
Axialbelastung axial load
Axiallager journal bearing, end-thrust bearing
Axiallüfter axial fan
Axialspiel end play, end clearance
Axialturbine axial-flow turbine
Azetylen acetylene

B

Backbord port side, port
Bagger dredger
baggern to dredge, to excavate
Bahn path, track, trajectory
Bajonettfassung bayonet holder
Bajonettsockel bayonet cap, bayonet base
Bajonettverschluss bayonet joint, bayonet fixing
Bake beacon
Bakelit bakelite
Balg bellows
Balken beam, girder
Ballastwiderstand ballast resistance, loading resistance
Ballen bale, bundle
ballig crowned, cambered
Balligdrehen spherical turning
Ballon balloon
Ballung packing, package, bunching
Bananenbuchse banana jack
Bananendoppelstecker U-plug
Band tape, belt (conveyor)
Bandabtastung tape sensing
Bandage bandage, dressing
Bandaufnahme tape recording, tape record
Bandbreite bandwidth
Bandfilter band-pass filter
Bandflechtmaschine braid-plaiting machine
Bandförderer belt conveyor, band conveyor
Bandgenerator belt-type generator
bandgesteuert tape-controlled
Bandleiter strip conductor
Bandlesegerät tape reading unit
Bandsäge band saw
Bandschleifen abrasive-belt grinding
Bandspeicher tape store
Bandstahl strip steel, hoop steel
Bandstahlbewehrung tape steel armouring
Bandsteuersystem type control system
Bandwalzwerk strip rolling mill
Bank bench
Barkhausen-Sprung Barkhausen jump
Barren bar, ingot
Barriere barrier
Base base
basenbildend basifying, base-forming
Basis basis, base
Basisschaltung ground-base circuit
Basisschicht base layer
Basisstrom base current

Batterie battery
Batterieanschlussklemme battery connecting terminal
batteriebetrieben battery-powered, battery-operated
Batterieempfänger battery-operated receiver
Bau building, construction, structure
Bauabschnitt stage of construction
Bauart design, type of design
Bauaufsicht supervision of works
bauchig barrel-shaped
Baudichte packing density
Bauelement component
bauen to build, to construct
Bauform design
Bauingenieur civil engineer
Baukastenprinzip building-block principle, modular concept, unit principle
Baumantenne fishbone aerial
Bauplatz site, building site
Baustahl mild steel
Bauteil element, basic unit, structural unit
Bauvorschrift building regulations, specification(s)
Bauweise design, system of construction
Bauxit bauxite
Beachtung observance, consideration
beanspruchen to load, to stress, to impose stress on
Beanstandung rejection, complaint
bearbeitbar workable, machinable
bearbeiten to work, to machine, to process, to treat
Bearbeitung machining, working, processing, tooling
Bearbeitungsfolge sequence of operations
Bearbeitungsgenauigkeit machining accuracy
Bearbeitungsplan operation sheet
beaufschlagen to admit
Beaufschlagung admission (turbine)
beben to tremble, to vibrate
Beben trembling, vibration
Becher bucket, cup
Becherförderer bucket conveyor, bowl feeder
Becherkondensator encased capacitor, box-type capacitor
Becherwerk bucket conveyor, bucket elevator
Becken pool, basin, reservoir, tank
Becquerel-Effekt photovoltaic effect, Becquerel effect
bedachen to roof
Bedarf want, need, demand, requirement
bedecken to coat, to cover
bedienen to operate, to attend, to service
Bedienung operation, attendance
Bedienungsanleitung operating instructions, working instructions
Bedienungsbrücke operation area
Bedienungseinrichtung control device
Bedienungselement control element
Bedienungselemente controls
Bedienungsfehler error in operation
Bedienungsgriff operating handle, master handle
Bedienungshandbuch operator's manual
Bedienungshebel control lever, operating lever
Bedienungsmann operator
Bedienungspult control console, control desk
Bedienungsvorschrift working instructions
bedrahten to wire
Bedrahtung wiring
beeinflussen to influence, to affect
Beeinflussung influence
beeinträchtigen to impair, to interfere with
befähigen to enable
Befähigung qualification, capability
Befehl instruction, order, command
Befehlsauslösung command action
Befehlsbibliothek library
Befehlsfolge control sequence
Befehlsimpuls command pulse
Befehlsschema flow chart
befestigen to attach, to fasten, to fix, to secure, to apply
Befestigung attachment, mounting
befeuchten to dampen, to moisten, to humidify

Befeuchter humidifier
befeuern to light, to beacon
Befeuerung lights, lighting, beacons
beflechten to braid
Beflechtung braiding
befördern to convey, to haul, to transport, to ship, to handle
Beförderung transport, conveying, haulage, shipment, handling
Beförderungsmittel transport facilities, means of transport
befugt entitled
begiessen to sprinkle, to water, to moisten
Beginn beginning, commencement, start, initiation
beginnen to start, to begin, to commence, to initiate
Begleitbrief covering letter
Begleitmetall foreign metal, impurity
begradigen to straighten out
begrenzen to limit, to terminate, to clip
Begrenzer limiter, clipper
Begrenzerdiode limiter diode
Begrenzerkreis clipping circuit, clipper
begrenzt durch bounded by
Begrenzung limitation, limit, clipping
Begriff term, notion, concept
Behälter container, bin, reservoir, tank, case, receiver
behandeln to treat, to process, to handle
Behandlung treatment, handling, processing, working, manipulation
beharren to persist, to remain,
Beharrungskraft inertia
beheizen to heat, to fire
Beheizung heating
beidseitig on both sides, both-ways
Beilegescheibe packing washer, shim
Beiluft admixed air
beimengen to add, to admix
Beimengung addition, additive, admixture
beiordnen to coordinate, to assign
Beiwert coefficient, factor
Beizanlage pickling plant
beizen to pickle
beladen to load, to charge, to burden
Beladestelle loading station
Belag lining, coating, plating, surface
belasten to load, to stress
Belastung loading, load
belaufen auf, sich to amount to
Beleg voucher, document
belegen to occupy, to cover
beleuchten to light, to illuminate
Beleuchtung lighting, illumination
Beleuchtungskörper lighting fixture
Beleuchtungsstärke intensity of illumination, illuminance
bemängeln to criticise, to find fault with
bemassen to dimension
Bemassung dimensioning
bemessen to dimension, to determine safe dimensions
benachbart adjacent, neighbouring
benennen to term, to denominate, to designate
benetzen to wet, to moisten
benutzen to use, to utilise, to employ
Benutzer user
Benutzung use, utilisation
Benzin petrol, gas, gasoline, benzine
Benzinmotor petrol engine
beobachten to watch, to observe, to view
Beobachtung observation, viewing
berechnen to compute, to calculate
Berechnung computation, calculation
Berechnungsfehler mistake (error) in computation
Bereich area, region, field, range
Bereitschaft standby, readiness
bereitstellen to place at disposal, to make available
bergen to salvage
Bergung salvage
Bericht report, information, message, notice
berichtigen to correct, to rectify, to adjust, to set right
Berichtigung correction, rectification, adjustment

berieseln to sprinkle, to spray, to wet
Berieselung sprinkling, watering
Berieselungsanlage sprinkler system
bersten to burst, to rupture
berücksichtigen to consider, to take into consideration
Berücksichtigung consideration
Beruf occupation, trade, craft, profession
Berufsschule trade school, apprentice school
beruhigen to stabilise, to steady, to smooth, to quieten
Beruhigung smoothing, damping, quietening
Berührung contact, touch, tangency
berührungslos non-contacting
berührungssicher shock-proof
Beryllium beryllium
beschaffen to supply, to provide
beschäftigen to employ, to engage
Bescheid, amtlicher official notice
beschichten to coat
Beschichtung coating
beschicken to load, to feed, to charge, to stoke
Beschickung loading, feeding, charging, stoking
beschiessen to bombard
beschleunigen to accelerate, to speed up
Beschleuniger accelerator
Beschleunigung acceleration, speeding up
beschreiben to describe, to specify
beschreibend descriptive
Beschreibung specification
beschriften to letter
Beschriftung lettering
beseitigen to dispose of, to remove, to eliminate
Beseitigung disposal, removal, elimination
besetzt occupied, filled, busy (tel.)
Besetztanzeige visual busy signal
Besetztsignal blocking signal
bespritzen to sprinkle
bespulen to load with coils
bespult coil loaded
Bessemerbirne (acid) Bessemer converter
Bessemerstahl (acid) converter steel
Bestand stock
beständig stable, continuous
Beständigkeit stability, continuity, durability, resistance
Bestandaufnahme stock taking
Bestandteil component
Bestandteile accessories, components
Bestätigung eines Signals acknowledgement of a signal
bestehen aus to consist of
bestellen to order
Bestellformular order form
Bestellung order
bestimmen to determine to analyse
Bestimmung determination,
Bestimmungen, gesetzliche statutory requirements
Bestimmungsort destination
bestrahlen to irradiate, to expose to rays, to bombard
Bestrahlung irradiation, exposure to rays, radiation treatment
Bestrahlungsanlage irradiation plant
Bestrahlungsapertur beam aperture
bestreichen to spread, to coat with
Bestreichungswinkel maximum traverse
bestücken to tip (cutting tools)
bestückt mit Hartmetall carbide-tipped
Bestückung complement (tube), tipping (cutting edge)
Beta-Strahlung beta radiation
betanken to refill, to refuel, to fuel
Betankung refuelling, fuelling
betätigen to actuate, to operate, to control
betätigt, indirekt remotely controlled
Betätigung operation, actuation
Beton concrete
Betrag amount, rate
betragen, durchschnittlich to average
betreiben to run, to operate, to conduct
Betrieb factory, mill, operation, service

Betrieb, ausser out of operation, out of service, out of use
Betrieb, in in service, working, in use, in running order
Betriebsanleitung operating instructions, instruction leaflet, operator's manual, instructions book for operator
Betriebsart mode of operation, duty, method of operation
Betriebsaufwand operating expenditure
Betriebsausrüstung factory equipment
Betriebsbereitschaft readiness for operation, standby
Betriebsdauer time of operation, working time
Betriebsdrehzahl operation speed, normal speed, service speed
Betriebseinrichtung operating facilities
Betriebserde earth bus
Betriebserfahrung operating experience, technical know-how
betriebsfähig ready for service, in running order
Betriebsforschung operational research
Betriebskosten operating expenses, operating cost, running cost (ship)
Betriebsleitung factory management
Betriebssicherheit reliability, reliability in service
Betriebsspannung operating voltage, service voltage, operational voltage
Betriebsstörung breakdown, shut-down, breakdown of service
Betriebsstunden working hours
Betriebsüberwachungsanlage monotoring system
betriebsunfähig unserviceable
Betriebsunfall factory accident, industrial injury
betriebswarm warmed up
Betriebszustand working condition, service condition
Bett bed (machine tool)
beugen to bend, to deflect, to crank, to diffract (ray)
Beugung diffraction
Beugung des Lichts diffraction of light
Beule bulge (sheet metal)
beulen to buckle
Beutel bag, sack
bevollmächtigt authorised
bevorzugen to prefer
bewässern to irrigate
Bewässerung irrigation
bewegbar movable
bewegen to move
Beweglichkeit mobility, flexibility
Bewegung movement, motion, swirling (melt), travel (sliding machine part)
Bewegungsablauf sequence of motions
Bewegungsänderung change in motion
Bewegungsfreiheit freedom of motion, (or: movement)
Bewegungsgeometrie geometry of kinematics
bewegungslos motionless
bewegungsmässig kinematic
bewehren to reinforce, to armour
Beweis proof, evidence, demonstration
beweisen to prove
bewerben, sich to apply for
Bewerber applicant
Bewerbung application
bewerkstelligen to accomplish
bewickeln to wind, to wrap
Bewilligung licence
bewirken to effect, to cause
bezahlen to pay, to settle
Bezahlung payment, settlement
bezeichnen to mark, to notate, to designate, to specify, to indicate
bezetteln to affix a label, to label
Bezettelung labelling, tagging
Beziehung relation, relationship
Beziehung zueinander, in interrelated with each other
Bezirk area, district, zone
bezüglich with reference to
Bezugsdaten reference data
Bezugsebene reference plane
Bibliothek library
biegbar bendable
Biegbarkeit bendability, flexure
Biegemaschine rod bender, bar bender, rod bending machine

Biegemoment flexural moment, bending moment
biegen to bend, to crank
Biegepresse bending brake
biegsam flexible, bendable
Biegung flexure, bending
bifilar bifilar
Bifilarantenne two-wire aerial
bikonkav biconcave, double-concave
bikonvex biconvex, double-convex
Bild image (opt.), picture, graph
Bildablenkschaltung vertical time-base generator, frame time-base circuit
Bildablenkspule frame coil
Bildablenkstufe scanning stage
Bildabtaster scanner, scanning device
Bildabtaströhre scanning tube
Bildabtastung scanning
Bildverstärker video-frequency amplifier
Biluxlampe bilux lamp
Bimetall bimetal
Bimetallauslöser bimetallic trip
binär binary
Binäraddierglied binary adder
Binärdarstellung binary notation
Binärelement binary cell
binärkodiert binary-coded
Binärziffer bit, binary bit
Bindeglied connecting link
Bindemittel binding medium, binding agent
binden to bind, to tie, to bond, to fix
Bindfaden twine
Bindung bond, linkage, tying
Binom binomial
Binomialreihe binomial series
binomisch binomial
bipolar bipolar
Bitumen bitumen
blähen to swell, to inflate
blanchieren to whiten
Blanchierstahl whitening steel
blank bright, bare, naked, uninsulated
Blankdraht blank wire
Blankglühen bright annealing
Blankstahl bright steel, cold-drawn steel
Blankziehen bright-drawing
Blasebalg bellows
Blasenstahl blister steel
blasig blown (casting)
Blasmagnet magnetic blow-out, arc deflector, blowing magnet
Blasofen wind furnace
Blasspule blow-out coil
Blatt sheet, leaf, blade
Blättchen lamina
blättchenförmig lamellar
Blattfeder leaf spring
Blattgrösse sheet size, size of drawing paper
Blatthalter blade holder (saw)
Blattkupfer copper sheet
Blaublech blue steel plate
Blauerz brown ore
blauglühen to open-anneal
Blauglühen open annealing
Blaupause blue print
Blausäure hydrocyanic acid, prussic acid
Blech plate, metal sheet, sheet metal
Blechausstechbohrer fly cutter (for holes in thin metal)
Blechbearbeitung sheet metal working
Blechbearbeitungsbetrieb plate shop
Blechbiegemaschine plate-bending machine
Blechbohrer sheet drill, hole cutter
Blechbüchse tin, tin can, tin canister
Blechduo two-high plate mill
Blechfalzmaschine seaming machine
Blechhandschere sheet metal workers' hand shears, snip
Blechhülse sheet metal tube
Blechlehre plate gauge, sheet metal gauge
Blechschere snip, plate cutting machine, plate shears
Blechwalzstrasse sheet rolling mill
Blei lead
Bleiabschirmung lead screen, lead shielding
Bleiarbeit lead smelting, plumbing
Bleiauskleidung lead lining
Bleibarren lead pig
bleiben to remain, to stay, to be permanent
Bleibenzin leaded petrol
bleichen to bleach
Bleierz lead ore

bleifrei unleaded
bleihaltig plumbiferous
Bleikabel lead-covered cable
Bleimantel lead sheath, lead sheathing, lead coating
Bleimantelkabel lead-covered cable
Blende aperture (opt.), diaphragm, orifice, stop
blenden to dazzle
Blendlicht glare
Blickfeld field of vision, range of vision
Blindanflugverfahren blind approach procedure
Blindflansch blank flange
Blindflug blind flight, instrument flight, blind flying
Blindlast reactive load
Blindleistung reactive power
Blindleitwert susceptance
Blindstecker dummy plug
Blindstrom reactance current
blinken to flash, to blink, to flicker
Blinkfeuer flashing light
Blinkgeber flasher unit
Blitz lightning, flash
Blitzableiter lightning conductor, lightning rod
Blitzgerät flash-light unit
Blitzgespräch lightning call
Blitzlichtaufnahme flash-light photograph
Blitzschlag lightning stroke
Blitzschutz lightning protection
Blockdiagramm block diagram
blockieren to block, to interlock, to jam
Blockierung blocking, locking, interlocking
Blockkokille ingot mould
Blockschaltbild block diagram
Blockwalzwerk blooming mill, cogging mill
Bockkran gantry crane, gantry hoist, portal crane
Boden ground, bottom, floor, earth
Bodenabstand ground clearance
Bodenanalyse soil analysis
Bodenfunkanlage ground radio installation
Bodenfunkstelle ground station
Bodenpersonal ground crew
Bodenprobe soil sample
Bodenuntersuchung soil investigation
Bogen arc, bow, curve, bend
bogenförmig arched, arch-like, bent, curved
Bogenmass circular measure
Bogenrohr bent pipe
Bogensäge web saw, bow-type saw
Bohranlage drilling rig (mine)
Bohrarbeit drilling, boring
bohren to drill (from the solid), to bore
Bohren drilling (from the solid), boring
Bohrer drill
Bohrfutter drill chuck
Bohrknarre ratchet brace
Bohrmaschine drilling machine, boring machine
Bohrung bore, bore hole, drilled hole
Bohrvorschub drill feed
Boje buoy
Bolzen bolt, pin
Bor boron
Bordanlage airborne system
Bördelmaschine flanging machine
bördeln to flange, to bead
Bordfunkstelle ship radio station, aircraft radio station
Bordkran deck crane
Bordküche galley
Bordpeiler ship direction finder, airborne direction finder
Bordradar airborne radar
Bordsender airborne transmitter
Bowdenzug Bowden wire, Bowden cable
Boxermotor opposed piston engine
Brand fire, burning
Brandbekämpfung fire-fighting
Brandgefahr fire hazard
Brandschaden fire damage
brechen to break, to fracture (material), to crush
Brechstange crowbar, wrecking bar
Brech- und Siebanlage crushing and screening plant
Brechung refraction
Brechungsindex refractive index
Brechwalzwerk crushing mill
breit wide, broad
Breitband wide band
Breitbandantenne wide-band aerial, broad-band aerial

Breitbandverstärker wide-band amplifier
Breite width, breadth
breiten to flatten
Breitstahl paring chisel
Bremsanlage braking system
Bremsausgleich brake compensation
Bremsbacke brake shoe
Bremsbelag brake lining
Bremsbelastung load on a brake
Bremse brake
bremsen to brake, to apply the brake
Bremsfeldröhre brake-field valve, retarding-field valve
Bremsflüssigkeit brake fluid
Bremskraft braking force
Bremsleistung brake horse-power
Bremsmotor braking torque
Bremsung braking, slowing down
Bremszaun Prony brake, brake dynamometer
brennbar combustible, inflammable, burnable
Brennelement fuel element
brennen to burn, to fire
Brenner blowpipe, torch, cutting blowpipe, burner (Bunsen)
Brennkammer combustion chamber
Brennofen firing kiln, calcinator, roasting furnace
Brennspiritus methylated spirit
Brennstoff fuel
Brennwert heating power
Brett deal, board, plank, panel
Bretter boards, timber
Brikett briquette
Brille spectacles, glasses
Brinellhärte Brinell hardness
Brinellhärtezahl Brinell hardness number
bröcklig crumbly, brittle, friable
Brom bromine
Bromsilber silver bromide
Bronze bronze
bronzieren to bronze
Broschüre booklet, brochure,
Bruch fraction, fracture
Bruchbildung fracturing, rupturing
Bruchdehnung strain at failure
bruchfest unbreakable, fracture-proof
Bruchfestigkeit ultimate breaking strength
brüchig brittle
Brüchigkeit brittleness
Bruchprobe specimen for strength test
Bruchstelle spot of rupture
Bruchteil fraction
Brücke bridge
Brückenabgleich bridge balance
Brummen hum (radio)
brummen to hum (radio)
Brummspannung hum voltage, ripple voltage
brünieren to burnish, to bronze
Brüter breeder reactor
Bruttoleistung gross output
Buchführung mit Lochkartenmaschine punched card machine accounting
Buchholz-Relais Buchholz relay
Buchse bush, bushing, sleeve
Büchse tin, can
Buckelschweissung projection welding
Bug bow
Bügel bow, frame (saw)
Bügelsäge hacksaw, web saw
Bugpropeller bow propeller
Bugstrahlruder bow jet rudder
Bühne platform, stage
Bund collar, flange
Bündelbreite beam width (radar)
Bündelkabel bank cable
Bündelleiter bunch conductor (overhead lines)
Bündelung beaming (rays), narrowing
bündig flush
Bundstahl faggot steel
Bunker bunker, silo, bin, hopper
Bunkerkohle bunker coal
bunkern to bunker, to coal, to fuel
Bunsenbrenner Bunsen burner
bunt many-coloured, varicoloured
Buntmetall non-ferrous metal
Büroarbeit office work, paper work
Büroklammer letter clip
Bürste brush
bürsten to brush
Bürstenabhebeeinrichtung brush lifter
Bürstenanschlusskabel pigtail
Bürstenbrücke brush yoke, brush rocker
Bürstenhalter brush holder

Butan butane
Butylalkohol butyl alcohol
Butylen butylen
Byte byte (quantity from 1 to 8 bits)

C

Cambridge-Walze Cambridge roller
Candela candela, cd
Celsiusgrad centigrade, degree centigrade
Charakteristik characteristic
charakteristisch characteristic
Charge (furnace) charge, melt, batch, heat, load
Chargenprozess batch process
Chargierlöffel charging spoon
Chargiermulde charging box
Chargiertrichter feed hopper
Chassis chassis, base, frame
Chefingenieur chief engineer
Chefkonstrukteur chief designer
Chemie chemistry
Chemieanlage chemical processing plant
Chemieingenieur chemical engineer
Chemikalien chemicals
Chemiker chemist
chemisch chemical
Chiffre cipher, code
chiffrieren to (en)code, to cipher
Chiffriernummer code number
Chiffrierschlüssel key, code
Chlor chlorine
chloren to chlorinate
Chlorstörstellen Cl impurities (semiconductor)
Chrom chromium
Chromat chromate
Chromnickelstahl chromium-nickel steel, nichrome
CO_2-Schweissen carbon-dioxide welding
Coolidge-Röhre Coolidge (ray) tube

D

Dach roof, top
Dachantenne roof aerial, over-house aerial
Dachziegel roof tile, pantile
Damm dam, ridge, bank, barrier
Dämm-Material insulating material
dämmen to insulate, to dam up
Dämmerlicht twilight, dusk (abends), dawn (morgens)
Dämmung (sound and heat) insulation
Dämmzahl (sound) damping factor
Dampf steam, vapour, fume
Dampfabgabe steam utilisation (for heating purposes)
Dampfablasshahn blow cock
Dampfabscheider steam separator
Dampfbetrieb steam service, driving by steam
dampfdicht steamproof, steam-tight
Dampfeinlassbüchse steam chest
Dampfeinlasshahn admission steam cock
Dampfeinlaßseite steam admission (port) side
Dampfeintritt steam admission
dampfen to steam
dämpfen to dampen, to attenuate, to absorb
Dampfer steamer, steamship
Dampferzeugungsanlage steam-generation plant
dampfförmig vapourous
Dampfheizungsanlage steam heating installation
Dampfkasten steam chest
Dampfkessel steam boiler
Dämpfung damping, attenuation, absorption
Dämpfungsdiode damping diode
Dämpfungsentzerrer attenuation equaliser
Dämpfungsentzerrung attenuation equalisation
Dämpfungsglied attenuator pad, resistance pad

darstellen to display, to represent, to obtain, to plot
darstellend descriptive (geometry)
Darstellung description, representation, display
Daten data, information
Datenantastsystem data sampling system
Datenaufzeichnung data logging
Datenauswertung data handling
Datenblock data block
Datenerfassung data gathering, data logging
Datenfluss data flow
Datenflussplan data flow chart
Datenspeicher data storage unit
Datenspeicherung data storage
Datenübertragung data transmission
Datenumsetzer data translator
Datenverarbeiter data processor
Datenverarbeitung data processing, data handling
Datenverarbeitungsanlage data processing equipment
Datenverschlüssler data-coding unit
Dauer period, duration
Dauerbelastung continuous rating(s), continuous load, sustained loading
Dauerbetrieb continuous duty (operation, use, work)
Dauerbiegefestigkeit bending stress fatique limit
Dauerfestigkeit fatique strength
Dauerform permanent mould
Dauerhaftigkeit durability
Dauerkurzschlußstrom sustained short-circuit current
Dauerlast continuous load, sustained load
Dauerleistung continuous rating
Dauermagnet permanent magnet
dauernd permanent
Dauerschlagfestigkeit impact fatique limit
Dauerschmierlager self-lubricating bearing
Dauerstrichmodulation continuous-wave modulation
Dauerstrichsender continuous-signal transmitter
Daumen thumb, lift, tappet, cog, cam
Daumenrad tappet wheel
Daumenregel thumb rule
Deblockierung unblocking
Decca-Navigationsverfahren decca navigation
dechiffrieren to decipher, to decode
Deck deck, top-side
Deckel cover, lid, cap
Deckfarbe finishing paint
Deckplatte cover plate, deck cover
Decksbalken beam
Deckung coincidence, congruence
Deflektor deflector
deformieren to deform
Deformierung deformation
dehnbar strainable, extensible, elastic, ductile
Dehnbarkeit strainability, extensibility, elasticity
dehnen to strain, to expand, to stretch
Dehnung strain, extension, expansion, elongation, stretch
Dehnungsfestigkeit tensile strength
Dehnungsmeßstreifen foil strain gauge, wire strain gauge
Dehnungszahl coefficient of expansion
Deionierungsschalter deion circuit-breaker
Dekadenimpulsgeber decade pulse generator
Dekadenschaltung decade connection
Dekadenzählröhre decade counter tube
dekodieren to decode, to decipher
Dekodierer decoder
Dekodierung decoding
Demodulation demodulation, detection
demodulieren to demodulate, to detect
demontieren to dismantle, to dismount
Demontierung dismantling, dismounting
demulgieren to de-emulsify, to demulsify
Depot depot, storage facility
Destillation distillation, distilling
Dezibel decibel

Dezimeterwelle decimetric wave, decimetre wave
Diagramm diagram, chart, graph, plot
Diamant diamond
Diamantabrichter diamond wheel dresser
diamantbestückt diamond-tipped
Diamantbohrer diamond drill
diamanthart adamantine
Diamantsplitter diamond particle
Diaprojektion slide projection
dicht leakproof, dense, close, tight
Dichte density, compactness,
dichten to pack, to seal
Dichtung packing, sealing
Dichtungsfläche packing surface, seal face
Dichtungsscheibe sealing washer
dick thick
Dicke thickness
dickflüssig viscous, consistent
dickwandig thick-walled
Dielektrikum dielectric
dielektrisch dielectric
Dielektrizität dielectricity
Dielektrizitätskonstante dielectric constant, permittance
dienen to serve, to be of use, to be utilised
Dienst service, duty
Dienstreise journey on duty, business trip
Dienstvorschrift service regulations
Dieselaggregat diesel-driven generating set
dieselelektrisch diesel-electric
Dieselkarren diesel truck
Dieselkraftstoff diesel fuel
Dieselmotor diesel engine
Differential differential
Differentialgetriebe differential gear
Differentialgleichung differential equation
diffundieren to diffuse
diffus diffuse
Digital-Analog-Wandler digital-to-analogue converter
Digitalablesung digital readout
Digitalanzeige digital display
dimensionieren to dimension
Dimensionierung dimensioning, proportioning, sizing, calculation of dimensions
dimensionslos nondimensional, dimensionless
Diodenaussenwiderstand diode load resistance
Diodenbegrenzer diode limiter
Diodenstromeinsatzpunkt diode-current starting point
Diodentorschaltung diode gate
Diplom diploma
Diplomingenieur professional engineer, graduate of technology
Dipol dipole
Dirac-Gleichung Dirac equation
Doppler duplicator, duplicating, punch, reproducing puncher
Doppler-Radar Doppler radar
Dorn arbor, mandrel
dornen to indent, to pierce, to expand
Dornpresse piercing press, mandrel press
Dose can, tin, cell
Dosenbarometer aneroid barometer
Dosieranlage batching plant
dosieren to dose, to batch, to proportion, to gauge
Dosis dosage, dose
Dotierung doping, dosage
Dotierungsmittel doping agent
Draht wire
Drahtbruch rupture of wire, wire break
Drahtbürste wire brush
drahten to wire, to cable
Drahtfunk line radio, carrier transmission
drahtlos wireless
Drall twist, helix, spin
Drehmaschine turning lathe, lathe
Drehmeissel lathe (turning) tool
Drehmelder synchro, selsyn
Drehmoment torque
Drehofen rotary kiln
Drehpfanne centre casting, pivot bearing
Drehpunkt centre of rotation, fulcrum, pivot
Drehrichtung direction of rotation
Drehrichtungsanzeiger phase-sequence-indicator
Drehschalter rotary switch
Drehschieber rotary slide valve

Drehsinn direction of rotation
Drehspindel lathe spindle
Drehspule moving coil
Drehstab torsion bar
Drehstabfeder torsion bar (or: rod) spring
Drehstrom three-phase current
Drehstromnetz three-phase system
Drehstromnetz mit Nulleiter three-phase four-wire system
Drehstromnetz ohne Nulleiter three-phase three-wire system
Drehsupport swivel head
Drehtisch rotary table, revolving table
Drehtür revolving door
Drehung rotation, revolution
Drehzahl speed
Drehzahlabfall speed drop
Drehzahländerung variation in speed, speed variation
dreiadrig three-core, three-wire
Dreibackenfutter three-jaw chuck
Dreibein tripod
Dreieck triangle
dreieckgeschaltet delta-connected
dreieckig triangular
Dreieckschaltung delta connection
Dreikantfeile triangular file, three-square file
dreikantig three-edged, three-cornered, three-square
Dreiphasenwechselstrom three-phase alternating current
Drifttransistor drift transistor
Drossel choke, throttle, ballast (of fluorescent lamps)
drosseln to throttle, to choke, to restrict
Drosselklappe choke control
Drosselspule choke coil, reactor
Drosseltransformator constant-current transformer
Drosselung throttling, choking
Druck pressure, thrust, compression
druckabhängig pressure-dependent, pressure-controlled
druckdicht pressure-tight, pressure-sealed, pressurised
drucken to print
drücken to force, to push, to depress
Druckgefälle pressure gradient, drop of pressure
Druckgefäss pressure vessel, pressure tank
Druckguss die casting
Druckknopf push-button
Druckluft compressed air
Druckluftbohrer pneumatic drill
Druckluftbremse air brake, pneumatic brake
Drucköl pressure oil, hydraulic oil
Dübel dowel, nog, plug
Dübelbohrer brad-awl
dübeln to dowel
Dunkelkammer dark-room
dünn thin, slender
Dünnfilm thin film
Dünnfilmschaltung thin film circuit
dünnflüssig light-bodied, thin-bodied, thin-fluid
dünnwandig thin-walled
Dunst fume, vapour
Duoschaltung lead-lag connection
Duowalzwerk duo mill
Durchbiegung deflection, sagging
Durchblasen bubbling,
durchbohren to through-drill
durchbrennen to blow, to fuse, to burn out
Durchbruch breakthrough, breakout
durchdrehen to crank (engine)
durchdringen to penetrate, to intersect
Durchdringung penetration, intersection
Durchfahrt passage, passing
durchfliessen to flow through, to pass
Durchflussanzeiger flow indicator
Durchflusskühlung straight-through cooling
durchführen to carry out, to perform, to accomplish, to pass through
Durchführung lead-through bushing
Durchführungshülse grommet
Durchführungsisolator insulated bushing, bushing insulator
Durchführungsklemme bushing clamp
Durchführungsperle beading
Durchführungsstromwandler bushing-type current transformer

Durchhang sagging, sag
durchhängend slack, sagging
Durchlassbereich pass-band width
durchlässig penetrable, transparent, permeable, porous, spongy
Durchlässigkeit permeability, penetrability
Durchlauf run, passage
durchlaufen to pass
Durchlauferhitzer flow heater
Durchlaufofen continuous furnace
Durchleuchtung X-raying
Durchmesser diameter
Durchsatzzeit time of passage
durchschalten to connect through
Durchschlag rupture (insulation), puncture, breakdown
durchschlagen to puncture, to break down, to spark through, to perforate, to punch
durchschlagsicher puncture-proof
Durchschlagskraft penetration voltage, puncture voltage,
Durchschlagsspannung disruptive breakdown voltage
Durchschnitt average, mean
Durchschnittslast average load
Durchschnittswert average value
Durchsicht overhaul
durchsichtig transparent
durchsickern to percolate, to leak through
durchspülen to flush
Durchsteckwandler bar-type current transformer, single-turn current transformer
durchsteuern to fully modulate
durchstossen to punch
Durchstossofen continuous discharge furnace
Durchtreiber round punch
durchverbinden to connect through, to put through
Durchwahl through-dialling
durchwählen to dial through
Durchziehglühofen continuous annealing furnace
durchzwängen to force through
Düse nozzle, spout (waveguide), tuyere (furnace)
Düsenantrieb jet propulsion
Düsenflugzeug jet-propelled aircraft
Düsengruppe group of nozzles
Düsenkammer nozzle box
Düsenkraftstoff jet fuel
Düsensatz nozzle ring segment, set of nozzles
Düsenstock pen stock
Dyn dyne
Dynamik dynamics
dynamisch dynamic

E

eben plane, flat, smooth
Ebene plane
Ebenheit evenness, flatness, smoothness
ebnen to level, to flatten, to smooth
Echoanzeige blip (radar)
Echobild echo pattern
Echolot echo sounder, depth sounder
Echolotung reflection sounding, echo depth sounding
Echoortung echo ranging
Echosignal blip
Echospannung return voltage
Echosperre echo suppressor
Echoweg echo path
echt real, genuine, pure, authentic
Echtheit trueness, purity
Echtzeit real time
Ecke corner, vertex
eckig angular, cornered, square
Edelgas inert gas, inactive gas, noble gas
Edelgaslaser noble gas laser
Edelguss high-strength cast iron
Edelmetall precious metal, noble metal
Edelstahl high-quality steel, high-grade steel
Edelstein precious stone, gem
Edison-Fassung Edison screw holder, screwed-type socket
Effekt effect
Effektivleistung actual power
Effektivspannung effective voltage, virtual voltage, r.m.s. voltage
eichen to calibrate, to gauge
Eichleitung attenuator, standard line, calibrating line
Eichmarke calibration mark
Eichmass calibration standard, gauge

Eichung calibration
Eierisolator egg-type insulator, globe strain insulator
eigenbelüftet self-cooled
eigenerregt self excited, self-oscillatory
Eigenfrequenz natural frequency
Eigengenauigkeit intrinsic accuracy
Eigenrauschen inherent noise, noise background
Eigenschaft property, condition
Eigenschwingung natural vibration
eigensicher intrinsically safe
Eigensicherheit intrinsic safety
Eigenstrahlung self-radiation, natural radiation, characteristic radiation
Eigentum ownership, property
Eigenverbrauch internal consumption
Eigenwiderstand inherent resistance
Eiisolator globe insulator
Eilbewegung rapid traverse, rapid motion
Eilgang fast traverse, quick traverse, rapid traverse
Eilrücklauf rapid return, quick return, rapid idle movement
Eimer bucket
Eimerbagger bucket dredger, bucket excavator
Eimerkette bucket chain
einadrig single-core
Einankerfrequenzwandler single-unit frequency converter
einarbeiten to recess, to sink (machining), to learn to train
Ein-Aus-Schalter single-throw circuit-breaker
Ein-Aus-Tastung on-off keying
Einbackenbremse single-block brake
Einbau mounting, insertion, assembly, installation, placement, placing
Einbauantenne built-in aerial
einbauen to mount, to insert, to assemble, to install, to place, to fit
Einbauinstrument panel instrument
einbetten to embed
Einbettung embedding
einblenden to fade in
Einbrand penetration (welding)
Eindampfer evaporator
eindeutig unique, unambiguous
eindringen to ingress (fluid), to intrude, to penetrate
Eindringen entry, ingress, penetration
Eindruck impression
eindrücken to press into, to force into
Eindrucktiefe indentation
einebnen to level
einengen to neck down, to reduce
einfach simple, single, plain
Einfachimpuls single pulse
Einfall incidence
einfallend incident (rays)
Einfallswinkel incidence angle, angle of incidence
einfangen to collect (electron), to trap
einfassen to border, to edge, to enclose, to line
Einfluss influence
einfrieren to freeze in
einfügen to insert, to fit in, to place between
einführen to introduce, to enter
Einführung introduction, insertion
einfüllen to fill in, to pour in, to charge
Einfüllstutzen filler neck
Eingabe input (data), entry
Eingabespeicher input memory, input store
Eingangsleistung input power, power input
Eingangsstufe input stage
Eingangswiderstand input resistance
eingebaut built-in, inbuilt
eingeben to enter, to feed
eingegossen cast-in
eingehen to shrink
eingelassen sunk, flush, recessed
eingelegt inlaid
eingepasst fitted-in
eingepresst mould-in
eingerastet engaged
eingerissen ragged
eingreifen to engage, to mate
eingerückt thrown in, engaged (gear)

Eingrenzung von Fehlerquellen localisation of error sources
Eingriff engagement, mesh
einhaken to catch, to hook, to dig (tool)
einhalten to hold, to maintain
einhängen to hang, to place on, to hinge
Einheit unit, item
einheitlich standard, standardised, uniformly
Einheitsbaustein standard modular unit
einhüllen to envelop
einkaufen to purchase, to buy, to shop
Einkäufer purchasing executive
Einkaufsabteilung purchasing department
einkehlen to channel
einkerben to notch, to indent, to score, to serrate
Einkerbung notch, indentation, serration
Einkesselschalter single-tank switch
einkleben to paste in
einklemmen to jam
einklingen to catch, to pawl
Einkreisempfänger single-circuit receiver
Einkristall single crystal
einkuppeln to release the clutch, to engage the clutch
Einlage insert, insertion
einlagig single-layer
Einlass intake, inlet, admission
Einlassdosenschalter recessed switch
einlassen to admit
Einlasskrümmer inlet manifold
Einlaßschalter recessed switch
Einlassventil admission valve
einlaufen to wear in (bearing), to run in
Einlaufen wearing-in (bearing)
Einlaufzeit run-in period, break-in period
Einlegekeil sunk key
einlegen to put in, to insert
Einlegen loading (work)
einleiten to trigger, to start, to pass into, to initiate
Einlenkrechner computer for guidance
einlesen to read in
einlippig single-lip, single-flute
einlöten to solder in
einnehmen to occupy (space), to receive, to take
einordnen to classify, to file
Einordnung classification, filing
einorten to position, to locate
Einortung locating, positioning
einpacken to pack up, to wrap
einpassen to fit in
einpolig single-pole, unipolar
einprägen to impress
einpressen to force in, to press into place
einrasten to engage, to insert, to click
einreihen to sequence
einrichten to set up (machine tool), to position, to arrange
Einrichten setting-up (machine tool)
Einrichter machine setter, tool setter
Einrichtung setting-up (machine tool), installation, establishment, arrangement, attachment
Einrichtungen facilities
einrollen to roll up
einrosten to grow rusty, to rust in
einrücken to engage, to clutch
Einrücken engagement, throwing
Einsatz application, use, utilisation, insert (tool)
einsatzbereit ready for service
einsatzgehärtet case-hardened
einsatzhärten to case-harden
Einsatzhärten case hardening
Einsatzmeissel bit, insert
einsäumen to hem, to border, to edge, to trim
einschaben to fit in by scraping
Einschalt-Ausschalt-Zeit make-break time
einschalten to switch on, to turn on, to connect, to close, to make a circuit
Einschalter single-throw switch, closing switch
Einschaltfolge switching-on sequence
Einschaltleistung making capacity
Einschaltstoss make impulse
Einschaltung starting, circuit closing

Einschaltvermögen making capacity
Einschaltzeit make time
einschiffen, sich to ship, to embark, to go on board, to take ship
einschleifen to grind in
einschliessen to include, to comprise, to lock in
Einschluss inclusion, entrapping
einschnappen to catch
Einschnürung constriction, restriction, necking, reduction
einschränken to restrict, to restrain, to reduce
einschrauben to screw, to screw into
Einschreibung entry
Einschubeinheit plug-in unit
Einschwingvorgang transient state, transient building-up, transient condition
einseitig unilateral, in one direction
einsetzen to insert, to install
einspannen to chuck, to clamp, to hold (work)
Einspannfutter chuck jaws
einspeisen to feed, to supply
Einspindelautomat single-spindle automatic (screw machine)
einspritzen to inject, to prime
Einspritzen injection, priming
Einspritzmotor fuel-injection engine
Einspritzpumpe injection pump
Einspritzversteller injection advance device
Einspritzverstellung injection timing
Einständerbauart open-side construction, single-column construction
Einständerexzenterpresse gap frame eccentric press
Einständerfräsmaschine single-column milling machine
einstechen to recess, to plunge
Einstechschleifmaschine plunge grinding machine
Einstechverfahren infeed process
einstecken to plug in, to insert
Einsteckschlüssel face spanner
einstellbar adjustable
Einstellbereich setting range
Einstellehre setting gauge
einstellen to set, to adjust, to tune in (radio), to align
Einstellen setting, set-up, adjusting, adjustment
Einstellgenauigkeit setting accuracy, adjustment accuracy, positioning accuracy
Einstellglied setting element
Einstellmarke timing mark, setting mark
Einstellscheibe dial
Einstellung setting, adjustment
Einstellwinkel plan angle (tool), working angle, setting angle
Einstich recess (machining)
Einstieg manhole
einstufen to grade
einstufig single-stage
einstürzen to collapse, to fall in
eintauchen to immerse, to dip, to steep, to submerge
Eintauchpumpe submersible pump
einteilen to divide, to classify, to grade, to calibrate
einteilig one-piece
eintragen to plot, to make an entry
eintreten to emerge into (fluid), to enter
Eintritt intake, inlet, entrance
Eintrittsgeschwindigkeit admission velocity
Ein- und Ausfahren in-and-out travel (machine))
einweichen to steep, to soak
einwickeln to wrap
einwirken to act upon, to affect, to attack
Einwirkung action
einzeichnen to plot
Einzelanfertigung single-piece production, individual production, single-piece job
Einzelantrieb individual drive
Einzelaufspannung single set-up
Einzelstück single-piece
Einzelteil one-off part
einziehbar retractable
einzig unique
Eisen iron
Eisendrossel iron-cored choke coil
Eisenerz iron ore
eisenhaltig ferruginous, ferriferous

Eisenhüttenkunde ferrous metallurgy
Eisenkern iron core
Eisen-Kohlenstoff-Diagramm iron-carbon diagram
Eisenoxyd ferric oxide, ferrous oxide
Eisenverlust iron loss
elektrifizieren to electrify
Elektrifizierung electrification
Elektriker electrician
elektrisch electric, electrical
Elektrizität electricity
Elektrizitätswerk power station
Elektrode electrode
Elektroenergie electric energy
Elektrohandbohrmaschine electric hand drill
Elektroingenieur electrical engineer
Elektrokarren electric truck, storage battery truck
Elektrolyse electrolysis
Elektrolyt electrolyte
Elektromotor electric motor
elektromotorisch electromotive
Elektron electron
Elektronenstrahl electron beam, cathode ray
Elektronikindustrie electronics industry
Elektrozug electric hoist
elementar elementary
Ellipse ellipse
elliptisch elliptic
Eloxalschicht anodised coating
eloxieren to anodise
Email enamel
Emitteranschluss emitter contact
Emitter-Basis-Diode emitter-base diode
Emitterfolger emitter follower
Emitterschaltung grounded-emitter circuit
Emitterschicht emitter junction
Emitter-Tor-Abstand emitter-to-gate spacing
Empfang reception
Empfänger receiver, receiving set, responder (radar), consignee
Empfängerseite receiving end (transport)
Empfangsanlage receiving installation
Empfangsgerät receiving set, receiver unit
Empfangskreis receiving circuit
empfehlen to recommend
empfindlich sensitive
Empfindlichkeit sensitivity, response
Endabnahme final examination
Ende end, termination, tip
endlos endless
Endmontage final erection, final assembly
Endstellung end position, final position
Energie energy
Energiebedarf energy demand, power consumption
Energieversorgung power supply
Energieverteilung energy distribution
Energiezufuhr power feed
eng narrow, close, tight, closely spaced
enger werden to narrow
Engpass bottleneck
entaschen to remove the ash
entdecken to discover
Entdeckung discovery
enteisen to de-ice, to defrost
Enteisung de-icing, defrosting
entfernen to remove, to clear
Entfernung distance, range, removal
entfetten to degrease
entflammbar inflammable
Entflammbarkeit inflammability
entgasen to degas, to dearate
Entgaser feed dearator, dearator
Entgasung degreasing, degasifying
entgegengesetzt opposite
entgraten to deburr, to burr
Entgraten deburring
enthalten to comprise, to contain
enthärten to soften
entkalken to declime, to decalcify
entkohlen to decarburise
entkoppeln to decouple, to isolate
Entkopplungsglied stopper circuit
Entkopplungskreis anti-resonant circuit
Entkopplungsschaltung decoupling circuit
entkuppeln to declutch, to uncouple
entladen to unload, to discharge
Entladung unloading, discharge
entlasten to relieve (of load), to unload, to release

Entlastung unloading, relieving, easing
entleeren to empty, to evacuate
entlüften to vent, to ventilate
Entlüfter exhauster, ventilator, dearator
Entlüftung ventilation, dearation, airing
entnehmen to take, to sample, to take out
Entriegelung unlocking, release
entrosten to remove the rust
entsalzen to desalinate
entscheiden to decide
Entscheidung decision
entsichern to unlock
entspannen to release, to unload
entsprechen to correspond to, to conform to
entsprechend corresponding
entstehen to develop, to occur, to result, to rise
entstören to suppress interference, to clear faults
Entstörung interference suppression, noise suppression, fault clearance
Entstörungsdienst fault clearing service
entwässern to dewater, to drain, to dehydrate
entweichen to escape, to disengage
entwerfen to design
entwickeln to develop, to evolve, to set free
Entwicklung development
entwirren to disentangle, to untwist
Entwurf design, draft, plan, project
entzerren to equalise, to rectify, to correct, to compensate
Entzerrer equaliser, corrector, compensator
Entzerrung equalisation, correction, compensation
entzünden to ignite, to initiate
entzundern to descale, to scale off
Entzundern descaling, scaling-off
Entzündung ignition, inflammation
Epoxydharz epoxy resin
Erdbauarbeiten earthworks
Erde earth, ground, soil
erden to earth, to ground
Erden, seltene rare earths
Erdklemme earth terminal, earthing terminal, ground terminal, ground clamp
Erdöl mineral oil, crude petroleum
Erdschluss earth fault, earth leakage, line-to-earth fault
Erdschlussschutz earth-fault protection
Erdstromkreis earth circuit
Erdtaste earthing key
Erdung earth lead, earth connection, grounding
Erdungsanlage earthing system
Erdungsstange earthing hook, switch-pole
Erdungswiderstand earthing resistance, discharging resistor
Ereignis event
erfassen to log, to detect
Erfassung logging, measuring, coverage, acquisition
erfinden to invent
Erfinder inventor
Erfinderschutz protection of inventors
Erfindervergütung award to the inventor
Erfindung invention
erfordern to require
Erforschung exploration, investigation
erfüllen to meet (specifications)
ergänzen to complement, to supplement, to complete
Ergänzung completion, completement, supplement
ergeben to provide, to result, to yield
Ergebnis result, yield
erhaben convex, embossed
erhalten to receive, to obtain,
Erhaltung conservation, preservation, maintenance
Erhebung elevation
erhitzen to heat up
erhöhen to increase, to raise, to step up, to elevate
erholen to recover
Erholung recovery
erkalten to cool, to chill
erkennen to detect, to recognise, to perceive
erleichtern facilitate

erleiden,Schaden to suffer damage
erlernen, einen Beruf to aquire a craft
ermächtigen authorize
ermässigen to reduce
Ermässigung reduction
ermitteln to detect, to determine, to find out
Ermittlung determination
ermöglichen to enable
ernennen to appoint, to nominate, to assign, to designate
erneuern to renew, to recondition, to reface, to face-lift
Erneuerung renewal, reconditioning, restoration, overhauling
erodieren to erode
Erosionsschutz erosion control
erproben to try, to test
erregen to excite, to energise
Erreger exciter, exciter set
Erregung excitation
erreichen to reach, to obtain, to arrive at
errichten to erect, to build, to set up
erringen to win, to achieve, to score
Errungenschaft achievement, result, advancement
Ersatz replacement, substitute, spare
Ersatzstoff substitute material
Ersatzteil spare part, replacement part, spare
erscheinen to appear, to be published (book)
Erscheinung appearance, phenomenon
erschliessen to develop (site)
erschöpfen to exhaust, to deplete
Erschöpfung der Batterie exhaustion of the battery
erschüttern to shatter, to rock, to vibrate
Erschütterung vibration shock, percussion, concussion
erschütterungsfrei vibrationless
erschütterungssicher shockproof
ersetzen to substitute, to replace, to exchange, to displace
erstarren to congeal, to gel, to solidify
erstatten to repay, to refund
Erteilung allowance
Ertrag yield
erwärmen to heat up, to warm up
Erwärmung warming, heating
erweitern to enlarge, to extend, to expand, to widen
Erweiterung extension, expansion, widening
erwerben to acquire, to obtain, to gain
Erz ore
erzeugen to produce, to generate
Erzeugnis product, article, make
Erzeugung generation
Erzeugungswälzkreis generating pitch circle
Esse chimney, stack, funnel
Etage floor, storey
Etat budget
Etikett label, tag
etikettieren to label
evakuieren to evacuate, to pump down, to pump out, to empty
Evolute evolute
Evolvente involute
Examen examination
experimentell experimental
explodieren to explode, to burst, to detonate
Explosion explosion, detonation, blast
explosiv explosive
exponentiell exponential
extrapolieren to extrapolate
extrudieren to mould by extrusion
Exzenter eccentric cam

First I thought you were ZA 10769422. You look very much like him.

F

Fabrik factory, works, mill, shop
Fabrikat make, manufacture
Fabrikation manufacture, making, production
fabrizieren to manufacture, to produce, to make
Fach subject, field, branch, line, profession
Facharbeiter skilled worker, craftsman
Facharbeiterprüfung qualifying examination
Fachausdruck technical term
Fachbuch, technisches technical book
Fachgebiet field, subject, branch, line
Fachliteratur technological literature, scientific literature
Fachmann expert, specialist
Fachrichtung special field
Fachzeitschrift, technische technical periodical, technical journal, technical magazine
Faden thread, filament, fathom (measure)
Fadentransistor filamentary transistor
fähig capable
Fähigkeit capability, ability, capacity
Fahrdraht contact wire, bus line
fahren to drive, to traverse (machine), to travel, to run
Fahrgestell base frame, chassis
Fahrt drive, driving, riding, run, running, passage, voyage, tour, cruise
Fahrwerk alighting gear, undercarriage
Fahrzeug vehicle (land), vessel (ship), craft (air)
Fall case, fall, drop
fallen to fall, to drop, to decrease
fällen to precipitate
Fallschirm parachute
Falltank header tank, gravity tank
Fällung precipitation

fälschen to duff, to adulterate, to forge, to conterfeit
falten to fold
Falz fold, seam, joint
falzen to fold, to seam
Falzfräser rebating cutter, notching cutter
Falzhobel grooving plane, fillister plane
Fangeinrichtung interception circuit
Fangelektrode target, collecting electrode
Farbe colour, color (am.), paint, dye
farbecht colour-fast
färben to dye, to colour
Fase bevel, chamfer, land (twist drill)
fasen to chamfer
Fasen chamfering
fassen to grip, to hold, to take
Fassung socket, lamp holder, holder
faul rotten, decaying
Faustregel rule of thumb, hard-and-fast rule
Feder spring, feather (key)
Federspannfutter chuck spring collet
fehlanpassen to mismatch
Fehler error, fault, flaw (material)
Feile file
Feinabgleich fine adjustment
Feingewinde fine screw thread, fine pitch thread
Feinstahlwalzwerk small section rolling mill
Feinvorschub fine feed
Feld field, bay, panel
Feldeffekttransistor field effect transistor
Feldschwächung field weakening, flux shunting
Feldspat felspar, feldspar
Feldstärke field strength
Feldtransistor field effect transistor
Feldwicklung field coil, field winding
Felge rim, felloe
Fell skin, fell
Fels rock
Fenster window
fern far, remote, distant
Fernamt trunk exchange, toll exchange
Fernanruf trunk call
Fernanschluss trunk subscriber's line
Fernanzeige remote indication
Fernbedienung remote control, telecontrol
fernbetätigt remotely operated
Ferngespräch trunk call, long-distance call
Ferngreifer manipulator
Fernheizung district heating
Fernmeldeanlage communication facility
Fernmeldetechnik telecommunication engineering, telecommunications
Fernrohr telescope
Fernsehempfänger television set
fernsehen to televise
Fernsehen television
Fernvermittlung trunk exchange
Ferrit ferrite
Ferritplatte ferrite slab
Ferritstab ferrite rod
fertig ready, finished, complete, completed
fertigbearbeiten to finish, to finish-machine
fertigbearbeitet finish-machined
fertigen to manufacture, to produce
Fertigfabrikat finished product
Fertigkeit skill, art
fertigstellen to complete, to finish
Fertigstellung completion, finishing
Fertigung manufacture, production, output, yield
Fertigungsingenieur production engineer
fest strong, solid, fixed, stationary, tied, firm
festbinden to lash
Festfahren sticking (machine)
festfressen to jam, to seize
festgefahren stuck (tool)
festhalten to hold in place, to arrest
festigen to strengthen, to stabilise, to compact
Festigkeit strength, rigidity, stability, compactness

Festigkeitslehre theory of the strength of materials
Festigkeitsprüfung strength test
festkeilen to key
festklemmen to clamp, to jam, to stick
Festkörper solid, solid body
Festkörperelektronik solid state electronics
Festkörperspeicher solid state memory
festsitzen to stick (tool)
feststellen to clamp, to lock
Festtreibstoff solid propellant
Festwertspeicher permanent memory, permanent store
festziehen to tighten
Festziehen tightening
Fett grease (lubrication)
Fettbüchse grease cup, greaser
fetten to grease, to lubricate
Fettnippel grease nipple
feucht humid, moist, damp
Feuchte humidity, moisture
feuchten to wet, to moisten, to dampen
feuchtigkeitsbeständig damp-proof, moisture-proof
Feuchtigkeitsschutz damp-proofing, moisture-proofing
Feuer fire, beacon (navigation), light
feuerbeständig fire-resistant, fire-proof
feuerdämmend flame-retardant, fire-retardant
Feuergefahr fire hazard
Feuerlöschanlage fire extinguishing plant
Feuerlöscher fire extinguisher
Feuermelder fire alarm
Feuerschweissen forge welding
feuersicher fire-proof, flame-proof
Feuertür furnace door
Feuerung firing, furnace, fuel
Feuerversicherung fire insurance
Feuerverzinnung hot tinning
Feuerwehr fire brigade
Fiber fibre
fieren to lower
Figur figure, pattern
Filmarchiv film library
Filter filter, strainer
Filtereinlage filter cone
filtern to filter
filtrieren to filtrate, to filter, to strain
Filz felt
Filzöler felt pad lubricator
Fingerfräser slot drill, shank-type slotting end mill, keyseating cutter
Firma firm
Firnis oil varnish, boiled oil
Fittings fittings
fixieren to fix, to set
flach shallow, flat, inclined
Flachbahnanlasser face-plate starter, disk-type starter
Fläche area, surface, face
Flacheisen flat iron
flächen to spot-face
Flächendichte surface density
Flächendiode junction diode
Flächenfräsen surface milling
Flächeninhalt area
Flächenschliff surface grinding
Flächenträgheitsmoment areal moment of inertia
Flächentransistor junction transistor
Flachgewinde square thread, flat thread
Flachwicklung plane winding
flackern to flash, to flicker
Flakon phial
flammbeständig flame-resistant
Flamme flame
Flämmen flame scarfing
flammengehärtet flame-hardened
Flammenschutz flame trap
flammofenfrischen to puddle
Flammpunkt flash point
Flammrohrkessel flue boiler
Flanke slope, flank, wing
Flankenform flank profile
Flankenspiel backlash
Flankenwinkel thread angle
Flansch flange
flanschen to flange
Flanschmotor flange-mounted motor
Flasche bottle, flask, jar
Flaschenzug pulley block, chain block, tackle
flechten to twist, to braid
Fleck spot, stain, blotch (of oil)
fleckenfrei stainless
flexibel flexible

Flickarbeit patch work, repair work
Fliehkraft centrifugal force
Fliehkraftregler centrifugal governor
Fliehkraftschalter centrifugally operated switch
Fliessarbeit flow-line operations, continuous production
Fliessband flow-line, conveyor belt
Fliessdehnung yield strain
fliessen to flow, to creep
Fliessen flow
Fliesspressen extrusion
Fließspan flow chip
flimmerfrei flickerless
flimmern to flicker
flink quick-acting (fuse)
Flip-Flop-Schaltung flip-flop circuit
Flucht row, alignment, play
fluchtend in line, aligned
Flugasche fly ash, flue dust
Flugblatt leaflet, pamphlet
Flügel wing
Flügelmutter wing nut
Flügelpumpe vane pump
Flügelrad impeller
Flugkörper missile
Flugmotor aero-engine
Flugstaub flue dust, smoke dust
Fluoreszenzlicht fluorescent light
fluoreszieren to fluoresce
Flussdiagramm flow chart
flüssig liquid, fluid
Flüssigkeit liquid, fluid
Flussmittel flux
fluten to flood, to flow
Flutlicht floodlight
Flutlichtbeleuchtung floodlighting
Fokus focus
Folge sequence, succession
Folgeabtastung sequential scanning
folgen to follow, to succeed, to track, to result
Folgesteuerung sequence control, sequential control
Folie foil, lamina, film
Förderanlage conveyor equipment, conveying plant
Förderer conveyor
Förderhöhe delivery head
Fördermittel handling device
fördern to convey, to handle, to transfer, to hoist, to deliver
Förderrinne conveying chute
Form shape, form, appearance, mould
Formabweichung error of form
Formänderung change of shape, deformation
Formätzen contour etching, chemical milling
Formbeständigkeit dimensional stability
Formdiamant diamond wheel dresser
Formdrehen form turning
Formel formula
formen to shape, to form, to mould
Formfräser form cutter, formed cutter
Formgebung shaping, forming
Formgenauigkeit truth of shape, accuracy of shape
Formgesenk shaped die
Formpressen pressure moulding
formrichtig properly formed
Formstahl structural steel shape, sectional steel
Forscher researcher
Forschung research
Forschungsarbeit research activities, research work
fortdauern to continue
fortfahren to go on
fortlaufend continuous
fortpflanzen, sich to propagate
Fortpflanzung propagation
Fortschaltrelais accelerating relay
Fortschaltwerk stepping mechanism
fortschreiten to advance
Fortschritt progress
Fotodiode photodiode
Fotoelement photovoltaic cell
fotoempfindlich photosensitive
Fotograf photographer
Fotografie photograph, photo, photographic picture
fotografieren to photograph
Fotozelle photocell
Fracht freight, cargo, load
Fräsarbeit milling job, milling operation
Fräsdorn cutter arbor
fräsen to mill

Fräsen milling
Fräser milling cutter
Fräskopf milling head
Fräsmaschine milling machine, miller
frei free, vacant, open, clear
Freifläche open-air space
Freigabe release
freigeben to release, to clear
freigelegt exposed, uncovered
freilegen to uncover
Freileitung overhead transmission line, open-wire line
Freileitungsbau open-line construction
Freileitungskabel overhead cable
Freiluftanlage open-air plant, outdoor plant
Freiluftstation outdoor station, outdoor substation
freisetzen to set free, to liberate
freistehend detached, free-standing
Freistrahlturbine free jet turbine
Freiwinkel angle of clearance,
freitragend self-supporting
Freizeichen free-line signal
Fremdatom impurity atom, foreign atom
Fremdbelüftung forced ventilation
Fremdkörper foreign matter
Fremdspeicher external memory
Frequenz frequency
fressen to seize (mach.), to fret (corrosion)
frisch fresh
Frischdampf live steam
frischen to blow, to fine, to oxidise
Frischluftgerät fresh-air respirator
fritten to frit, to sinter
Fritter coherer
Frontseite front, front side
frostbeständig frost-proof, frost-resistant
Fuge groove
fühlen to sense, to feel, to trace
Fühler sensor, sensing element, tracer
fühlergesteuert tracer-controlled
Fühlerstift stylus
führen to guide, to conduct, to lead, to carry
Führung guidance, run
füllen to fill, to pour in, to charge
Füllfaktor space factor
Füllstandsmesser level indicator, level gauge
Füllung filling, loading, charge
Fundament foundation
fundieren to found
Fünfeck pentagon
fünfeckig pentagonal
Funk radio, wireless
Funkapparat radio set
Funkbake radio beacon
Funkbefeuerung system of radio beacons
Funkbetrieb radio service
Funke spark
Funkeleffekt flicker effect
funkeln to sparkle
Funkempfang radio reception
funken to transmit, to radio
Funkenbildung sparking
Funkenfänger spark catcher
funkenfrei non-sparking
Funkenkammer arc chute
Funkenlöscher spark quencher, spark arrester
Funkenlöschspule blow-out coil
Funkenlöschung spark quenching, spark extinguishing
Funkenstrecke spark gap
Funkenstörung radio noise suppression
Funker radio operator
Funkfeuer radio beacon
Funkgerät radio set
Funkmeßsender radar transmitter
Funkmesstechnik radar engineering
Funkpeiler radio direction finder
Funkpeilung radio direction finding, wireless bearing
Funkruf radio call
Funkspruch radio message
Funktion function
funktionieren to work, to function, to operate, to be in operation
Funktionsprüfung functional gauging
funktionstüchtig serviceable
Furche ridge
Furnier veneer
Fuss foot (30,48 cm), leg, base
fussbetätigt pedal-operated
Fussboden floor

Fussbodenheizung underfloor heating
Fusshebel pedal, foot lever
Fusshöhe dedendum (gear)
Fussmotor foot-mounted motor
Futter chuck (clamping), lining (furnace)
Futteral case, sheath

G

Gabel cradle (tel.) yoke (milling arbor)
Gabelschlüssel open-end wrench
Gabelung split connection, Y-junction, bisection
Galerie gallery, platform
galvanisch galvanic
galvanisieren to electrodeposit, to electroplate
Gang gangway, aisle, alleyway, passage, passageway, gear (car), speed (mach.)
Ganzmetallgehäuse all-metal box, all-metal housing
Garantie guarantee, warrant
garantieren to guarantee, to warrant
gären to ferment
Garn yarn
Gasabscheider gas separator
gasartig gaseous
gasdicht gas-proof
gasen to evolve gas
Gasofen gas oven
Gatterschaltung gate circuit
Gaze gauze
Gebäude building
geben to give, to yield, to send, to result in
Geber transmitter
Geberwicklung pick-up winding
Gebiet area, field, zone, sector
gebietsweise sectional
Gebläse blower, fan
Gebläsekühlung forced-draught cooling
Gebläsemotor forced-induction engine
Gebläsewind blast
gebrauchen to use, to utilise, to apply
Gebrauchsanweisung instructions for use, operating instructions
Gebühr charge, rate, fee
gedopt doped
gedrängt packed closely together
geerdet earthed, grounded
Gefahr danger
Gefahrenklasse dangerous-materials class
Gefahrenzulage hazard bonus, danger money
Gefälle gradient, pitch, slope, fall
Gefäss vessel, cup, container
Gefüge structure, grain texture
Gefügebild micrograph
Gegendruck counter pressure
Gegendruckturbine reaction turbine, back-pressure turbine
Gegen-EMK counter e.m.f., back e.m.f.
Gegenfeld opposing field
Gegenkraft counterforce
Gegenlauffräsen up-cut milling, conventional milling
Gegenmutter jam nut, lock nut
gegenphasig in phase opposition, inversely phased, oppositely phased
Gegensprechanlage inter-communication system, talk-back system
Gegenstand object
Gegentaktgleichrichter push-pull rectifier
gegenüberliegend opposed, opposite
gegliedert articulated
gegossen cast
Gehalt salary (money), content
Gehäuse housing, case, casing, enclosure
Gehrung mitre
gekapselt enclosed, encapsulated, totally enclosed
geklebt glued
Geländer rail, hand rail, railing
Gelenk hinge, hinged joint, pinned joint
Gelenkrohr articulated pipe
gelten für to hold for, to be true for
gemässigt moderate, temperate
gemessen measured
genau accurate, exact, true, correct
Genauigkeit accuracy, precision
genehmigen to grant, to license
Genehmigung licence
Generalplan general plan, general arrangement plan

Generator generator
Generatorschutz generator protective system
genügen einer Bedingung to meet (or: to fulfil, to satisfy) a condition
Geometrie geometry
Gepäck luggage, baggage
Gerade straight line
geradlinig straight-lined
geradzahlig even-numbered
Gerät device, appliance, implement, equipment apparatus
Geräusch noise, sound
geräuschdämpfend silencing, sound absorbent
geräuschlos noiseless, quiet
geräuschvoll noisy
geruchlos odourless, inodorous, scentless
Gerüst scaffold, framing, staging
gesamt total, entire, whole
Gesamtansicht full-view illustration
Gesamtausschaltdauer total break time
Gesamtbereich overall range
Gesamtbetrag total, total amount
geschlossen closed, totally inclosed (apparatus)
geschraubt screwed, bolted
Geschwindigkeit speed, velocity, rate (process)
Geselle journeyman
Gesellschaft company
Gesenk die, cavity block
Gesenkfräser die-sinking cutter
Gesetz law
gesintered cemented, sintered
Gespräch (telephone) call
gesprungen cracked (ceramics)
Gestalt shape, form, design
Gestänge leverage, linkage
gestanzt pierced
gestapelt racked
Gestell frame, bed, stand, rack
gestört faulty, disturbed, troubled
gestrichelt dotted (line), broken
Gesuch application, request
geteilt divided, split
Getriebe gear, gear box
Getriebeturbine geared turbine
Gewicht weight
gewickelt wound
Gewinde thread
Gewindebohren tapping, thread tapping
Gewindebohrer tap, thread tap
Gewinn gain
gewinnen to win, to get, to extract
gezahnt toothed
Gicht top (blastfurnace), charge
Gichtbühne charging platform
giessen to cast, to pour
Giesserei foundry
Giessform mould
Giessharz cast resin
Gift poison
Gipfel summit, top, peak
Gips gypsum, gypsum plaster
Gitter lattice (crystal), grid (tube)
gittergesteuert grid-controlled
Glanz glaze, gloss
glänzen to glisten
glänzend glossy, bright, polished
Glas glass
Glasfaser glass fibre, fibre glass
glasieren to glaze
glatt even, smooth
Glätte smoothness
gleich equal (math.), like (similar)
gleichbleibend constant, uniform
gleichen to equal
gleichförmig uniform
gleichgerichtet unidirectional, equal in direction
Gleichgewicht equilibrium, balance
Gleichlauffräsen down milling, climb-cut milling
gleichphasig in phase
gleichpolig unipolar, homopolar
gleichrichten to rectify
Gleichrichter rectifier
Gleichrichtung rectifying, rectification, demodulation
gleichsetzen to equate
Gleichspannung direct current voltage, d.c. voltage
Gleichstrom direct current (d.c.)
Gleichung equation
gleichwertig equivalent
gleichzeitig simultaneous
Gleichzeitigkeitsfaktor simultaneity factor
Gleichzeitigkeitslogik concurrency logic

Gleis track, line
Gleitbahn slideway, glide, path
Gleitband slip band
gleiten to glide, to slip, to chute (work)
Gleitfeder sliding key, sliding feather
Gleitlager plain bearing, slide bearing
Gleitpassung sliding fit
Gleitschutz non-skid device
Glied joint, member, link, component
Gliederung sub-division, grouping
Gliederwelle articulated shaft
glimmen to glow, to smoulder
Glimmen glow
Glimmentladung glow discharge, corona discharge
Glimmer mica
glimmfrei corona-free
Glimmlampe glow tube
Glocke bell, glass jar, bubble cap
glühen to glow, to anneal (heat treatment)
Glühfarbe annealing colour
Glühkerze glow plug
Glühkopfmotor hot-bulb engine
Glühlampe incandescent lamp
Glühspirale glow plug filament
Gosse gutter
graben to dig, to trench
Graben ditch, drain, trench, dike
Grad degree (scale), order (equatation)
Graetz-Schaltung Graetz rectifier, bridge-connected rectifier
grafisch graphic(al)
Granulate granule, chips
granulieren to granulate, to corn, to shot
Graphit graphite
Grat burr
gravieren to engrave
Gravur engraving
Greifer claw, gripper
Grenzbeanspruchung limiting load
Grenze limit, boundary
Grenzfall borderline case
Grenzfläche interface, boundary surface
Grenzschalter limit switch
Grenzschicht boundary layer
Griff handle, hand
grob coarse, rough
Grobabstimmung coarse tuning
grobkörnig coarse-grained
gross big, large, bulky, great
Grösse size, magnitude, amount
Grössenordnung order of magnitude
Grosskraftwerk high-power plant
grosstechnisch large-scale, industrial
Grossversuch full scale test
Grube pit
Grund bottom, ground
gründen to found
Grundfläche base surface, floor area
Grundform basic design, basic form
Grundgebühr fixed charge
grundieren to prime, to bottom
Grundlast base load, normal load
grundlegend fundamental, basic
Grundmetall base metal, parent metal
Grundplatte baseplate, bed-frame, mounting plate
Grundregel basic rule
Grundriss plan view, layout
Grundschicht primary layer
Gründung foundation
Gruppe group
Gruppieren grouping
gültig bis valid up to
Gummi rubber
gummibereift rubber-tyred
Gurtbandförderer flat-belt conveyor
Gurtförderer belt conveyor
Guss casting, founding, pouring
Gusseisen cast iron
Gussform mould
Gusskern core
Gussstück casting, cast member
Güte quality
Güteklasse grade, quality
Güter goods, freight
gutheissen to approve
Gutlehre go-gauge

H

Haarriss hair crack, capillary crack
haften to adhere, to stick
Hahn tap, faucet, cock
Hahnküken taper plug
haken to hook
Haken hook, peg, clamp
Hakenmeissel hook tool
halbautomatisch semi-automatic
Halbleiter semiconductor
Hals neck, collar
haltbar durable, solid
halten to hold, to carry
Hammer hammer
Hand, von manually, by hand
handbetätigt manually operated
Handel trade, commerce
handhaben to handle
handwarm lukewarm
Handwerk trade, craft
Handwerker craftsman
Hanf hemp
Hängekatze overhead hoist
hängenbleiben to stick, to get stuck, to adhere, to seize
hängend suspended
hart hard
Härte hardness
härten to harden
Hartlot brazing solder
hartlöten to hard solder, to braze
Hartmetall hard metal, cemented carbide
Hartmetalleinsatz cemented insert, carbide insert
Harz resin
Haube hood
häufen to accumulate
Häufung accumulation
hauptamtlich full-time
Hauptantrieb main drive
Haustechnik domestic engineering, domestic services
Haut skin
Hebebühne lifting platform, rising platform
Hebel lever
Hebemagnet lifting magnet
heben to lift, to elevate, to raise, to hoist, to wind
Hebe- und Fördergeräte handling equipment
Heck stern, tail
Heft handle
Heissdampf superheated steam
heissen to hoist
Heissleiter thermistor
Heizanlage heating plant, heating equipment
heizen to heat, to fire, to stoke
Heizfaden filament, heater
Heizung heating, heating system
Heizzentrale central heating plant
hell bright, light
Helligkeit brightness, brilliance
Helling slipway
hemmen to arrest, to stop, to retard
herabdrücken to force down
herablassen to lower
herabsetzen to reduce, to decrease, to lower
heranführen to bring close to, to approach
herausführen to bring out
Herd hearth (furnace), range
Herkunft origin
herleiten to deduce, to derive
herstellen to make, to produce, to prepare
Hersteller manufacturer, maker
Hertz cycles per second (Hz, cps)
herumdrehen to turn around
herumschalten to index around
hervorstehen to project, to protrude
hervorstehend projecting
heulen to howl
Heuler howler
Heulton howl
Hilfe help, aid, assistance, support
Hilfseinrichtung auxiliary attachment
hin und her there and back
hindern to impede, to hamper, to restrain
Hindernis obstacle, obstruction
hindurchgehen to pass through, to reach through
hineinreichen in to extend into
hineinschlagen to drive home
hinten at the rear side, at the rear end, in the rear
hinterdrehen to relieve, to back off
hintereinandergeschaltet connected in series, series connected

hinterfräsen to relief mill
hinterschliffen relief-ground, backed-off
Hintertür backdoor
Hinundherbewegung reciprocation
hin- und herfahren to shuttle, to move back and forth
Hin- und Hergang to-and-fro motion
hin- und hergehen to reciprocate
hinzufügen to add
Hitzdrahtinstrument hot-wire instrument
Hitze heat
hitzebeständig heat-resistant, heat-proof, fire-proof
Hobel plane
hoch high
hochbeansprucht highly stressed
Hochbehälter elevated tank, overhead tank, gravity tank
Hochdruck high pressure
hochfahren to run up, to start up
Hochfrequenz high frequency (h. f.), radio frequency (r. f.)
hochheben to lift, to raise, to hoist
hochkant edgeways, edge-wise, on edge
Hochleistungs . . . heavy-duty, high-duty, high-power, high-speed
Hochofen blast furnace
hochohmig high-resistive, high-impedance
Hochschule, technische technological university, polytechnic
Hochspannung high voltage, high tension
Höchstbelastung maximum permissible load
hochwertig high-grade
Höhe height, altitude
höhengleich at the same level
hohl hollow
Hohlleiter waveguide
Hohlraum cavity, void, flaw
Hohlspiegel concave mirror
Honahle honing tool
honen to hone
hörbar audible
Hörbereich range of audibility
Hörschärfe acuity of hearing
Hub stroke
Hubraum displacement of piston, swept volume
Hufeisenmagnet horseshoe magnet
Hülle casing, sheath, shell, jacket
Hülse sleeve, tube, case, bush
Hupe horn, hooter
Hutmutter acorn nut
Hydraulik hydraulics
Hydraulikaggregat power pack, pump and motor unit
Hyperbel hyperbola
Hypothese hypothesis
Hystereseverlust hysteresis loss

I

Idealbedingung ideal condition
Identifizierung identification
Ilgner-Umformer Ilgner system
illuminieren to illuminate
imaginär imaginary
Impedanz impedance
Impedanzrelais impedance relay, distance relay
implodieren to implode
imprägnieren to impregnate, to preserve
Imprägniermittel impregnating matter, coating varnish
Impuls pulse, impulse
Impulsreihe pulse train
Impulsrelais time pulse relay
Impulstaktgeber clock-pulse generator
Impulsvorwahl preset counts
Inbetriebnahme commissioning, putting into operation, starting service
Indexklinke index latch
indirekt indirect
Induktion induction, flux density (magn.)
Induktionsschutz anti-inductive protection
Induktivität inductance
Industrieanlage industrial plant
Industrie- und Handelskammer Chamber of Industry and Commerce
Industriezweig branch of industry
induzieren to induce
ineinandergreifen to mesh, to mate
Informationsbild information pattern
Informationsfluss information rate (computer)

infrarot infrared
Infraschall infrasound
Ingenieur engineer
Ingenieurschule technical college
Inhalt contents, content, volume, capacity
Innenansicht interior view
Innenbeleuchtung interior lighting
Innendrehen turning internal surfaces
Innenraum interior space
Innenraumschaltanlage indoor switchgear, indoor switching system
Innentaster inside caliper(s)
Innenverkleidung interior lining
instabil unstable
Installateur fitter, plumber
Installationstechnik domestic engineering
installieren to install, to wire (el.)
instandhalten to maintain, to keep in order
Instandhaltung maintenance, servicing
instandsetzen to recondition, to repair, to reservice
Instrument instrument, device
Instrumententafel instrument panel, meter panel
Instrumentierung instrumentation
Integralrechnung integral calculus
Intensität intensity
intensivieren to intensify
intermittierend intermittent
Inventar inventory, stock
Inventaraufnahme stock taking
invers inverse
Inzidenz incidence
Ionenausbeute ion yield
Ionenaustausch ion exchange
Ionosphäre ionosphere
Irisblende iris diaphragm
irreversibel irreversible
Irrtum error, mistake
Irrungszeichen erasure signal
Is-Begrenzer short circuit current-limiting device
I-S-Diagramm Mollier diagram
Isolationsabstand insulation distance
Isolationsbemessung insulation rating
Isolator insulator
Isolierband insulating tape
isolieren to insulate, to isolate
Isotope isotope
Istmass actual size
Iststellung actual position
Istwert actual value

J

Joch yoke
justierbar adjustable
justieren to adjust
Justierfehler maladjustment
Justierung adjustment

K

Kabel cable
Kabelabschluss cable terminal, cable head
Kabelabzweiger cable box
Kabelader cable conductor, cable core
Kabelbruch cable break
Kabine cabin
Kadmium cadmium
Käfig cage
Käfigankermotor squirrel-cage motor
kalandern to calander
Kalanderwalze calander roll
Kaliber calibre, roll pass (rolling)
Kaliberdorn mandrel, sizing pin
kalibrieren to calibrate, to bring to size, to size, to adjust, to scale, to gauge
Kalium potassium
Kalk lime
kalken to whitewash
Kalkulator cost accountant
Kalorie calorie
Kalotte cap, spherical surface
Kaltband cold rolled strip
Kaltbiegeversuch cold bend test
Kälteanlage refrigerator
Kältebeständigkeit anti-freezing property
Kältediagramm psychometric chart
Kälteeinheit frigorific unit
kälteerzeugend cryogenic, frigorific
Kälteerzeugung refrigeration
Kaltprofil cold rolled section

Kammer chamber
Kanal canal, channel
Kante edge, corner, border
kanten to edge, to cant, to tilt
Kanüle cannula, tubule
Kanzel cockpit
Kapazitanz capacitance, capacitive resistance
Kapazität capacity, capacitance
kapillar capillary
Kappe cover, hood, cap
Kapsel capsule
kapseln to encapsulate
Karbid carbide
Karbidhartmetall cemented carbide
Karborund carborundum
Kardangelenk cardan joint, Hooke's coupling
Kardanwelle cardan shaft
Karosserie car body
Karte card, map, chart
Kartei card file
Kartenstanzer card punch
Kartenstapler card stacker
Karton cardboard, carton, pasteboard
Karusselldrehmaschine vertical boring and turning mill
Kaskade cascade
Kaskadenschaltung cascade connection
Kaskodenschaltung cascode circuit
Kasse cash register, cash office
Kasten case, box, bin
Kastenbett box-like bed
Kastenglühen flask annealing
Kastenständerbohrmaschine box column drill
Katalog catalogue
Katalysator catalyst, catalyzer
Katalyse catalysis
Kathete leg (math.)
Kathode cathode
Kathodenstrahloszillograph cathode-ray oscillograph
Kauf purchase, buying
Käuferwunsch customer requirement
Kaufpreis purchase price
Kautschuk caoutchouc
Kavitation cavitation
Kegel cone, taper
Kegeldrehen taper turning
Kegelrad bevel gear
Kegelradantrieb bevel gear drive
Kehle chamfer
kehlen to chamfer, to channel, to groove
Kehlkopfmikrophon necklace microphone, throat-type microphone
Kehlnaht fillet weld
Kehricht refuse, litter
Kehrwert reciprocal
Keil wedge, key, spline
Keilnut keyway
Keilnutenfräser keyseating cutter
Keilriemen vee belt
Keilriemenscheibe vee belt pulley
Keller cellar, basement
Kelvin-Grad degree Kelvin, °K
Kenndaten characteristics
Kennfeuer identification light, marker-beacon light
Kennimpuls label
Kennlinie characteristic line, characteristic curve
Kennwert parameter
kennzeichnen to mark, to designate, to characterise
Keramik ceramics
keramisch ceramic
Kerbe groove, notch, slot, kerf
kerben to groove, to notch, to slot
Kerbschalfestigkeit notch impact strength
Kern core, centre
Kernblech core lamination
Kernbohren trepanning
Kernbrennstoff nuclear fuel
Kernen coring
Kernenergie nuclear energy
Kernkraftwerk nuclear power station
Kernspaltung nuclear fission
Kernspeicher magnetic core memory, ferrite memory, ferrite store
Kerosin kerosene
Kerze candle, spark plug
Kessel boiler
Kesselbau boiler construction
Kesseldampf live steam
Kette chain
Kettenfahrzeug crawler-type vehicle
Kettenförderer chain conveyor
Kiel keel
Kies gravel
Kiesel pebble, flint

kieselsauer siliceous
Kinematik kinematics
Kinetik kinetics
Kinoprojektor cine projector
Kipp sweep (el. tubes)
Kippanhänger tipping trailer
Kippaufzug skip hoist
Kippe tip (site)
kippen to tip, to tilt, to cant
Kipper tipper, tip-truck, dumper
Kippgenerator sweep generator
Kipphebel valve rocker, rocker arm, rocking lever
Kiste box, case, chest
Kistenware box stock

Klammer clamp, clip, cramp, bracket (math.) parenthesis
klammern to clamp, to clip, to brace
Klangfarbe tone quality, tone colour, timbre of sound
Klangregler tone control
Klappe flap, lid, hatch
klappen to flap, to tilt
Kläranlage sewage disposal plant, sewage treatment plant, water treatment plant
Klärapparat clarifier, settling apparatus
klären to clarify, to settle, to clear, to purify
Klarmeldelampe all-clear signal light, all-ready signal light
klassifizieren to classify, to sort, to grade, to screen
Klaue jaw, claw
Klebeband adhesive tape
kleben to adhere, to stick to glue, to paste
klebrig sticky, adherent, gluey
Kleinbetrieb small factory, small-scale manufacturing
Klemmbacke clamping jaw
Klemme clamp, clip, terminal (connection)
klemmen to clamp, to jam, to seize
Klemmenbezeichnung terminal marking
Klemmleiste terminal block, terminal board, connecting block, connecting strip
Klempner plumber, pipe fitter
klettern to climb
Klima climate
Klimaanlage air-conditioning plant, air-conditioning system
klimatisieren to air-condition
Klinge blade
Klingel bell
klingeln to ring, to ring the bell
klingen to sound
Klinke latch, dog, spring jack
Klirrfaktor distortion factor
Klischee block cliché
klischieren to make blocks
klopfen to knock, to pink
klopffest antiknock, knockless, knockproof
Klopffestigkeit anti-knock quality
Klumpen clod, lump
Kluppe die-stock

knacken to click
Knagge dog, trip, catch, cam
Knall crack, bang
knallen to crack
knapp short (of . . .), insufficient
Knappheit shortage
Knarre ratchet
knarren to creak, to rattle
knattern to crackle, to sizzle (radio)
Knebel lock
Kneifzange pincers
Knick kink, bend
Knickbeanspruchung buckling stress
Knickbelastung buckling load
knicken to buckle, to fold
knickfest non-buckling
Knickfestigkeit buckling strength
Knicklast crippling load
Knickspannung critical stress
Knie elbow, angle, bend, knee
Kniehebel toggle lever
Kniestück bend, elbow, knee
Knopf knob, button
Knoten knot, kink, joint
Knotenamt tandem exchange, centre exchange, junction centre
Knotenamtsbereich tandem area
Knotenblech gusset plate
knüpfen to tie
Knüppel billet (met.)
Knüppelwalzwerk billet mill

koaxial coaxial
Kobalt cobalt
kochen to boil

Kocher cooker, boiler
Kochherd range
Kode code
Kodeumsetzer code converter, code translator
kodieren to code, to encode
Kodierung coding
Koerzitivkraft coercive force
Koffer trunk, box
kohärent coherent, single-frequency
Kohäsion cohesion
kohäsiv cohesive
Kohle coal, carbon
Kohlebürste carbon brush
kohlen to soot, to coal
Kohlenbergwerk colliery
Kohlenwasserstoff hydrocarbon
Kokille ingot, mould
Koks coke
Kolben piston, plunger
Kolbenbolzen piston pin, gudgeon
Kolbendampfmaschine reciprocating steam engine
Kollektor commutator, collector
kollidieren to collide, to run into, to run down
Kollision collision, boarding
Kolonne crew, gang, team, column (typing)
Kombizange combination pliers
Kombüse gallery, caboose
Kommanditgesellschaft limited partnership
Kommandoanlage control unit
kommen to come, to arrive, to approach
kompakt compact, self-contained
Kompensation compensation, balancing
kompensieren to compensate, to balance
Komponente component
kompoundieren to compound
Kompressionsraum compression chamber
Kondensat condensate
Kondensator capacitor (el.), condenser
Kondenshahn drain cock
kondensieren to condense
Konduktanz conductance
konform conformal
Kongruenz congruence
konkav concave
Konkurrent competitor
konkurrenzfähig competitive
Konossement bill of lading
konservieren to preserve, to tin, to can, to bottle
Konsole knee, bracket, console
Konsolfräsmaschine knee-and-column type milling machine
Konstante constant
Konstantgleichrichter stabilised rectifier
konstruieren to design, to build
Konstrukteur designer, design engineer
Konstruktion design, structure
Konstruktionsabteilung design department, designing department
Konstruktionsleiter chief designer
konstruktiv structural
Kontakt contact
Kontaktbahn contact path, contact bank
Kontaktbock contact jaw
kontaminieren to contaminate
Kontermutter lock nut, jam nut
kontinuierlich continuous
Kontinuität continuity
Konto account
Kontor office, bureau
kontrastreich high-contrast
Kontrollampe indicating light, signal lamp, tell-tale lamp, pilot lamp
Kontrolle check-up, checking, survey, monitoring
Kontroller controller, control device
kontrollieren to check, to monitor, to control
Konus cone
Konuseinsatz taper adapter
konvex convex
konzentrieren to concentrate
konzentrisch concentric
Konzession licence
Koordinatenachse coordinate axis
Kopf head, top, upper end
Kopfhöhe addendum (gear)
Kopfhörer headphone, headset, earphone
Kopflastigkeit nose heaviness
Kopfschraube cap screw
Kopfseite face
Kopfstütze head rest

Kopie copy, duplicate, print
kopierdrehen to copy-turn
kopieren to copy, to duplicate, to contour, to trace
koppeln to couple
Koppelnetzwerk interstage network
Koppelstufe buffer stage
Kopplung coupling
Kord cord
Kordel diamond-shaped knurling
kordeln to knurl
Kork cork
Körner punch mark, centre punch
Korona corona
Körper body, substance, hull
Korrektor proof-reader
korrodieren to corrode
Kosinus cosine
Kosten costs, expense(s), charge(s)
kostenlos free of charge
Krackanlage cracking plant
Kraft force, power, strength
kraftbetätigt power-operated
Kraftfahrzeug motor vehicle, motor car
kräftig vigorous, powerful
Kraftlinie line of force, field line
Kraftlinienverlauf path of lines of force
Kraftmaschine prime mover
Kraftquelle power source
Kraftstoff fuel, propellant
Kraftwerk power station
Kran crane
Krängungsmesser inclinometer
Kranportal gantry
Kranz gear rim, row of blades
kratzen to scratch, to mar, to scrape
Kratzer scratch, mar
Kräuselung rippling
Kreide chalk
Kreis circle, circuit
Kreisel gyroscope
Kreiselkompass gyro-compass
Kreiselpumpe centrifugal pump
Kreiseltochter repeater compass
kreisen to rotate, to circulate, to gyrate
kreisförmig circular
Kreisfrequenz angular frequency, gyro-frequency
Kreiskolbenmotor rotary engine (Wankel)
Kreislauf circuit, cycle, circulation
kreisläufig cycloididal
Kreisprozess cycle (Carnot Rankine)
Kreisquerschnitt circular cross section
kreisrund circular
Krempel carding engine (machine)
Kreuz cross
Kreuzgelenk swivel joint, universal joint
Kreuzkopf crosshead
Kreuzschlüssel four-way rim wrench
kriechen to creep, to leak
Kriechgang creep feed, creeping feed, inching
Kriechstrom creepage, leakage, leakage current
Kristall crystal
Kristallschwinger crystal resonator
Kristallschwingung crystal vibration
Kriterium criterion
kritisch critical
Krokodilklemme alligator clip
Krone crown, crest, top, summit
Kronenmutter castle nut
kröpfen to crank
Kröpfung offset, gap
krumm curved, bent, buckled, crooked
Krümmer elbow, bend, knee
Krümmung curvature, curve, bend
Kübel bucket, bowl, skip, trough
kubisch cubic
Kugel sphere, ball
Kugeldrehen ball turning
kugelförmig spherical, ball shaped, globular
Kugelfräsen cherrying
Kugelfräser cherry
Kugelgelenk ball-and-socket joint
Kühlanlage refrigeration plant, cooling plant, cooling system
kühlen to cool, to refrigerate
Kühler radiator
Kundendienst service
kündigen to give notice
Kündigungsfrist term of notice
künstlich artificial, synthetic
Kunststoff synthetic material,
Kühlmittel coolant, cooling agent
Küken stopcock

Kulisse main driving link
Kunde customer
Kühlluftmantel cooling-air jacket
Kunststoffkabel plastic cable
Kupfer copper
Kupolofen cupola furnace
Kuppe summit, crest
Kuppel cupola, dome
kuppeln to couple, to connect,
Kurbel crank
Kurbelgehäuse crankcase
Kurbelwelle crankshaft
Kurs course, heading
Kurve curve, graph, plot, bend
Kurvenbahnfräsen track milling
Kurvenleser graph reader, trace reader
Kurvenschar family of curves
kurz short, concise (book)
Kurzarbeit short time working
kürzen (math.) to cancel (a number out of a fraction)
Kürzen cancelling
Kurzfassung abridgement
Kurzhobelmaschine shaping machine
kurzhobeln to shape
kurzlebig short-lived
Kurzschluss short circuit
Kurzspan broken chip
Kurzwelle short wave
Kurzwellensender short-wave transmitter
kurzzeitig short-time, short-term
Küstenradar shore-based radar

Post Apollo

"It's amazing how peaceful it is — now the tourist season is over."
(New Scientist, London)

L

labil labile, instable
Labor laboratory
Lack lacquer, varnish
Lackanstrich varnish coat
Lackdraht enamel wire
Ladeaggregat charging set, battery charger
Ladegeschirr cargo handling gear
laden to load, to fill, to charge
Laderampe loading platform
Ladung load, loading, charge, charging
Lage position, location, layer
Lageeinstellung positioning
Lagen sections
Lagenwicklung layer winding
Lageplan layout plan, layout, site plan
Lager bearing (mach.), store (goods)
Lagerbock pillow block
Lagerhaltung stock keeping
Lagerkäfig bearing cage
Lagerkosten storage expenses
lagern to store
Lagerschale bearing shell, bearing bush
lahmlegen to paralise
lamellar lamellar
Lamelle lamina
lamelliert laminated
Lampe lamp, light, light fixture
Landungsbake landing beam beacon
Langdrehen plain turning, sliding
Langdrehmaschine sliding lathe
Länge length
längen to lengthen, to stretch, to elongate, to extend
langfristig long-term
längs longitudinal, along, lengthwise
langsam slow
Langsamläufer low-speed motor
Langspan long continuous chip
Längsspiel axial play, end play, play in longitudinal direction
Lappen rag, piece of cloth
läppen to lap
Läppen lapping
Lärm noise
Lärmbekämpfung noise control, noise reduction
Lasche strap, butt, splice, tie block, fish-plate
Laschennietung butt riveted joint
Laser light amplification by stimulated emission of radiation
Last load, weight, burden
Lastschalter load-breaking switch
Lastverteilung load distribution
Latte butten, strip
Lauf running, movement, operation
Laufbuchse bush, liner
laufen to run (mach.)
Läufer rotor
Läuferanlasser rotor starter, rotor-resistance starter, rotor rheostat
läufergespeist rotor-fed
Laufruhe smooth running
Lauge lye, leach
laut loud, noisy
Laut tone, sound
läuten to ring
Lautsprecher loudspeaker
Lautstärke loudness level, loudness, volume of sound, sound level
Lebensdauer life, working life
lecksicher leakproof
Leder leather
leer empty, blank (paper)
Leere vacuum, emptiness
leeren to empty
Leerlauf idle running, idling, no-load operation
legen to lay, to place, to deposit
Legierung alloy
Lehrdorn plug gauge
Lehrenbohren jig boring
Lehrling apprentice
Lehrmittel teaching aid
leicht light in weight, easy
Leim glue, size
leisten to perform, to carry out, to give
Leistung power, performance, output, yield, efficiency
leistungsfähig efficient, high-performance, high-capacity
Leistungsfaktor power factor
Leitblech guide plate, chute, baffle plate

leiten to conduct (current), to guide, to direct, to pass, to control
Leiter conductor (el.), manager
Leiterquerschnitt conductor cross section
Leitfaden textbook
Leitfähigkeit conductivity
Leitkarte master card
Leitlineal guide bar
Leitlinie directrix
Leitlochstreifen pilot tape
Leitung mains, line, circuit
Leitwalze guide roller
Leitwert admittance, conductance
Leitzahl routing code, guide-number
Lenkrolle steering idler, steering roller
Lenkung steering, guidance
Lenzpumpe bilge pump
lenzpumpen to free a vessel
Lenztank drain tank
Lesekopf reading head
Leselocher punch reader
lesen to read, to gather
Leuchtdichte luminance, illuminating power per square metre, brightness
Leuchte lamp, lighting fixture
leuchten to light, to emit
Leuchtfaden filament
Leuchtschaltbild illuminated mimic diagram
Leuchtskala illuminated dial
Leuchtstärke luminous intensity
Leuchtstofflampe fluorescent lamp, gas discharge lamp
licht clear
Licht light
Lichtanlage lighting installation
Lichtausbeute luminous efficiency
Lichtbogen electric arc
Lichtbogenlöschung arc quenching
Lichtbogenschweissen arc welding
lichtbrechend refractive, refringent
Lichtbrechung refraction of light
Lichteinfall light incidence
Lichteinheit photometric unit
lichtelektrisch photoelectric
Lichtpause blueprint, print
Lichtquelle light source
Lieferant supplier
Lieferfrist term of delivery
liefern to supply, to deliver, to provide
Lieferschein bill of delivery, invoice
Liefertermin delivery date
Lieferung delivery, supply
Lieferwagen delivery van
liegen to lie, to rest, to be situated
Lineal ruler, edge
Linie line
links left, on the left, to the left, left-hand, anti-clockwise (Rotation)
linksgängig left-hand thread
Linksgewinde left-hand thread
Linse lens
Liste list, register, catalogue, roll
Liter litre
Litze lace, strand, stranded wire, flexible wire
Litzenleiter stranded conductor
Lizenz licence
Lizenzinhaber licensee
Loch hole, aperture, gap, hollow, pit
Lochband punched tape
lochen to punch, to pierce
Löcherleitung p-type conductivity, hole conduction
Lochfrass pitting, localised corrosion
Lochkarte punch card, punched card
Lochkreis hole circle
Lochmitte centre point of hole
Lochschreiber punching recorder
Lochstreifen punched tape, perforated tape
locker loose, slack
lockern to loosen, to slacken
Logarithmus logarithm
Logik logic
Lohn wages, salary, payment, hire
Lohnerhöhung increase in pay, rise in wages
Loran loran (long range navigation system)
losbinden to untie
Löschdauer arcing-time
löschen to extinguish, to quench
Löschgerät erase unit, eraser (tapes)
Löschkammer quenching chamber, arc chute, arc-control device

Löschzeit arcing time, erase-time
lose in bulk (mat.), loose, free
lösen to unscrew, to slacken, to loosen, to unlock, to release, to unclamp, to unlatch
löslich soluble
loslöten to unsolder
losreissen to tear off, to pull off
losschrauben to unbolt, to undo, to unscrew
Lösung solution
Lot solder, hard solder
Lötdraht solder wire, soldering wire
loten to sound, to cast
löten to solder
Löten soldering
Lötfahne soldering tag, soldering lug, pigtail
Lötkolben soldering bit, soldering iron
Lötlampe blow lamp for soldering, blow torch for soldering
Lötmittel solder
Lötzinn soldering tin, tin-base solder
Lücke gap, spacing
Luft air, play (mach.)
Luftabscheider de-aerator, air release (separator)
Lufteinblasung air injection
lüften to ventilate, to ease (valve), to air, to aerate
Lüfter fan, blower, ventilator
Luftkabel overhead cable, aerial cable
Lüftung ventilation, venting, aeration
Luke hatch

M

machen to make, to produce, to manufacture, to perform
Magnetbandspeicher magnetic tape store
magnetisieren to magnetise
Magnetisierung magnetisation
Magnetventil solenoid valve
Magnetverstärker transductor
Magnetzünder magneto
mahlen to crush, to pulverise, to grind, to mill
Mangan manganese
Mangel lack, deficiency, shortness, lack of, shortage of
mangelhaft defective, imperfect, faulty
Mannschaft crew, team
Manometer pressure gauge, manometer
Mantel jacket, shell, sheath, outer cover
Marke mark, brand, label
markieren to mark, to label, to tag
Masche mesh (wire)
Maschine machine, engine
maschinell by machine, mechanical
Maschinenanlage engine plant, propelling machinery (ship)
Maschinenbau mechanical engineering, machine building
Maschinenraum engine room, machinery space
Maschinenschaden machine defect, breakdown of the machine
Maser maser (microwave amplification by stimulated emission of radiation)
Mass dimension, size, measure, gauge
Masse mass, weight
Massengutfrachter bulk carrier
massig bulky, massive
Massstab scale, rule
massstäblich full scale
Mast pole, tower, pylon, mast
Material material, stock, matter, substance
Materie matter, substance
Mathematik mathematics
Mathematiker mathematician
Matrix matrix
Matrize matrix, die, bottom die
matt mat, dim, dull
Mattglas frosted glass
Mauer wall
Maul jaw (snap gauge)
Maulschlüssel open-end wrench
Mechanik mechanics
Mechaniker mechanician
mehradrig multicore, multi-core
mehratomig polyatomic
Mehraufwand extra expenditure
mehrfach multiple
Mehrpreis extra charge, extra cost
Mehrwert surplus

Mehrzweck . . . general-purpose, multi-purpose
Meissel chisel, tool
meisseln to chip
Meisterschalter master controller
Meldelampe indicating lamp, signal lamp, pilot lamp
melden to signal
Meldung message
Membran membrane, diaphragm
Menge amount, quantity
mengen to blend, to mix
Mengenmesser flowmeter
Mennige red lead
Merklampe telltale-lamp
Merkmal feature, criterion
messbar measurable
Messbrücke measuring bridge
Messbuchse test socket
Messe fair
messen to measure, to gauge, to determine
Messer gauge, meter (measuring instrument), knife blade (mach.)
Messergebnis reading, measurement result
Messfühler sensor, sensing element, measuring element
Messgerät measuring instrument, measuring device
Messgrösse measured quantity
Messing brass
Messkette measuring chain
Messplatz test rack
Messwandler instrument transformer, voltage transformer
Messwiderstand measuring rheostat
Metall metal
Metallwaren hardware
Metazentrum metacentre
Meter metre
Meterkilogramm kilogramme-meter
Methan methane
Methode method
mieten to charter, to hire, to rent (flat)
Mignonsockel miniature Edison screw cap, midget cap, midget socket
Mikrofon microphone, transmitting set
Mikrometerschraube micrometer screw
Mikroschalter miniature switch, micro switch
Mikroskop microscope
Militärtechnik military engineering
Minderbetrag deficit, shortage
minderwertig low-grade, poor, low-quality
minimieren to minimise
Minoritätsträger minority carrier
Minuend minuend
Minusleiter negative conductor
Mischbatterie mixer tap, mixing valve
mischen to mix, to blend, to mingle, to merge
Mischer mixer, blender
Mischpult master control console
Mischung mixture, mix, blend
Mitarbeiter co-worker, assistant, collaborator, contributor (book)
mitführen to carry along
Mitglied member
mithören to listen in, to enter a circuit
Mithörschaltung monitoring circuit, side-tone circuit
Mitnahme drive (work), driving of the work
mitnehmen to drive (the work), to take along, to carry along
Mitnehmer work driver, dog, carrier
Mitte centre
Mittel mean (math.), average, means
mittelgross medium-size, medium-sized
Mittelleiter neutral wire, middle wire, zero wire
Mittellinie centre line
Mittelpunkt centre, centre point, mid-point, neutral
Mittelwert mean, average
mitten to centre
Mitten centring
Modell model, pattern, form
Modelltischlerei patternmaker's shop
Modellversuch model experiment
moderieren to moderate, to slow down
modifizieren to modify
Modifizierung modification
Modul modulus (math.), metric module

Modulationsfrequenz modulation frequency
Modus mode
möglich possible, feasible, practicable
Möglichkeit possibility, feasibility, practicability
Mol mol, mole
Molekül molecule
Mollierdiagramm Mollier diagram
Molybdän molybdenum
Moment moment, momentum
momentan instantaneous, momentary
Momentschalter quick break switch, quick-make-and-break switch, high-speed switch
Monelmetall Monel metal
Monozelle single cell
Montage mounting, erection, installation, construction, fitting, assembling
Montageabteilung assembly section
Montageanweisung fitting instruction, assembling instruction
Montageband assembly line
montagefertig ready-to-assemble
Monteur fitter, assembler
montieren to mount, to fit, to install, to erect
morsch decayed, rotten
Morseapparat Morse telegraph
morsen to morse
Mörtel mortar
Motor motor (el.), engine (aut.)
Motorzylinder engine cylinder
Muffe sleeve, bush, muff, sealing, joint
Multiplikand multiplicand
Multiplikator multiplier
multiplizieren to multiply
Mundstück nozzle, neck (turb.)
Mündung aperture, mouth
Muntzmetall Muntz metal
Muster sample, specimen, model, prototype, standard, example
Musterschutz trade-mark protection
Mutter nut
Muttergewindebohrer nut tap

N

n-leitend n-conducting
n-Leitung n-type conduction
n-Zone n-region
Nabe hub, boss
nachahmen to copy, to duplicate, to simulate
nacharbeiten to rework, to correct, to refinish
Nachbarkanal adjacent channel
nachbauen to reproduce
nachbehandeln to aftertreat
Nachbehandlung additional treatment, aftertreatment
nachbestellen to reorder
nacheilen to lag
Nacheilen lagging
Nachformdrehen contour turning, duplicate turning, copy turning
nachformen to copy, to duplicate, to contour
Nachführung follow-up
nachfüllen to refill, to replenish,
nachlassen to slacken in speed, to slow down
Nachlaufregler follower, follow-up mechanism, servomechanism
Nachlaufschaltung tracking circuit
nachprüfen to verify, to check
nachrechnen to calculate again, to check again
Nachricht message
nachrichten to re-level, to re-align
Nachrichteneinheit bit
Nachrichtenverkehr communication
Nachspannen des Werkstücks reclamping the work
nachstellbar adjustable, readjustable
nachstellen to readjust, to reset
nachsteuern to follow-up
nachstimmen to retune
Nachweis proof, verification
nachwiegen to check the weight
Nadel needle, stylus, broach (mach.)
Nagel nail, spike
Näherungswert approximate value
Naht seam, weld, joint
nahtlos seamless
Nahtschweissen seam welding

Nahverkehr local traffic, short-distance traffic
Namensschild nameplate
Napf cup
narrensicher fool-proof
Nase nose, catch
nass wet, humid, moist
Natrium sodium
Nebelkammer cloud chamber
neben alongside, near, close to
nebeneinander side by side
Nebenschluss shunt, by-pass
Nebensprechen crosstalk
Nebenstelle extension
Nebenuhr slave clock, repeater clock, secondary clock
Nebenwirkung side effect
negativ negative
nehmen to take, to receive, to seize
neigen to incline, to rake, to tilt
Neigung inclination, rake, tilt
Nenn ... rated, nominal
Nenner denominator
Neopren neoprene
Nettomasse net weight
Netz network, mains, system, supply
netzen to wet
netzgebunden mains-borne
Netzteil power supply, supply unit
Netzwerk network
Neuerung innovation
nichtätzend non-corrosive
Nichtbefolgung non-observance
Nichteisenmetall non-ferrous metal
nichtfluchtend misaligned
nichtleitend non-conducting
Nickel nickel
Niederdruck low pressure
Niederfrequenz low frequency, audio-frequency
niedergehen to descend, to move downward
Niederhalter hold-down, toe dog
niederlegen to lower, to hinge down
niederohmig low-resistance, low-resistive
Niederschlag precipitate, deposit, sediment, fallout
niederschlagen to deposit, to precipitate
Niederspannung low voltage
Niederspannungsanlage low-voltage installation
Niet rivet
nieten to rivet
Nippel nipple
Nische recess, niche
Nitrat nitrate
Niveau level
nivellieren to level
Nocke cam, dog, boss
Nocken cam
nockenbetätigt cam-actuated
nockengesteuert cam-controlled
Nockenwelle cam shaft
Nonius vernier
Norm standard
normieren to standardise, to calibrate, to normalise
Normteil standard component
Notabschaltung emergency shut-down
Notaggregat emergency diesel-generator station, stand-by unit
Notausschalter emergency stop, emergency cutout
Notausschaltung emergency tripping
Notbatterie emergency battery
nuklear nuclear
Nullabgleich zero balance
Nulldurchgang zero passage
Nulleiter neutral conductor, zero conductor
numerieren to number
Numerierung numbering
numerisch numerical
Nummernwahl dialling
Nussisolator egg insulator
Nut groove, flute
nuten to groove
Nutenkeil slot wedge
nutzbar utilisable, usable
nutzbringend profitable
Nutzfahrzeug commercial vehicle
Nutzlast payload, load
Nutzleistung effective power, net efficiency, service output
Nutzungsdauer working life, service life
Nylon nylon

Post Apollo

"This one's fully automatic"

O

obenliegend overhead
Oberfläche surface
oberflächlich superficial(ly)
Oberleitung overhead line, overhead contact wire (bus)
Oberwelle harmonic wave, harmonic component, upper harmonic
Oberwellenanteil harmonic content
Objekt object, specimen
ODER-Glied OR-gate, OR-element
Ofen furnace (met.), stove, oven
öffentlich public
offerieren to offer
Offerte offer, tender
offiziell official
öffnen to open, to unlock, to break, to disconnect
Öffnung aperture, opening, orifice, port
Öffnungskontakt break contact, opening contact
Ohm ohm
Öhr ear, eye
Oktan octane
Oktanzahl octane number, octane rate, anti-knock value
Okular ocular, eyepiece
Öl oil
Ölabdichtung oil seal
Ölabscheider oil separator, settling tank
ölbeständig oil-resistant
Ölbohrinsel oil-drilling rig, off-shore drill rig
Ölkabel oil-filled cable
opak opaque, black
Optik optics
optimieren to optimise
ordnen to arrange, to collate
Ordner folder, collator
Organ member, element
Ort place, site, spot
orten to locate, to position
örtlich local(ly)
ortsbesetzt locally busy (tel.)
ortsfest stationary, fixed
Ortung positioning, locating, position finding
Öse eyelet, ear
Oszillator oscillator

Oszillograph oscillograph
Oszilloskop oscilloscope
Ottomotor Otto engine
Oxid oxide

P

p-leitend p-conducting, p-type
Pacht lease
pachtweise on lease
Pack pack, bundle, bale
packen to pack, to bundle
Packung package, packing, gasket
Paket parcel, packet, package, bunch
Palette palette, pallet, loading pallet
Panel panel
Panne puncture, breakdown, failure
panzern to shield, to screen
Panzerrohr steel conduit
Panzerschlauch armoured hose
Papierfabrik paper mill
Pappe cardboard, board, carton
Parabel parabola
parabolisch parabolic
Parallelbetrieb parallel operation
Parallelreisser surface gauge, scribing block
parallelschalten to connect in parallel
partiell partial
Passagier passenger
Passarbeit fitting
Passdorn setting plug
passen to fit, to suit
Passfeder feather, key
Passfläche mating surface
passgenau true to size
passieren to pass, to traverse
Passlehre gauge
Paßstift locating pin, fitting pin, set pin
Passteil fitting piece, mating part
Passung fit
Patent, angemeldetes patent pending, filed patent
Patentamt patent office
Patrize counter-die, negative matrix
Patrone catridge (fuse)
Pause break, interval, tracing (drawing), print
pausen to trace, to print
Pech pitch
Pegel level, gauge
peilen to take a bearing, to bear, to gauge (level), to sound
Peilgerät direction finder
Peilung direction finding, bearing, sounding
Pendel pendulum
Pendelbetrieb shuttle service
Pendellager self-aligning bearing
Pendelmotor swivel bearing motor
pendeln to move to and fro, to shuttle, to swing, to reciprocate
Pendelverkehr shuttle service
perforieren to perforate, to punch
Perforierung perforation
Periode period, cycle
Periodendauer cycle duration
periodisch periodic, cyclic
Peripherie periphery, circumference
Permeabilität permeability
Persenning tarpaulin
Personalbestand staff
Pfahl pile, pole, post
Pfanne ladle, pan
Pfeife whistle
pfeifen to whistle
Pfeil arrow
Pfeiler pillar, column
Pfeilverzahnung herringbone teeth, double-helical teeth
Pferdestärke horsepower
pflastern to pave
Pflege attendance, maintenance, service
pflegen to maintain, to service, to tend, to look after
Pflicht duty, obligation
Pflichtenheft specifications
Pfosten post, pillar
pfropfen to graft
Pfropfen grafting
Pfusch slipshod work
pfuschen to botch
pH-Wert pH value, hydrogen ion concentration
Phänomen phenomenon; Pl.: phenomena
Phantomkreis phantom circuit
Phase phase
Phasenabgleich phase adjustment
phasenabhängig phase-sensitive

phasenverschoben dephased, offset in phase
phasenvertauscht misphased
Physik physics
physikalisch physical
Physiker physicist
piezoelektrisch piezoelectric
Pinole quill, sleeve
Pinsel brush
Piste runway
Plan plan, schedule, layout
plan plane, flat
Planarbeit facing operation
Plandreharbeit surfacing operation
plandrehen to face, to surface
planen to plan, to project, to design, to schedule, to face (mach.)
Planetengetriebe planetary gears
Planfräsmaschine fixed-bed type milling machine, manufacturing bed-type milling machine
plangemäss as per schedule, according to the plan
Planhobelmaschine surface planing machine
planieren to level, to plane, to planish
Planierung levelling, grading
plankonvex plano-convex
Plankurve face cam
planmässig scheduled, systematic, planned
Planscheibe faceplate
Planspiel management game
Planung planning, design
Plasmabrenner plasma torch
Plastik plastic (material)
Platin platinum
Platine sheet bar (billet), circuit card, circuit board (el.)
Platinenwalzwerk sheet-bar rolling mill
platinieren to platinise
platt flat, level, even, plane
Plättchen chip (el.), tool tip (mach.), wafer, slab (crystals)
Platte plate, sheet, slab
plätten to iron, to smooth, to press
Plattenventil disk valve
Plattform platform
plattieren to clad, to bond, to plate
Plattierung cladding, plating
Platz place, site, space, room, seat, spot, position, point
Platzarbeiter yard man
Platzbedarf space requirement, space required
platzen to burst, to explode, to crack, to blow out (tyre)
platzsparend room-saving, space-saving
Pleuel connecting rod
Plombe seal
pn-Flächentransistor p-n junction transistor
Podest platform, landing
Pol pole
Polanker armature with salient poles, pole armature
polarisieren to polarise
Polarität polarity
polen to pole (molten metal)
Polier foreman
polieren to polish, to burnish
Poliergerüst planishing stand
Polierläppen buffing
Polierschleifen abrasive-belt polishing
Polierstahl burnisher
Polrad magnet wheel, field spider
Polschuh pole shoe
Polteilung pole pitch
Polumkehr pole reversal
Pore pore, pin hole (paint)
porenfrei non-porous
Portalbauweise portal design
porös porous
Portalkran gantry crane, portal crane
Porzellan porcelain
positiv positive
Potenz power (math.)
Potenzierung raising to a power
Prägeform coining die, stamping mould
prägen to coin, to stamp, to mould
Prägepresse embossing press
Prägestempel punch, die
Prallblech baffle plate, deflector
präparieren to prepare, to treat, to preserve
prasseln to crackle, to splutter
Praxis practice
präzis precise, exact
Präzision precision
Preis price, cost
Preisangabe quotation

prellen to bounce, to rebound, to chatter
Prellwand baffle
Presse press, brake, jack
pressen to press, to mould, to squeeze
Pressform die block, compression mould
Pressgasschalter compressed-gas circuit-breaker
Pressglas moulded glass
Pressluft compressed air
Pressnaht fin
Presspappe pressboard
Presssitz press fit
Pressstoff pressed material, compression moulding material
Prinzip principle
Pressung pressure, compression
primär primary
Prinzipschaltbild schematic, schematic diagram
Prisma prism
Probe sample, specimen
Probeabzug proof sheet (print)
Probebetrieb trial operation, trial run, test run
Probefahrt trial trip (ship), trial operation
Probelauf test run, trial run
Probeentnahme sampling
Probezeit trial period
probieren to try, to check, to assay
Produktion production, output, yield, manufacture
produzieren to produce, to manufacture, to yield
Professur professorship
Profil profile, contour, section, shape
Profile (Pl.) sections (steel)
Profilfräsen profile milling
Profilbild profile chart
Profilstahl sectional steel
Profilwalzblock shaped ingot
Programm programme, program, routine, instruction
Programmgeber timer
Programmschalter controller, sequence switch
Projekt project
Protokoll record, minute, minute of meeting
Prozent per cent
Prozentsatz percentage
Prozess process, operation
Prüfbericht test report
Prüfeinrichtung test equipment, testing facilities
prüfen to test, to check, to verify, to gauge
Prüfen testing, checking
Prüffeld test bay, test laboratory
Prüfmittel testing medium
Prüfstand test stand, test rig
Prüfung test, testing, inspection, examination, verification
Prüfungsfach subject of examination
Prüfvorschrift test specification
puffern to trickle (charge), to buffer
Pufferschaltung buffer circuit
Pulserzeugung pulse generation
pulsieren to pulsate
Pulskodemodulation pulse-code modulation
Pult desk, console (control)
Pulver powder
pulverig powdery
Pumpenkolben pump piston, pump plunger
Pumpenkorb pump strainer
Punkt point, centre, spot, dot
Punktbildung dot formation
punktförmig point, pointlike
punktgeschweisst spot-welded
punktieren to point, to dot
Punktschweissen spot welding
Punktzahl score
pupinisieren to pupinise, to coil-load
Pupinkabel coil-loaded cable, loaded cable
putzen to clean, to trim, to burr
Putzmittel cleaning material, polish
Pyramide pyramid

Q

Quadrat square
quadratisch square, quadratic
quadrieren to square
Qualifizierung qualification, training
Qualität quality
Qualm smoke, exhaust fumes
Quantenenergie quantum energy
Quantenmechanik quantum mechanics
Quarz quartz
quarzgesteuert crystal-controlled
Quecksilber mercury
Quecksilberdampflampe mercury-vapour lamp
Quelle source
quellen to swell
quer transverse, cross
Querbeanspruchung transverse strain
Querschnitt cross section, section
quetschen to squeeze
Quetschklemme spring clip
quietschen to squeal, to squeak
quittieren to sign, to acknowledge, to receipt
Quittung receipt

R

Rachenlehre snap gauge
Rad wheel
Radabstand wheel-centre distance
Radantenne cartwheel aerial
Radar radio ranging and detecting: Funkortung
Radarnetz radar link
Räderantrieb geared drive
räderfrei gearless
Rädervorgelege back gears
Radialspiel radial play
radieren to erase, to etch, to grind
Radioaktivität (radio)activity
Radiotechnik radio engineering
radizieren to draw roots, to extract a root
Radnabe wheel hub
Radstand wheel base
raffen to gather, to gather up
Raffinerie refinery, refining plant
raffinieren to refine
Rahmen frame, chassis, border, rim
Rakete rocket, missile
rammen to ram, to drive piles
Rampe ramp, platform
Rand edge, rim, border
Randeffekt edge effect, boundary effect, fringe effect
Randeinfassung edging
Rändel knurling
Rändelschraube knurled screw
Randintegral integral around a closed contour
Randlinie boundary line
Randschicht transition region, boundary layer, peripheral layer
Randwert marginal value, boundary value
Rang rank, rate, order, class, grade, degree
rangieren to shunt, to switch
Raster screen, mesh, grid
Rasterbild picture frame
rastern to screen
Rasterplatte grid board
rationalisieren to rationalise
rationell rational, economical
ratterfrei chatterfree
rattern to rattle, to chatter
Rauch smoke, fume (chem.)
Rauchentwicklung smoke development
Rauchgasanzeiger smoke detector
Rauchgaskanal waste gas flue
Rauchpilz mushroom cloud
rauh rough
Rauheit roughness
Raum room, chamber, space, zone
Räumarbeit broaching operation (mach.), clearing job
Räumdorn push broach
Räumdurchgang broaching pass
räumen to broach (mach.), to clear
Raumfahrt space travel, space flight
Raumformfräsen cavity milling
Raumgitter space lattice
Raumheizung space heating
Raumladungsdichte space-charge density
Räumnadel broach, internal broach

Raumsonde space probe
raumsparend space-saving
Raupenschlepper crawler tractor, caterpillar tractor
Rauschanteil noise component
Rauschbild noise pattern
Rauschen noise, background noise, random noise, hiss
rauschend noisy
RC-Kopplung RC coupling
Reagenzflasche reagent bottle
reagieren to react, to respond
Reaktanz reactance
reaktionsfähig reactive, responsive
Reaktionsturbine reaction turbine
Reaktor reactor, pile
Realzeit real time
Rechenanlage computer, computer system
Rechenfehler computing error
Rechenschieber slide rule
rechnen to calculate, to compute, to reckon
Rechnen arithmetic, computing
Rechner calculator, computer
Rechteck rectangle
rechteckig rectangular
Rechteckimpuls square-wave pulse, square pulse, rectangular pulse
rechts, nach to the right, towards the right, right-hand side
rechtsdrehend clockwise
Rechtsgewinde right-hand thread
rechtsgültig legal, valid
rechtwinklig right angular, rectangular
recken to stretch, to extend
redigieren to edit
reduzierbar reducible
reduzieren to reduce, to cut down
Reedrelais reed relay
Reflexion reflection
Regal rack
Regel rule, principle
Regelantrieb controlled drive
regelbar controllable
Regelgerät controller, control unit, governor
Regelkreis regulating circuit, automatic control loop
regellos random (math.)
regelmässig regular, uniform
regeln to control (closed loop), to govern
Regelung closed loop control, automatic control regulation
Regelungstechnik control engineering
Regelventil control valve
Regelverstärker variable-gain amplifier
Regelwiderstand rheostat
Regler controller, automatic controller
Reibahle reamer
Reiben reaming
Reibkupplung friction clutch
Reibung friction
reichen bis to reach to, to extend to
reichlich copious, abundant, ample
Reichweite reach, range of transmission
Reifen tyre
Reihe row, array, series, batch, set, number of
Reihenfolge order, sequence
Reihenklemme terminal block
Reihenschaltung series connection
Reihenwiderstand series resistance
rein pure, clean
Reinheit purity, cleanliness
reinigen to clean, to cleanse, to purify, to refine
Reinigung cleaning, cleansing, purification
Reise travel, journey, voyage
Reisegeschwindigkeit cruising speed
Reissbrett drawing board
Reissbrettstift drawing pin, thumbtack
reissen to fracture, to crack, to rupture, to tear
Reissnadel scriber
Reitstock tailstock
rekombinieren to recombine, to re-unite
Relais relay
relativ relative
Relativitätstheorie theory of relativity, relativity theory
Reling railing
Reluktanz reluctance
remanent remanent, residual
Rennstahl ore steel, bloomery iron

Renovierung face-lift, renovation, renewal, restoration
rentabel profitable
Rentabilität profitability
Reparatur repair, overhaul
reparieren to repair, to overhaul, to recondition, to mend
repetieren to repeat
reproduzieren to reproduce, to duplicate, to repeat
Reserve reserve
Reserveanlage stand-by plant
Reservebatterie spare battery
Resonanz resonance
Rest remainder, residue, remanent
Resultat result
Resultierende resultant
retten to save, to rescue
Retusche retouching
retuschieren to retouch
Revolverdrehmaschine turret lathe
Rezept recipe
reziprok reciprocal
Richtantenne directional aerial
richten to direct (radio), to align, to straighten, to put up
Richtfunkstation beam station
Richtfunkverbindung radio link
Richtmagnet control magnet
Richtplatte surface plate
Richtung direction, alignment (mach.)
Richtverstärker directional amplifier
Richtwert reference value, recommended value
riefen to corrugate, to flute, to ridge, to groove
Riemen belt
Rieselkühler spray cooler
rieseln to run,to trickle
Riffelblech chequer plate
Riffelstahl fluting tool
Riffelung fluting
Rille groove, scratch
Ring ring, annulus, washer
Ringbohren trepanning
ringförmig ring-shaped, annular, cyclic
Rippe rib, web
Rippenheizkörper radiator
Rippung ribbing
Riss fissure, fracture, crack
rissig cracky
Ritzel pinion
ritzen to scratch
Roboter robot
Rockwellhärte Rockwell hardness
roh crude, raw, blank, rough
Roheisen pig iron
Roherz crude ore, raw ore
Rohling blank, raw piece
Rohmass rough size
Rohmaterial raw material, stock
Rohr pipe, tube, conduit, duct
Rohrbruch pipe fracture
rohrförmig tubular
Rohrgewinde pipe thread
Rohrkrümmer pipe bend, elbow, knee
Rohrleitung piping, tubing, pipe conduit
Rohrmuffel pipe bell
Rohstahl crude steel
Rohstoff raw material
Rolle pulley, idler, block, reel
rollen to roll
Rollenförderer roller conveyor
Rollenlager roller bearing
Rollwagen wheeled truck
röntgen to X-ray, to radiograph
Röntgengerät X-ray unit, X-ray apparatus
Rost rust, grate, grating, grid, screen
rosten to rust, to corrode
rostfest rustproof, anticorrosive
rostig rusty
Rostschutz rust prevention, rust-proofing
Rostschutzanstrich anti-rust paint, rust-protecting paint
rotglühend red-hot
Rotguss red brass
Rubin ruby
Ruck jerk
Rückansicht rear view
ruckartig jerky
rückbilden to re-form, to reshape
Rückdruckturbine reaction turbine
rucken to jerk
rücken to shift, to jack, to throw (lever), to move over, to push
Rückfluss return flow
Rückführung recycling, recirculation, feedback
rückgängig machen to cancel
rückgewinnen to reclaim, to recover
Rückgewinnung recovery, reclamation

rückkoppeln to feedback
Rückkopplung feedback
rückkühlen to recool
Rücklauf reverse stroke, return stroke, non-cutting stroke (mach.)
Rücklauftank return tank
Rückleistung reverse power
rückleiten to reconduct
rückmelden to feedback
Rückmeldung feedback
Rückprall rebound, recoil
Rückschlagventil non-return valve
Rückseite rear, rear side, back, back side
Rückstand residue, deposit
Rückstau back pressure, back-surge, reflux
rückstellen to reset
Rückstellung resetting
rückstossfrei recoilless
rückstreuen to back-scatter
Rückstrom reverse current
Rückstromrelais reverse-power relay, reverse-current relay
Rückwand rear wand, back wall, rear panel
rückwärts backwards, back, reverse
ruckweise jerky
Rückweisung rejection
rückwirken to react
Rückwirkung reaction
Rückzugsfeder release spring
rückzünden to arc back
Ruf call
rufen to call, to ring
Ruhe rest, silence, quietness
Ruhekontakt normally-closed contact
Ruhelage rest position
ruhend stationary, at rest, static
Ruhespannung normal-operation voltage
Ruhestellung off-position, rest-position
Ruhestromanlage closed-circuit system
Ruhestromauslösung no-voltage release
Ruhestromkreis closed circuit
Ruhezeit rest period, dwell period
ruhig quiet, calm, smooth
rühren to agitate, to stir
Rumpf body, trunk, hull
rund round, circular, rounded
Rundfunk radio, broadcasting, wireless
Rundfunkgerät radio set
Rundfunktechnik radio engineering
Rundheit roundness
Rundlauf true running
rundschalten to index around
Rundschalttisch indexing rotary table
Rundschleifen cylindrical grinding
Rundstahl round bar, round bar steel
Rundstrahlantenne omni-directional aerial
Russ soot
Russbläser soot blower
rüsten to set up, to tool, to prepare
Rüstzeit tooling time, set-up time
rutschen to slip, to slide, to chute (workpiece)
rutschfest anti-slip, non-slip, non-skid
Rutschkupplung friction coupling, safety coupling
rutschsicher non-skid
rütteln to vibrate, to rock, to shake

S

Sanitärtechnik sanitary engineering
Saphir sapphire
Satellit satellite
satt saturated
sättigen to saturate
Sättigung saturation
Sättigungsdrossel saturable reactor
Satz set, batch
Satzfräser gang cutter
Sauberkeit cleanliness
sauer acid, sour
Sauerstoff oxygen
saugen to suck, to absorb
Sauggebläse exhauster
Saugkorb suction strainer, suction basket
Saugluft intake air, indraft
Saugseite intake side (pump)
Saugventil suction valve
Saugzug suction, draft

Säule pillar, post, column
Säulenbohrmaschine upright drilling machine
Säulendiagramm bar chart
Säure acid
schaben to scrape
Schaben scraping
Schaber scraper
Schacht shaft (mine), well (elevator), stack (blast furnace), casing (ship), trunk
Schachtel box
Schachtelung lamination overlap (transformer cores)
schachten to excavate, to sink
Schachtwand manhole wall (cable)
Schaden damage, breakdown, defect
schadhaft faulty, defective, damaged
schädigen to damage
Schaft shank, shaft, handle, stem
Schaftfräser end mill cutter, shank cutter
Schale shell, pan, basin, bowl
Schall sound
schallabsorbierend sound-absorbing
Schallbekämpfung suppression of noise
Schalldämpfung sound damping, sound attenuation
Schalldämpfer silencer, exhaust silencer, sound absorber
schalldicht sound-proof
schallen to sound
schallisoliert sound-insulated
Schallpegel sound level
schallschluckend sound-absorbing
Schaltalgebra Boolean algebra
Schaltanlage switchgear, distribution system, switching system
schaltbar controllable, switch selected, indexing
Schaltelement switching element, circuit element, control element
schalten to switch, to connect, to control, to operate, to index, to wire
Schalter switch, circuit-breaker, interrupter
Schaltfolge switching sequence
Schaltgerät switchgear, control gear, switching device
Schalthebel switch lever, operating lever, control lever
Schaltimpuls switching pulse
Schaltklaue shift dog
Schaltkreislogik circuit logic
Schaltleiste connecting block
Schaltleistung breaking capacity, rupturing capacity
Schaltorgan switching element
Schaltplan circuit diagram, wiring diagram, wiring scheme
Schaltpult control console, control desk, control panel
Schaltschloss latch (el.)
Schaltschrank control cabinet, control box
Schaltspiel switching cycle
Schaltstück contact member
Schalttafel switchboard, control panel
Schaltuhr timer, time switch
Schaltung circuit, connection
Schaltzeichen wiring symbol
Schaltzelle control cubicle
scharf sharp, defined
Schärfe sharpness, keenness (edge), definition (opt.)
schärfen to sharpen
Schärfentiefe depth of focus
scharfkantig sharp-edged, angular, feather-edged, sharp-cornered
Schatten shade, shadow
schätzen to estimate, to appreciate, to value
Schätzung estimation, evaluation
Schaubild graph, mimic diagram, plot
Schaufel blade (turb.)
Schauglas sight glass, oil-flow indicator
Schaukasten display box, show-case
Schauloch sight hole, peep hole
Schaum foam
Schaumstoff foamed plastic
Schauzeichen indicator, annunciator
Scheibe disk,slice, plate, wafer
scheiden to separate, to part, to refine
Schein brightness, shine, appearance

Scheinleistung apparent power
Schellack shellac
Schelle clip, clamp, bracket
Schema block diagram, general layout
Schenkel arm, leg, side
Schenkelpolgenerator homopolar generator
Scherbelastung shear load, shearing load
Schere scissors, shear(s) (large scissors)
scheren to shear, to clip
Scherfestigkeit shear strength, shearing strength
scheuerfest wear-resistant
scheuern to gall (metals), to scrub, to scour
Schicht layer, coat, film, coating, lamina, deposit
Schichtarbeit shift work
Schichtbildung lamination
Schichttransistor junction transistor
schichtweise by layers, laminated
Schieber slide valve, sliding member (general)
Schiebung translation, displacement
schief oblique, skew, sloping, inclined
Schiefe obliquity
Schieflast load unbalance, asymmetrical load
schiefwinklig oblique-angled, skew
Schiene rail, beam, bar
schiessen to shoot, to blast
Schiff ship, vessel, boat, craft
Schiffsantrieb (ship's) propulsion, marine propulsion
schimmeln to mould
Schild protective shield
schimmern to shimmer, to glitter
Schirm protective screen, shield
schlackebildend slag-forming
schlackenreich rich in slag
schlaff slack, loose
Schlag blow, impact, beat
Schlagbohrer percussion drill, churn drill
schlagen to blow, to strike, to beat
schlagfest impact-proof
Schlagfräser fly cutter
Schlagseite lapside, lopside, list
Schlagseite haben to list
Schlamm mud, sludge, slurry, dredge, slime
Schlange coil
schlank slender, slight
Schlankheitsgrad slenderness ratio, coefficient of fineness, length constant
Schlauch hose, flexible tube
Schlechtlehre not-go gauge
Schleichgang creep feed, inching
Schleier fog, haze, mist
Schleife loop
schleifen to grind, to cut
Schleifring collector ring, collector
Schleifringanlasser slip-ring starter
Schleifscheibe grinding wheel, abrasive wheel
schleppen to tug (ship), to tow
Schleuderguss centrifugal casting
schleudern to centrifuge, to sling, to throw
Schleuderprüfung dynamic balance test
schlichten to finish-machine, to dress, to smooth
Schlichten finishing, dressing, sizing
Schlichtmeissel finishing tool
schlichträumen to finish-broach
schliessen to close, to lock, to shut, to close down (factory), to make (el.), to complete
Schliesser a-contact, closing contact, making contact
Schliff grind
Schlinge loop
Schlingerbewegungen corkscrew motion, corkscrewing
Schlingern roll, easy roll
Schlitten slide (mach.), carriage, saddle
Schlitz slot, kerf, groove, slit
Schlitzring slit ring
Schloss lock, padlock, latch (breaker)
Schlosser locksmith, mechanic, fitter
Schlosserei fitter's shop
Schlosskasten apron (lathe)
schlucken, Schall to absorb sound
schlupffrei non-slip

Schlüssel key, wrench, spanner, code, cipher
Schlüssel, verstellbarer adjustable open-end wrench
schlüsseln to code, to cipher
schmal narrow, thin
schmelzen to fuse, to melt
Schmelztiegel crucible, melting pot
schmiedbar forgeable, malleable
Schmieden forging
schmieren to lubricate, to grease, to oil
Schmiernippel grease nipple
Schmutz dirt, scud
schmutzig dirty
schnappen to snap, to catch
Schnecke worm
schneckenförmig spiral, helical
Schneckengetriebe worm gearing
Schneidbrenner blowpipe, cutting torch, flame cutter
Schneide cutting edge, bit
Schneideisen die
schneiden to cut, to clip, to shear, to blank, to notch, to trim
Schneideneinsatz set of inserts
schneidhaltig capable of retaining the cutting edge
schnell fast, quick, high-speed, speedy, rapid
Schnellabschaltung quick breaking, rapid interruption, high-speed breaking
schnellaufend high-speed
Schnellauslösung instantaneous tripping
Schnellbremse quick-acting brake
Schnitt cut, cutting, sectional drawing, section
Schnittkraft cutting force
Schnittpunkt intersection point
Schnüffelventil sniffing valve
Schnur cord, string, twine, flexible lead
Schornstein chimney, stock, funnel, flue
Schott bulkhead
schraffieren to shade, to section, to hatch
Schraffierung hatching, sectioning
schräg oblique, inclined, sloping, slanting, skew, tilted
schrägbohren to drill holes at an angle
Schräge slope, obliquity,bevel, slant
Schrägkante chamfer, bevel
Schrank cabinet, locker, board
Schranke barrier, gate
schränken to set (saw)
Schaube bolt, screw, propeller
schrauben to screw, to tighten
Schraubfassung screwed lamp-holder
Schraubkopf fuse-carrier
Schraubstock vice
schreiben to write, to type, to record, to plot
Schritt step, pitch
Schrittregler step-acting controller
schrittweise stepwise
Schrott scrap
schruppen to rough, to rough-machine
Schrupphobeln rough planing
Schub thrust
Schubkraft thrust force, thrust, shearing force, shear, pushing force
Schuh shoe, ferrule
Schukosteckdose socket outlet with earthing contact
Schukostecker plug with earthing contact
Schulung des Personals personnel training
schuppenartig scale-like
Schuppenbildung scaling
Schutt wastage, rubbish, debris
Schuttabladeplatz dump site
schütteln to shake, to agitate, to vibrate
Schüttelrinne shaking trough, shaking chute
schütten to pour (liquid, bulk), to dump
Schutz protection, safeguard, cover
Schütz contactor
Schutzanzug protective frogsuit, overall
Schutzart type of protection
Schutzbrille safety glasses (goggles)
Schutzeinrichtung protective equipment, protective gear
schützen to protect, to guard, to screen, to shield
Schutzerde protective earth

Schutzwandler isolation transformer
Schutzwiderstand protective resistance
schwachbeleuchtet dimly lit
schwächen to weaken, to diminish, to lessen
Schwächung attenuation, reduction, fading
Schwalbenschwanz dovetail
Schwallwasser splash water
schwallwassergeschützt splash-proof
Schwankung fluctuation, variation
schwärzen to blacken, to dark
schweben to float, to hover, to be suspended
Schwebstoff aerosol
Schwebung beat
Schwefel sulphur
schwefelarm low-sulphur
Schwefelsäure sulphuric acid
Schweissaggregat welding unit, welding set
Schweissarbeit welding, welding job
Schweissbrenner welding torch
schweissen to weld
Schweissen welding
Schweissraupe welding bead
schwelen to smoulder
Schwelle threshold, step, barrier, sill (house)
schwellen to swell, to increase, to rise
Schwellenwert threshold value
Schwengel handle, pendant lever
Schwenkachse swivelling axis
schwenkbar swivelling pivoting
schwenken to swing, to swivel, to slew, to revolve
schwer heavy (load, duty), difficult, hard
Schwere gravity
Schwerpunkt centre of gravity
schwimmen to swim, to float, to drift
Schwimmer float
Schwimmerschalter float switch
schwinden to contract, to shrink, to fade (radio)
schwingen to oscillate, to vibrate, to rock, to hunt
Schwingung oscillation, vibration
Schwitzwasserbildung deposit of moisture
Schwund shrinkage, fading (radio)
Schwundregelung automatic gain control
Schwungkraft centrifugal force
Schwungrad flywheel
Sechseck hexagon
sechseckig hexagonal
Sechskantschlüssel hexagonal spanner
Sechskantschraube hexagonal-head bolt (or: screw)
Seefunk marine radio communication
Sehfeld field of view
Seide silk
Seidenlackdraht varnished silk-braided wire
seigern to segregate, to liquate
Seil rope, cord, line, cable, string
Seilbahnförderer ropebelt conveyor
seilbetätigt cable-powered
Seite side page (book), wing (building)
Seitenband sideband
seitlich sidewise, lateral
sekundär secondary
Selbstabgleich automatic balancing
selbständig self-contained, independent, unaided
selbstansaugend self-priming
selbstätig automatic, self-acting, self-contained, independent
selektieren to pick out
Selen selenium
selten rare (metal)
Sende-Empfang-Schalter sending and receiving switch
Sendegerät transmitting set
senden to transmit, to emit, to send
Sender transmitter, emitter
Senke depression, hollow, sink
senken to lower, to counterbore (machine)
Senker counterbore, countersink
Senklot plummet, lead
senkrecht vertical, perpendicular
Serie series, run (production)
Serienarbeit repetition job
Servomotor servo-motor, pilot

Serienmotor series(-wound) motor, booster
setzen to set, to set up, to put, to place
sicher safe, proof, protected
Sicherheit safety
sichern to secure, to lock, to guard, to fasten
sicherstellen to ensure
Sicherung fuse (el.), locking
Sicht visibilty
sichten to classify, to sort
Sichtgerät display unit
Sickergrube soakaway
sickern to seep, to percolate, to ooze
Sickerung percolation, seepage
Sieb screen, sieve, riddle, filter, strainer
sieben to screen, to sieve,to riddle, to strain, to sift
Siedepunkt boiling point
Siederohrkessel water tube boiler
Siegel seal
siegeln to seal
Signaleingabe signal input
Signalflussdiagramm signal-flow graph, signal-flow diagram
Signallampe signal light, indicator light
Signalverfolger signal tracer
Silber silver
Silberfolie silver foil
Silberlot silver solder
Silikat silicate
Silikon silicone
Silizium silicon
Siliziumplättchen silicon chip, silicon wafer
Siliziumgleichrichter silicon rectifier
Simmerring oil-seal ring
Sockel pedestal, base, socket, holder
Sockelbuchse cap sleeve
Sockelstift contact pin, base pin
Soffittenlampe festoon lamp, tube lamp, strip-light
Sofortruf immediate ringing
Sog suction
Sole brine
Solleistung required output, nominal output
Sollwert nominal value, rated value, desired value
Sollwertgeber reference element
Sonde probe
Sonderausführung special design
sondieren to probe, to sound
Sonnenbatterie solar cell, solar battery
sortieren to classify, to sort, to grade, to size
Spalt gap, fissure
spalten to split, to break, to crack
Spaltfilter gap-type filter
Span chip, cut
Spanabfuhr chip disposal, chip removal
Spanfläche cutting face, chip-bearing surface
Spanfluss flow of chips
Spanndruck clamping pressure
spannen to load, to hold, to grip, to set up, to clamp
Spannen der Werkstücke setting up work
Spannfutter chuck
Spannung voltage (el.), stress, tension, load (mech.)
Spannung gegen Erde voltage to earth
spannungsführend live, active
Spannut chip groove, flute
Spannvorrichtung clamping fixture, holding fixture, clamping device
Spanplatte chipboard
Spant frame, rib
Spanung cutting
Spanwinkel rake angle, cutting rake
Spardiode efficiency diode
Speicher store, memory
Speicherung storage
Speisekabel feeder cable, feeder
speisen to feed, to supply, to load, to charge
Speisewasser feed water
Speisung feeding, feed supply, charge
Spengler pipe fitter, plumber
Sperre lock, interlock, block
sperrig bulky, awkward
Sperrschicht barrier, junction
Sperrstrom inverse voltage, gate voltage, biasing potential
Spiegel mirror, reflector

spiegelbildlich mirror-inverted
Spiel game (allg.); play, backlash, clearance, allowance, free travel (mach.)
Spindel spindle, screw
Spiralbohrer twist drill
spitz pointed, acute
Spitze point, tip
splitterfrei shatterproof
splittern to splinter, to split
Sprache, chiffrierte coded speech
spritzen to spray, to injection-mould, to extrude
sprengen to blast
Spritzgiessen moulding by injection, die-casting
Spritzguss die-casting
spröde brittle, short
sprühen to spray
sprudeln to bubble
Sprung crack, fissure, fault
Spule coil, bobbin, reel, spool
spulen to reel, to spool, to wind
Spülluft scavenging air
Spülung rinsing, cleansing
Spur trace, gauge
Stab bar, rod, column
Stabbatterie tube cell, cylindrical cell
stabil stable, rugged, rigid, robust
Stabilisator stabilizer, balancer
Stabilität stability, ruggedness, rigidity, balance
Stahlband steel band platform, position, level
Ständerblech stator lamination
Standortpeilung position fixing
Stand state, status, stand,
Stange stick, bar, rod, pole
Stanzblech punching sheet
Stanze puncher
stanzen to punch, to stamp, to blank
stapeln to stack, to tier, to pile
Stärke starch (chem.), strength, intensity
Starkstromanlage power plant
Starkstromtechnik electrical power engineering
Starrheit rigidity
Start take-off, start, start-up, launch, launching
Startzeit start-up time
Statik statics
stationär fixed, stationary
Stau congestion
Staub dust
staubförmig dusty, powdered
stauchen to compress, to upset, to head, to machine-forge, to crush
stauen to crowd, to stagnate, to choke, to stow
stechen to pierce, to hole, to engrave, to cut, to carve to punch
Stechmeissel parting tool
Steckbuchse receptacle
Steckdose socket, socket outlet, plug connector
Stecker plug, plug connector, attachment plug
Steckverbindung plug connection, plug and socket connection, connector
Stehbolzen stud, stay bolt
steif stiff, rigid
steigen to mount, to climb, to ascend, to rise, to increase
steigern to increase
Steigung gradient, incline, inclination, slope
steil steep
Stein stone, brick
Steinkohle hard coal, mineral coal
Stellantrieb servo drive, actuator
stellbar adjustable
Stelle point, site, place
stellen to put, to set, to place, to adjust, to control
stellen auf Null to set on (or: to) zero, to reset
Stellung position, setting
Stellungnahme comment(s)
Stempel stamp, punch, die, puncheon, piston
Sternpunkt neutral point, zero point, centre point
Sternpunkterdung neutral earthing
stetig continuous, steady
steuern to control, to drive, to regulate
Steuerungsgerät control gear
Steuerung control, drive, timing
Stickstoff nitrogen
Stichleitung tie line, open feeder
Stift pin, stud, stick

stillegen to shut down, to put out of service, to close
Stillstand rest, stoppage
Stirnrad spur gear
stocken to stop, to end, to cease, to jam
Stoff substance, material, matter
Stopfbuchse stuffing box, packing box
Stöpsel plug, peg
stöpseln to plug in, to plug
stören to disturb, to interfere
Störfaktor interference factor, signal-to-noise ratio
stornieren to cancel
Störsignal spurious signal, parasitic signal
Störung breakdown, disturbance, trouble, failure, fault, interference
Stoss impact, percussion, shock, jerk, pulse
Stossdämpfer shock absorber, dashpot, shock mount
Stössel cam follower, ram (shaper)
stossen to push, to strike, to hit, to slot, to shape
Stoßstrom surge current, transient current
straff tight, tensioned, taut
Strahl ray, beam, jet
Strang rope, cord
strangpressen to extrude
Strebe brace, strut
Strecke distance, line, segment of a line
strecken to stretch, to lengthen, to elongate, to draw out
Streichung deletion
Streuecho scatter echo
Streuung scattering, spread, leakage
Strich line, stroke, dash
stricheln to dash, to shade
Strom current (el.), stream (flow)
Strömung flow, flux, motion of a fluid
Stromversorgung power supply, current supply
Stück piece, workpiece, part, lump
Stufe step, degree, level
Stütze support, pillar, post, column, rest, bracket, stanchion
Stützpunkt point of support, bearing surface
Styrol sterene
suchen to look for, to hunt for, to search, to find out
Sucher locator, detector
summen to hum
Summer buzzer
Summton buzzing sound
Sumpf sump (mach.)
synchron synchronous, in step

T

Tangens tangent
tanken to fuel
Tarif tariff, rate(s)
Tabelle table, chart
Tableau indicator board, panel
Takt stroke, cycle
tarnen to camouflage
Tarnung camouflage
Taschenlampe torch lamp, flashlight
Taste key, push-button
Taster push-button switch, tracer
Tau rope, hawser, cable
tauchen to submerge, to dive, to immerse, to dip
tauglich suitable, suited, usable, worthy, serviceable
Taumelscheibe wobbling disk, wobble plate
taxieren to taxe, to assess, to rate
Technik technology, engineering
Techniker technician, engineer
Teer tar
Teil part, section, portion, share, element, component, workpiece
Teilen eines Kreisumfanges circular spacing, circumferential indexing
Teilkreis pitch circle, angular scale
Teilmarke index
teilnehmen to participate, to take part
Teilnehmer subscriber (tel.)
Telefonapparat telephone set
telegrafieren to telegraph, to cable, to wire, to send a telegram
Temperaturabfall temperature fall, drop in temperature
Temperaturanstieg temperature rise, rise in temperature

Temperguss malleable cast iron
Tendenz tendency, trend
Terminplan schedule, time schedule
Theorie theory
Thermalhärtung time quench hardening
thermisch thermal
Thermometerkugel thermometer bulb
Thermoschalter temperature switch
Thomasstahl Thomas-Gilchrist steel, basic converter steel
tief deep
Tiefbau civil engineering
Tiefenmesser depth gauge, fathometer
Tiefenwirkung depth effect
Tiefgang draught
Tiefpassfilter low-pass filter
Tiegel crucible, melting pot
Tilgung deletion
Tinte ink
Tippbetrieb inching, jogging
Tippschaltung finger-tip control
Tisch platen, table
Titan titanium
Titel heading (drawing)
Toleranz allowance, tolerance
Ton clay (earth), sound, tone
Tonabnehmer sound pick-up
Tonband tape
Tonfrequenz audio frequency, voice frequency
Tonne barrel, cask
Topf jar, pot
Tor gate
Torschaltung gate circuit
Totalbetrag total amount, total sum
Totpunkt dead centre, slack point
Tourenregler speed controller
Tragbalken column, standard, girder
träge slow-acting
tragen to bear, to support, to carry
Trägerfrequenz carrier frequency
tragbar portable
Tragfähigkeit load-carrying capacity
Traggriff lifting handle
Trägheitsmoment moment of inertia
tränken to impregnate, to soak, to saturate
Tränklack impregnating varnish
Transformator transformer
Transistortechnik transistor engineering, transistor technology
transponieren to transpose
transportieren to transport, to convey, to transfer, to handle
Trapez trapezium
Traverse bridge beam, cross-beam, cross girth, cross bar
Treffplatte target
treiben to drive, to propel
Treibstoff fuel, propellant
Trenndiode buffer diode
trennen to separate, to disconnect, to interrupt
Trennschärfe channel selectivity, discrimination
Trennsicherung disconnecting fuse
Trennung separation
Trichter funnel, hopper
trichterförmig funnel-shaped
Triebwerk engine, propulsion unit, power plant
trocken dry, arid
Trocknung drying
Trog trough, tub, tray, vat
Trommel drum, barrel
tropenfest tropic-proof
Tröpfchen droplet
tropfen to drop, to drip, to trickle, to leak
Tropfen drop
trüb opaque, cloudy, turbid
Tube tube
Turbinenanlage turbine plant
Tür door
Turbolader turbo-supercharger
Turm tower
Tüte paperbag
Typ design, type
typisieren to standardise

U

Überbau superstructure
überlasten to superload, to overload
überbrücken to bridge, to shunt, to connect across

Überdrehzahl overspeed
Überdruckturbine reaction turbine
übereinstimmen to agree, to match together, to coincide
Überfluss abundance
Überflüssigkeit redundancy
überfluten to overflow, to flood
Übergabe delivery, handing-over, transfer
Überhitzer superheater, overheater
überholen to overhaul, to recondition
Überlagerung superposition, beat
überlappend overlapping
Überlastung overload, overloading
Überlauf overflow
übermässig excessive
überprüfen to verify, to examine, to check
überschlagen to flash over
überschreiten to exceed, to overshoot, to overtravel
Überschuss excess, surplus
Überspannungsableiter surge diverter, overvoltage arrester
überspringen to skip, to flash (spark)
übersteuern to overmodulate, to overdrive, to override
Überstunden overtime
Übertragung transmission, transfer
überwachen to monitor, to supervise, to inspect, to control
überweisen to transfer
Überweisung transfer
überziehen to coat, to plate, to laminate
Uhrzeiger clock hand
UKW v.h.f. very high frequency
Ultraschall ultrasonic sound, ultrasonics
Umarbeitung reworking
Umbau rebuilding, alteration, reconstruction
umbauen to reconstruct, to rebuild, to alter
Umdrehung revolution, turn
Umdrehungen pro Minute revolutions per minute
Umfang circumference, perimeter
Umflechtung braid
umformen to convert, to transform
Umformer transformer, converter, transducer
Umgang handling, manipulation
Umgebung environment, vicinity
Umgebungstemperatur ambient temperature
umgehen to by-pass
umgekehrt inverse, reverse, reciprocal
umgestalten to modify, to transform
Umgrenzung boundary
umhüllen to encase, to jacket, to wrap, to cover, to coat, to mantle
Umkehr reversal, turning
umkehren to reverse, to return
umkreisen to orbit, to circle
umlaufen to circulate, to rotate
Umschalter change-over switch
umsponnen braided, covered
umspulen to rewind
umstellen to shift from . . . to . . .
umsteuern to reverse
Umsteuerung reversal
Umwälzverfahren forced-air system
umwenden to turn over, to turn round, to invert
unbestimmt indefinite
undicht leaky, leaking
unfachgemäss inexpert
Unterlagen data, information, documentation
Unterlegscheibe washer, lock washer, plain washer
Untermass undersize
Unterschlitten lower slide
Unterseite bottom side, underside
Untersetzungsgetriebe reduction gear, step-down gear
unterstützen to support, to carry
untersuchen to examine, to investigate, to test, to analyse, to study, to check
unterteilen to subdivide
unverarbeitet raw, untreated,
Unfall accident
ungültig invalid, void
unhörbar inaudible
unterbrechen to break, to interrupt, to open, to disconnect, to cut off

V

Ventil valve
Ventilator ventilator, fan, blower
verändern to alter, to vary, to modify
verarbeiten to work, to process, to treat, to convert
Verarmung depletion
verbessern to improve
Verbesserung improvement
Verbesserungsvorschlag suggestion for improvement, improvement suggestion
verbinden to tie, to bind, to connect
Verblockung interlocking
Verbrauch consumption, dissipation
verbrauchen to consume, to dissipate
Verbrennung combustion, burning, incineration
Verbrennungsmotor internal combustion engine
Verbundnetz grid
verdampfen to evaporate, to vaporise
Verdampfer evaporator
Verdichter compressor
verdrahten to wire, to interconnect
Verdrängung displacement
Verdrehung torsion, twisting
verdunsten to evaporate, to vaporise, to volatilise
verdünnt diluted
vereinigen to unite, to combine
verengen to narrow, to contract, to reduce
Verfahren process, method, technique, treatment
verflüchtigen to volatilise
verflüssigen to liquify, to condensate
verfügbar available
Vergaser carburettor
Vergleich comparison
vergleichen to compare
vergrössern to enlarge, to increase, to magnify
Verhalten behaviour, response, performance
Verhältnis ratio, relation, proportion
verhindern to prevent
Verkauf sale
Verkaufsabteilung sales department
Verkehr traffic
verkehrt wrong, upside down
verkeilen to key, to wedge, to cotter
verketten to link, to correlate
Verkleidung lining, coating
verkleinern to reduce, to decrease, to diminish, to scale down
verklemmen to jam, to stick, to seize
verlagern to shift, to displace
verlängern to lengthen, to prolong (time)
Verlust loss
verriegeln to bolt, to block,
versagen to fail
verschieben to shift, to displace
Verschleiss wear, abrasion
Versicherung insurance
versickern to seep
versiegeln to seal
versorgen to supply, to provide
Versorgung supply
verstärken to strengthen, to reinforce (mech.), to amplify, to gain, to boost (el.)
Verstärker amplifier, repeater, follower, booster (pneumatics)
versteifen to stiffen, to brace, to strengthen, to reinforce
verstellen to adjust, to shift (lever)
Verstellgetriebe variable speed gear
verstopfen to clog, to blind, to plug
Versuch trial, test-run, attempt
vertauschen to transpose, to permute, to exchange
verteilen to distribute
Verteilung distribution
Vertrag contract, agreement
Vertreter representative, agent
Vertrieb sale, distribution, trade
verunreinigen to pollute, to contaminate
verursachen to cause
vervielfachen to multiply

verwenden to use, to utilise, to apply
verwischen to blur
verzerren to distort
Verzerrung distortion
verzinken to galvanise, to coat with zinc
verzögern to delay, to retard, to lag
Verzögerung delay, lag, retardation
verzweigen to branch
Verzweigung network, branch
vielfach multiple
vielseitig versatile
Vierkant square
vollelektrisch all-electric
vollelektronisch full electronic
voltaisch voltaic
Vorarbeit preliminary work
vorbeifahren to pass
vorbereiten to prepare
vorbohren to predrill
Vorderseite front face
voreilen to advance, to lead (phase)
Voreilen advance, lead, leading
vorfristig ahead of schedule
vorführen to demonstrate
Vorgang process, operation, action, event
vorgespannt preloaded
Vorhang curtain
vormagnetisieren to premagnetise
Vormontage sub-assembly
Vorwahl preselection
Vorwähler preselector
Vorwärmer preheater, economiser
vorwärts forward, forth
Vorzeichen sign
vulkanisieren to vulcanise, to cure

W

Waage balance, scale
wabenförmig honeycomb
Wachs wax
wachsen to grow, to increase
Wachstum growth
wägen to balance, to weigh
Wagen car, vehicle, coach, carriage
Wahl selection, dialling (tel.)
Wahrscheinlichkeit probability
walzen to roll, to mill
Walzwerk rolling mill
Wanderfeldröhre travelling-wave tube
wandern to migrate, to shift
Wandler transformer, transducer, converter, transmitter
Wanne tank, tub, oil sump
Ware goods, ware
Wärme heat
Wärmeausbeute heat yield
wärmeisoliert thermally insulated
Wartung maintenance, servicing
Waschmittel washing agent, detergent
Wasserbau hydraulic engineering
Wasserstoff hydrogen
weben to weave
wechseln to change, to alternate, to exchange
Wechselrichter inverter
Wechselspanung alternating voltage
Wechselstromgenerator alternator
Wechselwirkung interaction
wegleiten to carry off
weich soft
Weichlöten (soft) soldering
Weite width, distance, breadth
Welle wave, shaft (mech.)
Wendegetriebe reversing gear
werben to advertise
werfen to throw, to cast
Werk works, mill, factory
Werkstoff material
Werkstück workpiece, work, part, component
Werkzeug tool
Werkzeugmaschine machine tool
wickeln to wind, to coil, to reel
Wicklung winding
Widerstand resistance
wiederholen to repeat
Windung turn, twist
Winkel angle
Wirbelstrom eddy current
wirken to act, to operate
Wirklast active load, actual load
Wirkungsgrad efficiency
Wismut bismuth
Wortgeber word generator
Würfel cube
Wurzel root

X Y Z

x-Achse x-axis
y-Achse y-axis
z-Achse z-axis
Z-Schneide Z-bit
Zacke jag, notch
zäh tough
Zahl number, quantity
zählen to count, to register
Zählen counting
Zahn tooth (gear)
Zahnstange gear rack
Zange pliers, tongs
Zapfen journal (bearing)
Zeichen sign, signal, mark, symbol
Zeichnung drawing, graph, plan
zeigen to show, to point, to indicate
Zeiger pointer, hand, needle, indicator
Zeile line
Zeilenabtasten line scanning
Zeitgeber timer, timing element
Zeitrelais time relay, time-lag relay
Zement cement
Zentralheizung central heating system
zentrieren to centre, to true, to centralise
Zentrum centre
Zerfall decay
Zerhacker chopper
zerspanen to destroy
Ziegel brick
ziehen to pull, to draw
Ziel target, numeral, digit
Zifferblatt dial
Zinc zinc
Zinn tin
Zoll inch
Zone zone, area
Zubehör accessories, attachments, fittings
Zufuhr supply, admission
Zug tension (mech.)
Zugriff access
Zunahme increase, rise, growth
zünden to ignite, to fire
Zungenfrequenzmesser reed-type frequency meter
zuordnen to coordinate, to associate
zurückführen to recycle, to restore
zurücklaufen to return
zurückziehen to retract, to withdraw
zusammenbauen to assemble
Zusatz addition, additive
Zustand state, condition
zwängen to force
Zweig branch
zweistufig two-stage, two-step
Zweitakter two-stroke engine
Zwischenraum clearance, interspace, gap
Zwischenstück adapter, connecting piece
zyklisch cyclic
Zylinder cylinder
zylindrisch cylindrical

"It's opened my eyes to a whole new world of incomprehensible mumbo-jumbo."

incomprehensible unfassbar, unbegreiflich
mumbo(-)jumbo Quatsch, Blödsinn

Pictorial Machine Tool Vocabulary

Englisch - Deutsch

A

ability to maintain cutting power (or: to retain cutting edge) Schneidhaltigkeit
abrasion Schleifwirkung, Verschleiss
abrasive wheel cutting-off machine Trennschleifmaschine
abutment Widerlager
abutting end Stossfläche
accumulated tooth spacing error Summenteilfehler
accuracy of dimension Massgenauigkeit
accuracy of measurement Messgenauigkeit
accuracy of positioning Zentriergenauigkeit, Einstellgenauigkeit
accuracy of reproduction Nachformgenauigkeit
accuracy of shape Formgenauigkeit
accuracy of spacing Teil(ungs)-genauigkeit
accuracy to size Masshaltigkeit
active cutting edge Hauptschneide
active portion of a broach Schneidenteil eines Räumwerkzeugs
actual size Istmass
acute angle spitzer Winkel
adapt, to anpassen, einrichten
adaptability Anpassungsfähigkeit, Verwendungsbereich
adapter Zwischenstück, Hülse, Übergangshülse
adapter (Räumwerkzeug) Führungsbüchse
adjust, to einstellen, anpassen, nachstellen
adjustability Einstellbarkeit, Nachstellbarkeit

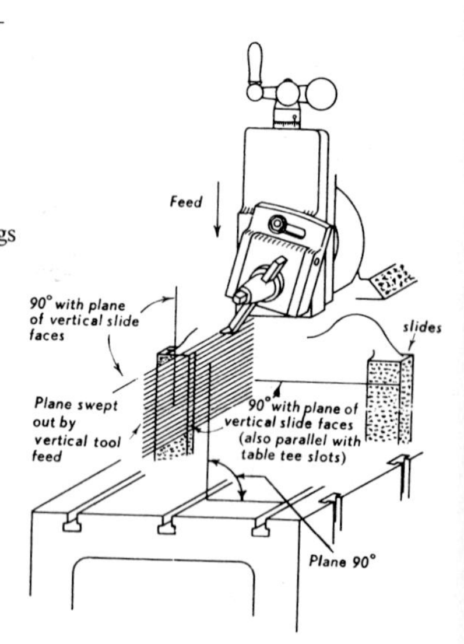

accuracy relationships of a surface Genauigkeit der Bezugsfläche

adjustable speed gear
Verstellgetriebe, Motor und Getriebe für Drehzahlregelung

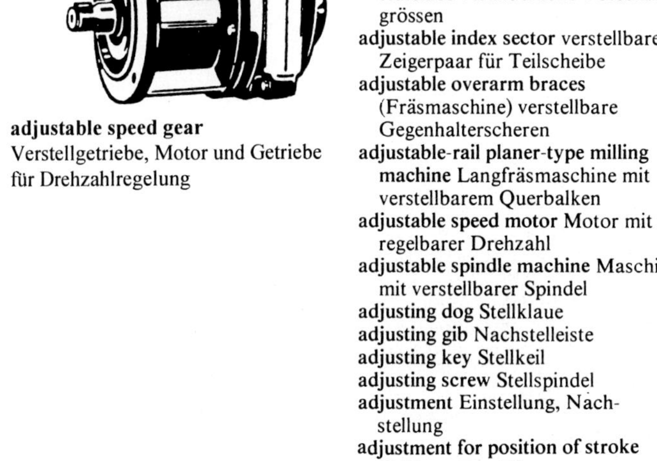

adjustable block Gleitschieber, Klemmplatte, Klemmstück (Waagrechtstossmaschine)
adjustable dog verstellbarer Anschlag
adjustable feed trip dog einstellbarer Anschlag zur Vorschubausrückung
adjustable head horizontal milling machine Planfräsmaschine mit verstellbarem Spindelkopf
adjustable hydraulic flow control valve for infinitely variable feed rate einstellbares Drosselventil für stufenlos veränderliche Vorschubgrössen
adjustable index sector verstellbares Zeigerpaar für Teilscheibe
adjustable overarm braces (Fräsmaschine) verstellbare Gegenhalterscheren
adjustable-rail planer-type milling machine Langfräsmaschine mit verstellbarem Querbalken
adjustable speed motor Motor mit regelbarer Drehzahl
adjustable spindle machine Maschine mit verstellbarer Spindel
adjusting dog Stellklaue
adjusting gib Nachstelleiste
adjusting key Stellkeil
adjusting screw Stellspindel
adjustment Einstellung, Nachstellung
adjustment for position of stroke Hublagenverstellung
adjustment for wear Nachstellung zum Ausgleich für Verschleiss, Verschleissausgleich
admission Beaufschlagung, Eintritt

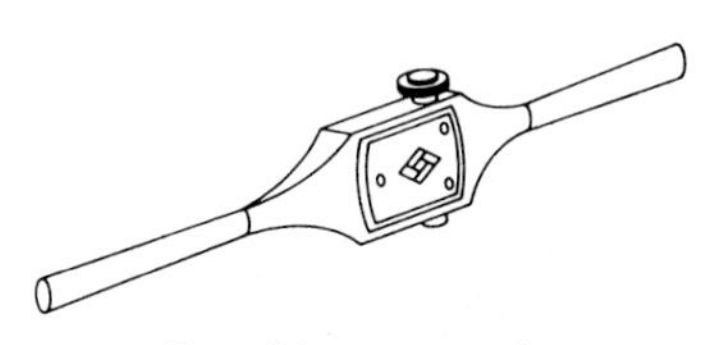

adjustable tap wrench
verstellbares Windeisen

advance, to zustellen, vorbringen, verschieben
air-actuated druckluftbetätigt
air chuck Pressfutter
air-cooled luftgekühlt
airdrain petcock Entlüftungshahn
air-hardening Lufthärtung
air-operated druckluftbetätigt
air-relief valve Entlüftungsventil
air vice pneumatisch betätigter Schraubstock
align, to ausrichten, ausfluchten, in eine gerade Linie bringen
alignment Ausrichtung, Flucht, Axialität
alignment section (Räummaschine) Führungsstück
all-angle milling head auf jeden Winkel verstellbarer Fräskopf
allowable maximum pull höchstzulässiger Zug
allowance Toleranz, Spielraum, Abmass
alloy for cutting tools Schneidlegierung
alloy steel legierter Stahl, Sonderstahl
alternate angle side and face cutter Kreuzzahn-Scheibenfräser
alternate angle staggered tooth type tee-slot cutter kreuzverzahnter T-Nutenfräser
alternate gash plain mill kreuzverzahnter Walzenfräser (auch: Scheibenfräser)
alternate helical tooth cutter kreuzverzahnter Fräser
alternate nicking gegeneinander versetzte Kerbung, auf Lücke stehende Spanbrechernuten
alternate tooth milling cutter kreuzverzahnter Fräser
aluminium-oxide tool tip Aluminiumoxidschneide
amount of angularity Schrägstellung, Grösse des Winkels
amount of back-off (Räummaschine) Grösse des Freiwinkels

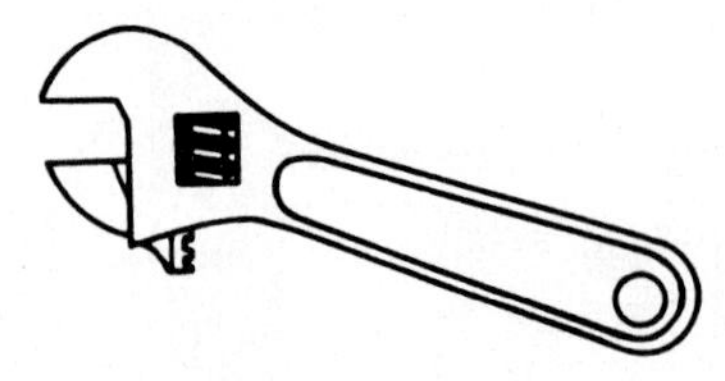

adjustable wrench
verstellbarer Schraubenschlüssel

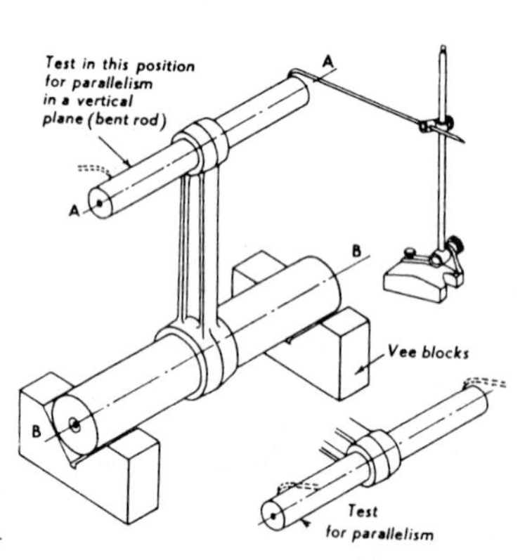

alignment tests on a connecting rod
Fluchtungsprüfung an einer Pleuelstange

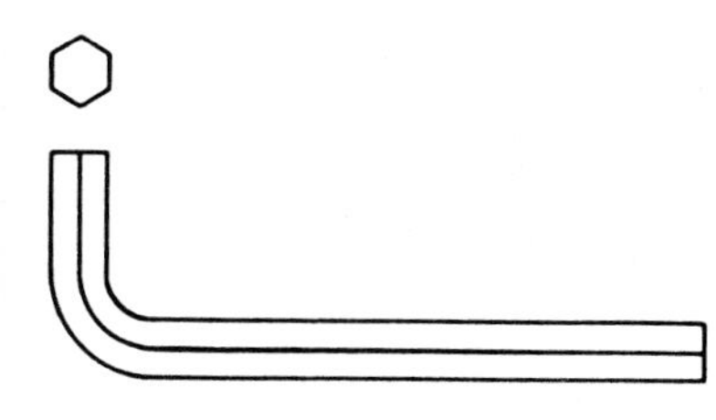

Allen key (or: wrench) Imbusschlüssel, Schlüssel für Innensechskantschrauben

alternating current motor Wechselstrommotor

amount of chip space Grösse des Spanraums
amount of eccentricity Exzentrizität
amount of feed Vorschub, Vorschubgrösse
amount of metal removed Spanmenge, Spanabnahme
amount of rake Grösse des Spanwinkels
amount of stock left for . . . Zugabe für . . ., Übermass als Bearbeitungszugabe für . . .
amount of stock removed abgenommenes Werkstoffvolumen
angle Winkel, Knie
angle milling cutter Winkelstirnfräser
angle end mill Winkelfräser mit Zylinderschaft
angle of approach (Fräser) Spanumfangswinkel, Eingriffswinkel
angle of clearance Freiwinkel
angle of inclination Neigungswinkel
angle of keenness Keilwinkel
angle of relief Freiwinkel (an der Fase, hinter der Schneide)
angle of the tool helix Drallwinkel
angle of the thread Flankenwinkel
angle of tilt Kippwinkel, Drehwinkel
angle plate Aufspannwinkel, Winkeleisen
angle tool Prismenmeissel
angular winklig, Winkel . . .
angular adjustment Schrägstellung
angular cut schräger Schnitt
angular cutter Winkelfräser
angular displacement Winkelverschiebung
angular division of work(piece) Teilung des Werkstücks nach Winkelmass
angular downfeed schräger Tiefenvorschub
angular drive Winkelantrieb
angular end milling schräges Fräsen in axialer Richtung des Schaftfräsers
angular gashing Fräsen geneigter Schlitze

angular half side mill Winkelstirnfräser
angular indexing Teilen nach Winkelmass
angular milling operation Fräsarbeit an schrägen Flächen
angular parallels Keilpaar, Keilstücke
angular planing Schräghobeln
angular positioning Schrägstellung
angular range Winkelbereich
angular setting Schrägeinstellung, Winkeleinstellung
angular spacing Winkelteilung
angular teeth Schrägverzahnung
angular tilt Schrägstellung, Schrägkippen
anneal, to (aus)glühen
annular ringförmig
annular clearance Ringspalt
annular groove ringförmige Auskehlung, Ringnut
annular spring Ringfeder
anti-backlash device Spielausgleicheinrichtung, Gleichlauf-Fräseinrichtung
anti-backlash longitudinal table feed nut Mutter der Tisch-Längsvorschubspindel mit Spielausgleich
anti-clockwise im entgegengesetzten Sinne des Uhrzeigers, links
anti-clockwise rotation Linksdrehung
anti-friction bearing Wälzlager
anti-wear properties Verschleissbeständigkeit, Verschleissfestigkeit
anvil Amboss
aperture Öffnung, Spalt, Mündung
apex Scheitel, Spitze
appendix Anhang, Zusatz
appliance Gerät, Apparat
application Anwendung, Einsatz, Verwendung
apply, to anwenden, anbringen, befestigen
approach, to zuführen, sich nähern
approach Heranführung, Vorlauf
approach of cutter Anschnittweg des Fräsers, Anlaufweg

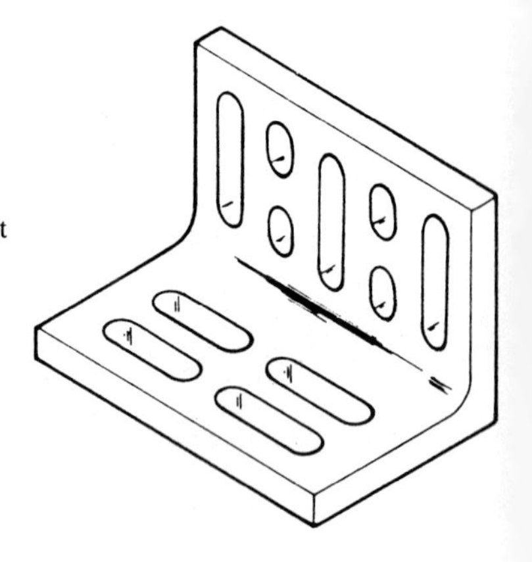

angle plate
Aufspannwinkel

approach side (Fräser) Anschnittseite
apron (Drehmaschine) Supportschlossplatte; auch: Sammelbegriff für Meisselhalter, Klappe, Klappenträger
apron (Waagrechtstossmaschine) Querschlitten des Tisches
apron clamping bolt Klemmschraube des Klappenträgers
apron slide Querschlitten des Tisches
arbor Dorn
arbor brace Gegenhalterschere
arbor bracket Dorntraglager
arbor cutter Aufsteckfräser
arbor nut Dornmutter
arbor support Traglager des Dornes
arbor supporting bracket Dornstützlager
arbor yoke (Fräsmaschine) Traglager, Führungslager
arbor type mill Aufsteckfräser für Aufsteckdorn
arc Bogen
area of contact Berührungsfläche
area of cut Schnittfläche
arm Arm, Ausleger, Stössel
arm brace (Fräsmaschine) Gegenhalterschere
arrange, to anordnen
arrangement Anordnung, Aufbau
arrest, to arretieren, festhalten, sperren
articulated pipe Gelenkrohr
ascend, to aufwärts bewegen, ansteigen
assemblage Zusammenbau, Montage
assemble, to zusammenbauen, montieren
assembly line Montageband
attach, to befestigen, anbringen
attachment Befestigung, Zusatzeinrichtung
attachment for taper milling Kegelfräseinrichtung
attendance Wartung
augment, to vergrössern (Fördermenge)
auto-cycle mit automatischem Arbeitsablauf

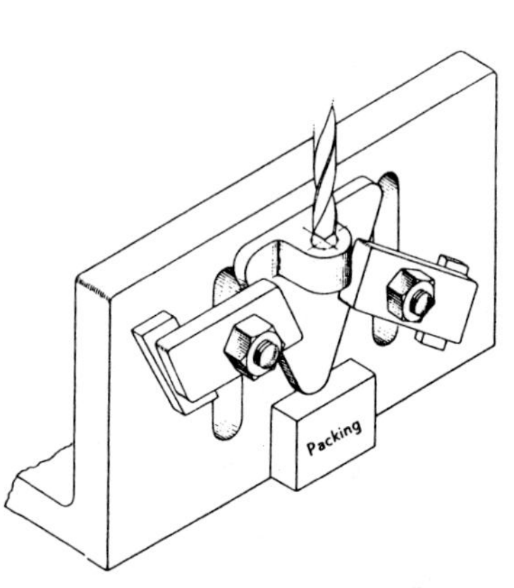

angle plate applied for drilling work
Aufspannwinkel bei der Verwendung von Bohrarbeiten

automatic backlash elimination automatischer Ausgleich des Totganges
automatic chip disposal unit Einrichtung zur automatischen Späneabfuhr
auxiliary Hilfs .., Zusatz .., zusätzlich
auxiliary equipment Zusatzausrüstung, Sonderausstattung
auxiliary rolling table (Einständer-Hobelmaschine) Hilfslaufbahn, Rollenbahn
axial rake (or: angle) (Fräser) axialer Spanwinkel
axis (Pl.: axes) Achse
axis of abscisses X-Achse
axis of inclination Neigungsachse
axis of ordinates Y-Achse
axis of rotation Drehachse
axis of work Werkstückachse

arrow staple gun Kabelmarkierer, Spezialhefter

B

backed off hinterdreht, hinterschliffen, mit Freiwinkel
back end piloting (Räumwerkzeug) Führung am Endstück
back gearing Zahnradvorgelege
backing block Stützblock
backing off Hinterdrehen, Hinterschleifen
backlash toter Gang, Spiel(raum)
backlash free ohne toten Gang, spielfrei
back-off angle (Räumwerkzeug) Freiwinkel
back-off clearance Hinterschliff, Freiwinkel
back pin (Teilscheibe) rückwärtiger Raststift
back rake angle Spitzenspanwinkel
back rest (Räumwerkzeug) Gegenhalter
back slope Spitzenspanwinkel
back-up ring (or: washer) Stützring, Stützscheibe

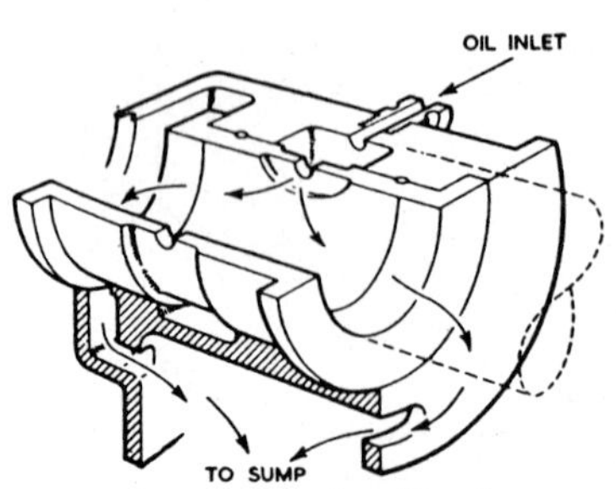

arrows showing oil path Pfeile, die den Ölverlauf zeigen

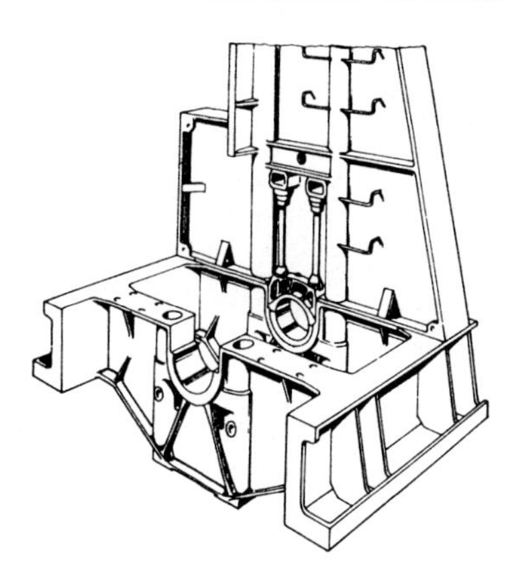

bedplate of an engine
Grundplatte eines (Diesel-)Motors

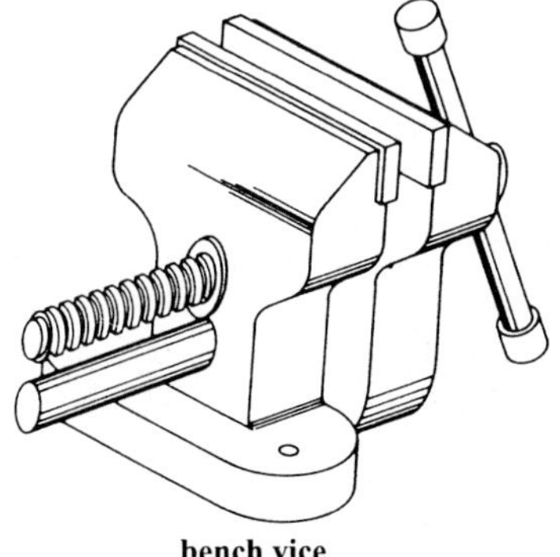

bench vice
Schraubstock

baffle plate Trennungsblech, Zwischenplatte, Stauscheibe
balanced ausgeglichen
balance weight Gegengewicht, Ausgleichsgewicht
ball bearing Kugellager
ball cup Kugelpfanne
ball-end cutter Fräser mit runder Stirn
ball-end mill Schaftfräser mit runder Stirn
ball joint Kugelgelenk
ball-nose end mill Schaftfräser mit runder Stirn
bar Stange, Stab, Traverse
bar stock Stangenmaterial
base Grundplatte, Basis, Fundament, Gestell
baseplate Grundplatte
basic design Grundform, typische Ausführung
basic rating (Fräser) Spanvolumen
batch Satz, Menge, Reihe, kleine Stückzahl
batch production Reihenfertigung
batch milling Fräsen in Reihenfertigung
bay Abteilung in einer Halle
beam Ausleger, Balken, Träger
bearing Lager, Lagerung
bearing bush Lagerbüchse
bearing support (Fräsdorn) Traglager
jam, to klemmen
bed (Maschinen-)Bett
bed slides Bettführungsbahnen
bed-type milling machine Planfräsmaschine, Fräsmaschine mit beweglichem Spindelstock
bed ways Bettführungsbahnen
bell Muffe
bell-mouthed mit trichterförmiger Öffnung
bellows type cover harmonikaähnlicher Schutz
belt Riemen
belt conveyor Bandförderer
bench Werkbank, Bank, mit Füssen versehenes Maschinenbett

bench milling machine Planfräsmaschine
bench plate Anreissplatte
bend cold, to in kaltem Zustand biegen
bent finishing tool gebogener Schlichtmeissel
between-grind broach life Standzeit des Räumwerkzeugs (zwischen zwei Nachschliffen)
between-grind life Standzeit
bevel, to abschrägen, abkanten, ausschärfen, abfasen, brechen
bevel Abschrägung, Fase
bevel schräg, abgefast
bevel cut Schrägschnitt
bevel gear cutter Kegelradfräser
bevel gear formed cutter Kegelrad-Formfräser
bevel gear shaper Kegelrad-Hobelmaschine
bevel protractor Anlegewinkelmesser
billet Knüppel
bin Behälter, Fülltrichter
binding bolt Klemmbolzen, Feststellschraube
binding nut Gegenmutter
bit insert Einsatzwerkzeug
bit tool Einsatzmeissel
blade Messer (Fräser)
blade inserted cutter Fräser mit eingesetzten Messern
blade life Lebensdauer (Standzeit) des Messers
blade regrinding Messernachschliff
blade renewal Messerauswechslung
blank Rohling
blend, to mischen, vermengen, verschmelzen
blend Mischung
block, to blockieren, verriegeln, absperren, arretieren
block Unterlegeklotz, Klemmstück
block indexing Schritteilung, Sprungteilung
block method of spacing Schritteilmethode
blunt stumpf

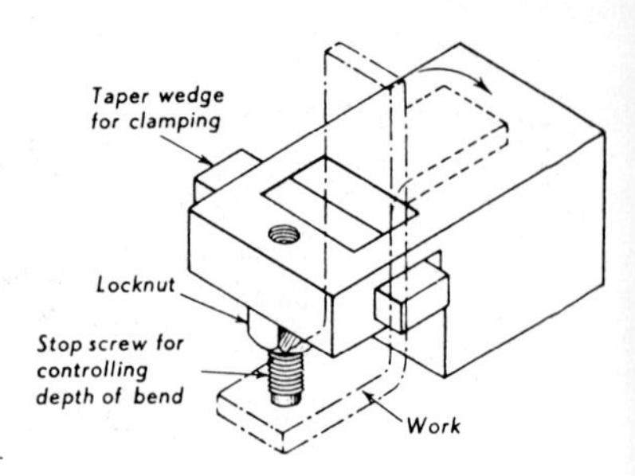

bending fixture
Biegevorrichtung

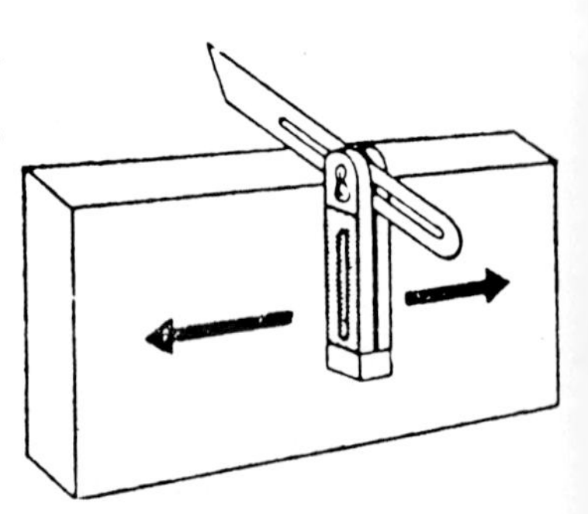

bevel for testing of trueness
Prüfen der Geradheit mittels Gehrungswinkels

bevel-wheel gear
Kegelradgetriebe

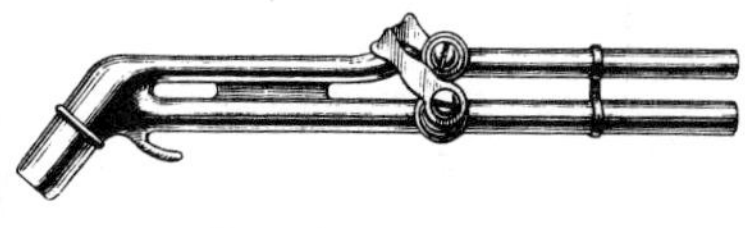

blowpipe
Schneidbrenner

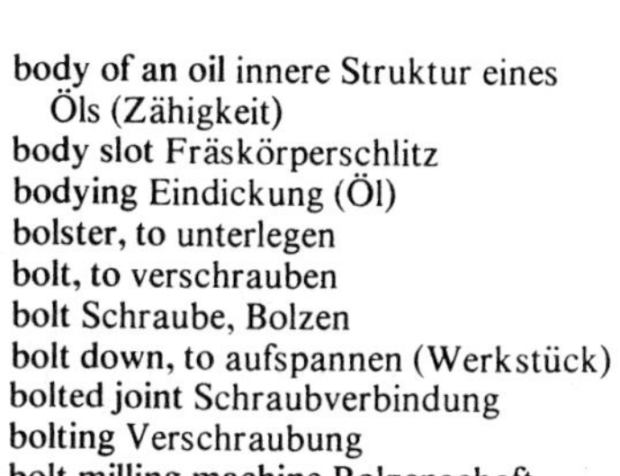

body of an oil innere Struktur eines Öls (Zähigkeit)
body slot Fräskörperschlitz
bodying Eindickung (Öl)
bolster, to unterlegen
bolt, to verschrauben
bolt Schraube, Bolzen
bolt down, to aufspannen (Werkstück)
bolted joint Schraubverbindung
bolting Verschraubung
bolt milling machine Bolzenschaftfräsmaschine
bolt, nut and tube threading machine Schrauben-, Mutter- und Rohrgewinde-Schneidmaschine
bolt thread cutting machine Bolzengewinde-Schneidmaschine
bolt washer Bolzenscheibe, Unterlegscheibe für die Mutter
bore, to aufbohren, bohren
bore Bohrung, lichte Weite
bore diameter Bohrungsdurchmesser
bore of pipe lichte Weite eines Rohres
bore size Bohrungsdurchmesser, lichte Weite
boring Aufbohren
boring bar Bohrstange
boring head (Hobelmaschine) Bohrsupport
boring head (Fräsmaschine) Bohreinrichtung
boss Nabe, hervorstehendes Stück
bottom Boden, unteres Ende, Basis, Unter . . .
bottom die Untergesenk
bottom plate Bodenplatte, Bodenblech
bow Bogen, Bügel, Krümmung
box Kasten, Gehäuse, Behälter
box nut Überwurfmutter
box section overarm Gegenhalter des kastenförmigen Querschnitts
box spanner Steckschlüssel
box table (or: box-type table) Kastentisch
box-type column Kastenständer
brace, to durch Stützen unterbauen, absteifen, verstreben, verklemmen

brace Stütze, Spannstift, Steife, Strebe, Brustbohrer, Klammer
brace (Fräsmaschine) Gegenhalterschere
bracing Verstrebung, Versteifung
bracing rip Rippe zum Versteifen
bracket Stütze, Schelle, Träger, Klammer (Mathematik)
bracket plate Befestigungsplatte
brake Bremse
brake, to bremsen
brass Messing
braze, to hartlöten
brazed-on tip hart aufgelötetes Plättchen
brazed-tip carbide tool Werkzeug mit hart aufgelötetem Hartmetallplättchen
brazed-tip tool Werkzeug mit hart aufgelöteter Schneide
brazing alloy Hartlot
brazing spelter Messinghartlot
break, to brechen, zerbrechen, unterbrechen
break Bruch, Sprung, Unterbrechung
breakdown Betriebsstörung, Versagen
breaking Bruch, Zerreissen, Bruchstelle
breaking elongation Bruchdehnung
break-proof bruchsicher
breast planer Blechkanten-Hobelmaschine
breather line Entlüftungsleitung
breather Luftventil, Entlüftungsventil, Schnüffelventil, Atmungsventil
breech cover Verschluss, Verschlusskappe
bridge Brücke, Portal (Fräsmaschine, Hobelmaschine)
brittle spröde
broach, to räumen
broach Räumwerkzeug, Räummaschine
broach and centre machine Räum- und Zentriermaschine
broach assembly zusammengesetztes Räumwerkzeug
broach carrier Räumwerkzeugträger

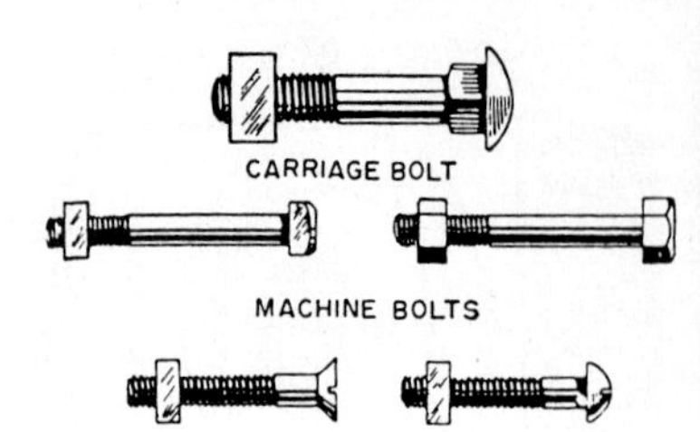

bolts
Schrauben

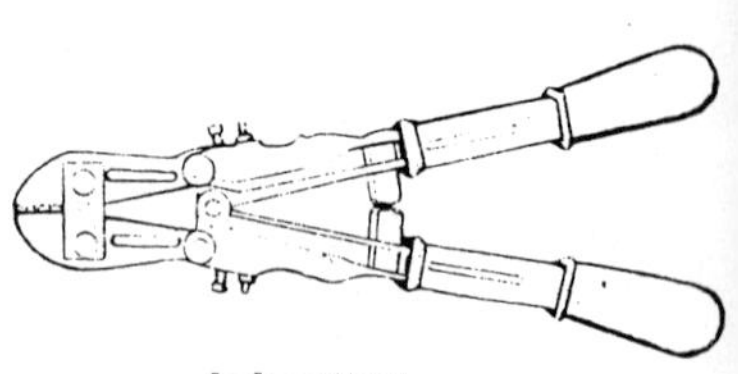

bolt cutters
Bolzenschneider

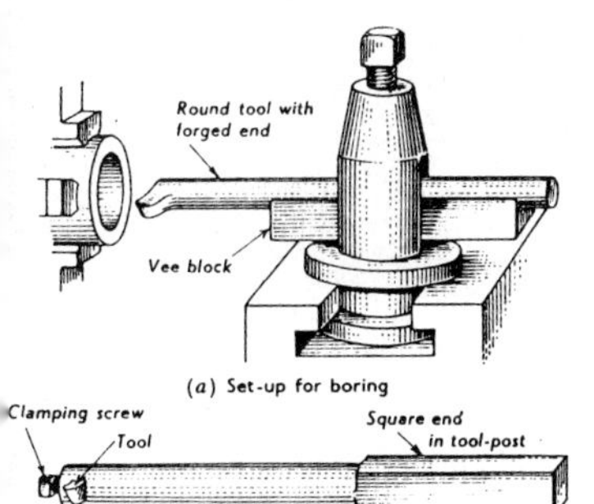

(a) Set-up for boring

(b) Boring bar

boring work
Aufbohrarbeit

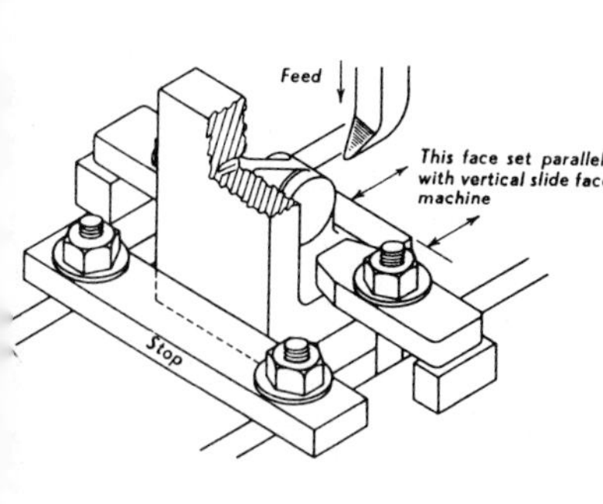

brackets used for shaping operation
Stützen (oder: Klammern) bei der Verwendung von Kurzhobeln

broach cutting section Schneidenteil des Räumwerkzeuges
broached surface geräumte Fläche, Arbeitsfläche beim Räumen
broacher Räummaschine
broach front pilot Führungsstück an Aufnahme des Räumwerkzeuges
broach handling Zubringung des Räumwerkzeuges
broach-handling slide Anhebeschlitten, Räumwerkzeug-Zubringeschlitten
broach handling unit Zubringeeinrichtung an der Räummaschine
broach holder Räumnadelhalter
broaching job Räumarbeit
broaching machine Räummaschine
broaching pass Räumdurchgang
broaching rate Stückzahl beim Räumen
broachings Räumspäne
broaching setup Räumaufspannung
broaching tool Räumwerkzeug
broach insert Schneideinsatz für das Räumwerkzeug
broach lifter cylinders Räumnadel-Anhebezylinder
broach pull down machine Ziehräummaschine mit nach unten gehendem Arbeitsgang
broach puller Ziehwerk (an der Räummaschine)
broach pull head Werkzeughalter des Zugorgans der Räummaschine
broach ram Räumstössel, Räumschlitten
broach slide Räumschlitten
broad cut finishing Breitschlichten
broad finishing tool Breitschlichtmeissel
broad nose finish tool Breitschlichtmeissel
brokes (USA) Ausschuss, Abfälle bei der Verarbeitung
buckle, to knicken, krümmen, verbiegen, verziehen
buckling strength Knickfestigkeit
build, to bauen

«building brick» machine tool Werkzeugmaschine nach Baukastenbauweise
built-in eingebaut
built-on aufgebaut
built-up cutting edge Aufbauschneide
built-up teeth (Räummaschine) Zahnungseinsätze
built-up type cutter Fräser mit eingesetzten Zähnen
bulk production Fertigung grösserer Mengen
bulky sperrig
burnish, to polieren, bräunen, glätten
burnisher (Räummaschine) Glättzahn
burnishing Polierdrücken, Polierrollen
burnishing broach glättende Räumnadel
burr, to abgraten, entgraten
burr Grat
burring chisel Abgratmeissel
burr-removing device Entgratungseinrichtung
burst, to bersten, zersprengen, platzen
bush Buchse, Büchse, Muffe
butt, to stumpf aneinanderfügen, anstossen
button Knopf
butt-weld, to stumpfschweissen
by-pass, to umgehen, vorbeiströmen
by-pass line Umgehungsleitung

Butt weld

butt-welded lathe tool
stumpfgeschweisster Drehmeissel

C

calibrate, to eichen, kalibrieren
calipers Taster
cam Nocke, Kurve, Knagge, Daumen
cam miller Kurvenfräsmaschine
cam milling Kurvenfräsen, Nockenfräsen
cam milling attachment Kurvenfräseinrichtung

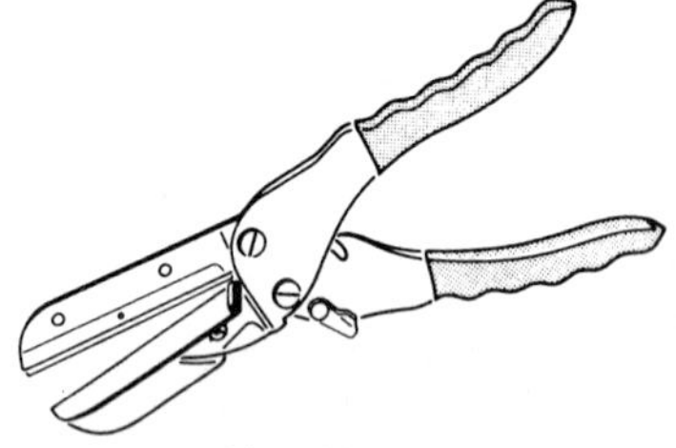

cable cutter
Kabelschneider

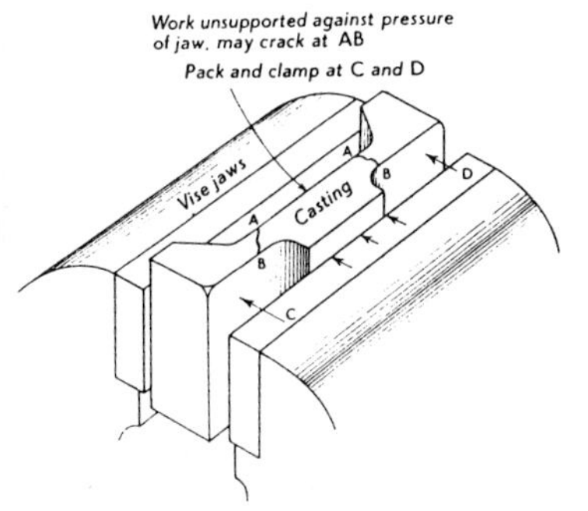

casting not supported under point of clamping
Gussprofil, nicht abgestützt an der Spannstelle

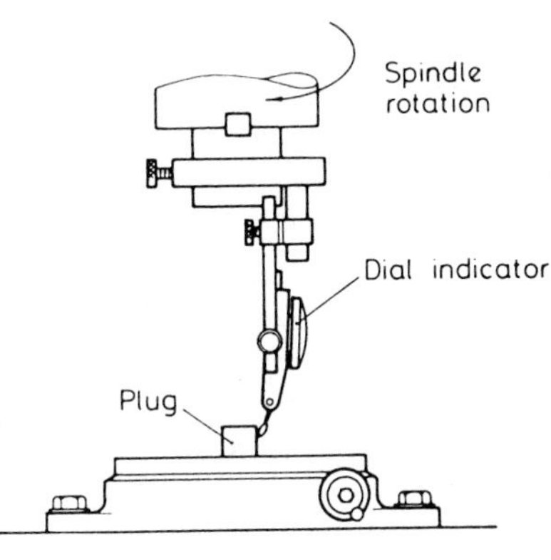

centring a rotary table
Zentrieren enes Drehtisches

cam-roller Kurvenroller
camshaft Nockenwelle
cancel out, to aufheben, ausgleichen
cant, to kanten, kippen
cantilever Ausleger
capacity Kapazität, Aufnahmefähigkeit, Leistungsfähigkeit
capacity of vice jaws Spannweite der Schraubstockbacken
cap nut Überwurfmutter
capstan Drehkreuz
carbide cutter Hartmetallfräser
carbide cutting edge Hartmetallschneide
carbide cutting section Hartmetallschneidenteil
carbide edge Hartmetallschneide
carbide insert Hartmetalleinsatz
carbide milling Fräsen mit Hartmetallwerkzeug
carbide planing Hobeln mit Hartmetallmeissel
carbide shaping Kurzhobeln mit Hartmetallmeissel
carbide-tipped cutter hartmetallbestückter Fräser
carbide-tipped tool hartmetallbestücktes Werkzeug
carbide tool Hartmetallwerkzeug
carbide-tooled mit Hartmetallwerkzeugen ausgerüstet
carbon steel Kohlenstoffstahl
carriage Schlitten, Bettschlitten
carrier Mitnehmer, Drehherz
carrier plate Mitnehmerscheibe
carry, to tragen, unterstützen, halten, aufnehmen, führen, lagern
carry off, to ableiten, wegleiten
case Gehäuse, Behälter, Kasten
case-hardened einsatzgehärtet
case study Arbeitsbeispiel
casing Gehäuse, Hülle, Kapsel
casing wall Gehäusewand
cast, to giessen
cast-alloy cutter Fräser aus Gusslegierung
castellating Schlitzen von Kronenmuttern

castel (or: castle) nut Kronenmutter
casting Gussstück, Giessen
catch, to einklinken, einschnappen, einhaken
catch Sperrhaken, Knagge, Arretierhebel, Sperrklinke
catch plate Mitnehmerscheibe
cavity milling Raumformfräsen
cemented carbide Hartmetall, Sinterkarbid
cemented carbide tip Hartmetallplättchen
cemented carbide tool Hartmetallwerkzeug
cemented insert Hartmetalleinsatz
cemented tungsten carbide gesintertes Wolframkarbid
center, to zentrieren, ankörnen
center Mitte, Spitze
center drilling Zentrierbohren
centering Zentrieren
centerless spitzenlos
centerline Mittellinie
central pintle Mittelzapfen
center punch Körner
ceramic bit keramischer Einsatzmeissel
chamfer Abschrägung, Anfasung, gebrochene Kante
chamfering tool Anfasmeissel
change-gear box Wechselrädergetriebekasten
change-gear calculation Wechselräderberechnung
change gears Wechselräder
change-over and resetting Umrüsten für andere Serie
change-over time Umrüstzeit
channel, to auskehlen
channel Kanal
chasing Gewindestrehlen
chatter, to rattern
chatter Rattern
chatter mark Rattermarke
check nut Gegenmutter
cheese-head screw Rundkopfschraube
cherry Kugelfräser, Gesenkfräser

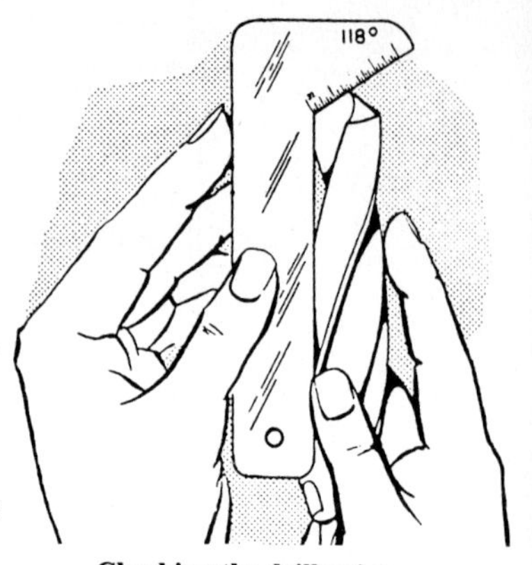

Checking the drill point angle and cutting edge
Prüfen des Bohrerspitzenwinkels und der Schneidkante

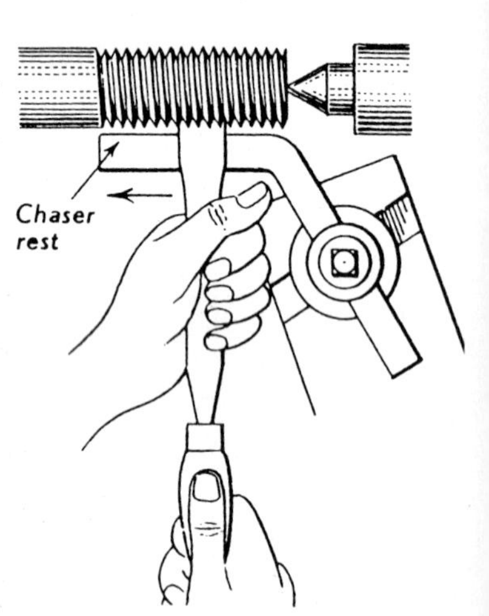

chasing a thread in the lathe
Gewindestrehlen auf einer Drehmaschine

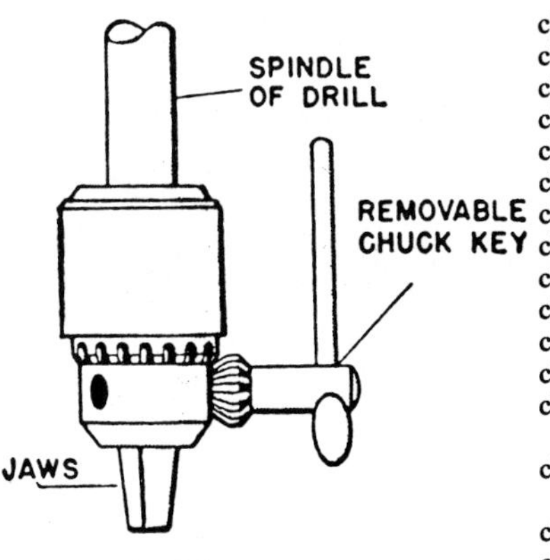

chuck key
Futterschlüssel

chip Span
chip-bearing surface Spanfläche
chip breaker Spanbrecher
chip-breaking groove Spanbrechernut
chip chute Spanleitblech, Spänefang
chip clearance Spanraum
chip conveyor Spanförderer
chip disposal Spanabfuhr
chip flow Spanabfluss
chip formation Spanbildung
chip groove Spannut
chipless process spanloses Verfahren
chip load per tooth (Fräsen) Spantiefe je Zahn
chipping (Schneide) Ausbruch, Ausbrechen
chip pocket Spannut
chisel Meissel
choke, to drosseln, stauen, verstopfen
chuck, to aufspannen, einspannen (in Spannfutter)
chuck Spannfutter
chucking Einspannen, Aufspannen
chucking device Spannvorrichtung
chuck jaw Einspannbacke
chuck jaws Einspannfutter
chuck spring collet Federspannfutter
chuck work Futterarbeit
chute, to rutschen, gleiten
chute Leitblech, Rutsche
chute feed Zuführung von Werkstükken über Leitblech
circle of holes Lochkreis
circuit Kreislauf, Stromkreis, Schaltung
circular rund, rundlaufend, kreisförmig, Rund . . .
circular broach Räumwerkzeug mit kreisförmigem Querschnitt
circular cut Kreisschnitt
circular dividing table Rundtisch mit Teileinrichtung
circular division of work Rundteilung des Werkstücks
circular feed Rundvorschub
circular forming tool Formscheibenstahl
circular milling Rundfräsen

clamp for screw threads
Spannstück für
Schraubengewinde

circular milling attachment Rundfräseinrichtung
circular milling operation Rundfräsarbeit
circular milling table Rundfrästisch
circular motion attachment Rundhobeleinrichtung
circular recessing Fräsen kreisförmiger Auskehlungen
circular slotting Rundstossen
circular spacing Rundteilen
circular table Rundtisch
circular work rundes Werkstück, Bearbeitung mit Rundvorschub
circulate, to umlaufen, kreisen, umwälzen, umlaufen lassen
circumference Umfang, Peripherie
circumferential indexing Teilen eines Kreisumfangs
circumferential surface Umfangsfläche, Mantel, Zylinderfläche
clamp, to klemmen, festklemmen, aufspannen, feststellen
clamp Klemmung, Klemmschraube, Spannschraube
clamp bed (Fräsmaschine) Rundführung
clamp bolt Spannbolzen
clamp coupling Schalenkupplung
clamp dog Spannkloben
clamping Aufspannung, Festklemmung
clamping area Aufspannfläche
clamping bolt Spannbolzen
clamping device Klemmeinrichtung, Aufspannvorrichtung
clamping fixture Spannvorrichtung
clamping screw Feststellschraube
clamping table Aufspanntisch
clamp-type toolholder Klemmhalter
clapper Meisselklappe
clapper clamping nut Klemmschraube für den Klappenträger
claw coupling Klauenkupplung
clean, to reinigen
cleanliness Sauberkeit, Reinheit
cleanse, to spülen, reinigen

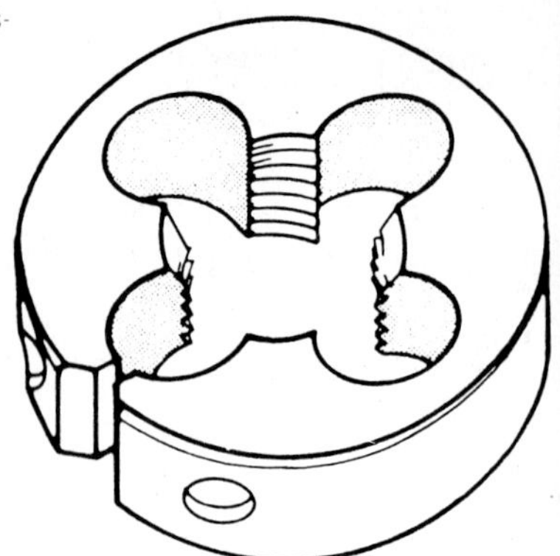

circular split die
Schneideisen, geschlitzt

circulating pump for heating system
Heizungsumwälzpumpe

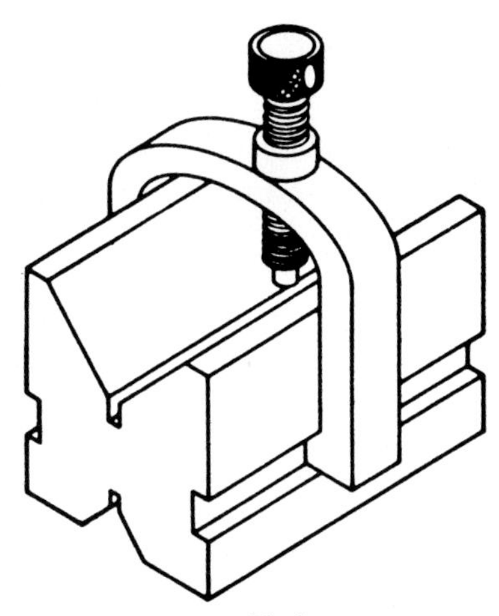

clamp on vee-block
Spannstück auf
Prismenblock

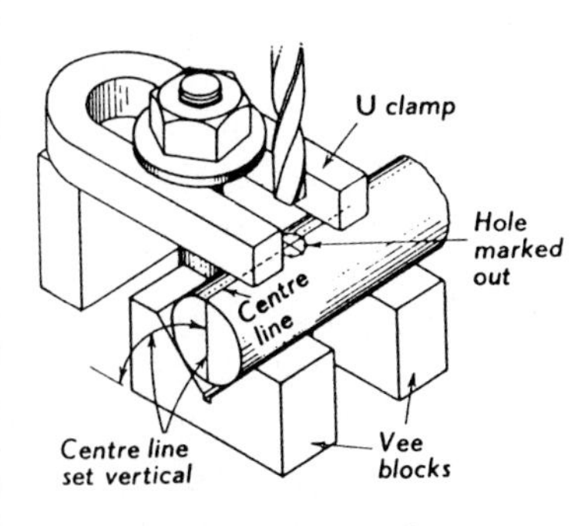

clamping and setting for drilling a shaft
Spannen und Einrichten zum Bohren einer Welle

clearance Zwischenraum, Spiel, Abstand, Spielraum, Freiwinkel
clear of entfernt von, mit Abstand von, ohne Berührung mit
clevis Schäkel
climb-cut milling Gleichlauffräsen
climb cutting Gleichlauffräsen
climbfeed method Gleichlauf-Fräsverfahren
cling, to haften an
clip Schelle, Klemme
clockwise im Uhrzeigersinn, rechtsdrehend
clockwise rotation Rechtsdrehung
clogging of chips Spanstauung
close, to schliessen
close limit production Fertigung unter Einhaltung enger Toleranzen
close limits Massgrenzen
closely measured genau bemessen
close-meshed engmaschig
clutch, to einrücken, schalten, kuppeln
clutch Kupplung, Schaltkupplung
coarse grob, rauh
coarse-feed series Grobvorschubreihe
coarse-pitch cutter Fräser mit grober Zahnteilung
coated überzogen, umhüllt
coating Überzug, Umhüllung
cock Hahn (ventil)
coil, to aufrollen, wickeln
coil Spule
coil spring Schraubenfeder
collar Manschette, Buchse, Hülse, Skalenring, Skalentrommel
collar (Fräsmaschine) Zwischenring, Abstandsring
collet Spannzange, Reduziereinsatz
collet chucking Zangenspannung
column Ständer
column and knee type machine Konsolmaschine
column and knee type milling machine Konsolfräsmaschine
column ways Führungsbahnen am Ständer
combination broach-burnisher tool Räum- und Glättwerkzeug

comma-shape kommaförmig
commencement chip thickness Spanstärke am Anfang des Schnittes
compensate for wear, to Verschleiss ausgleichen
complete, to fertigstellen, vollenden
component Teil, Bestandteil, Werkstück, Komponente
component of force in cutting Schnittkraftkomponente
composed of zusammengesetzt aus
compound indexing Verbundteilen
compound slide Kreuzschlitten
compress, to verdichten, komprimieren, zusammendrücken
compression coupling Klemmkupplung, Schalenkupplung
compressive clamp Federspanneisen
comprise, to einschliessen, enthalten, umfassen
compute, to berechnen, ausrechnen
concave milling cutter Fräser für Halbkreisprofil
concentric chuck Universalfutter
conduct, to leiten, führen
conduit Leitung, Rohr, Kanal
cone Konus, Kegel
cone drive Stufenscheibenantrieb
cone pulley Stufenscheibe
cone-type face milling cutter with inserted blades Fräskopf mit eingesetzten Messern
conical konisch, kegelig
conical clutch Konuskupplung, Kegelkupplung
connect, to verbinden
connecting rod Verbindungsstange, Pleuelstange
connection sleeve Verbindungsmuffe
constriction Verengung, Einschnürung, Verjüngung
construct, to bauen, ausführen
construction Bau, Ausführung
contact area Berührungsfläche, Traganteil
contact pressure Anpressungsdruck
contact roller Führungsrolle

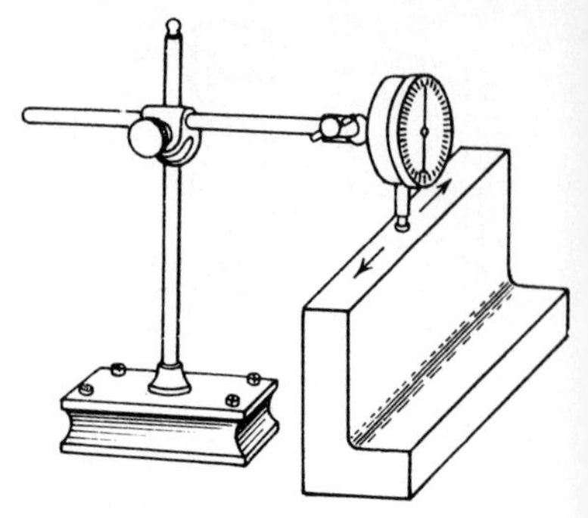

clock indicator (also: dial gauge) checking for parallelism
Messuhr zum Prüfen von Parallelität

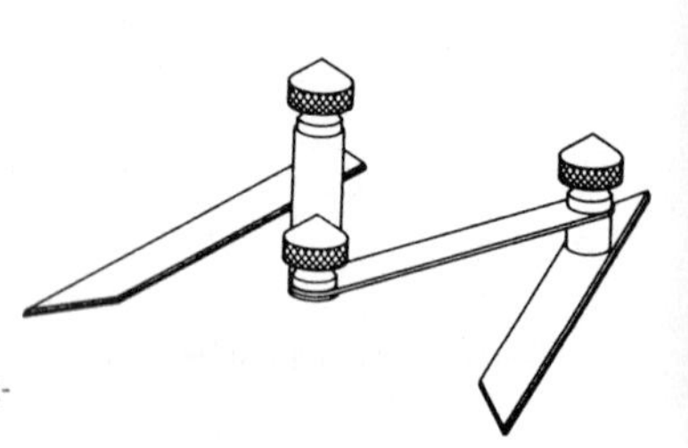

combination bevel
Stellwinkel

control box
Steuerkasten

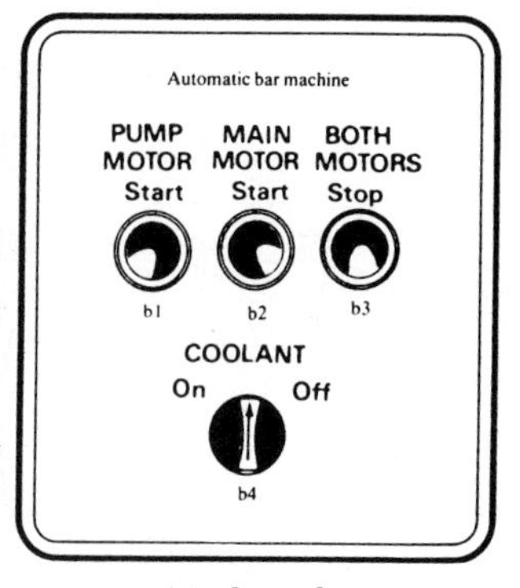

control panel
Steuertafel

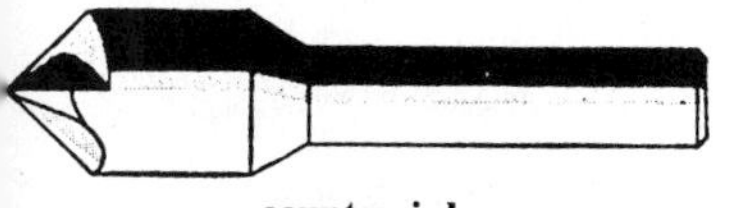
countersink
Senkbohrer

contaminate, to verunreinigen, verschmutzen
contamination Verunreinigung, Verschmutzung
content Gehalt, Inhalt
continuous broaching ununterbrochenes Räumen
continuous curly chip Wendelspan, Lockenspan
continuous rating Dauerleistung
contour, to im Umriss fräsen, profilieren, kopieren, nachformen
contour Aussenlinie, Umriss
contouring Nachformen, Kopieren
contour milling Umrissfräsen
contour planing Umrissnachformhobeln
contour shaping machine Kehlhobelmaschine
contraction Einengung
control, to steuern, schalten, kontrollieren
control Steuerung, Schaltung, Regelung
control desk Steuerpult, Schaltpult
controller Steuerorgan, Steuergerät
control lever Schalthebel, Bedienungshebel
control panel Schalttafel, Schaltfeld
controls Schaltelemente
control slide valve Steuerschieber
control station Kontrollstation, Steuerstand
control worm Schaltschnecke
conventional cut Gegenlauffrässchnitt
conventional milling Gegenlauffräsen
conversion Umwandlung
convert, to umwandeln, umbauen
convex milling cutter Fräser für Halbkreisprofil
convey, to fördern
conveyance Förderung, Beförderung, Transport
conveyor Fördereinrichtung
coolant Kühlmittel
coolant supply Kühlmittelzuführung
cooling slot Kühlungsnut

co-ordinate table Koordinatentisch
copper, to verkupfern
copy, to nachformen, kopieren
copy and die milling cutter Kopier- und Gesenkfräser
copying Kopieren, Nachformen
copying attachment Kopiereinrichtung, Nachformeinrichtung
copying shaper Kopier-Waagrechtstossmaschine
copy milling Kopierfräsen, Nachformfräsen
copy milling machine Kopierfräsmaschine
core Kern
corner-rounding cutter Radiusfräser
corrosion-preventive korrosionshindernd
corrugated gewellt, gerieft
cotter mill Langlochfräser
counterbalance, to mit Gegengewicht versehen
counter-clockwise im entgegengesetzten Uhrzeigersinn
countershaft Vorgelegewelle
couple, to kuppeln
coupling Kupplung
crack Riss, Bruch, Sprung
crank, to kröpfen, beugen, biegen
crank Kurbel, Kulisse
crank arm Kurbelschwinge
cranked planing tool gebogener Hobelmeissel
crank planer Hobelmaschine mit Kurbelantrieb
crank shaper Waagrechtstossmaschine mit Kurbelschleifenantrieb
crater, to auskolken
crater Auskolkung
creep feed Kriechgang, Schleichgang
crooked gekrümmt
cross adjustment Querverstellung
cross arm Ausleger, Querbalken
cross bar Querbalken
cross beam Traverse, Querhaupt
cross feed Quervorschub, Querverschiebung
cross feed screw Querschubspindel

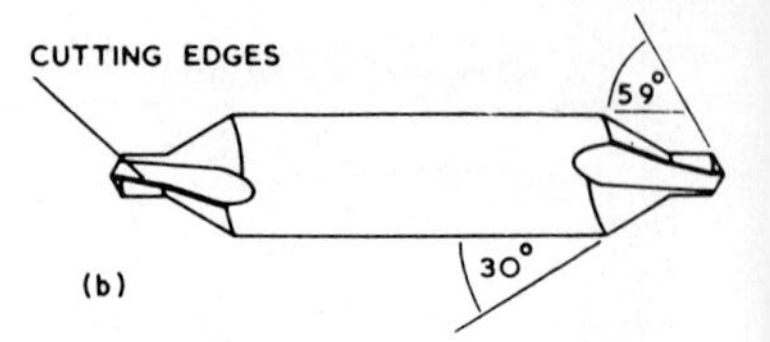

countersink or centre drill
Senk- oder Zentrierbohrer

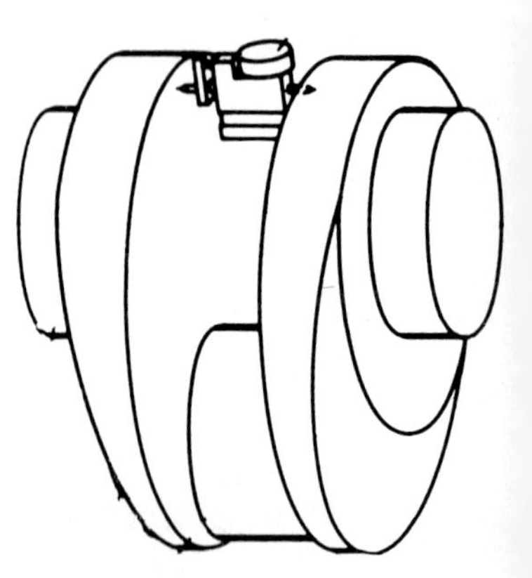

crankweb deflection gauge
Kurbelwellenatmungs-Messgerät

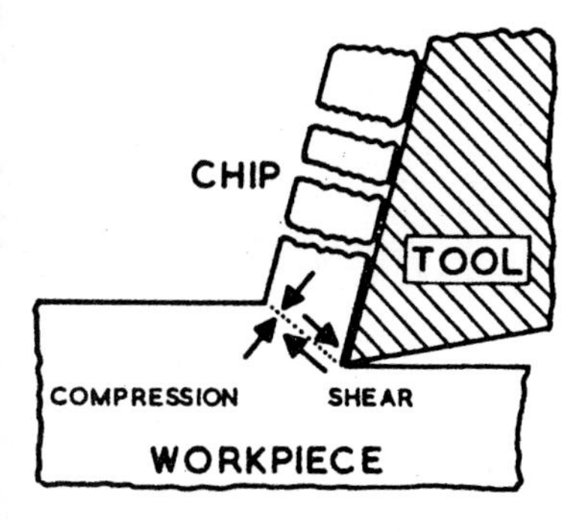

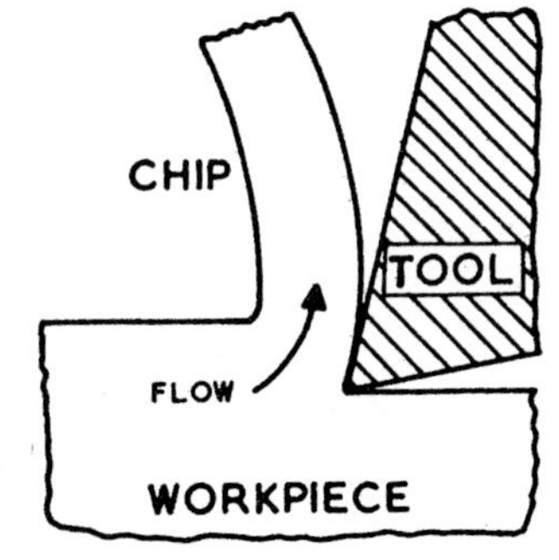

cutting action: brittle (above) and ductile material
Schneidvorgang: sprödes (oben) und zähes Material

cross girth Traverse, Querhaupt
crosshead Traverse, Querhaupt
cross milling Fräsen in Querrichtung
cross planing Querhobeln
cross rail Querbalken
cross rail planing head Querhobel-support
cross rib Querrippe
cross section Querschnitt
cross slide Querschlitten
cross travel Querbewegung
cross slot Quernut
crumble, to abbröckeln
crumbly bröcklig
cup-shaped cutter Topffräser
current Strom
cushion, to dämpfen, puffern
cut, to schneiden, zerspanen
cut Schnitt, Span
cutting-off tool Stechmeissel
cut-off saw Trennsäge
cut-out-milling Ausschneidfräsen
cut per tooth Schnittiefe je Zahn
cutter Fräser, Schneidwerkzeug
cutter arbor Fräsdorn
cutter clamp Messerklemme
cutter for stub-arbor mounting Fräser zum Aufstecken auf fliegend angeordnetem Dorn
cutter head Messerkopf
cutter life Fräserstandzeit
cutter pitch Fräserzahnteilung
cutter relief Fräserabhebung
cutter speed Fräserdrehzahl
cutter spindle Frässpindel
cutter spindle quill Frässpindelhülse
cutter tooth Fräserzahn
cutter wear Fräserverschleiss
cutting Schneiden, Zerspanen, Zerspanung
cutting angle Schnittwinkel
cutting blade Messer
cutting blade life Messerstandzeit
cutting capacity Zerspanungsleistung
cutting depth Spantiefe, Schnittiefe
cutting down Gleichlauffräsen
cutting edge Schneidkante, Schneide

cutting edge angle Komplementwinkel des Einstellwinkels
cutting efficiency Wirkungsgrad bei Zerspanung, Zerspanungsleistung
cutting face Spanfläche
cutting force Schnittkraft
cutting from the solid Schneiden aus dem Vollen
cutting life Standzeit, Lebensdauer
cutting load Schnittkraft
cutting medium Zerspanungsmittel, Schneidwerkstoff
cutting-off machine Trennmaschine
cutting-off tool Stechmeissel, Abstechwerkzeug
cutting operation Zerspanungsarbeit, Zerspanungsvorgang
cutting path Fräsweg
cutting pressure Schnittkraft
cutting rake Spanwinkel
cutting section Schneidenteil
cutting speed Schnittgeschwindigkeit
cutting spindle Frässpindel
cutting stroke Arbeitshub
cutting technique Zerspanmethode, spanabhebendes Verfahren
cutting teeth (Räumwerkzeug) Schneidenzähne
cutting tool Schneidwerkzeug, zerspanendes Werkzeug
cutting tool tip Schneidwerkzeugplättchen, Schneidplatte, Werkzeugschneide
cutting tooth Schneidzahn
cutting up Gegenlauffräsen
cutting-wedge angle Keilwinkel
cycle Zyklus, Arbeitszyklus, Arbeitsgang, Arbeitsablauf
cycle setting Einstellung des Arbeitsablaufs
cylindrical rotary valve Drehkolbenventil

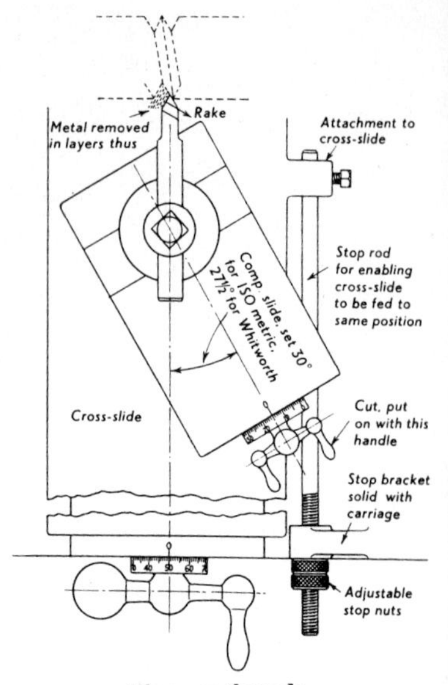

cutting vee threads
Schneiden von Spitzgewinde

D

damage, to beschädigen
damage Schaden, Beschädigung

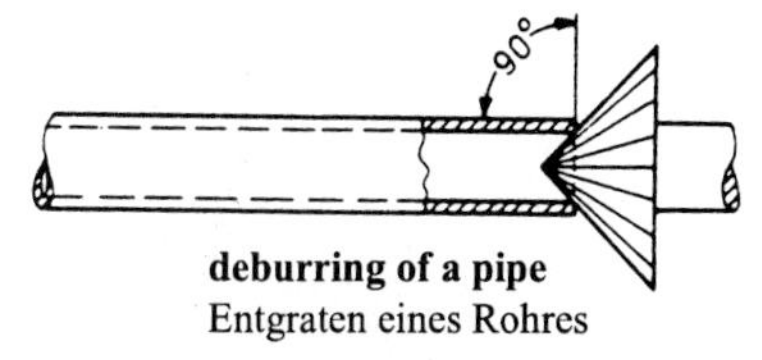

deburring of a pipe
Entgraten eines Rohres

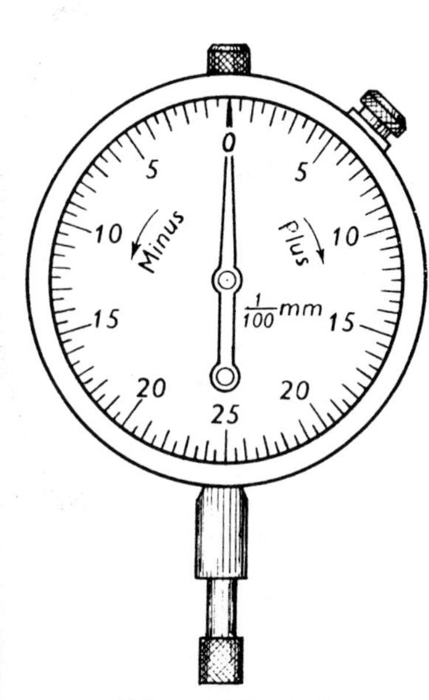

dial gauge (also: clock indicator)
Messuhr

dashpot Bremszylinder, Stossdämpfer, Puffer
dead centre feststehende Spitze, Reitnagel
deburr, to entgraten
deburring Entgraten
declutch, to entkuppeln, ausrücken
decrease of volume Volumenabnahme
dedendum (Pl: dedenda) Fusshöhe, Zahnfuss
dedendum circle Fusskreis
defect Defekt, Fehler, Beschädigung
deflect, to ablenken, durchbiegen
deflection Ablenkung, Durchbiegung
degreasant Entfettungsmittel
degrease, to entfetten
degree of accuracy Genauigkeitsgrad
degree of taper (Räumwerkzeug) Vorschub je Zahn, Verjüngung
deliver, to liefern, abgeben, fördern
delivery Förderung, Lieferung
delivery control member Regelorgan für die Fördermenge
delivery head Förderhöhe
delivery rate Fördermenge
demesh, to ausrücken
demount, to abnehmen, demontieren, ausbauen
depress, to niederdrücken, senken
depression Vertiefung, Druckminderung
depth cutting type of broach senkrecht zur Werkstückoberfläche schneidendes Räumwerkzeug, Räumwerkzeug mit Tiefenstaffelung der Zähne
depth feed Tiefenvorschub, Senkrechtvorschub
depth micrometer Tiefenmass, Tiefenmikrometer
depth of cut Spantiefe, Schnittiefe
depth of gap Ausladung
depth stop Tiefenanschlag
derive from, to ableiten von
de-scaling Entzundern
descend, to abwärts bewegen, niedergehen

design, to konstruieren, entwerfen, bestimmen
design Konstruktion, Entwurf, Bauart, Ausführung
designer Konstrukteur
desired value Sollwert
detachable abnehmbar
detent Auslöser, Arretierung, Klinke
detergent Reinigungsmittel, ablösendes Mittel, Detergent
develop, to entwickeln
deviate, to abweichen
deviation Abweichung
device Vorrichtung, Gerät, Einrichtung, Apparat
diagram Schaubild, Plan, Diagramm
dial Skalenscheibe, Wahlscheibe, Skalenring, Einstellscheibe
dial indicator Anzeigeskala, Messuhr
diameter Durchmesser
diametrically opposed diametral gegenüberliegend
diamond-knurled kordiert
diamond-tipped diamantbestückt
diaphragm Membrane
die Gesenk, Matrize, Form
die and mold copy miller Gesenk- und Formen-Nachformfräser
die-casting Spritzgiessen
die making Herstellung von Gesenken
die mill Gesenkfräser
die milling machine Gesenkfräsmaschine
die sinking machine Gesenkfräsmaschine
dig in, to einhaken (Meissel in Werkstück)
dimensional accuracy Massgenauigkeit
direct indexing head einfacher Teilkopf
directional in Bewegungsrichtung
direction of cutting Schnittrichtung
direction of travel Bewegungsrichtung
direct motor drive Einzelantrieb
disassemble, to zerlegen, auseinandernehmen

dividers
Teil-, Stechzirkel

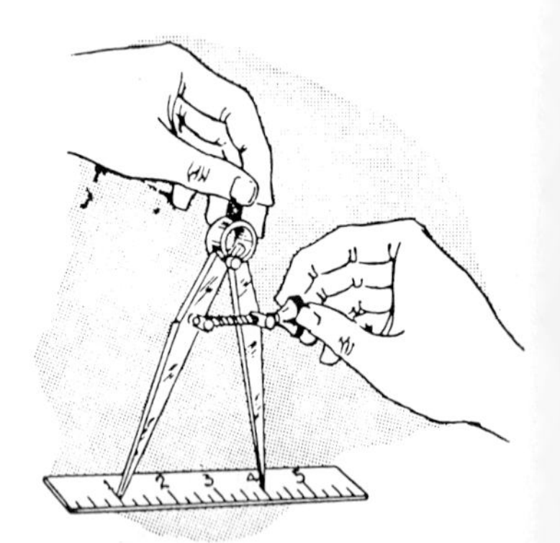

dividers being set
Teilzirkel wird eingestellt

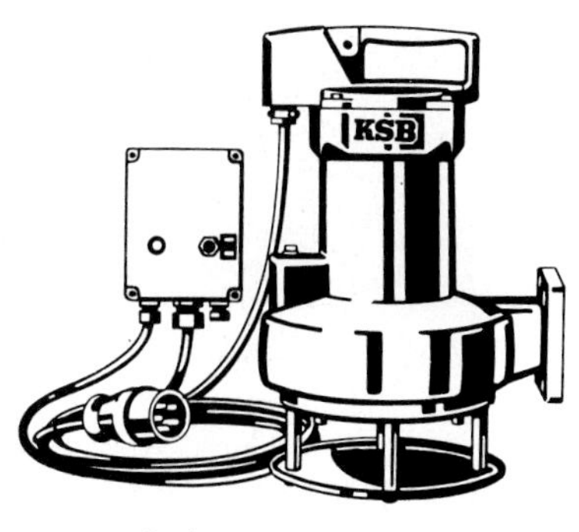

drain pump
Entwässerungspumpe

disc cam Scheibenkurve
discontinuous chip Bruchspan, Reissspan
disengage, to ausrücken, aus Eingriff bringen
dismantle, to auseinandernehmen, demontieren
displace, to verdrängen, fördern
disposal Beseitigung
dissipate, to verstreuen, verteilen
distortion Verwindung, Verziehen, Verdrehung
diverge, to sich erweitern, zerstreuen
divided-table planer Hobelmaschine mit zwei Arbeitstischen für kontinuierliches Arbeiten
dividing apparatus Teilapparat
dividing attachment Teileinrichtung
dividing head Teilkopf
dividing-head plate Teilscheibe
division Teilung
dog Drehherz, Knagge, Klaue, Mitnehmer, Anschlag
dog-actuated anschlagbetätigt, nokkenbetätigt
dope, to dem Öl chemische Zusätze beifügen
double-acting doppeltwirkend
double-angle cutter Winkelfräser für Werkzeuge mit gefrästen Zähnen für Spannuten mit Drall
double-column planer Doppelständer-Hobelmaschine
double-cut broach zweiseitig schneidendes Räumwerkzeug
double-cut planing Hobeln mit Spanabnahme in beiden Richtungen
double-cutter Schlagzahnfräser mit zwei Meisseln
double-ended doppelseitig
double-end mill zweiseitiger Schaftfräser, doppelseitiger Langlochfräser
double-end milling machine Zweispindel-Abflächmaschine zum gleichzeitigen Fräsen der Stirnflächen von Wellen
dovetail, to Schwalbenschwanznuten ausarbeiten

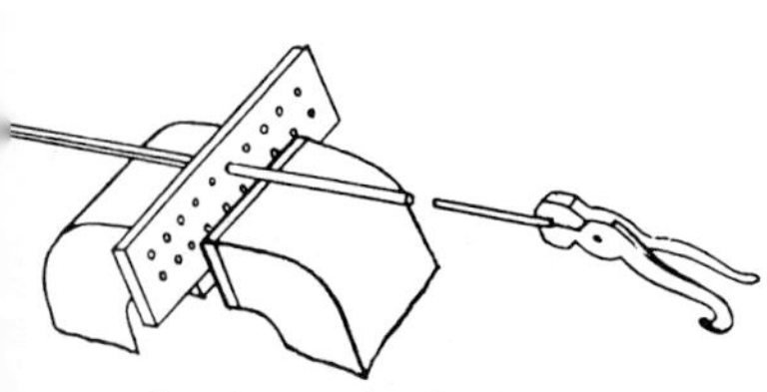

drawing a tube through a plate
ein Rohr wird durch eine Platte gezogen

dovetail Schwalbenschwanz(führung)
dowel, to dübeln
dowel Dübel
down-cut milling Gleichlauffräsen
downfeed Tiefenvorschub
downstroke Abwärtshub
downward travel Abwärtsbewegung
draftsman Zeichner
drag Zwischenschneide
drain, to ablassen, zurückfliessen lassen
drain Abfluss, Rückfluss
drain cock Ablasshahn
drawbolt Spannschraube
drawbar Zugstange
draw-cut shaper Waagrechtstossmaschine mit ziehendem Schnitt
draw grooves, to Nuten ziehen
draw head Ziehkopf
draw home, to anziehen, hineinziehen
draw rod Zugstange
drill, to bohren (aus dem Vollen)
drill flute Bohrernut
drilling Bohren (aus dem Vollen)
drive, to treiben, antreiben, mitnehmen
drive Antrieb, Mitnahme
driver Antriebsrad, Drehherz, Mitnehmerbolzen
driving dog Mitnehmer, Drehherz
driving pinion Antriebsritzel
drop, to fallen (lassen), abfallen
drop Tropfen, Abfall
drop-forging die Schmiedegesenk
drum Trommel
drum-type milling machine Trommelfräsmaschine
dry grinding Trockenschleifen
dry milling Trockenfräsen
dual ram broaching machine Doppelstössel-Räummaschine, Zwillingsräummaschine
dual table planer type miller Langfräsmaschine mit zwei Tischen
duct Kanal, Röhre
ductile zäh, geschmeidig, dehnbar
dull, to abstumpfen, stumpf werden, stumpf machen

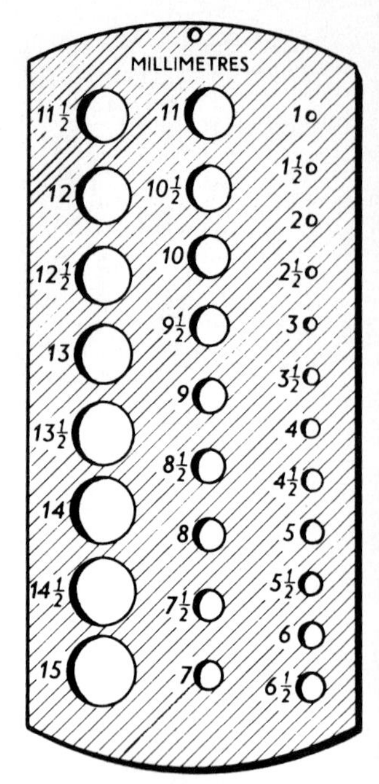

drill gauge
Bohrlehre

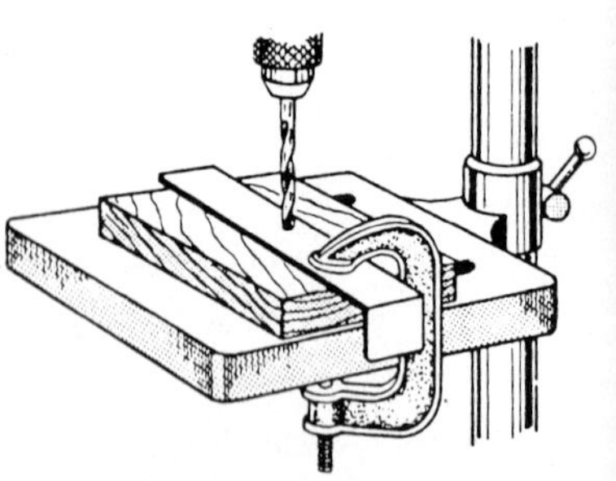

drilling thin-gauge metal
Bohren von dünnem Material

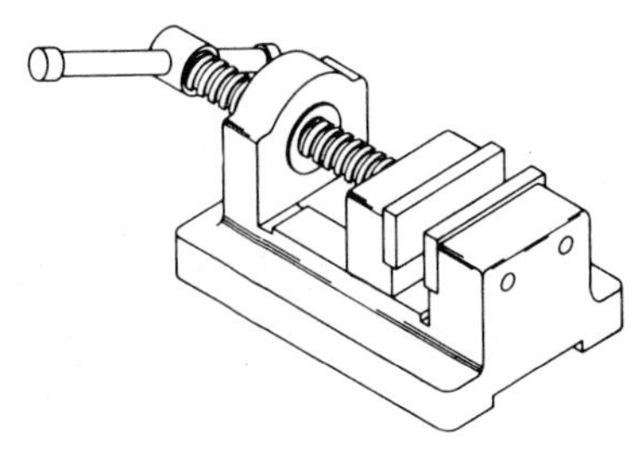
drill vice
Bohrschraubstock

duplex-head milling machine Doppelspindel-Planfräsmaschine
duplex milling gleichzeitiges Fräsen mit zwei Fräsköpfen
duplex planer Doppelhobelmaschine
duplicate, to nachformen, kopieren
duplicate in three dimensions, to raumnachformen, dreidimensional nachformen
duplicate machining Nachformen, Nachformarbeit
duplicate part Serienteil
duplicating attachment Nachformeinrichtung
duplicator Kopiereinrichtung, Nachformeinrichtung
dust guards Staubschutz
dwell Stillstand, Bewegungspause
dwell teeth (Räumwerkzeug) Führungszähne
dynamic seal dynamische Dichtung, Abdichtung beweglicher Teile gegeneinander

E

ear Öse
easy machining gut zerspanbar, leicht bearbeitbar
eccentric aussermittig
eccentric shaft Exzenterwelle
eddy, to wirbeln
eddy Wirbel, Turbulenz
edge Kante, Schneide, Rand
edge chipping Ausbrechen der Schneide
edge-machine, to abfasen, an Kanten bearbeiten
edge wear Verschleiss an der Werkzeugschneide
effect, to leisten, bewirken
effect Leistung, Wirkungsgrad
effective wirksam, effektiv
efficiency Wirkungsgrad, Aufnahmefähigkeit
eject, to auswerfen

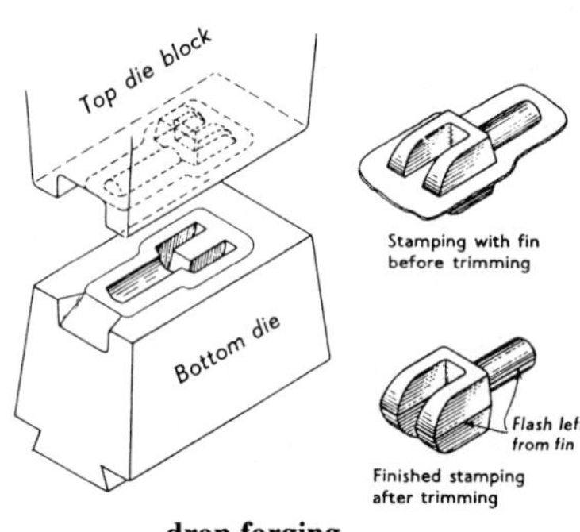

drop forging
Gesenkschmieden

ejector Auswerfer
elbow Knie, Kniestück (Rohr)
elephant machine tool Schwerstwerkzeugmaschine
elevate, to heben, anheben
elevating slide (Räummaschine) Anhebeschlitten
elongate, to ausdehnen, verlängern, strecken
emboss, to prägen, hohlprägen
emergency stop Nothalt
employ cutting speeds, to Schnittgeschwindigkeiten verwenden
empty, to leeren, entleeren
empty leer
emulsifier Emulgierzusatz zu Öl
end-brazed carbide tool an der Stirn gelötetes Hartmetallwerkzeug
end clearance Axialspiel
end-cut single point tool Kopfmeissel
end-cutting in axialer Richtung schneidend, mit der Stirnseite schneidend
end face mill Walzenstirnfräser
end flank Freifläche der Nebenschneide (Stirnseite)
end journal Stirnzapfen
end mill Schaftfräser, Fingerfräser
end mill adapter Zwischenstück für Schaftfräser
end stop Längsanschlag
end thrust axialer Druck
endwise axial, in Achsrichtung
engage, to einrücken, eingreifen, einrasten, einschalten
engaged im Eingriff, eingerastet, eingeschaltet
engaging dog Mitnehmerklaue
engineering service technischer Dienst
engrave, to gravieren
engraving cutter Gravierfräser
engraving machine Graviermaschine
ensure, to sichern, garantieren, sicherstellen
entering angle Einstellwinkel
entering end Anschnittseite
enter the cut, to anschneiden, Spanabnahme beginnen

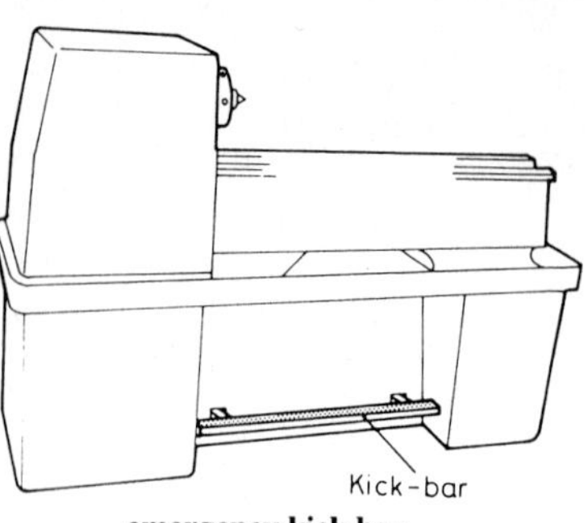

emergency kick bar
Leiste für Fussnotabschaltung

energy cost indicator
Energiekostenanzeiger

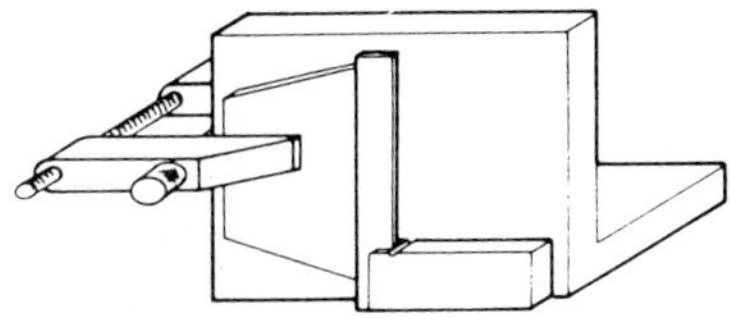

engineer's square
Mess-, Anschlagwinkel

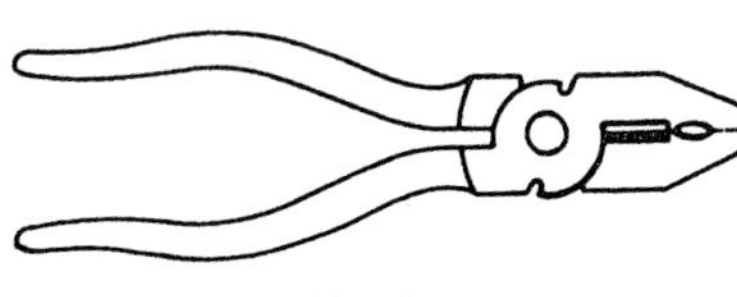

engineer's pliers
Kombizange

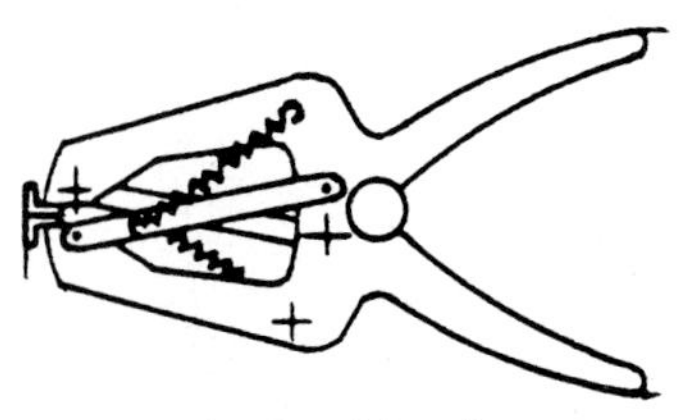

expander for piston rings
Kolbenringzange

equal angle cutter Prismenfräser
equalize, to ausgleichen
equip, to ausrüsten, ausstatten, versehen
equipment Ausrüstung, Ausstattung, Einrichtung
erection Aufstellung, Montage
error Fehler, Abweichung
evaporation Verdampfung
even eben, glatt
evenly spaced in (mit) gleichem Abstand
exceed, to überschreiten
exert, to ausüben (Druck)
exhaust line Rückflussleitung
expansion Erweiterung
expansion arbor Spreizdorn
expansion joint Ausdehnungsfuge
expel, to ausstossen
explorator Nachformfräser
ex stock vom Lager
extension piece Verlängerungsstück
external broach Aussenräumwerkzeug
external broaching machine Aussenräummaschine
external gear aussen verzahntes Rad
extrude, to fliesspressen
extrusion Strangpressen, Fliesspressen
eye Öse

F

fabricate, to fertigen, herstellen
fabrication Fertigung, Herstellung
face Stirnfläche, Vorderfläche
face angle Komplementwinkel des Einstellwinkels
face cam Plankurve
face cutting edges Stirnschneiden
face-ground formed milling cutter Formfräser (an der Spanfläche nachgeschliffen)
face land Fase der Stirnschneiden (beim Scheibenfräser)

face mill Stirnfräser
face milling Flächenfräsen, Walzenstirnfräsen
face milling cutter Stirnfräser
face of the tooth Spanfläche eines Fräserzahnes
face plate Planscheibe
face ridge fasenartiger Anschliff auf der Spanfläche
facilities Einrichtungen
facing Plandrehen, Flächenfräsen
facing cutter Stirnfräser
facing operation Bearbeitung einer Fläche, Planarbeit
factory Fabrik, Betrieb
fail, to versagen
failure Versagen
false jaws Einsatzbacken
fasten, to befestigen
fastener Verbindungselement
fast traverse Eilgang
fatigue Ermüdung
fatigue strength Dauerfestigkeit
faulty design Konstruktionsfehler, Fehlkonstruktion
feathered aufgekeilt
feather edge zugeschärfte Kante
feather key Schiebekeil
feed Vorschub, Zustellung
feedback Steuerungssystem mit Rückmeldung
feed change gear Wechselräder für den Vorschub
feed control Vorschubregelung, Vorschubsteuerung
feed dial Vorschubeinstellscheibe
feed driving disc Nutenscheibe der Querschubeinrichtung
feed gear box Vorschubräderkasten
feed hopper trichterförmige Zuführungseinrichtung
feeding mechanism Vorschubgetriebe, Einrichtung für den Vorschub
feed mark Vorschubmarke
feed-pawl mechanism Sperrgetriebe für den Vorschub

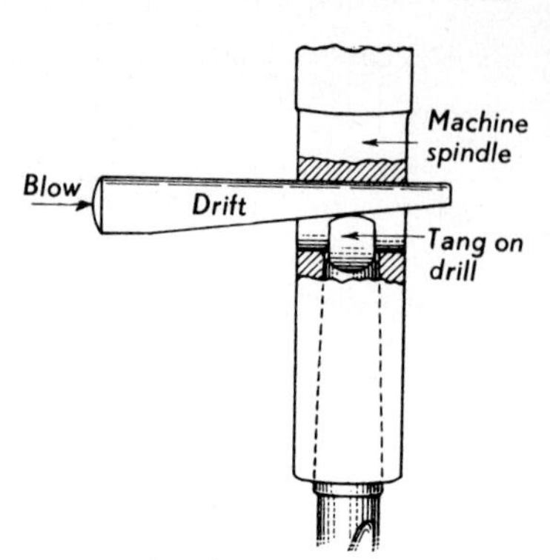

extracting drill with taper drift
Herausnehmen des Bohrers mittels Austreibers (oder: Keils)

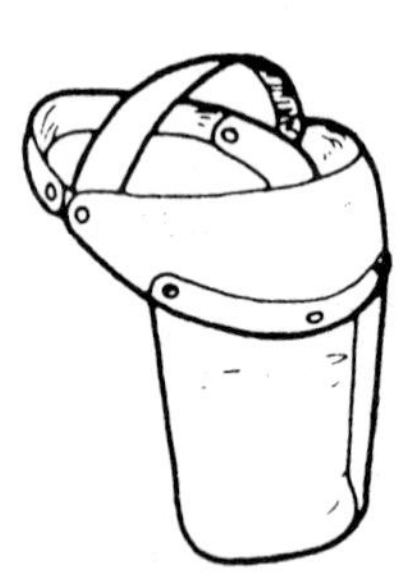

facescreen
Gesichtsschutz

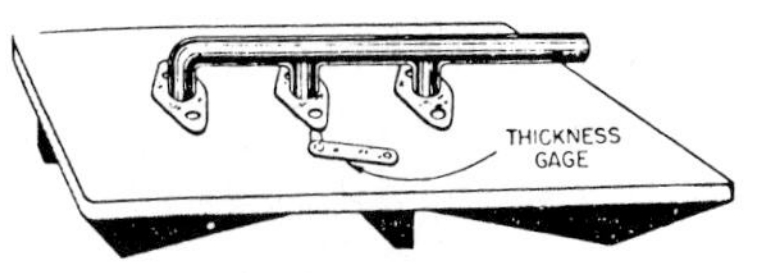

flat surface being checked of conformity
Planfläche wird auf Ebenheit geprüft

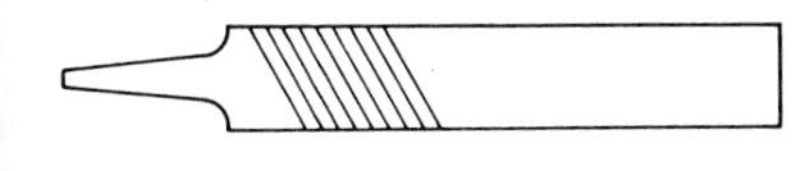

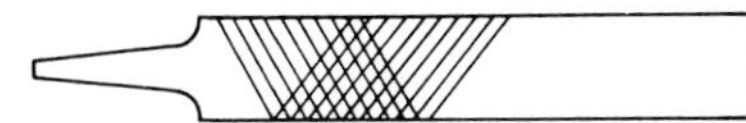

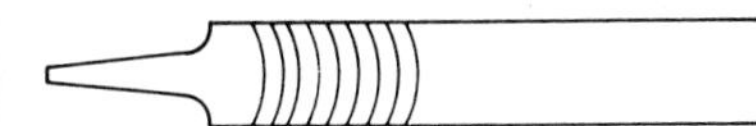

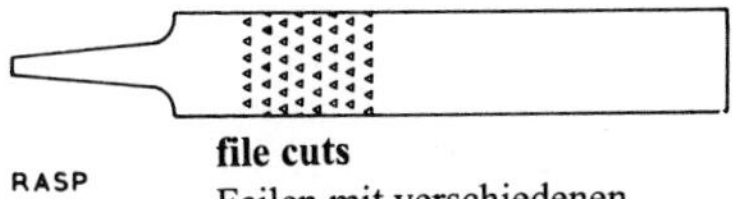

file cuts
Feilen mit verschiedenen Hiebarten

feed ratchet wheel Schaltrad, Vorschubklinkenrad
feed rate Vorschub, Vorschubwert, Vorschubgrösse
feed rod Vorschubwelle, Zugspindel
feed screw Vorschubgewindespindel
feed selector Vorschubwähler, Vorschubwahlschalter
feed setting Vorschubeinstellung
feed shaft Vorschubwelle
feed trip Vorschubausrückung
feel, to tasten, fühlen
feeler Taster, Taststift
feeler gauge Fühlerlehre
feeler pin Taststift
felt packing Filzdichtung
female piece Matrize
female thread Innengewinde
ferrous scrap Eisenschrott, Alteisen
ferrule Ring, Zwinge
file, to feilen
filing and sawing machine Feil- und Sägemaschine
filings Feilspäne
filler cap Eingussverschluss
fillet, to auskehlen
fillet Ausrundung, Abrundung
fillister head runder Schraubenkopf
filter, to filtern
filter Filter
filter element Filtereinsatz
fin Grat, Naht, Rippe (Rohr)
final position Endeinstellung
final sizing Fertigkalibrieren
fine adjustment Feineinstellung, Feinnachstellung
fine feed Feinvorschub
fine-grained feinkörnig
fine mesh gauge engmaschig
fine pitched cutter Fräser mit feiner Zahnteilung
fine setting Feineinstellung
fine-tooth cutter Fräser mit Feinverzahnung
finger clamp Spanneisen mit Stift
finish, to schlichten, fertigbearbeiten
finish Endbearbeitung, Oberflächengüte

finish broach, to schlichträumen, fertigräumen
finished part fertigbearbeitetes Teil
finished size Fertigmass
finishing Endbearbeitung, Schlichten
finishing broach Schlichträumwerkzeug
finishing cutter Schlichtfräser, Nachfräser
finishing dimensions Fertigmasse
finishing tool Schlichtmeissel
finish-machined fertigbearbeitet
finish milling Schlichtfräsen, Fertigfräsen
finish size Fertigmass
finish to size, to auf das Fertigmass bringen
finless gratlos
fire proof feuersicher
first cycle machining Bearbeitung im ersten Arbeitsgang
fir tree profile Tannenbaumprofil
fit, to passen, montieren, anbringen
fit Passung, Sitz
fit into, to einbauen, einpassen
fitted with versehen mit, ausgerüstet mit
fitter Monteur, Installateur
fitting instruction Montageanweisung
fitting pin Passstift
fittings Armaturen, Fittings
fix, to anbringen, befestigen, aufspannen, festmachen, feststellen
fixed pulley Festscheibe
fixture Aufspannvorrichtung
flange, to flanschen
flange Flansch
flanged connection Flanschverbindung
flank Freifläche (der Hauptschneide)
flank of tooth Zahnflanke
flap, to klappen
flap Klappe
flare, to ausdehnen, weiten
flash, to abgraten
flash Grat
flat eben (Fläche)

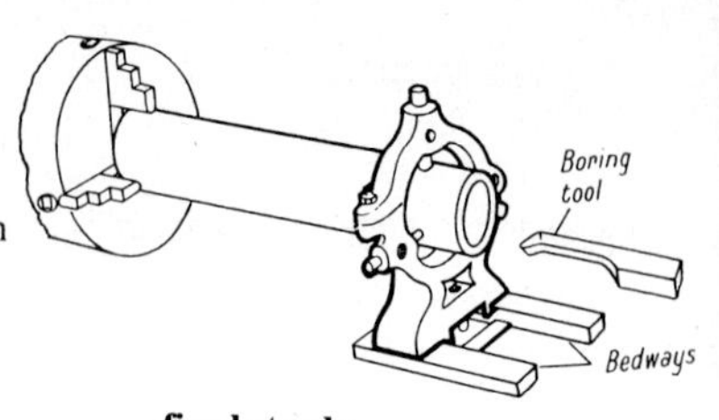

fixed steady
feststehender Setzstock

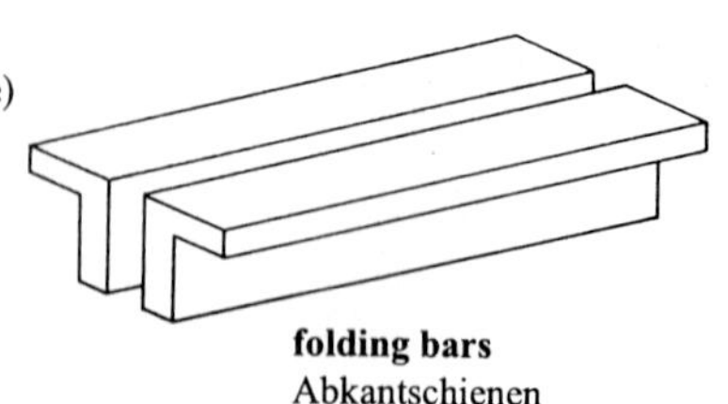
folding bars
Abkantschienen

flatness Ebenheit
flat slideway Flachführung
flat-surface shaping Kurzhobeln ebener Flächen
flat thread Flachgewinde
flat way Flachbahn, Flachführung
flaw Fehler (im Werkstoff), Blase
flex, to (sich biegen), beugen
flexibility Biegsamkeit, Anpassungsfähigkeit
flexible anpassungsfähig, vielseitig, elastisch, biegsam, beweglich, flexibel
flexible shaft biegsame Welle
floating ring freibeweglicher Führungsring
flood, to überfluten, überströmen
floor Fussboden
floor plate Grundplatte
floor switch Fussschalter
floor-to-floor time Boden-zu-Boden-Zeit
flow, to strömen, fliessen
flow Strömung
flow chip Fliessspan
flow of the chips Spanabfluss
flow production Fliessfertigung
fluid flüssig
fluid Flüssigkeit
fluid-feed milling machine Fräsmaschine mit hydraulischem Vorschub
flush, to spülen
flush, to einfluchten, bündig machen
flute Nute, Spannut
flute miller Nutenfräsmaschine
flute milling Nutenfräsen
fluting Nuten, Fräsen von Nuten
fly cutter Schlagzahn, Schlagfräser
fly nut Flügelmutter
flywheel Schwungrad
foam, to schäumen
foam Schaum
follower roll Tastrolle
follow-up servo Folgeeinrichtung
fool-proof narrensicher
foot lever Fusshebel
foot-mounted in Fussbauform

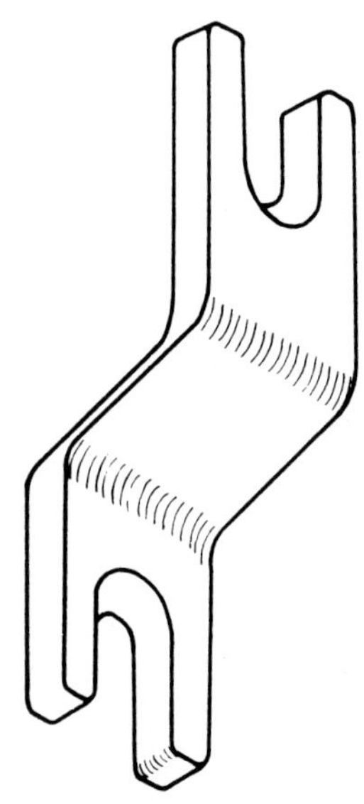

forked piece
Gabelstück

footstock Reitstock
footstock sleeve Reitstockpinole
foot switch Fussschalter
force, to zwängen, drücken
force Kraft
force down, to niederdrücken
forced lubrication Druckschmierung
force of the cut Schnittkraft
force of the stylus Fühlerdruck
forge, to schmieden
forged tool einteiliges Werkzeug, Vollmeissel
forging Schmiedestück
forging die Schmiedegesenk
form and punch shaping machine Form- und Stempelhobelmaschine
formation of burrs Gratbildung
form broach Formräumwerkzeug, Räumwerkzeug für Sonderformen
form copying and profile miller Formkopier- und Profilfräsmaschine
form cutter Formfräser
form-duplicating machine Nachformmaschine
formed milling cutter Formfräser
former pin Führungsstift
former plate Leitlineal
forming tool Formmeissel
form milling Formfräsen
form milling machine Formfräsmaschine
form profile cutter profilgeschliffener Formfräser
form-relieved cutter hinterdrehter Formfräser
form tracer Nachformfühler, Taster
foundation Fundament
four-jaw chuck Vierbackenfutter
fracture, to brechen
fracture Bruch
frail zerbrechlich
frame Rahmen, Ständer, Gestell
free cutting gut zerspanbar
free cutting steel Automatenstahl
free of play spielfrei
free from wear verschleissfrei
friction Reibung

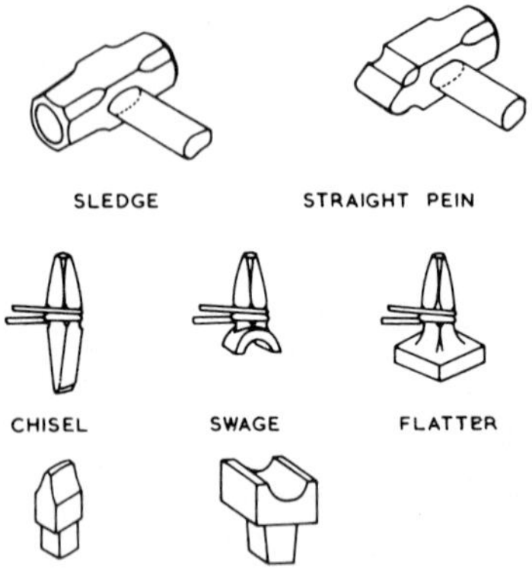

SLEDGE
STRAIGHT PEIN
CHISEL
SWAGE
FLATTER

forging tools used on the anvil
Schmiedewerkzeuge, wie sie am Amboss verwendet werden

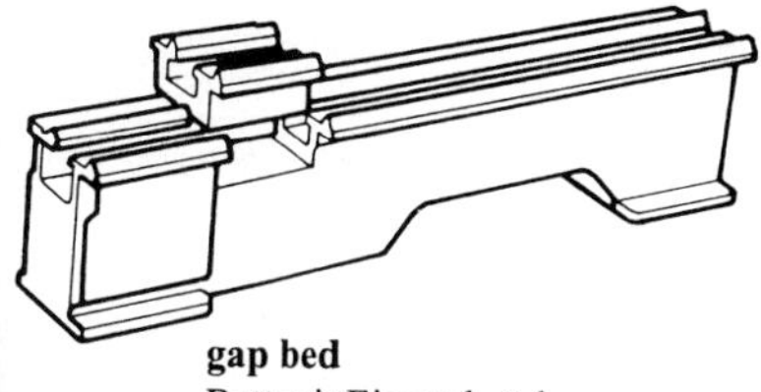

gap bed
Bett mit Einsatzbrücke

frictional reibungs . . ., durch Reibung
friction clutch Reibkupplung
friction saw Reibsäge
front angle Einstellwinkel der Nebenschneide
front brace Gegenhalterschere
front edge Vorderkante, Stirnschneide, Nebenschneide
front elevation Vorderansicht
front end Stirnseite, vorderes Ende
front face Vorderfläche, Vorderseite, Stirnseite
front rake Spitzenspanwinkel
front top rake angle Spitzenspanwinkel
front upright Vorderständer
fulcrum Drehpunkt, Drehachse
fume Rauch, Dampf
funnel Trichter
fuse, to schmelzen

G

gage, to (GB: to gauge) messen, vermessen, kalibrieren
gage (GB: gauge) Lehre, Eichmass
gall, to scheuern, reiben, sich festfressen
gang-cutter Satzfräser
ganged cutters zu einem Satz zusammengestellte Fräser, Satzfräser
gang-mill, to mit einem Satzfräser bearbeiten, mit Doppelspindelstock bearbeiten
gang-mill Satzfräser
gang milling Satzfräsen, Fräsen mit Werkzeugsatz
gang planing Hobeln mit mehreren Meisseln gleichzeitig
gang tool Werkzeugsatz, Meisselsatz
gang tooling setup Mehrmeisselaufspannung
gang-type toolholder Mehrmeisselhalter, Mehrstahlhalter
gang up cutters, to einen Satz Fräser aufspannen

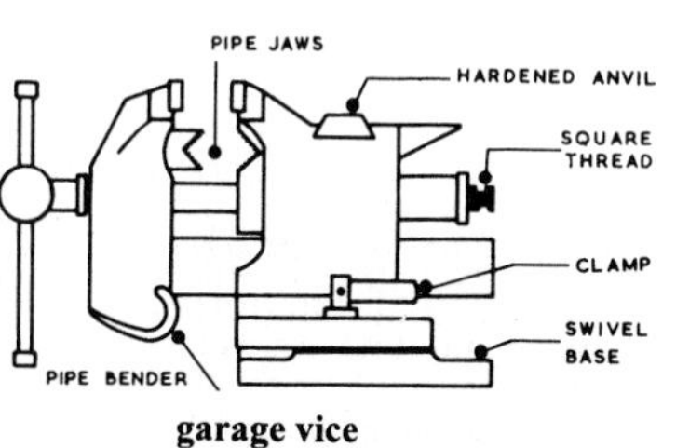

garage vice
Werkstatt-Schraubstock

gantry Gerüst, Portal, Kranportal
gap Spalte, Lücke, Kröpfung, Zwischenraum
gap-type filter Spaltfilter
gash, to Vorfräsen von Zahnlücken
gash Spanlücke, Zahnlücke
gash angle Lückenwinkel
gashing Fräsen von Schlitzen, Vorfräsen von Zahnlücken
gashing cutter Zahnformvorfräser für Fräserzahnlücken
gasket Dichtung (zwischen ruhenden Flächen)
gasket surface Dichtfläche
gate valve Schieber, Absperrventil
gauge, to (USA: to gage) messen, vermessen, kalibrieren
gauge (USA: gage) Lehre, Eichmass
gauge block Endmass
gauge block measuring Messen mit Parallelendmassen
gear, to durch Zahnräder verbinden, übersetzen, eingreifen, ineinandergreifen
gear Zahnrad, Getriebe, Übersetzung, Eingriff
gearbox Getriebe, Getriebekasten, Räderkasten
gear cutter Zahnradfräser
gear cutting attachment Zahnradfräseinrichtung
gear cutting machine Zahnradfräsmaschine
geared drive Rädertrieb
geared shaper Waagrechtstossmaschine mit Stösselantrieb durch Zahnstange
geared slotting machine Senkrechtstossmaschine mit Zahnräderantrieb
geared to über Zahnräder verbunden mit
gear generating machine Zahnrad-Wälzfräsmaschine
gear hob Zahnradwälzfräser
gear hobbing machine Zahnrad-Wälzfräsmaschine
gear hobbing technique Zahnrad-Wälzfräsverfahren

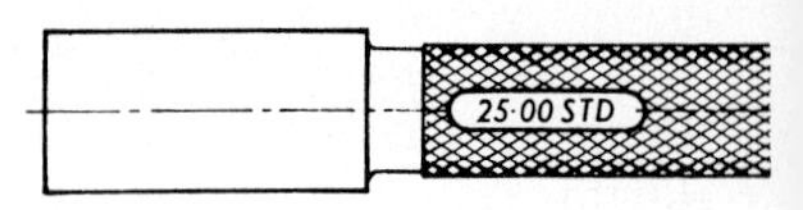

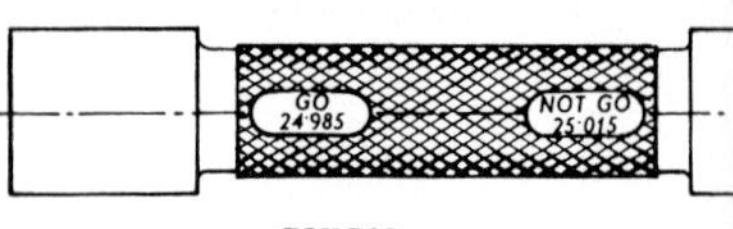

gauges
(Bohr-) Lehren

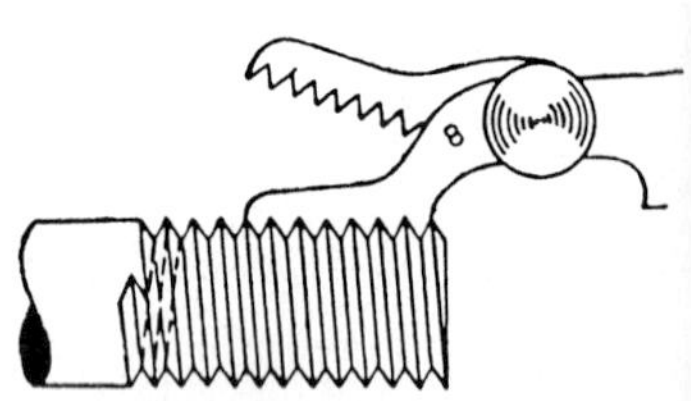

gauging external thread
Messen von Aussengewinde

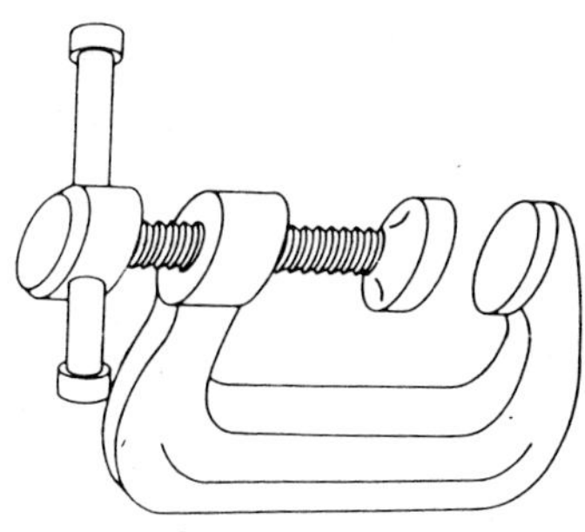

G-cramp
Schraubzwinge

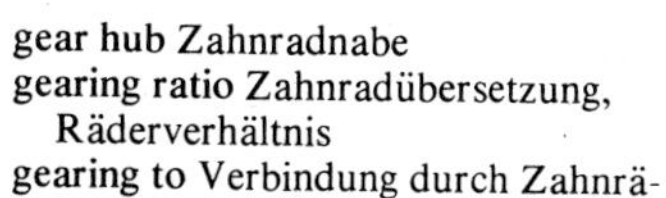

gear hub Zahnradnabe
gearing ratio Zahnradübersetzung, Räderverhältnis
gearing to Verbindung durch Zahnräder mit
gear milling machine Zahnradfräsmaschine
gear of the single helical type schrägverzahntes Rad
gear rack Zahnstange
gear reduction Zahnraduntersetzung
gear scroll chuck Universalfutter
gear stocking cutter Zahnformvorfräser
gear together, to miteinander kämmen
gear-tooth number Zähnezahl des Zahnrades
gear tooth profile Zahnprofil
gear train Rädergetriebe
general utility tool allgemeines, zu einer Maschine gehörendes Werkzeug
generate, to erzeugen, durch Wälzfräsen herstellen
generating Wälzfräsen
generating method Wälzverfahren
Geneva stop Malteserkreuz
geometry of the cutting edge Geometrie der Schneide
gib Leiste, Nachstelleiste
gibbed mit einer Leiste versehen
gibbed surface broach assembly mit Keilleisten nachstellbares Räumzeug
gill Rohrrippe
gilled tube Rippenrohr
girder Träger, Tragbalken
gland Brille (Stopfbüchse)
glass pane Glasscheibe
globe valve Kugelventil
goose neck finishing tool gekröpfter Schlichtmeissel
govern, to regeln
governor Regelgerät, Regulator
grade, to einteilen, abstufen, einstufen, sortieren
grade Klasse, Gütegrad, Härtegrad

goggles
Schutzbrille

graded in size der Grösse nach abgestuft, in Grösse verschieden
gradient Neigung, Steigung, Gefälle
graduate, to einteilen, mit Teilstrichen versehen
graduated collar Skalentrommel, Büchse mit Gradeinteilung, Skalenring
graduated disk Skalenscheibe
graduating Anbringen von Masseinteilung, Graduieren
graduation Gradeinteilung, Abstufung, Teilstrich
graduation in degrees Gradeinteilung
grain Körnung, Korn
graph Schaubild, graphische Darstellung
graphite lubricant Graphitschmiermittel
gravity Schwerkraft
grease, to fetten
grease Schmierfett
grease cup Fettbüchse
grease gun Druckschmierpresse
greasy fettig, ölig
grind, to schleifen
grind Schliff, Anschliff
grinding allowance Schleifzugabe
grinding head Schleifsupport
grinding life Standzeit (zwischen zwei Anschliffen)
grip, to fassen, klemmen, spannen
gripping jaw Spannbacke
gritty spröde, körnig
groove, to nuten, auskehlen
groove Nut, Rille, Aushöhlung
grooved mit Nut versehen, genutet
grooved ball bearing Rillenkugellager
groove-ring collar Nutringmanschette
grooving Nuten
grooving cutter Nutenfräser
ground (to grind) geschliffen
ground clearance Bodenabstand
ground-in chipbreaker eingeschliffener Spanbrecher
ground-in step type chipbreaker Spanleitstufe

grinding wheel dresser
Schleifscheibenabrichter

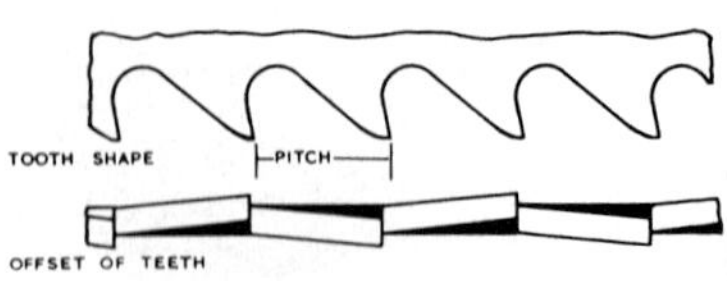

hacksaw teeth
Zähne einer Bügelsäge

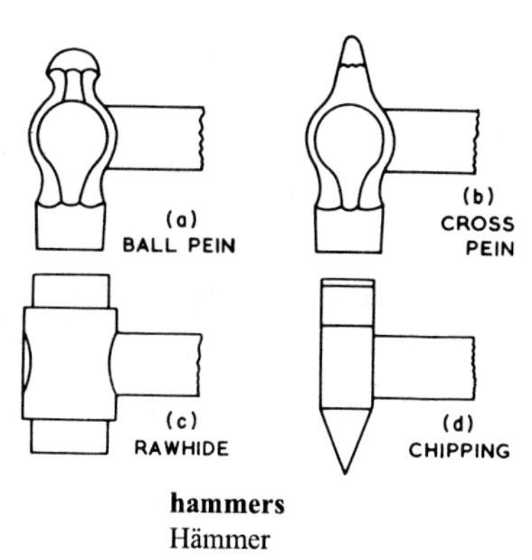

hammers
Hämmer

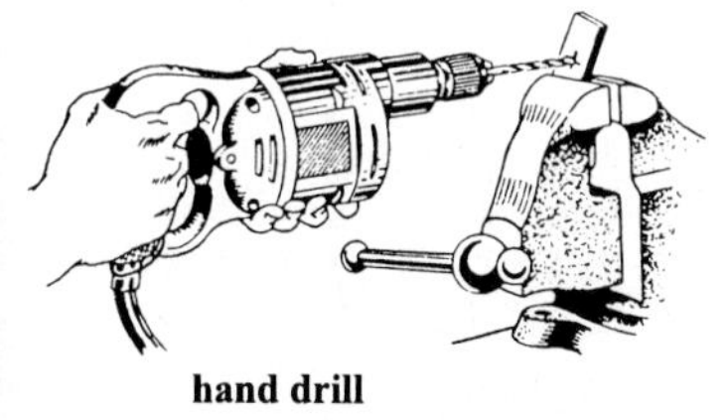

hand drill
Handbohrmaschine

ground to give clearance hinterschliffen, mit angeschliffenem Freiwinkel
group, to anordnen, zu Gruppen zusammenfassen
group Gruppe
guard, to schützen, sichern, sperren
guide, to führen, leiten
guide bar Führungsstange, Leitschiene
guide bar Aufnahmedorn (Räumwerkzeug)
guide block Stufenblock
guide bush Führungsbüchse
guide pin Führungsstift, Kopierstift
guide plate Kopierlineal
guide roller Führungsrolle
guide screw Leitspindel
guide way Führungsbahn
guiding pulley Spannrolle, Führungsrolle
gullet Zahnlücke, Rundung am Zahnfuss, Zahngrund
gummed with old oil mit Ölrückständen behaftet
gyrate, to umlaufen, sich drehen
gyration Drehung, Kreisbewegung, Umlauf

H

hack-saw machine Bügelsäge
half-side milling cutter Scheibenfräser mit einseitiger Verzahnung
halve, to halbieren
hammer, to hämmern
hammer Hammer
hand adjustment Handeinstellung
hand control Steuerung (oder: Schaltung) von Hand
hand crank Handkurbel
hand feed Handvorschub
handle, to handhaben, behandeln, bearbeiten, bedienen
handle Griff, Hebel
hand lever Handhebel
handling Handhabung, Bedienung, Transport, Förderung, Zubringung

handling equipment Transporteinrichtung
hand of cut Schneidrichtung, Schnittrichtung
hand of helix Drallrichtung
hand of rotation Drehrichtung
hand reamer Handreibahle
hand scraped handgeschabt
hand setting Handeinstellung
hand tool Handwerkzeug
harden, to härten
hardened steel gehärteter Stahl
hardening Härten
hardening shop Härterei
hard facing Bestücken mit Hartmetall
hard metal tipped milling cutter hartmetallbestückter Fräser
hardness Härte
hard solder, to hartlöten
hard solder Hartlot
have a bearing on, to Einfluss haben auf
head, to anstauchen
head Spindelstock, Support (Hobelmaschine), Stösselkopf (Waagrechtstossmaschine)
header Sammelrohr
head of delivery Förderhöhe
head slide Schieber, Senkrechtschlitten, Meisselschlitten
headstock Spindelstock
head swivel Drehteil
heat, to erwärmen, erhitzen, heizen
heat Wärme, Hitze
heat conducting wärmeleitend
heat exchange Wärmeaustausch
heat insulation Wärmeschutzisolierung
heat resistant wärmebeständig, hitzebeständig
heat treated wärmebehandelt
heat treatment Wärmebehandlung
heat up, to erhitzen, erwärmen
heel of the tooth Schneidenrücken (Kante zwischen Fase und Zahnrücken)
height gauge Höhenlehre
height of centres Spitzenhöhe

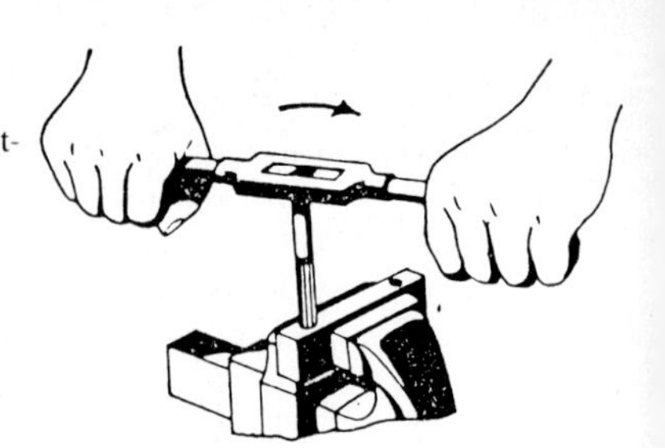

hand reamer being turned by a tap
Reibahle, betätigt durch Windeisen

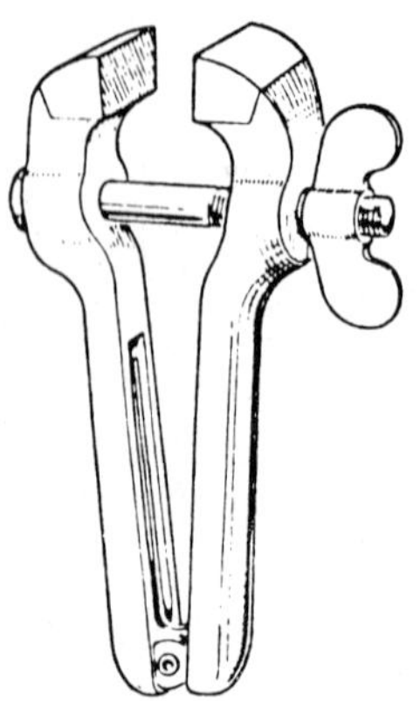

hand vice
Feil-, Handkloben

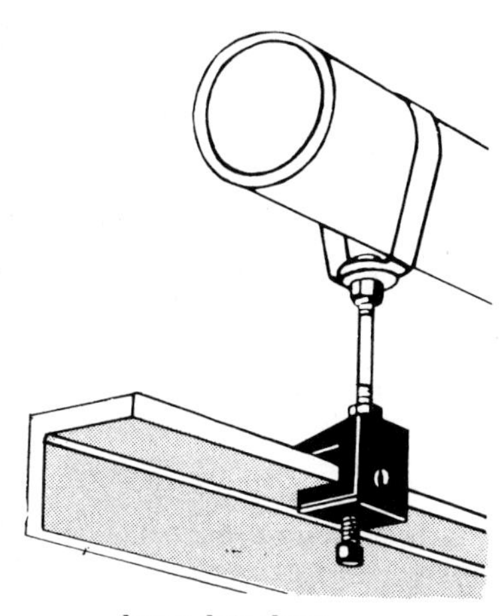

heavy-duty clamp Rohrschelle für schwere Lasten

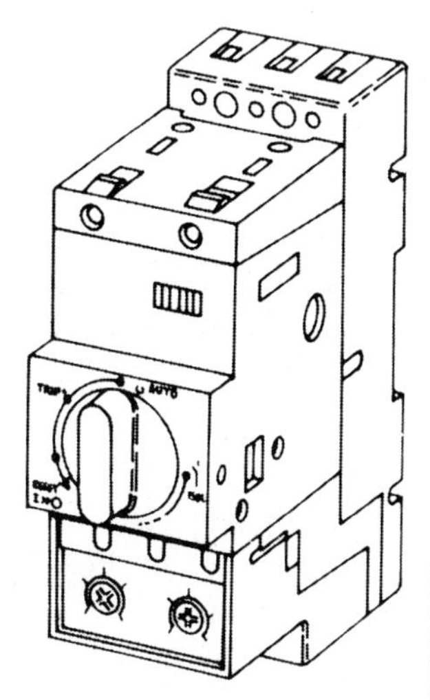

heavy-duty contactor Hochleistungsschütz

helical schraubenförmig, drallförmig, spiralförmig
helical broach Räumwerkzeug für drallförmige Nuten
helical broaching Räumen von Drallnuten
helical carbide-tipped end mill hartmetallbestückter Schaftfräser mit spiralförmiger Verzahnung
helical curve Schraubenkurve, Schraubenlinie
helical drive Tischantrieb über Schrägzahnräder
helical form milling Formfräsen von Schraubennuten
helical gear schrägverzahntes Rad, Schraubenrad
helical groove schraubenförmige Nut
helical mill Fräser mit spiralförmiger Verzahnung
helical milling Spiralnutenfräsen, Fräsen einer schraubenförmigen Nut
helical mill with pilot spiralverzahnter Fräser mit Führungszapfen
helical plain milling cutter Hochleistungs-Walzenfräser
helical rack schrägverzahnte Zahnstange
helical slab mill Hochleistungs-Walzenfräser
helical spline broach Räumwerkzeug für drallförmige Nuten
helical spring Schraubenfeder
helical tooth cutter Fräser mit Spiralverzahnung
helix (Pl.: helices) Schraubenlinie
helix angle of threads Steigungswinkel von Gewinden
helix tooth cutter spiralverzahnter Fräser
hermaphrodite caliper Tastzirkel
herringbone gear Pfeilrad
hex, to einen Sechskant fräsen
hex Sechskant
hexagon Sechskant
hexagonal sechseckig
high-alloy steel hochlegierter Stahl

high-division index plate Teilscheibe für weiten Teilbereich
high-feed milling Fräsen mit grossem Vorschub
high-rake milling Fräsen mit grossem Spanwinkel
high-rake cutter Fräser mit grossem Spanwinkel
high-speed carbide tool machining Schnellzerspanung mit Hartmetallwerkzeugen
high-speed milling attachment Schnellfräseinrichtung
high-speed screw slotter Schnellschraubenschlitzmaschine
high-speed stock removal Schnellzerspanung
high-velocity turning Schnelldrehen
hinge, to gelenkig befestigen, schwenken, klappen
hinge Gelenk, Angel
hinged gelenkig befestigt, aufklappbar
hinged knee Kniegelenk
hinged pin Scharnierstift
hob, to wälzfräsen
hob Wälzfräser
hobber Wälzfräsmaschine
hobbing Wälzfräsen
hobbing cutter Walzfräser
hobbing machine Wälzfräsmaschine
hold a tolerance, to eine Toleranz einhalten
hold-down Niederhalter
holder Schneidenhalter (Räumwerkzeug)
holding device Aufspannvorrichtung
holding of size Masshaltigkeit
holding strap Spanneisen
hole circle Lochkreis
hole operation Bearbeitung einer Bohrung
hollow grinding Hohlschliff
hollow milling cutter Hohlfräser
honeycomb wabenartiges Gitter
hood Kappe, Haube, Verdeckung
hook Haken
Hooke's joint Kardangelenk
hook tool Hakenmeissel

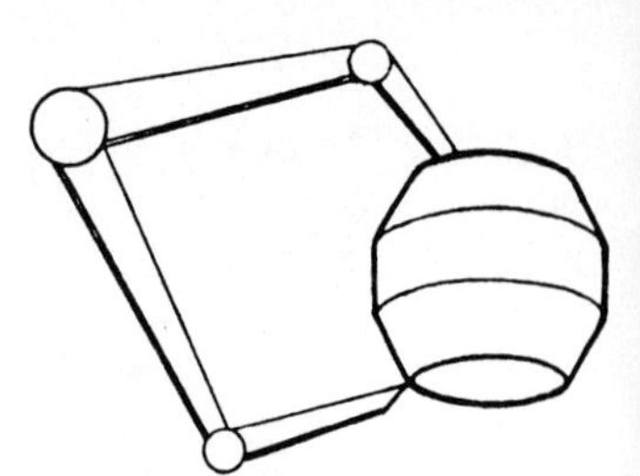

hinged-leg dividers
Tastzirkel mit Gelenkgliedern

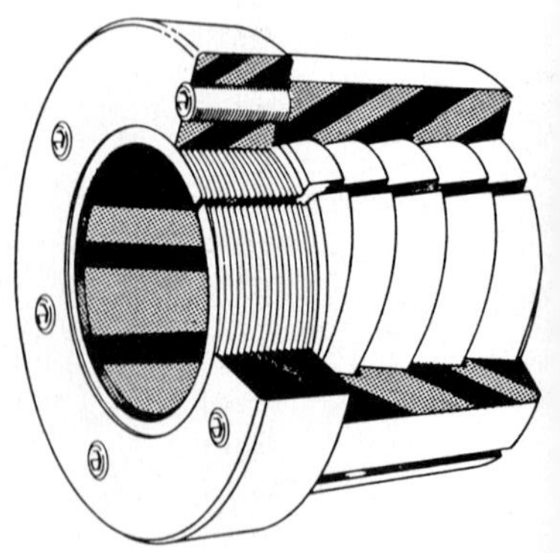

holding bush
Spannbüchse

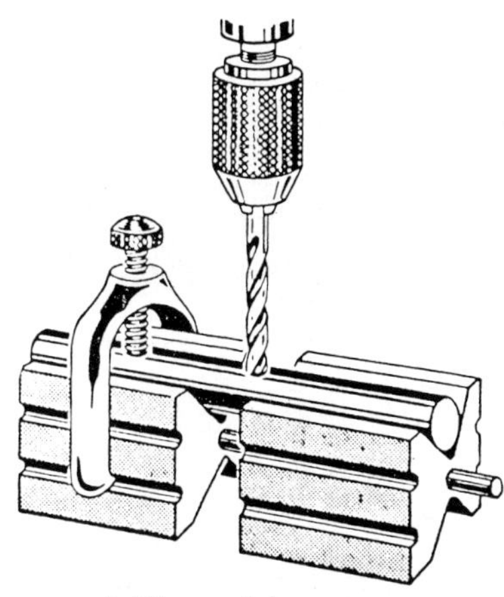
holding work in vee-blocks
Halten von Werkstücken auf Prismenstücken

hopper Beschickungstrichter, Füllkasten, Teilzuführungseinrichtung
horizontal front clearance Einstellwinkel der Nebenschneide
horizontal head Seitensupport
horizontal internal broaching machine Waagrecht-Innenräummaschine
horizontal miller Waagrechtfräsmaschine
horizontal rail Querbalken (Hobelmaschine)
horizontal-type shaper Waagrecht-stossmaschine
horn Aufnahmedorn (Räummaschine)
hose Schlauch
house, to unterbringen, aufnehmen
housing Gehäuse
hub Nabe
hydraulically operated hydraulisch betätigt
hydraulic jack Hydraulikmotor mit geradliniger Bewegung, Kolben

I

idle, to leerlaufen
idle stroke Leerhub
idler gear Zwischenrad
idler pulley Riemenspannrolle
idle time Brachzeit, Nebenzeit
idling Leerlauf
immerse, to eintauchen
impact load Stossbelastung
impact strength Schlagfestigkeit
impair, to beeinträchtigen, beschädigen
impart pressure, to Druck ausüben, unter Druck setzen
impede, to festhalten, bremsen, arretieren
impeller Antriebsrad
inaccuracy Ungenauigkeit
inaccurate ungenau
inching (or: jogging) Rucken, Feineinstellung, Kriechgang, Langsameinstellung

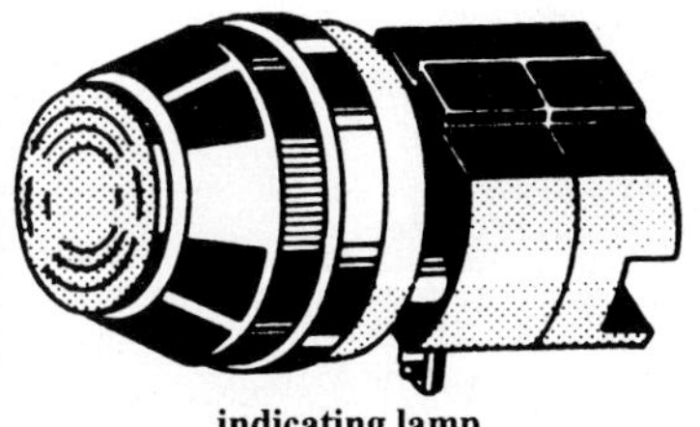
indicating lamp
Anzeigenleuchte

inclinable schrägstellbar, neigbar
inclination Schrägstellung, Neigung
incorporate, to einbauen
increase, to vergrössern, erhöhen, steigern, zunehmen, anwachsen
increase Vergrösserung, Erhöhung, Steigerung, Zunahme, Zuwachs
increment Stufe, Zunahme, Zuwachs
increment cut ungleiche Teilung
in-cut method Verfahren des Gleichlauffräsens
indent, to dornen, einkerben, verzahnen
indentation Einkerbung, Einschnitt, Verzahnung
independent four-jaw chuck Planscheibe
index, to anzeigen, alphabetisch ordnen, teilen, schalten
index Anzeiger, alphabetisches Verzeichnis, Beiwert
index base Rundtisch, Teileinrichtung
index base milling pausenloses Fräsen mit Rundschalttisch
index circle Lochkreis
index disc (or: disk) Teilscheibe
index head Teilkopf
index head operation Teilarbeit, Arbeit mit Teilkopf
indexing Teilen, Schalten
indexing attachment Teileinrichtung
indexing error Teilfehler, Fehler beim Teilen
indexing latch Indexklinke, Indexraststift
indexing rotary table Rundtisch mit Teileinrichtung, Rundschalttisch
index plate Teilscheibe
indicate, to anzeigen, indizieren
indicator Anzeigegerät, Messuhr
individual drive Einzelantrieb
inequality Ungleichheit, Abweichung
inertia Trägheit
inexpert unfachgemäss
infinitely variable feeds stufenlos regelbare Vorschübe
inflow Einfliessen, Einströmen
ingot planer Blockhobelmaschine

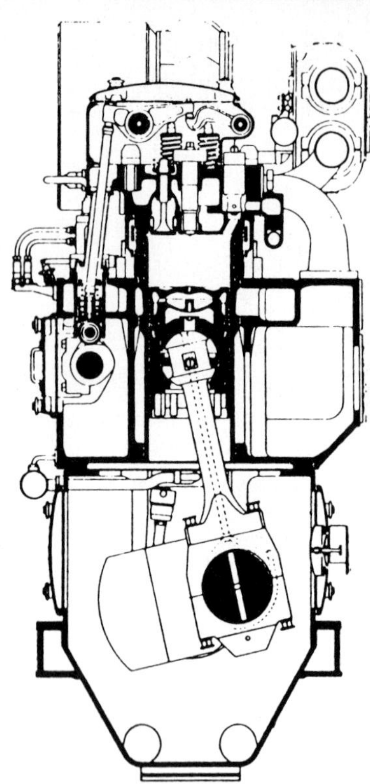

in-line engine
Reihen(diesel)motor

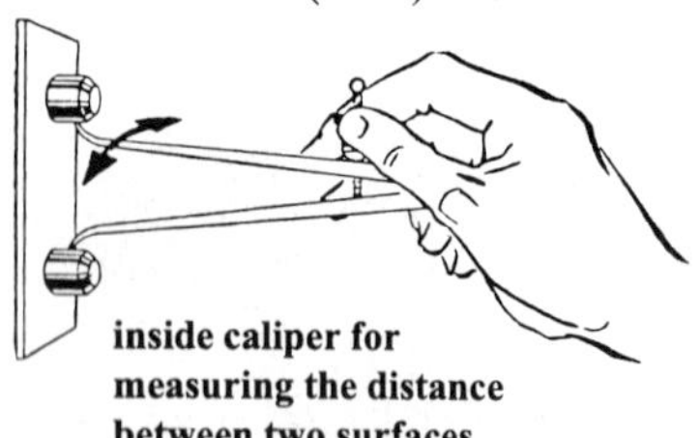

inside caliper for measuring the distance between two surfaces
Innentaster zur Messung des Abstandes zwischen zwei Flächen

inside diameter being measured with an inside caliper
Innendurchmesser wird mit einem Innentaster gemessen

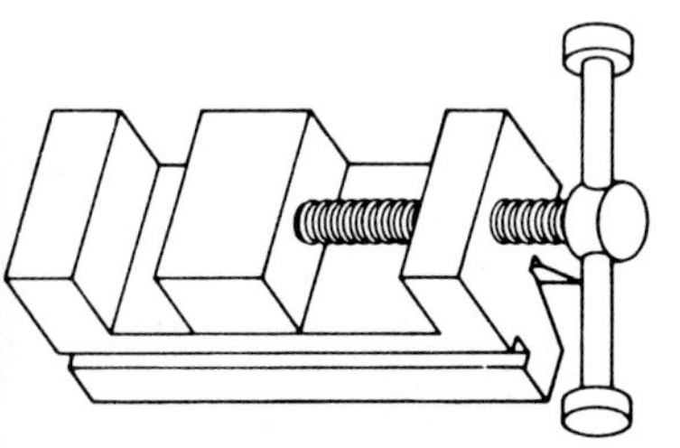

instrument vice
Instrumentenschraubstock

ingredient Bestandteil
ingress, to eindringen, einströmen
initial Anfangs . . ., anfänglich
initiate, to in Gang setzen
inlet port Eintrittskanal
in-line transfer machining zerspanende Bearbeitung auf Taktstrasse
inner diameter Innendurchmesser
in-process gauging of a workpiece Messen des Werkstücks zwischen den Bearbeitungsgängen
input Leistungsaufnahme
input shaft Antriebswelle
insert, to einsetzen
insert Einsatz
inserted blade Einsatzmesser
inserted teeth broach Räumwerkzeug mit eingesetzten Zähnen
inserted tooth fly mill Schlagfräser mit eingesetztem Meissel
insert the pin, to den Stift einrasten
inside calipers Innentaster
inside diameter Innendurchmesser, lichte Weite
install into, to einbauen in
installation of machines Aufstellung (Montage) von Maschinen
insulate, to isolieren
intake Eintritt
intake port Eintrittsöffnung
integral with ein Ganzes mit, fest an
interchangeable austauschbar
interconnection Zwischenglied, Verbindung
interfere with, to beeinträchtigen, stören
interlock, to verblocken, gegeneinander verriegeln
interlock Verblockung, gegenseitige Verriegelung
interlocked cutter set gekuppelter Fräsersatz
interlocked with each other miteinander verriegelt
intermediate arbor support Zwischentraglager für den Dorn
intermediate size Zwischengrösse
intermediate stage Zwischenstufe

intermittent feed Ruckvorschub, Sprungvorschub
internal broach **Innenräumwerkzeug, Räumnadel**
internal broaching Innenräumen
internal broaching machine Innenräummaschine
internal copying attachment Innenkopiereinrichtung, Innennachformeinrichtung
internal gear innenverzahntes Rad
internal gear milling Innenradfräsen
internally ribbed mit Innenverrippung
internal thread Innengewinde
internal work Bearbeitung von Innenflächen
intersection Schnittpunkt, Stossstelle
intricate kompliziert, verwickelt, schwierig
invariable unveränderlich
inverted cup seal Hutmanschette
inverted Vee-guide Prismenführung
involute gear Evolventenrad
iron Eisen
iron scrap Alteisen, Eisenschrott
irregular spacing ungleiche Teilung
issue from, to ausströmen aus
issue Ausfluss

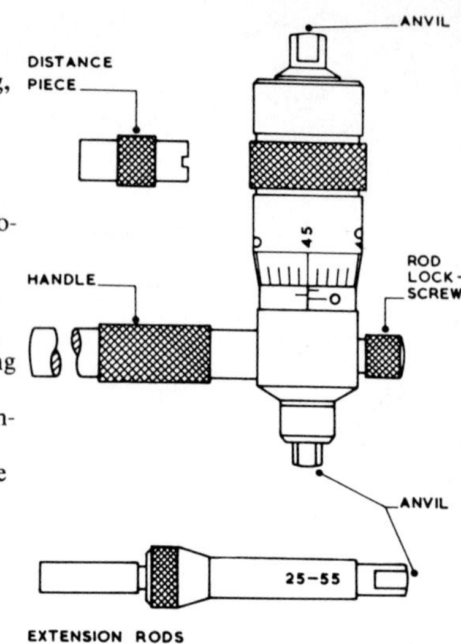

internal micrometer
Innenmikrometer

J

jack Hebezeug, Winde, Stützbock
jack the work, to das Werkstück unterbauen
jacket, to umhüllen, ummanteln
jack screw Hubspindel, Verstellspindel
jam, to klemmen, festfressen
jamming Festfressen, Verklemmen
jam nut Gegenmutter
jaw Backe, Backen, Klaue
jaw chuck Backenfutter
jaw-type pull head Ziehkopf mit Backenspannung
jenny Tastzirkel
jet Düse, Strahl

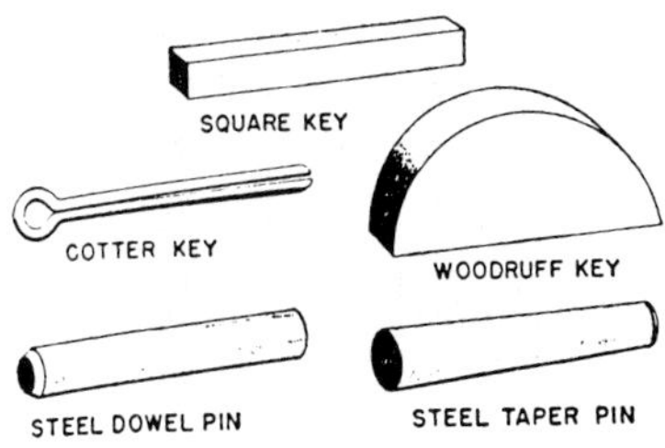

keys and pins
Keile und Stifte

jig Bohrschablone, werkzeugsteuernde Vorrichtung
jig bore, to Lehren bohren
jig boring Lehrenbohren
jig boring and milling machine Lehrenbohr- und Fräsmaschine
jig mill Lehrenfräsmaschine
jigs and fixtures Vorrichtungen (für Werkzeugmaschinen)
job change-over Übergang zu neuer Serie
job production Einzelfertigung
jog, to stossen, langsam bewegen, auf Einstellung bringen
join, to verbinden, zusammenfügen
joining pipe Anschlussrohr
joint Verbindungsstelle, Stoss, Gelenk
joint ring Dichtungsring für den Flansch
journal Zapfen, Achshals, Wellenzapfen
journal bearing Halslager, Traglager
jump feed Sprungtischvorschub

K

keen scharf
key, to aufkeilen, festkeilen
key and slot Feder und Nut
key drive Mitnahme durch Keil
keyed end Einspannteil
keygroove, to Keilnuten ziehen
keyseat, to Keilnuten einarbeiten
keyseater Keilnutmaschine
keyway Keilnut
keyway broach Räumwerkzeug für Keilnuten
keyway cutting Keilnutenfräsen
keyway tool Keilnuten-Stossmeissel
kickout Auslösung, Auslöseeinrichtung
knee Konsol(e) Winkeltisch
knee clamping Konsolklemmung
knee elevating screw Konsolhubspindel

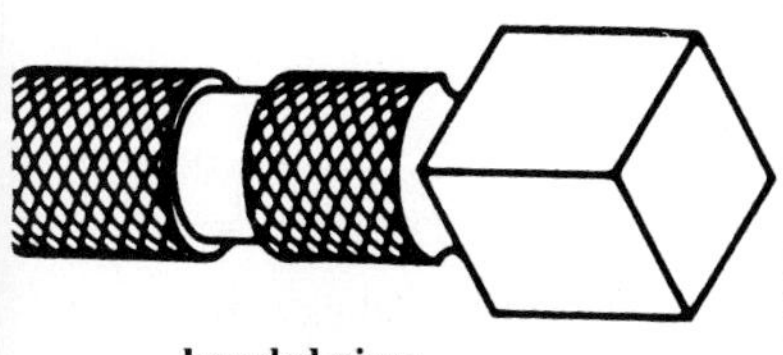

knurled piece
Rändelstück

kneeless type milling machine Planfräsmaschine, Fräsmaschine mit beweglichem Spindelstock
knee mill Konsolfräsmaschine
knee slides Konsolführungen
knob Knopf
knurl, to rändeln
knurling Rändeln
knurling tool Rändelwerkzeug

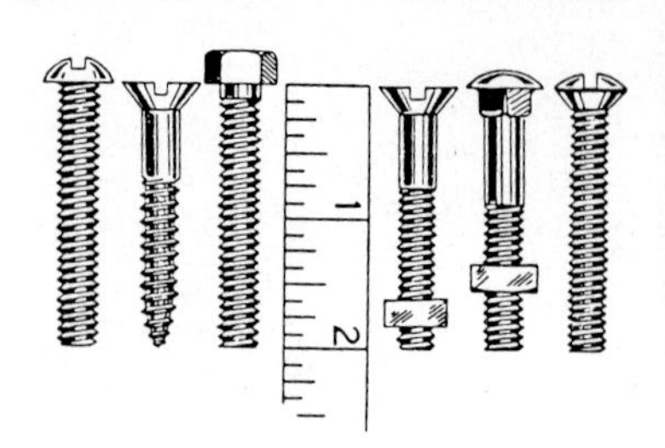

length of a bolt being measured
Länge eines Bolzens (einer Schraube) wird gemessen

L

labour saving arbeitssparend
lack, to ermangeln, nicht besitzen
lack of Mangel an, Fehlen von
lamina Schicht, Lamelle
land Fase
land and groove type of chipbreaker Spanleitrille mit Fase
land relief Hinterschliff der Fase, Freiwinkel
land wear Verschleissmarkenbreite
lap, to läppen, überlappen
latch Klinke, Schnapper
latch-and-fire mechanism federgespannter Auslösemechanismus
latch pin Raststift
lateral Seiten . . ., seitlich
lateral feed seitlicher Vorschub, Quervorschub
lathe carrier Drehherz
lathe tool Drehmeissel
layer Lage, Schicht
laying out Anreissen
laying-out plate (layout plate) Anreissplatte
lay out, to anreissen
layout Anreissen, Anriss, Grundriss
layout man Anreisser
lead, to führen, leiten
lead Blei
lead Steigung, Ganghöhe
lead angle Anschnittwinkel, Schrägungswinkel
lead bar Leitlineal
leading cutting edge Hauptschneide

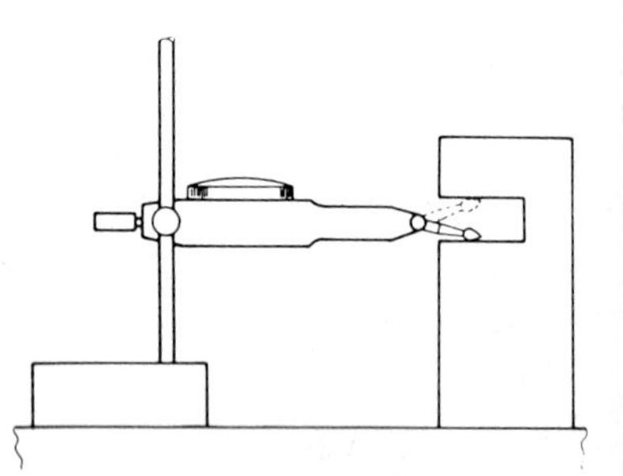

lever-type dial-indicator
Messuhr mit Gestänge

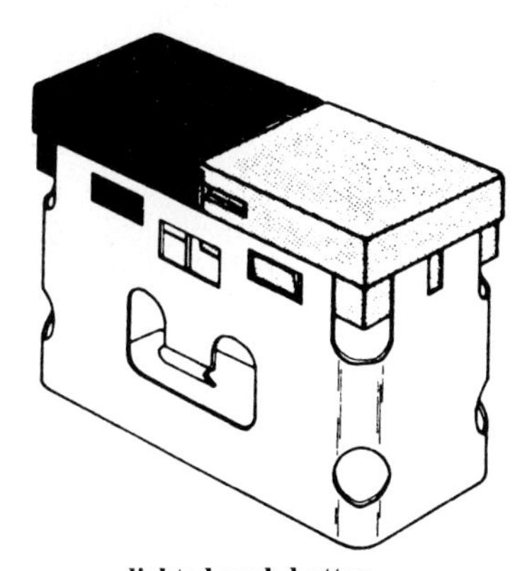

lighted push-button
Leuchttaster

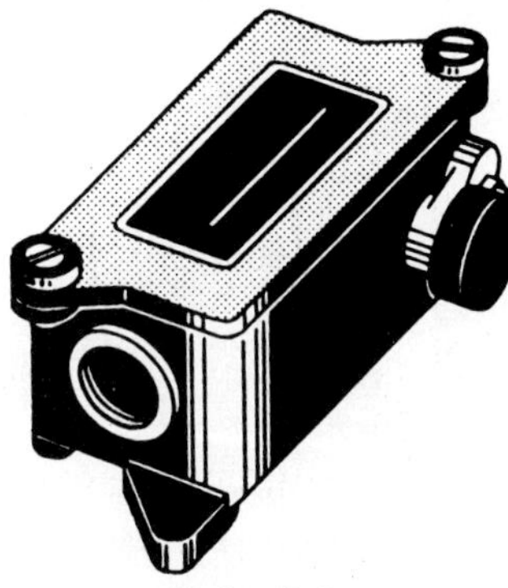

limit switch
Endschalter

leading tool edge abfallende Schneide, negative Überhöhung der Schneide
lead of feed screw Steigung der Tischvorschubspindel
lead of helix Steigung der Schraubenlinie
lead screw Leitspindel, Verstellspindel
leak, to lecken, undicht sein
leakage Lecken, Leckmenge
leak-proof leckdicht, abgedichtet
leather cup Ledermanschettendichtung
left-bent tool gebogener, linker Meissel
left-cut tool linker Meissel
left-hand cut linksschneidend
left-hand helical cutting edges Schneiden mit Linksdrall
left-hand nut Mutter mit Linksgewinde
left-hand parting tool linker Stechmeissel
left-hand rotation Linksdrehung
left-hand rail head linker Quersupport
left-hand thread Linksgewinde
length Länge
lengthen, to verlängern, strecken
length of cut Schnittlänge
length of stroke Hublänge
length stop Längsanschlag
lengthwise in Längsrichtung
lessen, to vermindern
level, to nivellieren, in Waage bringen
level in Waage, gleichmässig
lever Hebel
lever action Hebelwirkung
lever-controlled durch Hebel geschaltet
lid Deckel, Klappe
life Lebensdauer, Standzeit
life capacity of a tool Stückzahl, die das Werkzeug während seiner Lebensdauer fertigen kann
life of the cutting edge Standzeit der Schneide
lift, to anheben, lüften, abheben
light cut Schnitt mit geringer Spantiefe, leichter Schnitt

light-duty plain milling cutter normaler Walzenfräser
light metal Leichtmetall
light-slit method Lichtschnittverfahren
limit, to begrenzen, beschränken auf
limit Grenze, Begrenzung
limited quantity production Kleinreihenfertigung
limit stop Endanschlag, Anschlag zur Begrenzung der Bewegung
limit switch Endschalter
line, to ausfüttern, ausgiessen
line Leitung, Rohrleitung
linear indexing Längenteilung, geradlinige Längsteilung
line assembly work Fliessmontage
line-by-line milling Zeilenfräsen
liner out Anreisser
liner Unterlegstreifen
line setup Reihenaufspannung
line tracer for profiling operations Fühler für Nachformarbeit nach Zeichnung
line up, to ausrichten
lining out Anreissen
link Verbindungsglied, Glied
link drive Kulissenantrieb
link rod Kulissenstange
lip Schneide, Lippe
lip angle Keilwinkel
liquid flüssig
liquid Flüssigkeit
live centre umlaufende Körnerspitze
load, to laden, belasten, einführen, beschicken, auflegen
load Belastung
loading Belastung
loading time Aufspannzeit
locate, to plazieren, an einer Stelle anbringen, einstellen
locating plunger Raststift der Teilscheibe
location Aufspannung
lock, to festklemmen, feststellen, verriegeln
lock Verriegelung, Festklemmung
locking Verriegelung, Festklemmung

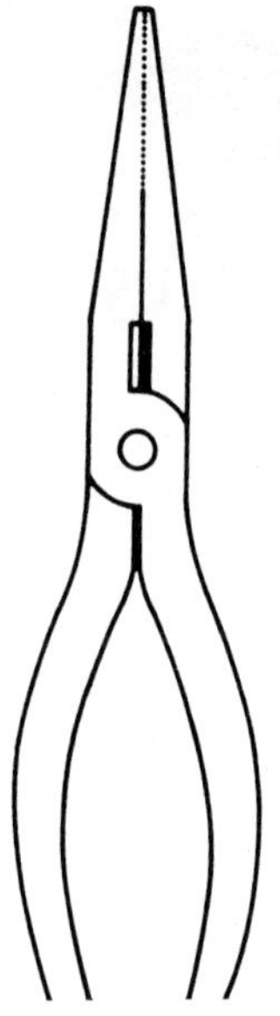

long-nosed pliers
Spitzzange

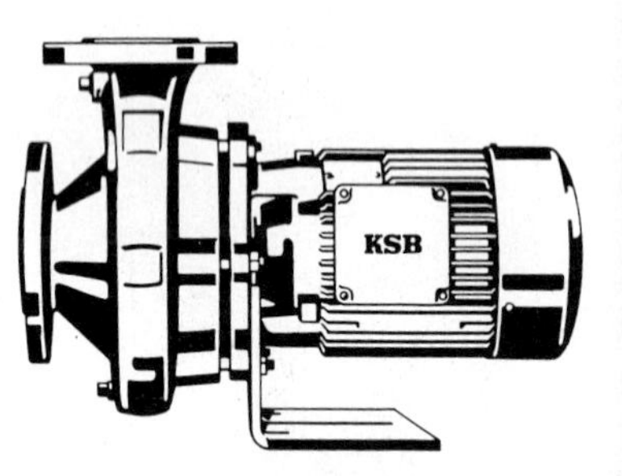

low-pressure pump
Niederdruck-Umwälzpumpe

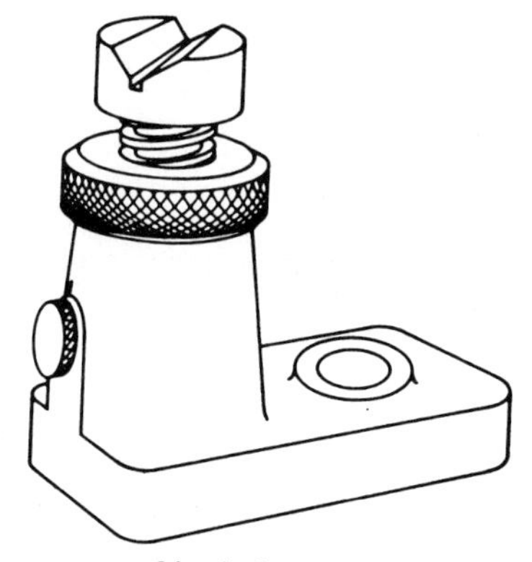

machine jack
Schraubbock mit Befestigung, Lünette, Schrauben-Hebebock

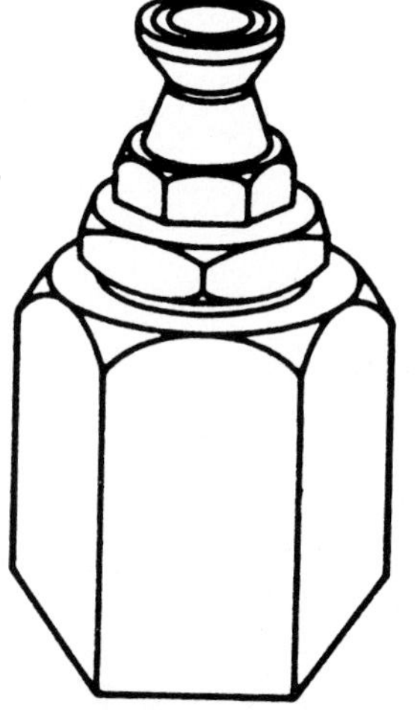

machinist's jack
Stützwinde, (Sechskant-) Hebebock, manchmal auch: Schraubbock

locking disk Rastscheibe
locking pin Verriegelungsstift
lock nut Gegenmutter
long-chip material langspanender Werkstoff
longitudinal Längs . . ., längs
longitudinal feed Längsvorschub
longitudinal force Vorschubkraft
longitudinal planing Längshobeln
long-run job grosse Serie
long-run repetition work grosse Serie
long-stroke broaching machine Langhub-Räummaschine
loose locker, lose, abnehmbar
loose coupling schaltbare Kupplung
loose gear Wechselrad
loosen, to lockern, lösen
loss Verlust, Abfall
loss due to slippage Schlupfverlust
loss of head Verlusthöhe, Druckverlust
loss of tool hardness Erweichen der Schneide
lost motion play toter Gang
lot Posten, Serie
low niedrig, gering, tief
low-carbon steel kohlenstoffarmer Stahl, Flussstahl
lower, to herablassen, senken
lower part Unterteil
lubricant Schmiermittel
lubricant return Schmiermittel-Rückführung
lubricate, to schmieren
lubricating Schmieren, Schmierung
lubricating oil Schmieröl
lubrication Schmierung
lubrication engineering Schmiertechnik
lug Nase, Ansatz
luting agent Dichtungskitt

M

machinability Bearbeitbarkeit, Zerspanbarkeit

machinable bearbeitbar, zerspanbar
machine, to bearbeiten, zerspanen
machine all over, to allseitig bearbeiten
machine attendant Maschinenwärter
machine bed Maschinenbett
machine builder Maschinenbauer
machine building Maschinenbau
machine column Maschinenständer
machine cycle Bearbeitungszyklus, Folge der Arbeitsgänge
machined from the solid aus dem Vollen herausgearbeitet
machined surface Arbeitsfläche, bearbeitete Fläche
machine-handling time Nebenzeit
machine operator Maschinenarbeiter
machine relieved hinterdreht
machinery Maschinenpark, Maschinen
machine shop Maschinenwerkhalle, Maschinenhalle
machine tool Werkzeugmaschine
machine tool building Werkzeugmaschinenbau
machine tool control Steuern von Werkzeugmaschinen
machine tool expert Werkzeugmaschinenfachmann
machine tool plant Werkzeugmaschinenfabrik
machine upright Maschinenständer
machine vice Maschinenschraubstock
machining (spanabhebendes) Bearbeiten
machining allowance Bearbeitungszugabe
machining operation Zerspanungsarbeit
machining rate Bearbeitungsgeschwindigkeit
magnetic chuck Magnetspannfutter
magnification Vergrösserung
magnify, to vergrössern
magnitude Grösse
main clearance Hinterschliff der Phase, Freiwinkel
main cutting edge Hauptschneide

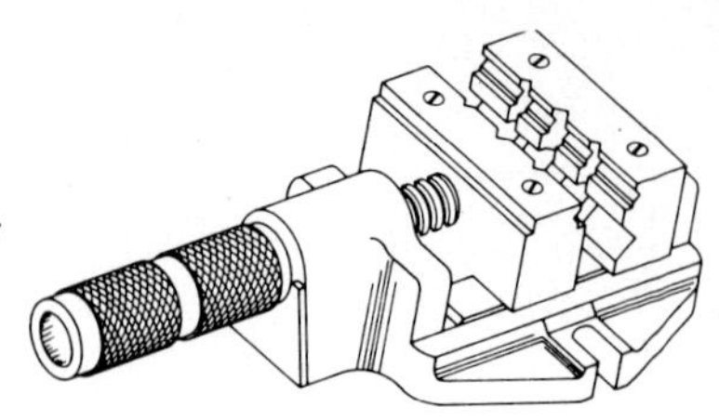

machine vice
Maschinenschraubstock

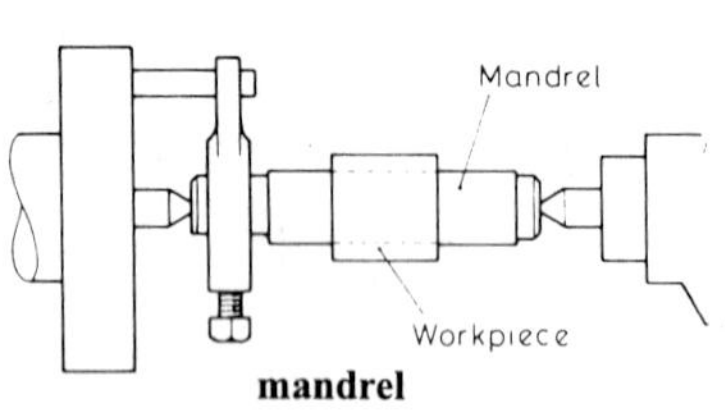

mandrel
Dorn

main dimensions Hauptabmessungen
mains Hauptleitung, Netz
mains voltage Netzspannung
maintain, to beibehalten, aufrechterhalten
maintenance Wartung
maintenance free wartungsfrei
major diameter Aussendurchmesser (Gewinde)
make Erzeugnis, Fabrikat
maker Hersteller
makeshift design Behelfskonstruktion
male thread Aussengewinde
mandrel Dorn
manipulator Wendeeinrichtung
manual control Steuerung von Hand, Handschaltung, Handregelung
manually von Hand
manufacture Fertigung, Herstellung
manufacturing programme Fertigungsprogramm
mar, to beschädigen
mark, to bezeichnen, markieren, anreissen
marker out Anreisser
marking Anreissen
marking-off table Anreissplatte
mass duplicate, to in Massen anfertigen
mass production Massenfertigung
master Bezugsstück, Muster
master plate Schablone, Kopierlineal
master switch Hauptschalter
master template Kopierschablone
mate, to ineinanderpassen, kämmen, eingreifen
material testing Werkstoffprüfung
mating faces aneinanderpassende Flächen
matter Materie, Stoff
mean reference line mittlere Bezugslinie
measure, to messen
measure Mass
measurement Messung
mechanical engineer Maschinenbauingenieur, Maschinenbauer
mechanical engineering Maschinenbau

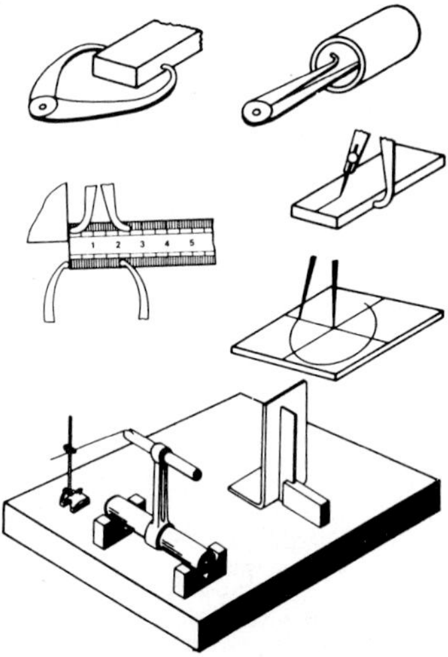

marking-out tools
Anreisswerkzeuge

medium size mittelgross
melting point Schmelzpunkt
member Glied, angetriebener Teil
mercury Quecksilber
mesh, to kämmen
mesh Eingriff
metal braid Metallgeflecht
metal-cutting research Zerspanungsforschung
metal forming spanlose Verformung
metal hose Metallschlauch
metallic Metall . . ., metallisch
metallurgy Metallurgie
metal-removal rate Spanmenge, spezifisches Spanvolumen
metalworking Metallbearbeitung
meter, to messen, zumessen
meter Mass, Messgerät
meter out, to dosieren
metric thread metrisches Gewinde
micrometer Mikrometer
microscopic eye-piece Okular
middle slide Mittelschlitten
mill, to fräsen
mill Fräser, Fräsmaschine
milling Fräsen
milling cutter Fräser
milling feed Fräsvorschub
milling head Fräskopf, Frässpindelstock
milling machine Fräsmaschine
milling quill Fräspinole
millings Frässpäne
minimize, to verringern, auf ein Mindestmass herabsetzen
minimum limit Kleinstmass
minor diameter Kerndurchmesser (Gewinde)
minute klein, sehr gering
mirror-copying spiegelbildliches Kopieren
misalignment Fluchtungsfehler
mist coolant Kühlmittel für Ölnebelkühlung
mist cooling Ölnebelkühlung, Sprühkühlung
mist-lubrication Ölnebelschmierung
mixture Mischung

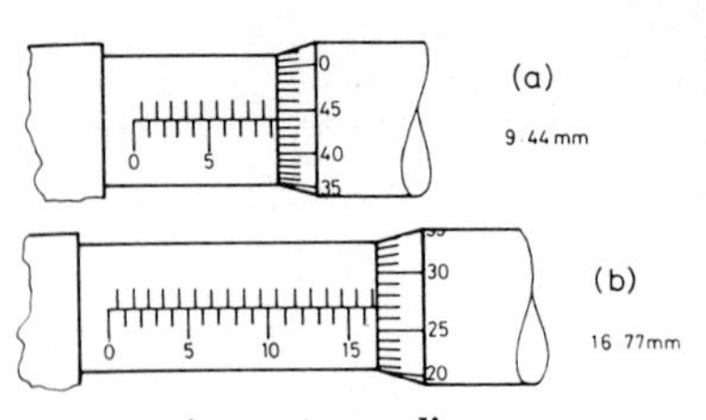

micrometer readings
Ablesewerte am Mikrometer

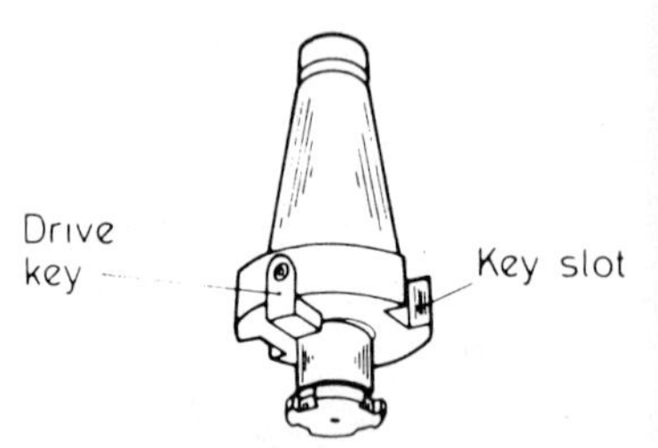

milling-machine stub arbor
fliegend aufgespannter Dorn für Fräsmaschine

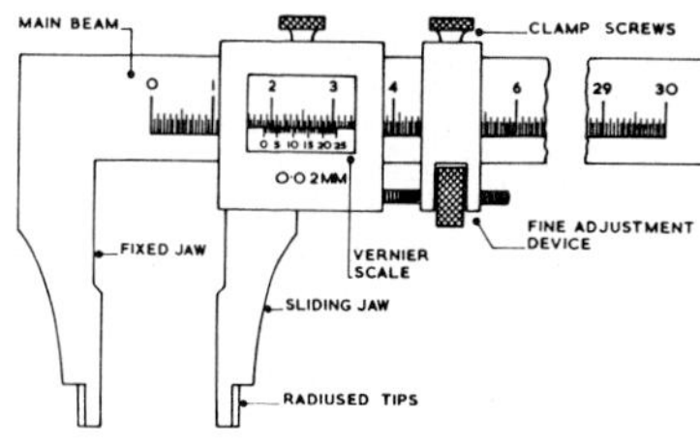

metric vernier calper
metrische Schieblehre

motor starter being tested
Motoranlasser wird geprüft

modify, to abändern, umgestalten
moist feucht
moisten, to benetzen
moisture Feuchtigkeit
molybdenum high speed steel Molybdän-Schnellstahl
moment of inertia Trägheitsmoment
mono-lever control Einhebelsteuerung, Einhebelschaltung
morphy caliper Tastzirkel
motion Bewegung
motive power Treibkraft, Triebkraft
mold (mould) Form, Giessform
mount, to aufspannen, montieren, anbauen
mounting Anbringen, Aufbringen, Montieren
mounting bracket Befestigungsschelle
movable beweglich
moving seal Abdichtung beweglicher Teile
muff Muffe
multi-disk clutch Mehrscheibenkupplung
multihead milling machine Mehrspindel-Fräsmaschine
multipart mehrteilig
multipoint tool mehrschneidiges Werkzeug
multi-purpose milling machine Mehrzweckfräsmaschine
multistage mehrstufig, Mehrstufen . . .
multi-toothed mehrzahnig, mehrschneidig
multi-toothed cutter mehrzahniger Fräser

N

nail Nagel
narrow eng, schmal
needle bearing Nadellager
negative-rake cutter Fräser mit negativem Spanwinkel
negligible error Abweichung, die man vernachlässigen kann

nick, to einkerben
nick Kerbe, Spanbrechernut, Schlitz
nickel Nickel
nipple Nippel
no-backlash spielfrei
noise Lärm, Geräusch
noiseless geräuschlos
nominal output Nennleistung
nominal rating Nennleistung
non-combustible nicht brennbar, nicht entzündbar
non-ferrous Nichteisen . . .
non-machining time Nebenzeit
non-vibrating schwingungsfrei
non-warping verzugsfrei
non-wearing verschleissfrei
numerical machining Bearbeitung mit numerischer Steuerung
nose Werkzeugspitze, Spindelnase, Spindelkegel
notch Raste, Kerbe
notched bar impact test Kerbschlagversuch
notched wheel Rastenrad
not-go gage (gauge) Ausschusslehre
nozzle Düse
number of teeth Zähnezahl
numerical control numerische Steuerung
nut Mutter

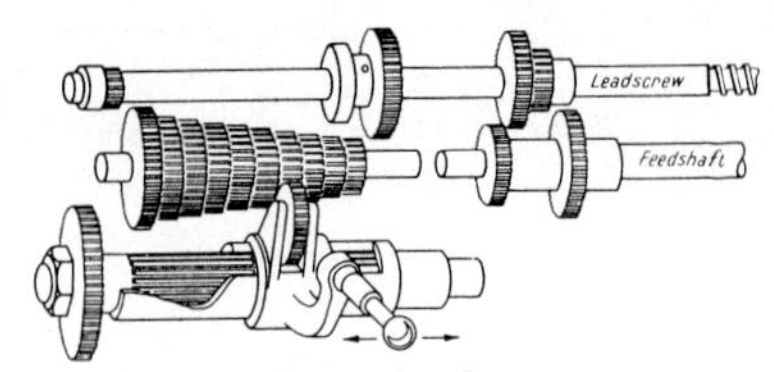

Norton gearbox
Nortongetriebe

O

oblique schräg, schief
oblique rake schräger Spanwinkel
oblong hole Langloch
obtain, to erhalten, erreichen
obtuse angled stumpfwinklig
occupy, to beschäftigen, besetzen, einnehmen, belegen
octagon Achtkant, Achteck
odd angle cutter Winkelfräser für Werkzeuge mit gefrästen Zähnen (für Spannuten mit Drall)
odd-fluted end mill Schaftfräser mit ungerader Nutenzahl

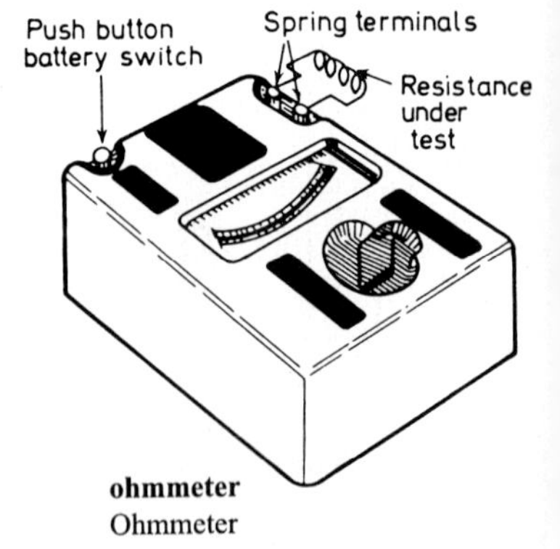

ohmmeter
Ohmmeter

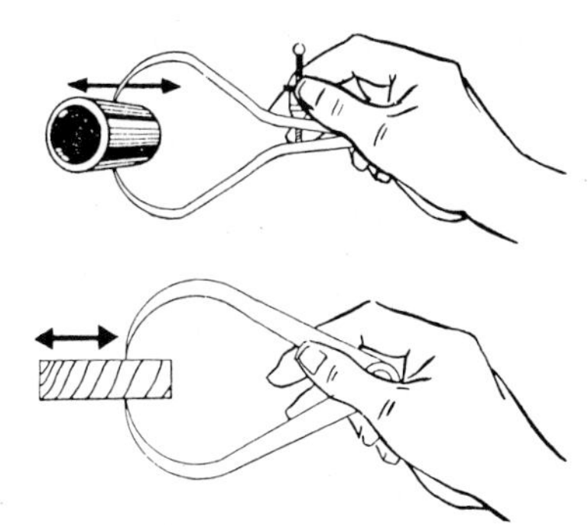

outside calipers in use
Aussentaster bei der Anwendung

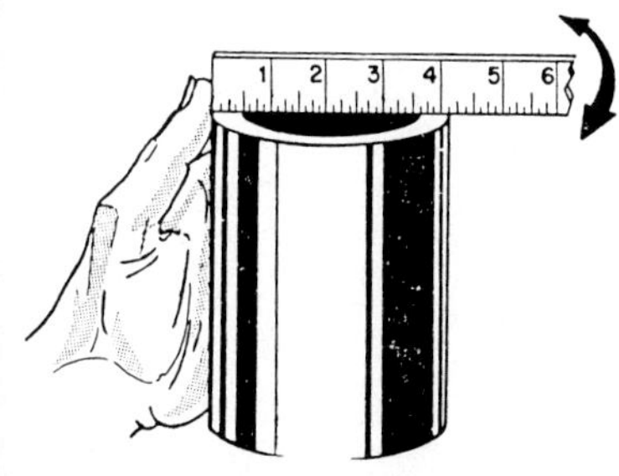

outside diamter of a pipe being measured
Aussendurchmesser eines Rohres wird gemessen

odd leg Tastzirkel (Werkstattausdruck)
off-centre aussermittig
off-centre position Aussermittigkeit
offset, to versetzen
offset versetzt, gekröpft
offset clamp verstellbares Spanneisen
offset milling machine Fräsmaschine mit aussermittig schwingender Spindel
offset tool gekröpfter Meissel, abgesetzter Meissel
oil, to ölen, schmieren
oil circuit Ölkreislauf
oil cup Öler, Ölbuchse
oil groove Schmiernut, Ölnut
oil-hardened ölgehärtet
oiling point Schmierstelle
oil-in-water emulsion Emulsion von Öl in Wasser
oilstone, to mit Ölstein abziehen
one-man control Einmannbedienung
one-off production Fertigung von Einzelstücken
one-pass cutting Schneiden in einem Durchgang
one-piece einteilig, aus einem Stück
one-way flow Strömung in einer Richtung
on-position Einschaltstellung
open-front design nach vorn offene Konstruktion, Einständerkonstruktion
opening capacity Spannweite (Schraubstock)
open-sided planer Einständer-Hobelmaschine
operate, to betätigen, bedienen, arbeiten
operating device Bedienungseinrichtung
operating head Nutzförderhöhe
operating instruction Bedienungsanweisung
operating lever Bedienungshebel, Schalthebel
operating time Nutzungszeit (Maschine)

operation Betätigung, Bedienung, Arbeitsweise
operator Bedienungsmann, Maschinenarbeiter
optical dividing head optischer Teilkopf
order Reihenfolge, Anordnung, Bestellung
ordinary milling Gegenlauffräsen
orifice Öffnung, Mündung
O-ring Dichtungsring mit rundem Querschnitt
orthogonal cutting Zerspanen ohne Überhöhung der Schneide
oscillating arm Kurbelschwinge, Schwingarm
outboard brace Gegenhalterschere (Fräsmaschine)
out-cut milling Gegenlauffräsen
outfit Ausrüstung, Gerät
outlet port Austrittskanal
outline Umriss
out of mesh ausser Eingriff, ausgerückt
out of round unrund
out of truth ungerade, unrund
output Mengenleistung, Ausstoss, Fertigungsleistung, abgegebene Leistung
output per annum Jahresproduktion
output shaft Abtriebswelle
outside calipers Aussentaster
outside diameter Aussendurchmesser
overall length Gesamtlänge
overarm Gegenhalter (Fräsmaschine)
overarm support Gegenhalterstütze
overhang Überhang
overhead obenliegend, darüber angebracht, Ober . . .
overlap, to überlappen
overload, to überlasten
overload Überlastung
overload clutch Überlastungskupplung
overload protection Überlastschutz
overshoot, to hinausgehen über, hinausgeraten über
oversize Übermass

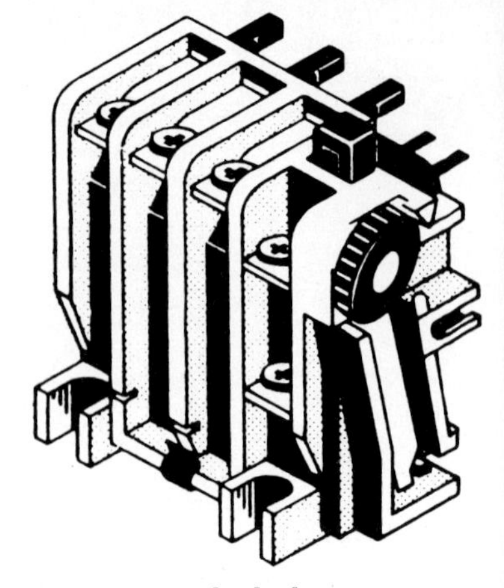

overload relay
Überstromrelais

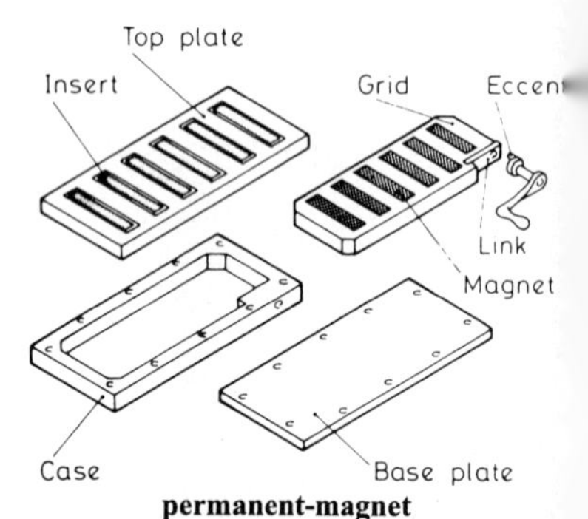

permanent-magnet chuck
Dauermagnetspannplatte

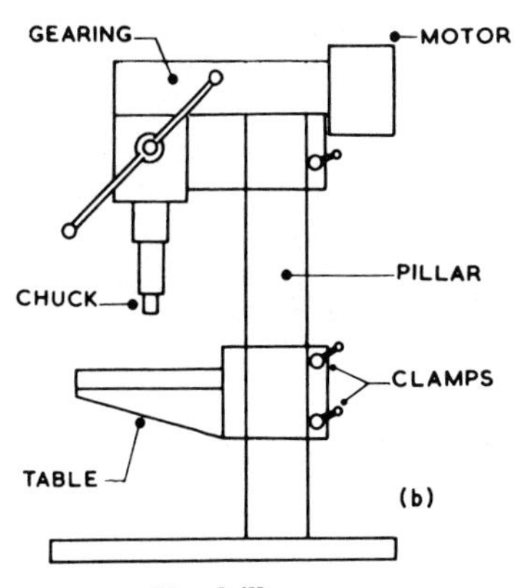

pillar drill
Säulenbohrmaschine

pin vice
Stiftkloben

overtighten, to zu fest anziehen
oxidize, to oxidieren
oxygen Sauerstoff

P

pack, to verpacken, abdichten, unterbauen
packing Packung, Abdichtung, Unterbauung
packing block Unterlegeblock
packing cord Dichtungsschnur
packing ring Dichtungsring
packing shim Packungsscheibe, Unterlegscheibe
pad, to ausfüttern, polstern
pad Polster, Kissen, Puffer
pan Schale, Wanne
panel Tafel
pantographic engraving machine Pantographengraviermaschine
pantographing Nachformfräsen
parallel clamp Parallelschraubzwinge
parallelism Parallelität
parallel planing machine Langhobelmaschine
parallel strock milling Zeilenfräsen
part Teil, Stück, Werkstück
part-ejection device Werkstückauswerfeinrichtung
parting Abstechen, Trennen
parting off Abstechen, Trennen
parting off tool Abstechmeissel
parting tool Trennwerkzeug, Stechmeissel
pass, to durchlaufen, durchgehen, Durchgang haben
pass Durchgang, Arbeitsgang
passage of a tool Werkzeugdurchgang
pattern Muster, Schablone
pawl-and-ratchet mechanism Sperrwerk, Sperrgetriebe
pawl arm Schwingarm
pedal operated fusshebelbetätigt
peel, to abschälen

peg, to anstiften, verdübeln
pendant herabhängend, schwebend
pendant Hängedruckknopftafel, Hängeschalter
pendant control panel Hängeschalttafel
pendulum milling Pendelfräsen
penetrate, to durchdringen, eindringen in
percussion Schlag, Stoss, Erschütterung
perforated tape Lochstreifen
perform, to ausführen, leisten
performance Leistung
perimeter Umfang, Umkreis
period of dwell Bewegungspause
peripheral cutting edges Umfangsschneiden
peripheral milling Wälzfräsen
permanent joint unlösbare Verbindung
permanent magnetic chuck Spannfutter mit Dauermagnet
permeable durchlässig
permissible zulässig
permit, to erlauben, zulassen, aufnehmen
perpendicular senkrecht, Senkrechte
piece Stück, Werkstück
piece of work Werkstück
pieces per hour Stundenleistung, Stückzahl je Stunde
pierced gestanzt, gelocht
pilot Führung, Führungsstück
pilot gear Führungszahnrad
piloting Führen, Führung
pin Stift
pin of the driving plate Mitnehmerbolzen
pincers Zange, Kneifzange
pinion Ritzel
pintle Stift, Zapfen, Achse
pin, to anheften, befestigen
pipe Rohr
pipe assembly Schlaucharmatur
pipe bend Rohrbogen
pipe thread Rohrgewinde
pipe wrench Rohrschlüssel, Rohrzange

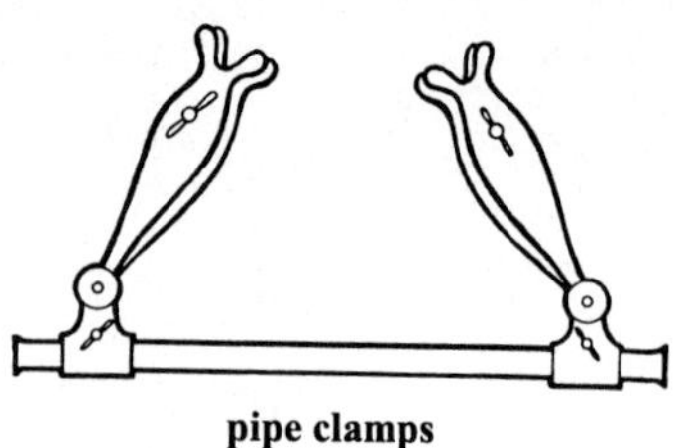

pipe clamps
Rohrschellen

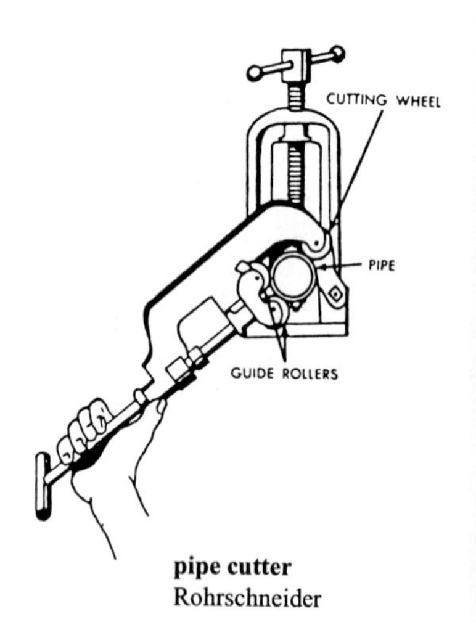

pipe cutter
Rohrschneider

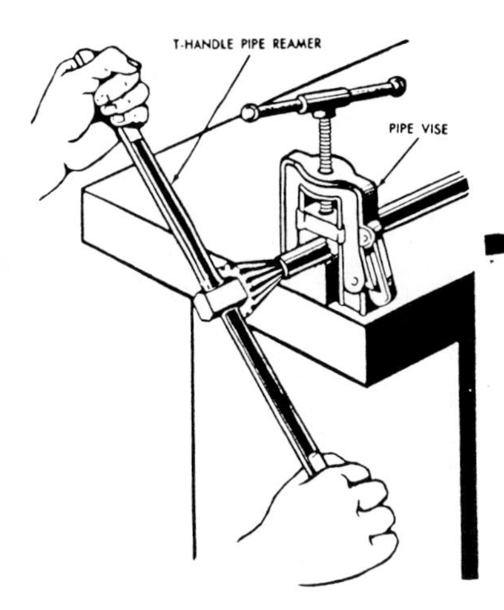

pipe reamer to remove burrs
Reibahle zur Entfernung von Grat

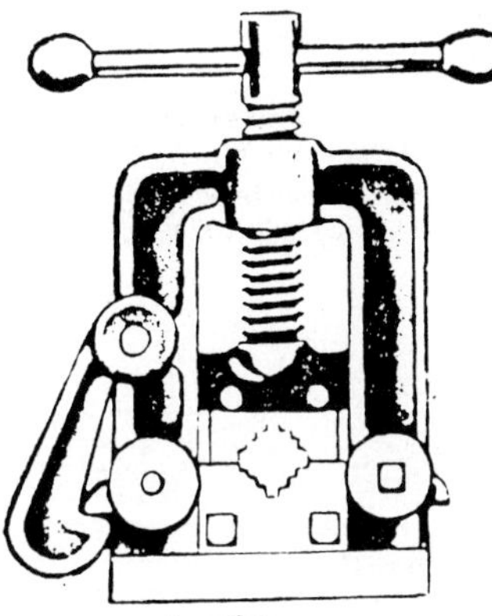

pipe vice
Rohrschraubstock

piston Kolben
piston ring Kolbenring
piston rod Kolbenstange, Pleuelstange
pit, to auskolken (Spanfläche)
pitch Teilung, Steigung (Gewinde)
pitch circle Teilkreis
pivot, to gelenkig aufhängen, sich um einen Zapfen drehen
pivot Zapfen, Drehbolzen, Drehpunkt
placed angeordnet
plain einfach, flach, eben
plain carbon steel unlegierter Kohlenstoffstahl
plain cutter Walzenfräser
plain dividing head einfacher Teilkopf
plain milling Planfräsen
plain milling machine Einfachfräsmaschine
plain slide valve Flachschieber
plain washer Unterlagscheibe
planamilling Fräsen mit Planetenspindel-Fräsmaschine
plan angle Einstellwinkel
plane, to hobeln
plane eben
plane Ebene
planer Hobelmaschine
planer head Hobelmaschinensupport
planer miller Langhobelfräsmaschine
planer platen Hobelmaschinentisch
planer poppet Gegenhalter, Spannbock
planer table Hobelmaschinentisch
planer tool Hobelmeissel
planer-type milling machine Langfräsmaschine, Portalfräsmaschine
planer vice Hobelmaschinenschraubstock
planetary gearing Planetengetriebe
planetary milling Fräsen mit Planetenspindel-Fräsmaschine
planing Hobeln
planing crosswise Querhobeln
planing machine Hobelmaschine
planing setup Aufspannung auf Hobelmaschinentisch

planing tool Hobelmeissel
plano-milling machine Langfräsmaschine, Portalfräsmaschine
plano-miller Langfräsmaschine, Portalfräsmaschine
plan relief angle Einstellwinkel der Nebenschneide
plant Anlage, Betriebsanlage, Werk, Fabrik, Betrieb
plant layout Aufstellung der Maschinen
plate Tafel, Platte, Blech
plate-bending machine Blechbiegemaschine
plate cam Scheibenkurve
plate-edge planing machine Blechkanten-Hobelmaschine
plate locking pin Haltestift der Teilscheibe
play Spiel
pliers Zange
plot, to aufreissen, einzeichnen, darstellen
plug Stopfen, Stift, Stecker (el.)
plunge cutting Tauchfräsen
plunge milling machine Tauchfräsmaschine
plunger Raststift, Tauchkolben, Plungerkolben
plunger pin Raststift
pneumatically operated druckluftbetätigt
pneumatic chucking Druckluftspannung
pneumatic conveying Druckluftförderung
pocketing dreidimensionales Fräsen, Gesenkfräsen
point Punkt, Stelle, Spitze
pointer Zeiger
pointer deflection Zeigeranschlag
point of gear intermesh Eingriffsstelle
polish, to polieren
polygon profile Mehrkantprofil
poor finish geringe Oberflächengüte
poorly clamped schlecht gespannt

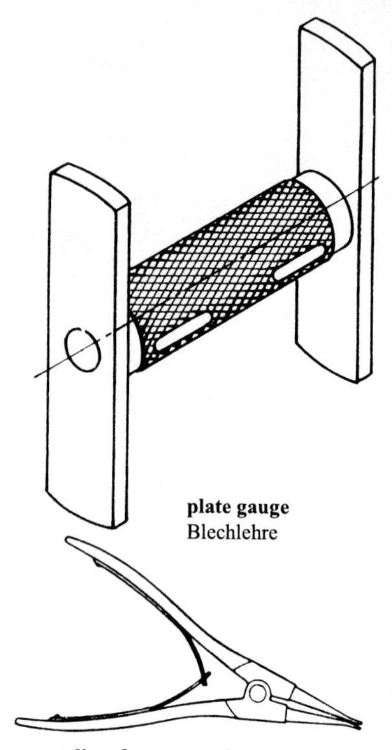

plate gauge
Blechlehre

pliers for contracting lockrings (shafts)
Zange für innenspannende Sicherungsringe (Wellen)

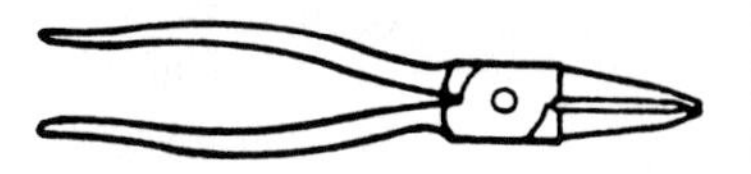

pliers for expanding lockrings (holes)
Zange für aussenspannende Sicherungsringe (Bohrungen)

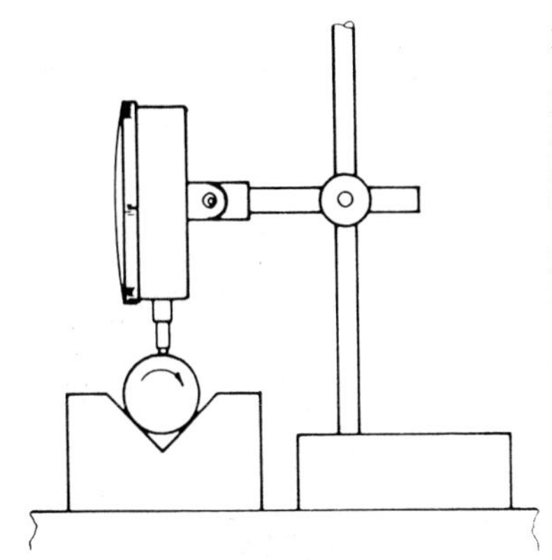

plunger-type dial indicator
Messuhr mit (Mess-)Bolzen

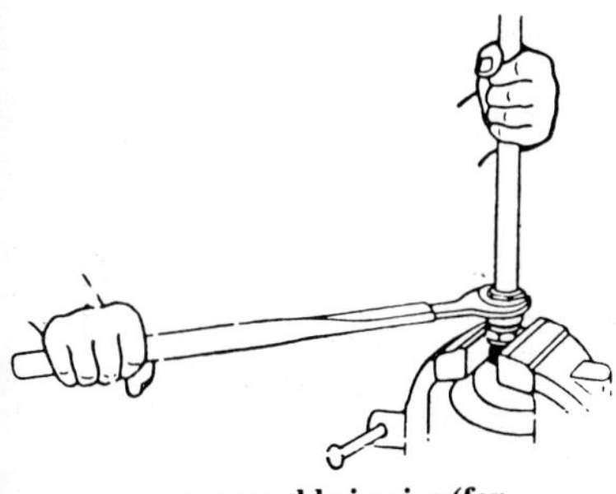

pre-assembly in vice (for pipe cutting)
Vormontage im Schraubstock

poppet Gegenhalter, Spannbock, Spannschraube
port, to leiten, führen (Flüssigkeit)
port Kanal
ported mit Kanälen versehen
position, to einstellen, in Stellung bringen, positionieren, zentrieren
position Stellung, Lage
positive drive zwangsläufiger Antrieb, zwangsläufige Mitnahme
positive rake milling Fräsen mit Positivschnitt des Werkzeugs, Fräsen mit positivem Spanwinkel
positive stop fester Anschlag
post-process gauging Messen nach Beendigung des Arbeitsgangs
pour point Stockpunkt
powder Pulver
power absorption Leistungsaufnahme
power consumption Kraftverbrauch
power of a machine Leistung einer Maschine
power down feed maschineller Tiefenvorschub, automatischer Tiefenvorschub
power input Leistungsaufnahme
power output abgegebene Leistung
power pack Hydraulikaggregat
power traverse Selbstgang
preadjusted vorher eingestellt
precaution Vorsichtsmassregel
precision job Präzisionsarbeit
predrill, to vorbohren
preloaded vorgespannt
preoptive mit Vorwahl
preselect, to vorwählen
preselector Vorwähler
preselective vorwählbar
preset, to vorher einstellen
press, to drücken, pressen
press Presse
press type broaching machine Stossräummaschine
pressure Druck
pressure-balanced mit Druckausgleich
pressure drop Druckabfall
pressure head Druckhöhe
pressure switch Druckschalter

prevent, to verhüten
prevention of accidents Unfallverhütung
prick punch Anreisskörner
primary clearance Fasenfreiwinkel, Freiwinkel an der Freiflächenfase, Hinterschliff der Fase
primary cutting edge Hauptschneide
primary land fasenartiger Anschliff an der Schneide
primary negative rake Spanflächenfase mit negativem Spanwinkel
prime, to ansaugen
prime mover Antriebsmotor, Antriebselement
principal cutting edge Hauptschneide
principal part Hauptteil
prior machining operations Vorbearbeitung
prismatic bearing surfaces prismatische Führungen
prismatic guide ways prismatische Führungen
procedure Verfahren, Arbeitsweise
process Verfahren, Prozess
produce, to herstellen, erzeugen, fertigen
produce mirror-image parts, to spiegelbildkopieren
production Produktion, Herstellung, Fertigung
production engineer Fertigungsingenieur
production engineering Fertigungstechnik
production increase Produktionssteigerung
production technique Fertigungsmethode
productivity Produktivität
profile Profil, Form
profile cutter Profilfräser, Formfräser
profile ground cutter profilhinterschliffener Fräser
profile milling Profilfräsen, Nachformfräsen
profiler Formfräsmaschine, Nachformfräsmaschine

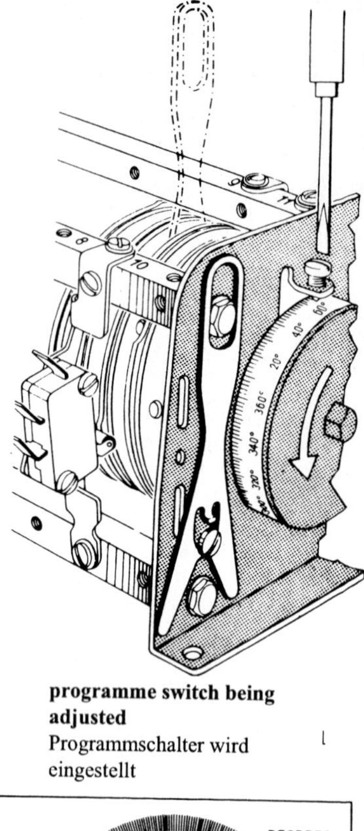

programme switch being adjusted
Programmschalter wird eingestellt

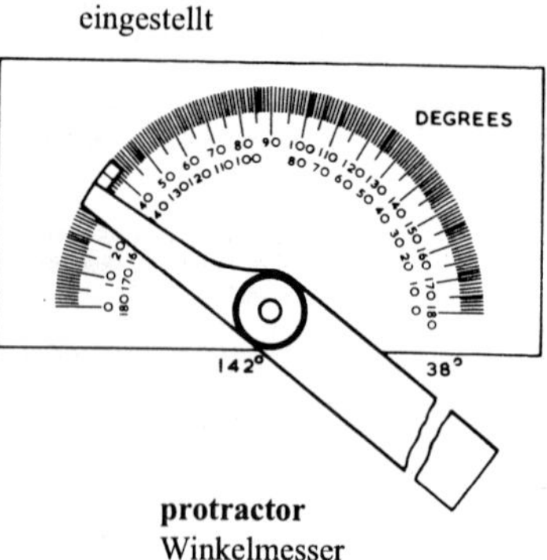

protractor
Winkelmesser

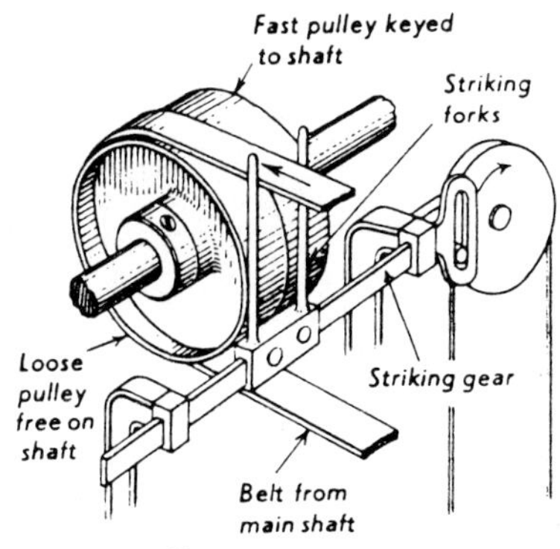

pulleys
Riemenscheiben und Rollen

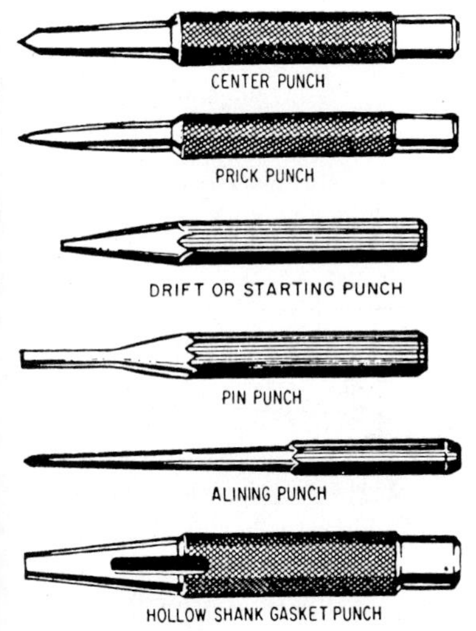

punches
Durchschläge und Körner

profile-relieved cutter profilhinterschliffener Fräser
profiling Profilfräsen
profiling machine (zweidimensionale) Nachformfräsmaschine
programme control Programmsteuerung
project, to hervorstehen, herausragen, planen, projizieren
prolong, to verlängern
property Eigenschaft
protect, to schützen
protective hood Schutzhaube
protractor Winkelmesser
protrude, to herausragen, hervorstehen
prove, to beweisen, sich erweisen als
provide, to vorsehen (sich), ergeben, bereitstellen, liefern
provided with versehen mit
pry Brecheisen, Hebezeug
pull, to ziehen
pull Zug, Zugkraft
pull broach Räumnadel
pull broaching Ziehräumen
pullers Ziehkopf, Spannkopf
pulley Riemenscheibe
pulling load Zugbelastung
pulsation-free pulsationsfrei, gleichmässig, stossfrei, schwingungsfrei
punch, to lochen, stanzen
punch Stempel, Patrize
punched card Lochkarte
punched card programming Programmsteuerung mit Lochkarte
punched paper type Papierlochband
punched tape control Lochbandsteuerung
punch and form shaper Form- und Stempelhobelmaschine
punch mark Körnungspunkt, Körner
push, to stossen, schieben, drücken, treiben
push broach, to stossräumen
push broach Räumdorn
push-button Druckknopf
put into operation, to in Gang setzen

Q

quadrant Wechselräderschere
quality control Qualitätskontrolle, Gütekontrolle
quantity Menge
quantity of heat Wärmemenge
quantity production Massenfertigung
quarter bend Rohrbogen von 90°
quench, to abschrecken
quick action chuck Schnellspannfutter
quick return Eilrücklauf
quick reverse Eilrücklauf
quietness in operation Laufruhe
quill Pinole, Hülse

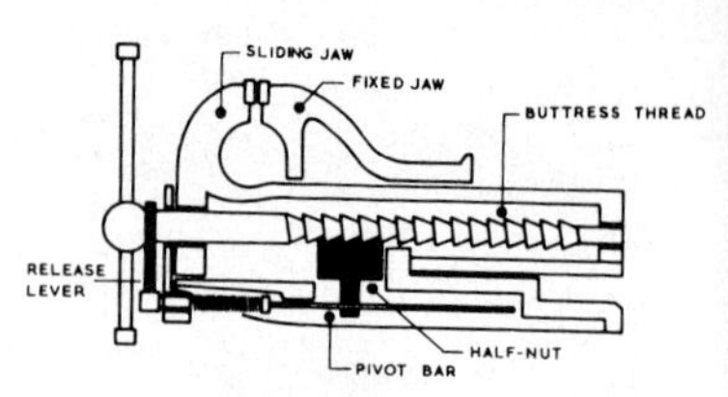

quick-release vice
Schraubstock für Schnellösung

R

race Laufring (Lager)
rack Zahnstange
rack milling Fräsen von Zahnstangen
radial radial, sternförmig
radially expanding clutch Spreizringkupplung
radial rake radialer Spanwinkel
radiused gerundet, mit Radius versehen
ragged eingerissen, zackig
rail Schiene, Querbalken, Querführung
rail head Quersupport
raise, to heben, anheben, abheben
rake Spanfläche
rake face Spanfläche
ram Stössel
ram bearing Stösselführung
range Bereich
range of feeds Vorschubbereich
rapid idle movement Eilrücklauf
ratchet Sperrad, Klinkenrad
ratchet-and-pawl mechanism Sperrwerk, Sperrgetriebe
rate, to bemessen, schätzen, veranschlagen
rate Grösse, Menge, Wert, Geschwindigkeit

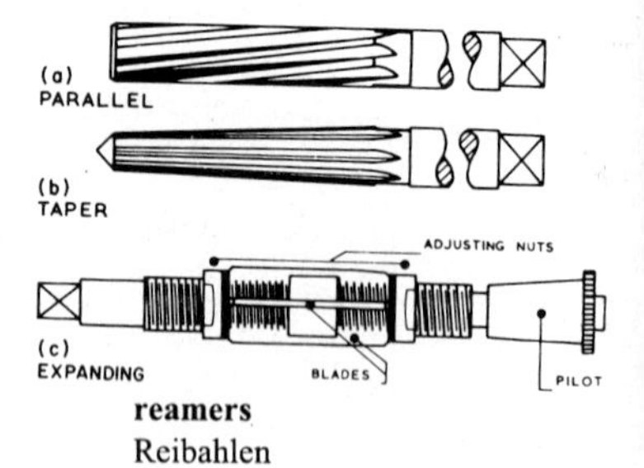

reamers
Reibahlen

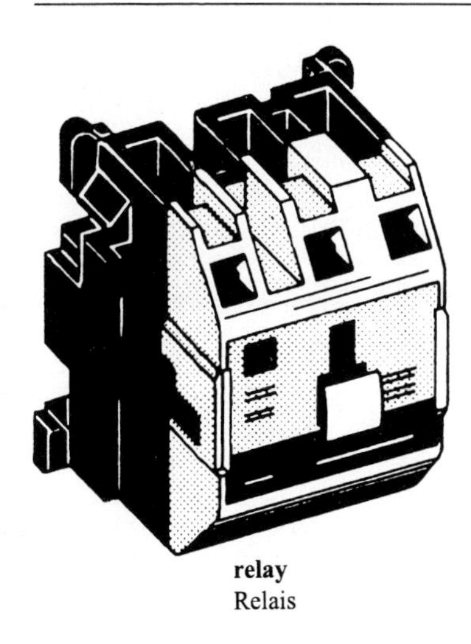

relay
Relais

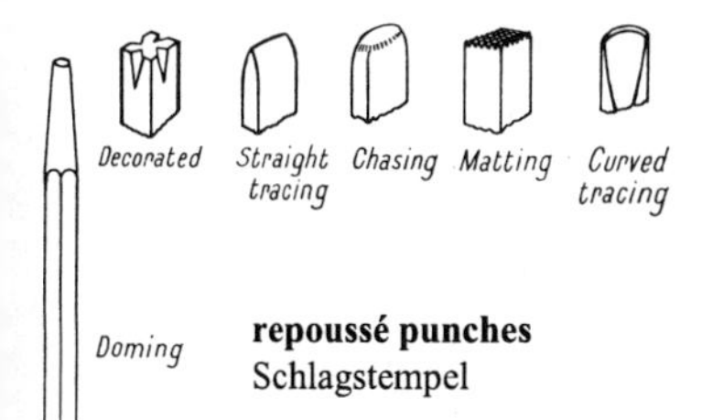

repoussé punches
Schlagstempel

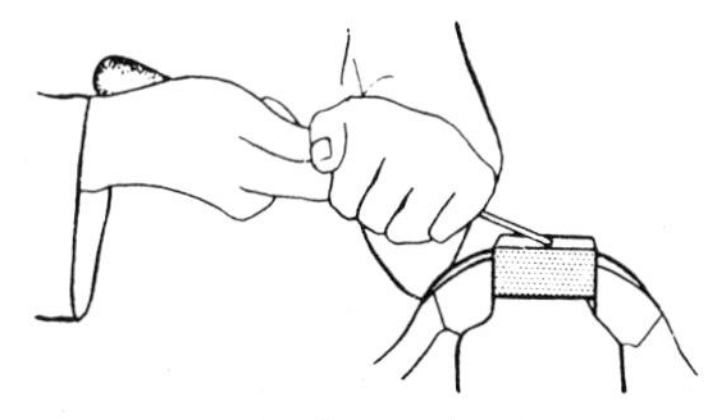

scraping work
Schabarbeit

rated capacity Nennleistung,
rated output Nennleistung
rate of cut Spantiefe
rate of feed Vorschubwert, Vorschubgrösse
rate of flow Strömungsmenge, Strömungsgeschwindigkeit
rate of tool wear Verschleissfortschritt beim Werkzeug
rating Leistung, Nennleistung
ratio Verhältnis
raw material Rohstoff
ray Strahl
reach, to reichen, erreichen
reach Reichweite
reader Lesegerät
reading Ablesung
reading error Ablesefehler
ream, to reiben, ausreiben
reamer Reibahle
reamer fluting cutter Reibahlennutfräser
rear hinterer, Hinter . . .
rear pilot Führungsstück, Endstück
rear side Rückseite
reassemble, to wieder zusammensetzen
recedling table Schiebetisch
receive, to aufnehmen, empfangen
receiving Aufnahme
reception Aufnahme
recess, to einstechen
recess Aussparung, Vertiefung
recessing tool Hakenmeissel
rechuck, to umspannen
reciprocal milling Pendelfräsen
reciprocate, to hin- und hergehen
reciprocating machinery Maschinen mit hin- und hergehender Hauptbewegung
reciprocation Hin- und Herbewegung
reclamping Neuaufspannung, nochmaliges Aufspannen
recondition, to reparieren, wieder instandsetzen
red hardness Warmhärte
redesign, to umkonstruieren

reduce, to reduzieren, vermindern
reduced scale verkleinerter Massstab
reduction gear Untersetzungsgetriebe
reference data Richtwerte
reflux Rückfluss
register, to registrieren, anzeigen
regrind Nachschliff
regrinding Nachschleifen
regulate, to regeln, regulieren
reject, to zurückweisen
reject rate Ausschussquote
rejects Ausschuss
release, to lösen, lockern, freigeben, loslassen
release Freigabe, Lösen, Auslösung
reliability Betriebssicherheit, Zuverlässigkeit
relief Hinterschliff der Fase, Freiwinkel
relief angle Freiwinkel
relief ground hinterschliffen
relief mill, to hinterfräsen
relief motion Abhebebewegung
relieved hinterschliffen, hinterdreht
relieving toolholder Meisselhalter mit Abhebung
relocate, to (Werkstück) umspannen
relocating work Werkstückumspannung
relocation Umspannung, Veränderung der Stellung
remain set, to in Einstellung bleiben
remote control Fernsteuerung, Fernbedienung
remotely controlled ferngesteuert
remotely operated fernbetätigt, indirekt betätigt
removable tool bit abnehmbarer Meisseleinsatz, Einsatzmeissel
removal Entfernung, Abnahme, Abfuhr (Span)
remove, to entfernen, (Span) abnehmen
remove metal, to zerspanen
renewal Erneuerung
repair, to reparieren
repair Reparatur

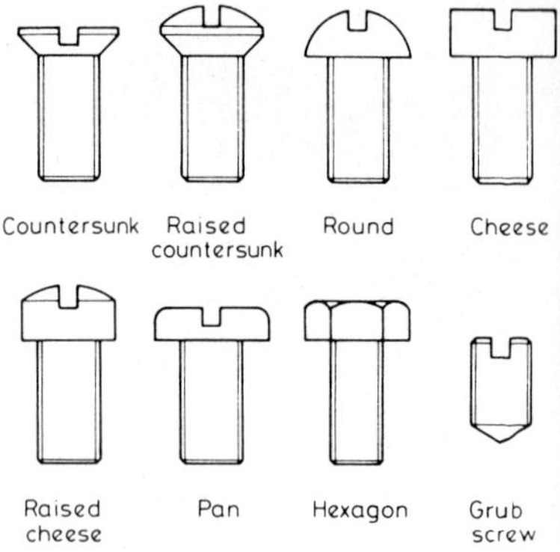

Countersunk
Raised countersunk
Round
Cheese
Raised cheese
Pan
Hexagon
Grub screw

screw heads (slotted) Schraubenköpfe (mit Schlitz)

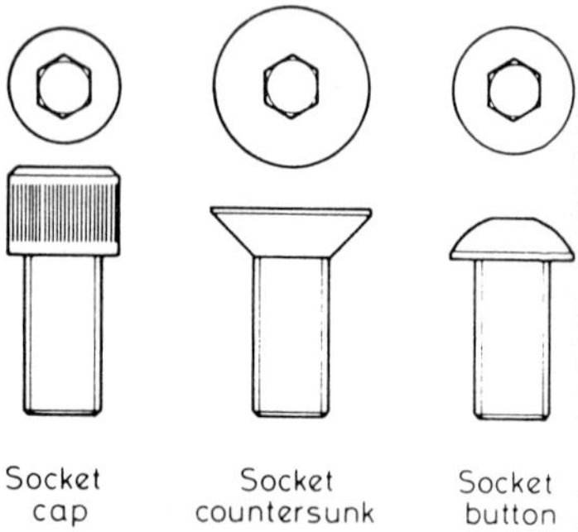

Socket cap
Socket countersunk
Socket button

screw heads (recessed) Schraubenköpfe (mit Innensechskant)

screw jack
Schraubwinde, Einstellschraube; manchmal auch: Maschinen-Hebebock

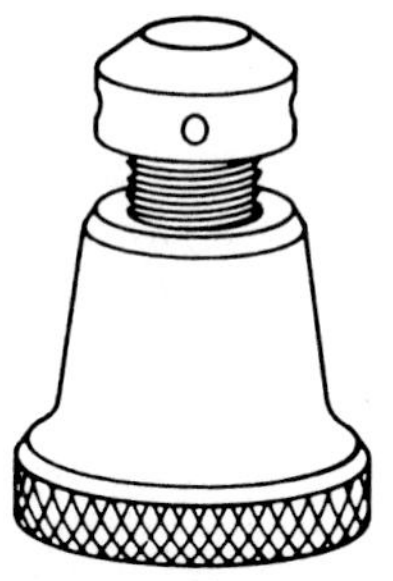

screw packing
Richtschraubbock, wenn mit Feingewinde; manchmal auch: Schraubensupport

repeatability Nachformfähigkeit, Möglichkeit der Wiederholung eines Arbeitsganges
repetition job Serienarbeit
repetition milling Serienfräsen
repetition production Serienfertigung
repetitive parts Serienstücke
replace, to ersetzen
replaceable auswechselbar
replacement blade Ersatzmesser
replacement part Ersatzteil
replica Ebenbild
represent, to darstellen
reproduce, to nachformen, kopieren, reproduzieren
reproduction Nachformung, Kopierarbeit
re-schedule, to umplanen
research Forschung
research and development division Forschungs- und Entwicklungsabteilung
reseat, to den Sitz neu schleifen, einschleifen
reservicing Wiederinstandsetzung, Reparatur
reset, to umrichten, umspannen, neu einstellen, neu einrichten, nachstellen
resetting Umrichten, Umspannung, Neuaufspannung
resharpen, to nachschleifen, nachschärfen
resharpening Nachschliff
residue Rückstand, Abfall
resilience Elastizität
resilient elastisch
resin Harz
resist, to widerstehen, standhalten, aushalten, aufnehmen
resistance Widerstand
resistance to wear Verschleissfestigkeit
resistant widerstandsfähig
resistant to wear verschleissfest
rest Ruhe, Stillstand
restrict, to einschränken, einschnüren, verengen, drosseln

restriction Verengung
resultant force (R) Gesamtresultierende (P), Schnittkraft
retain, to festhalten, zurückhalten
retard, to verzögern
retardation Verzögerung
retool, to einspannen anderer Werkzeuge, neurüsten, umrüsten
retract, to zurückziehen, sich zurückziehen
retractable zurückziehbar
retraction Zurückziehen
retraction of milling head Fräserabhebung
retriever head Zubringerkopf (Räumen)
return line Rückflussleitung
return speed Rücklaufgeschwindigkeit (Hobelmaschine)
return stroke Rücklauf, Leerhub
reversal Umsteuerung
reversal of the direction Richtungsumkehr
reverse, to wenden, umsteuern, umkehren, rückwärts laufen lassen, rückwärts fahren
reverse Rücklauf
reverse rückwärts
reverse dog Umsteuerknagge, Anschlag zum Umsteuern
reverse gear Wendegetriebe
reverse mirror duplicating spiegelbildliches Nachformen
reverse image attachment Spiegelbild-Kopiereinrichtung
reversible umsteuerbar
reversing lever Umsteuerhebel
revolution Umdrehung
revolve, to sich drehen
revolving table Drehtisch
rheostat Regelwiderstand
rib Rippe
ribbed verrippt
ribbon chip Bandspan
ridge Furche, Riefe, fasenartiger Anschliff
ridge left by the tool Bearbeitungsriefe

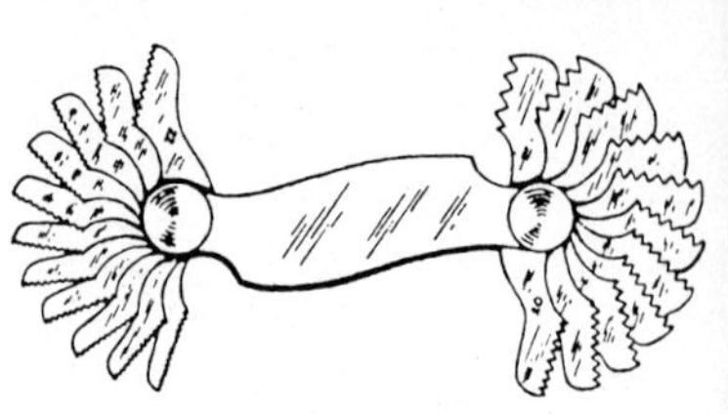

screw pitch gauge
Gewindelehre für Ganghöhen

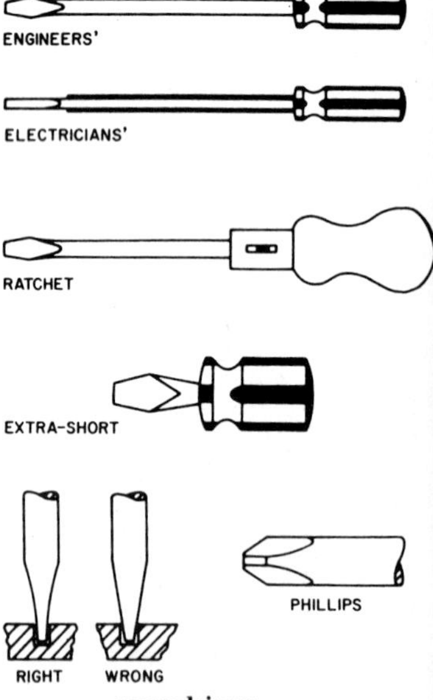

screwdrivers
Schraubenzieher

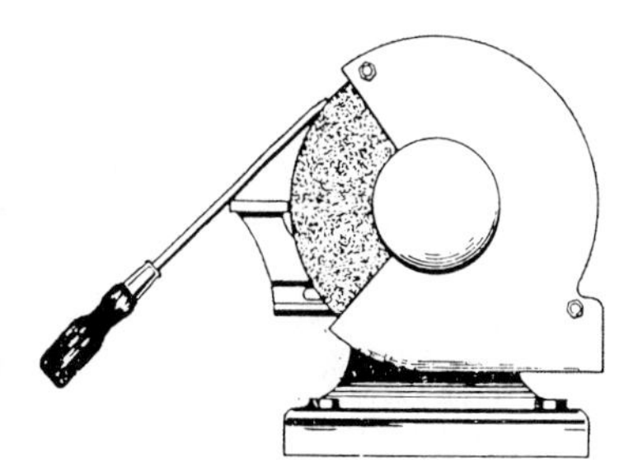

screwdriver tip on grinding wheel
Schraubenzieher an der Schleifscheibe

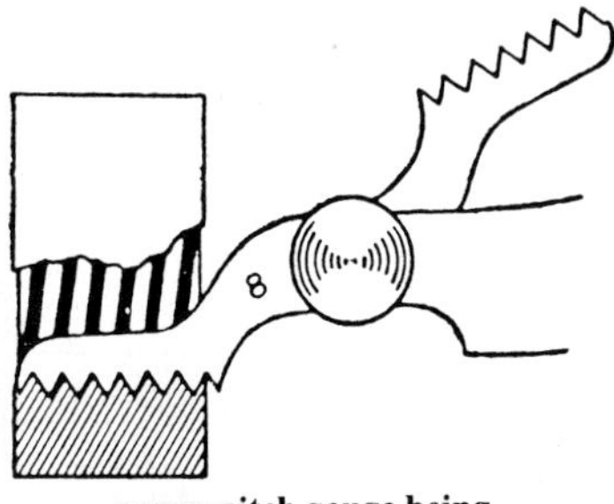

screw pitch gauge being applied
Gewindelehre wird angelegt

right-and-left-hand helix cutter Fräser mit Pfeilverzahnung
right-angle head Querhobelsupport
right angle rechter Winkel
right angularity Rechtwinkligkeit
right-bent tool gebogener rechter Meissel
right-cut rechtsschneidend
right-hand cut rechtsschneidend
right-hand helical cutting edges Schneiden mit Rechtsdrall
right-hand rotation Rechtsdrehung
right-hand spiral Rechtsdrall
right-offset tool abgesetzter rechter Seitenmeissel
rigid starr
rigidity Starrheit
rim Rand, Radkranz
ring gasket Dichtungsring
rinse, to spülen
rise and fall miller Planfräsmaschine mit Hebe- und Senkeinrichtung für die Spindel
rise in pressure Druckanstieg, Drucksteigerung
rise per tooth Überhöhung von Zahn zu Zahn (Räumen)
rise the tool, to den Meissel abheben
rock, to schwingen, schwanken, rütteln, erschüttern
rocking arm Schwingarm
rod Stange
roll, to walzen
rolled gewalzt
roller Walze, Kopierrolle
roller bearing Walzenlager, Rollenlager
roller follower Führerwalze
roll-over fixture Wendevorrichtung
roll-over station Wendestation
root Wurzel (Mathem., Zahngrund)
root circle Fusskreis
root clearance Spitzenspiel, Spiel an der Zahnwurzel
root diameter Kerndurchmesser, von Zahngrund zu Zahngrund gemessen (Räumwerkzeug)

root-mean square average (rms) Wurzel aus Mittelwert der Quadrate
rotary drehend, Dreh...
rotary attachment Rundfräseinrichtung
rotary feed Rundvorschub
rotary indexing machine Rundschaltmaschine
rotary milling Rundfräsen
rotary table Rundtisch
rotation axis Drehachse
rough, to aufrauhen, rauh machen, schruppen
rough rauh, grob
rough adjustment Grobeinstellung
rough-cut, to vorfräsen
rough cutter Vorschneider, Vorfräser
rougher Schruppräumwerkzeug
roughing Schruppen
roughing broach Schruppräumwerkzeug
roughing cherrying operation Vorfräsen runder Gesenkformen
roughing cut Schruppschnitt, Schruppspan
roughing cutter Schruppfräser, Vorfräser
roughing operation Schrupparbeit, Schruppschnitt, spanende Vorbearbeitung
roughing pass Schruppdurchgang
roughing tool Schruppwerkzeug, Schruppmeissel
roughing tooth Schruppzahn
rough-machine, to vorbearbeiten, schruppen
rough-machined geschruppt
rough-milling Vorfräsen, Schruppfräsen
roughness Rauheit
rough planing Schrupphobeln, Vorhobeln
rough positioning Grobeinstellung
round bar Rundeisen
round bar vice Schraubstock für rundes Stangenmaterial
round broach, pull type Ziehräumnadel mit kreisförmigem Querschnitt

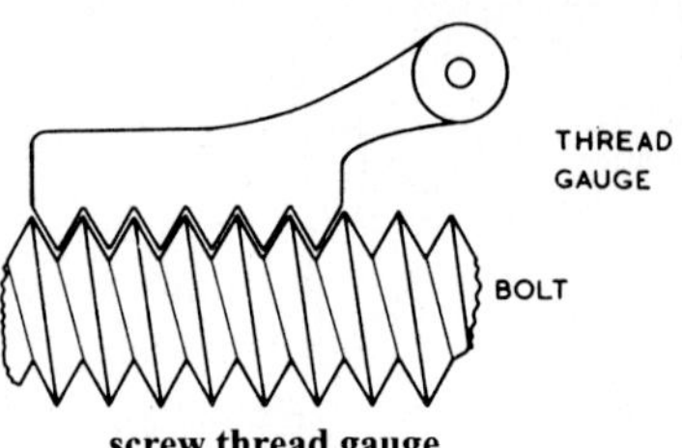

screw thread gauge
Schraubengewindelehre

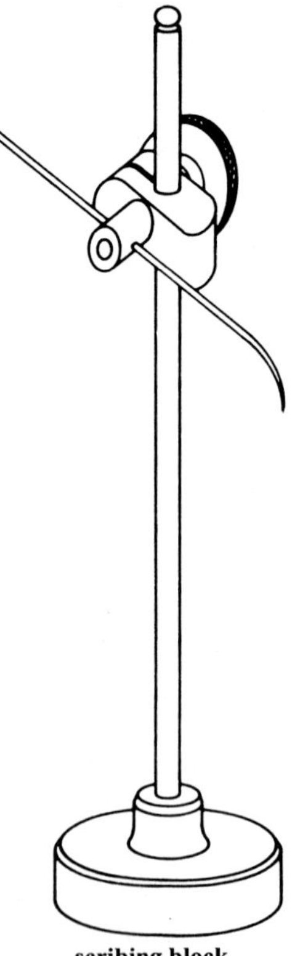

scribing block
Parallelreisser

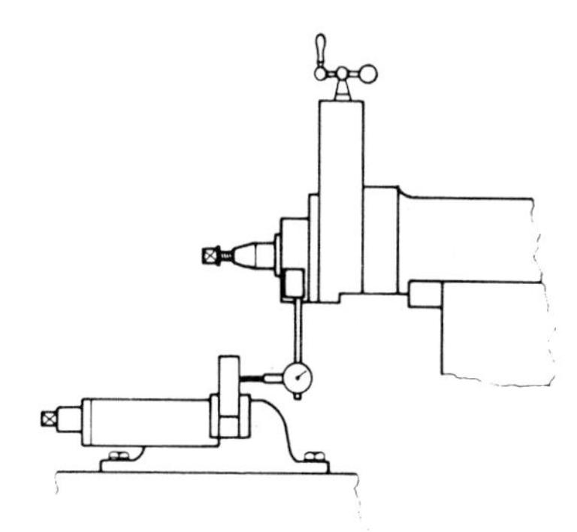

setting up the vice Aufspannen eines Schraubstockes

round-end cutter Fräser mit runder Stirn
round-hole broaching Räumen runder Durchbrüche
round-pilot at a square broach runde Aufnahme an quadratischem Räumwerkzeug
round-point tapering die-sinking cutter kegeliger Gesenkfräser mit runder Stirn
router Fräsmaschine zum Ausschneiden von Umrissen aus Blechen mit Handvorschub
routing Nachformfräsen mit Handvorschub, Arbeitsvorbereitung
routing card Laufkarte für das Werkstück, Arbeitskarte
routing cutter Gesenkfräser, Langlochvorfräser
row Reihe
rub, to reiben
rubber gasket Gummidichtung
rugged stabil
ruggedness Stabilität
ruler Lineal
rule of thumb Faustregel
run Lauf
running Lauf
rush job eilige Arbeit
rust, to rosten
rust Rost
rust preventive Rostschutzmittel
rust removal Entrosten
rusty rostig

S

saddle Sattel, Querschieber, Schlitten, Kreuzschieber
safeguard Schutz, Sicherung
safe load zulässige Last, zulässige Belastung
safety Sicherheit
safety guard Schutzvorrichtung
safety stop dog Sicherheitsanschlagnocke

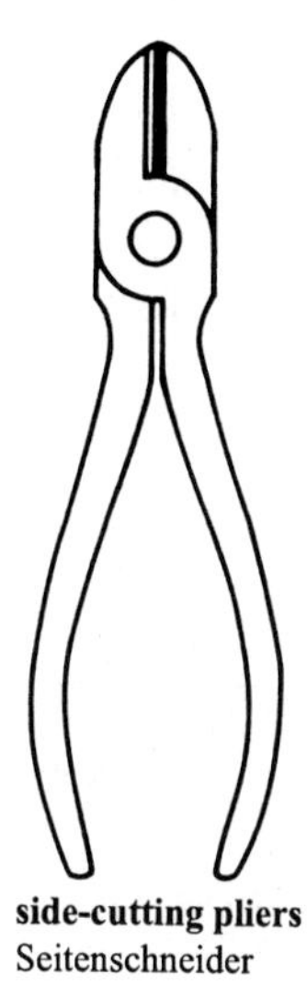

side-cutting pliers Seitenschneider

sag Durchhang (Riemen)
salvage, to rückgewinnen, reparieren, ausschlachten
sample Muster, Probe
sample-work-piece Musterstück, Bezugsstück
sandwich brazed hart eingelötet
saturate, to sättigen
saturation point Sättigungspunkt
saving in material Werkstoffeinsparung
saving in time Zeitersparnis
saw, to sägen
saw Säge
saw-tooth cutter spitzgezahnter Fräser
scale Massstab
scale Gusshaut, Zunder
scaled kalibriert
scale formation Zunderbildung
scarf, to schärfen
scarfing Anschärfen, Abschrägen
scavenge, to spülen, reinigen
scope of work Arbeitsbereich, Anwendungsbereich
score, to einkerben, beschädigen
scoring Einkerben
scrap Schrott
scrap component Ausschussteil
scrape, to schaben
scraper Schaber
scraping Schaben
scrap metals Schrott
scrap rate Ausschussrate, Ausschussquote
scrap work Ausschuss
scratch Kratzer
screw, to schrauben
screw Schraube, Gegenspindel
screw and nut table feeding Tischvorschub durch Spindel und Mutter
screw cutting Gewindeschneiden
screw-driver Schraubenzieher
screwed connection Schraubverbindung
screwing Verschraubung
screw into, to einschrauben in
screw on, to aufschrauben

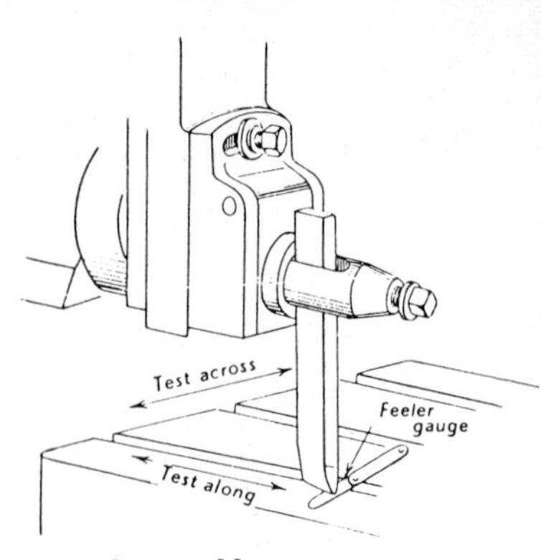

shaper table
Tisch einer Waagerechtstossmaschine

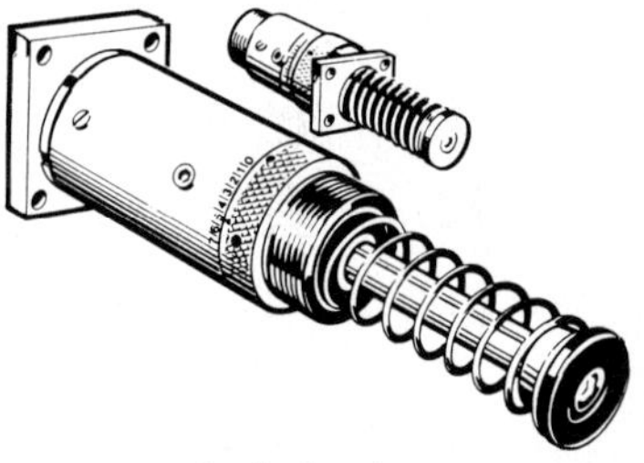
shock absorber
Stossdämpfer

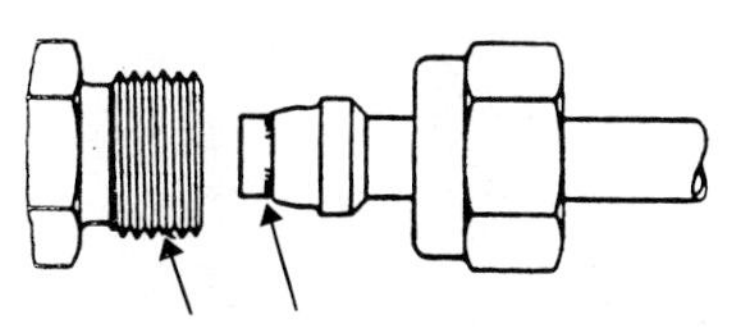
sleeve and nozzle
Hülse und Düse

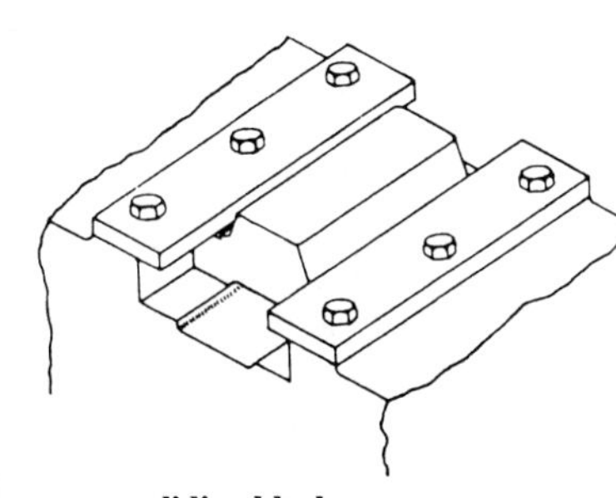
sliding block
Kulissen-, Führungsstein

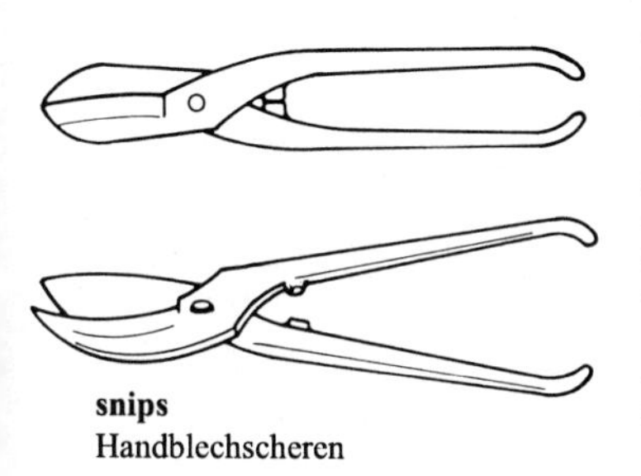
snips
Handblechscheren

screw-on cutter Aufschraubfräser
screw slotter Schraubenschlitz-maschine
screw slotting Schraubenschlitzen
screw slotting saw Schraubenschlitz-säge
screw within a telescopic cover Tele-skopspindel
scribe, to anreissen
scribed line Anrisslinie
scriber Reissnadel
scribing Anreissen
scribing block Parallelreisser
scroll Spindel (Schraubenpumpe)
scroll gear Triebkranz
seal, to abdichten, abschliessen, ver-siegeln
seal face Dichtungsfläche
sealing Abdichtung, Dichtung
seam Naht
seamless nahtlos
season, to altern
seasoning Alterung
seat, to aufnehmen, aufliegen (Werk-stück)
seat Sitz
seat dresser Ventilsitz-Fräseinrichtung
seat for key Keilsitz
seating Sitzfläche
secondary clearance sekundärer Frei-winkel
secondary cutting edge Nebenschneide
second cycle operation zweiter Arbeitsgang
section (or: cross section) Quer-schnitt
sectional broach mehrteiliges Räum-werkzeug
sectional drawing Schnitt (Zeichnung im Querschnitt)
sector arm Zeiger (Teilen)
secure, to befestigen, spannen
securing pin Befestigungsstift
seepage Versickerung, Durch-strömung
segmental chip Bruch- oder Reissspan
segment copying Nachformen im Teilumriss

segment gear Zahnsegment
seize, to fressen, festfressen, verklemmen, hängenbleiben
seizure Fressen, Festfressen
select, to wählen
selector switch Wahlschalter
self-acting selbsttätig
self-aligning ball bearing Pendelkugellager
self-centring selbstzentrierend
self-centring chuck Universalfutter
self-contained mit eigenem Antrieb, unabhängig
self-indexing selbstschaltend
self-priming selbstansaugend
semicircle Halbkreis
semi-skilled worker angelernter Arbeiter
sensitive feinstufig, gefühlvoll
sensitivity feinstufige Schaltbarkeit
separate, to trennen
separate getrennt
separation Trennung, Ablösung (Strömungslehre)
sequence of operations Arbeitsablauf, Folge der Arbeitsvorgänge
sequencing Folgesteuerung, Einstellung der Arbeitsfolge
serial number Fertigungsnummer
series Serie, Reihe
series of tests Versuchsreihe
serrated gerieft, geriffelt, verzahnt
serrated blade cutter Messerkopf mit auswechselbaren, geriffelten Messern
serration Kerbverzahnung
serve, to dienen
service life Nutzungsdauer, Lebensdauer
service stress Betriebsbeanspruchung
servicing Instandhaltung
servo-controlled mit Hilfssteuerung, mit Vorsteuerung
servomechanism Hilfssteuerelement, Vorsteuerelement
set, to einstellen, setzen
set at an angle, to schrägstellen
set Satz

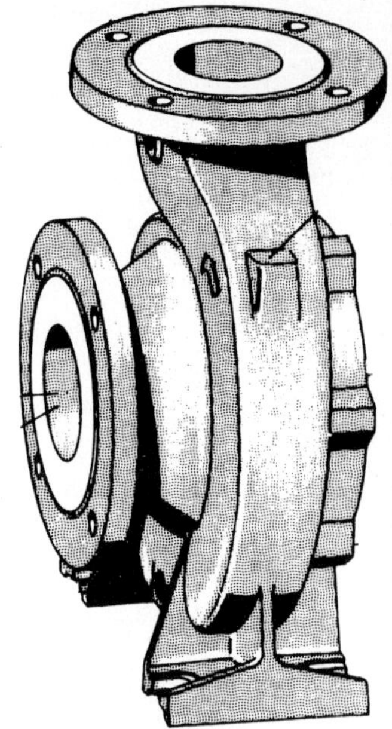

spiral housing of pump
Spiralgehäuse einer Pumpe

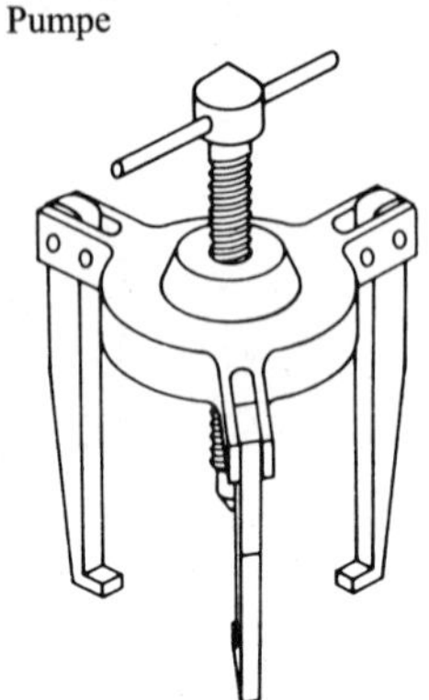

sprocket remover
Riemenscheibenzange

spur gear
Stirnradgetriebe

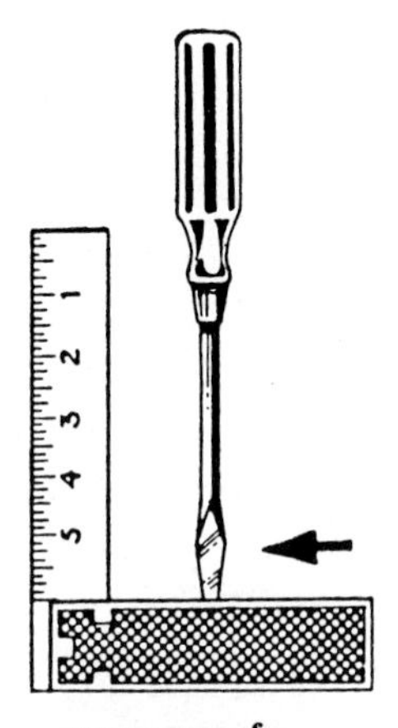

squareness of a screwdriver being checked
Winkligkeit eines Schraubenziehers wird geprüft

set of change gears Wechselrädersatz
setover of the swivel head of the shaper Verstellen des Drehteils der Waagrechtstossmaschine
setscrew Klemmschraube, Einstellschraube
setting Einstellen, Einspannen, Aufspannung
setting accuracy Einstellgenauigkeit
setting gauge Einstellehre
setting out Anreissen
setting-up time Aufspannzeit
settle, to sich setzen, sich senken, in Ordnung bringen
set to the full-depth position, to auf volle Schnittiefe einstellen
setup Aufspannung
setup fixture Aufspannvorrichtung
shaft Welle
shaft coupling Wellenkupplung
shallow flach
shank Schaft
shank back rake Spitzenspanwinkel
shank cutter Schaftfräser
shape, to formen, gestalten, profilieren, kurzhobeln, stossen
shape Form, Gestalt, Umriss
shaped profile cutter profilhinterschliffener Formfräser, spitzgezahnter Formfräser
shaper Waagrechtstossmaschine, Kurzhobler
shaper tool Hobelmeissel
shaping Formgebung
shaping Kurzhobeln
shaping machine Waagrechtstossmaschine, Kurzhobelmaschine
sharp-edged orifice scharfkantige Öffnung
sharpen, to schärfen, nachschärfen, nachschleifen
sharpening Schärfen, Schleifen
sharp-pointed mit scharfer Spitze, mit scharfer Schneide
shaving cut Schabeschnitt
shear, to scheren
shear angle Abscherwinkel, Scherwinkel

shear chip Scherspan
shear cut scherender Schnitt
shear cutting Abscheren, Zerspanen mit scherendem Schnitt
shearing Abscheren
shearing action Scherwirkung
shearing force Scherkraft
shear strength Scherfestigkeit
shear tool Schermeissel
sheave Scheibe, Rillenscheibe, Blockscheibe
sheet metal Blech
sheet metal gauge Blechlehre
sheet metal templet Blechschablone
sheet metal working Blechbearbeitung
shell type tool Aufsteckwerkzeug
shifter Steuerschieber
shim, to unterbauen
shim Unterlagstück, Unterlegplatte, Zwischenlegblech
shock Stoss, Erschütterung
shock absorber Stossdämpfer
shockless stossfrei
shock load Stossbelastung
shop Werkstatt, Werkhalle, Betrieb
shop term Werkstattausdruck
shortage Knappheit, Mangel
shortage of skilled workers Facharbeitermangel
short chip material kurzspanender Werkstoff
shorten, to verkürzen
short run job kleine Serienarbeit, kleine Stückzahl
short stroke ram kurzhubiger Stössel
short taper Steilkegel
shoulder Schulter, Stufe, Ansatz, Absatz
shoulder tool Seitenmeissel
shoulder of the cut Schnittfläche
shrinkage cavity Lunkerstelle
side Seite
side and face milling cutter Scheibenfräser
side angle Komplementwinkel des Einstellwinkels
side box Seitensupport

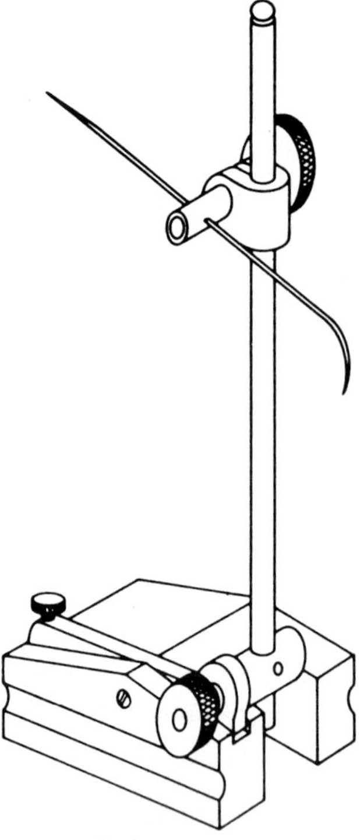

surface gauge
Höhenreisser, auch: Parallelreisser

side clearance seitlicher Freiwinkel, Hohlschliff (Schlitzsäge)
side cutting tool Seitenmeissel
side face Seitenfläche
side head Seitensupport
side head screw Spindel zur Verstellung des Ständerschlittens
side housing Seitenständer, Ständer
side mill Scheibenfräser
side milling Walzenstirnfräsen
side milling cutter Scheibenfräser
side-relieved seitlich hinterschliffen, hohlgeschliffen
side rougher Seitenschruppmeissel
side slope Seitenspanwinkel
side tool Seitenmeissel
sidewise seitlich, in seitlicher Richtung
silent geräuscharm, ruhig
silent running Laufruhe, ruhiger Lauf, geräuscharmer Lauf
silicone Silikon
simple gear train Rädergetriebe mit einfacher Übersetzung
simple indexing einfaches Teilen
simplex manufacturing type milling machine Einspindel-Planfräsmaschine
simplify, to vereinfachen
simultaneous gleichzeitig
simultaneous twin spindle milling gleichzeitiges Fräsen mit zwei Spindeln
sine bar Sinuslineal
single-acting einfachwirkend
single angle cutter Winkelstirnfräser
single edged tool einseitiges Werkzeug
sinter, to sintern
sintered carbide Sinterkarbid, Hartmetall
sintered carbide tip Hartmetallplättchen
sintered tool Hartmetallwerkzeug
six speed gearbox sechsstufiger Räderkasten
size Grösse, Abmessung, Massgenauigkeit

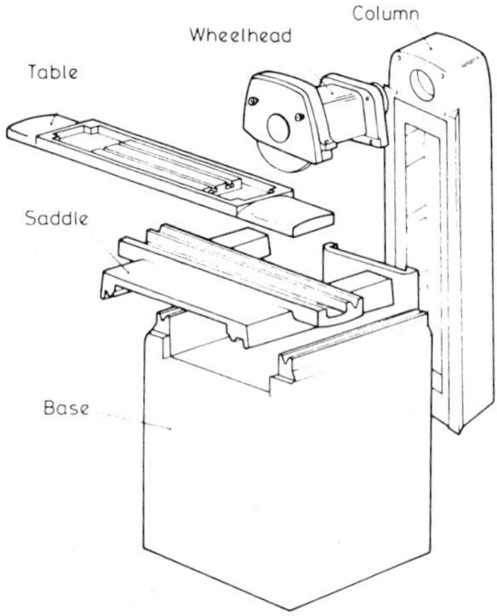

surface grinder, main elements
Flächenschleifmaschine, Hauptteile

sized kalibriert
sized tool Formmeissel
size loss Verlust der Massgenauigkeit, Verringerung der Grösse
size of land Fasenbreite, Zahnrückendicke
size to width, to auf Breite bringen
sizing broach Kalibrierräumwerkzeug
sizing teeth Kalibrierzähne
sizing work Kalibrierarbeit
skew geneigt, schräg, schief
skilled labour gelernte Arbeitskräfte, Facharbeiter
skilled worker Facharbeiter
skill test praktische Lehrlingsprüfung
skin Gusshaut
skip, to überspringen
skip-feeding Sprungvorschub
skip-feeding for rapid traverse between gaps Sprungvorschub für Leerwege im Eilgang
skip milling Fräsen unterbrochener Flächen
slabbing cutter Walzenfräser (der breiter als sein Durchmesser ist)
slabbing machine Langfräsmaschine zum Flächenfräsen
slab-mill, to mit breitem Walzenfräser fräsen, Flächen fräsen
slab mill breiter Walzenfräser
slab milling Planfräsen mit breitem Schnitt
slacken, to lockern, lösen (Schraube)
slacken the locking screw one turn, to die Klemmschraube um eine Umdrehung lösen
sleeve Hülse, Zwischenstück, Muffe, Pinole
sleeve joint Muffenverbindung
sleeve-lock Pinolenfestklemmung
slender schlank
slenderness ratio Schlankheitsgrad (s/i)
slice, to in Streifen schneiden
slide, to gleiten
slide Schlitten, Führung
slide bar Gleitschiene

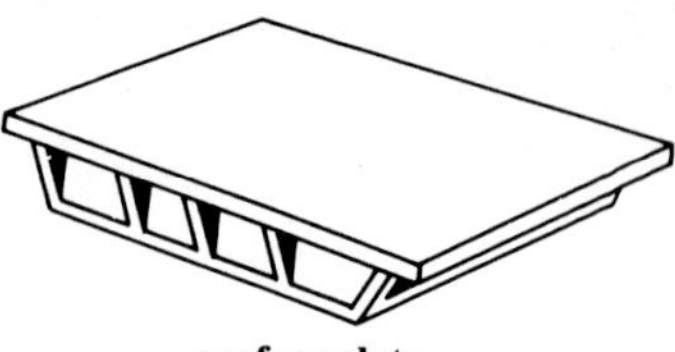

surface plate
Anreissplatte

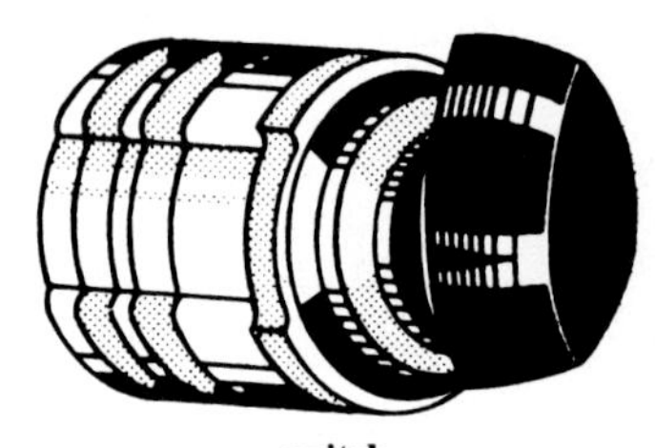

switch
Schalter

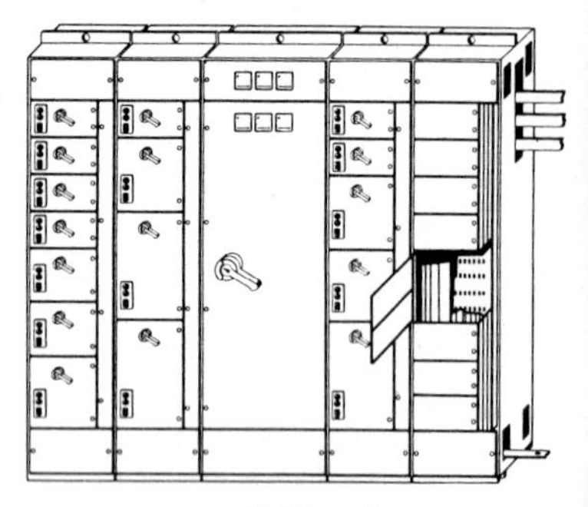

switchboard
Schalttafel

slide head Schlitten (Räummaschine)
slide rest Support
slide rule Rechenschieber
slideway Gleitbahn, Führung
sliding block Führungsstein, Kulissenstein
sliding cam plate Kurvenschieber
sliding fit Gleitsitz
sliding gears Schiebezahnräder
sliding surface Gleitfläche
slip, to rutschen, gleiten
slip Schlupf
slip gauge Parallelendmass
slip loss Schlupfverlust
slippage Schlupf
slitting cutter Schlitzfräser
slog schwere Arbeit
slogging Abnahme eines grossen Spans
slope Neigung
slot, to schlitzen
slot Schlitz, Nut
slot and keyway milling machine Langloch- und Keilnuten-Fräsmaschine
slot drill Langlochfräser, Fingerfräser, Nutenfräser
slot milling Schlitzfräsen, Schlitzen
slot milling cutter Nutenfräser
slotted arm Kurbelschwinge, Kulisse
slotter Senkrechtstossmaschine
slotter tool Stossmeissel
slotting Schlitzen, Langlochfräsen, Senkrechtstossen
slotting attachment Stosseinrichtung, Senkrechtstosseinrichtung
slotting bar Stossstange (Senkrechtstossmaschine)
slotting cutter Langlochfräser
slotting machine Senkrechtstossmaschine
slotting saw Schraubenschlitzfräser
slotting setup Aufspannung für Senkrechtstossarbeit
slotting tool Stossmeissel
slow down, to verlangsamen, bremsen
sluggish zäh

small batch production Kleinreihenfertigung
small lot milling Kleinreihenfräsen
small quantity production Kleinreihenfertigung
small run kleine Serienarbeit
smooth glatt, stossfrei, stufenlos
smoothing Glätten
socket Buchse, Fassung, Hülse, Muffe, Stutzen
socket pipe Muffenrohr
socket ring Taumelscheibe
soft packing Weichpackung, Weichdichtung
softening Erweichen
solder, to löten
solder Lot
soldered joint Lötverbindung
soldering iron Lötkolben
solderless joint lötlose Verbindung
solder wire Lötdraht
solenoid operation magnetische Betätigung
solid fest, massiv, einteilig, Voll . . .
solid bed type milling machine Planfräsmaschine
solid broach einteiliges Räumwerkzeug
solid carbide finishing shell Schlichtschneideeinsatz aus Hartmetall zum Aufstecken auf Räumwerkzeug-Tragkörper
solid cutter einteiliger Fräser, massiver Fräser
solid type end mill Schaftfräser
solid metal gasket Metalldichtung
soluble löslich, lösbar
solution Lösung
solvent Lösungsmittel
source off errors Fehlerquelle
space, to teilen, auf Abstand bringen
space Raum, Abstand
spacer Abstandsring
spacing Abstand, Teilung
spacing accuracy Teilgenauigkeit
spacing of teeth Zahnteilung
spall, to abbröckeln
span, to umspannen, einfassen

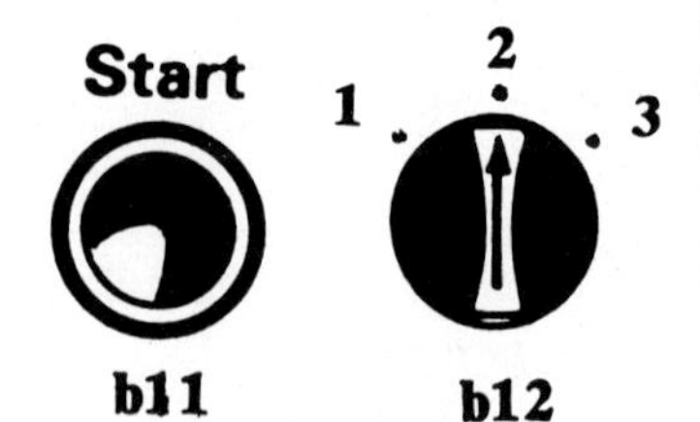

start and selector switch
Anlass- und Wahlschalter

spanner Schraubenschlüssel
spare part Ersatzteil
spark Funke
spark erosion machine Funkenerosionsmaschine
specification(s) technische Daten, Bauvorschrift, Hauptabmessungen, genauere Angaben
specified technisch festgelegt
specimen Musterstück, Probestück
speed Geschwindigkeit, Drehzahl
speeding-up Beschleunigung
speed of rotation Drehzahl
speed ratio Übersetzungsverhältnis
speed reducing gears Untersetzungsgetriebe
speed up, to schneller laufen lassen, beschleunigen
spherical end mill Schaftfräser mit Kugelstirn
spherical milling Fräsen von Kugel- oder Halbkugelformen
spigot joint Muffenrohrverbindung
spigot nut Überwurfmutter
spindle Spindel
spindle stock Spindelstock
spindle motor reversing switch Umsteuerschalter für den Spindelmotor
spindle nose Spindelkegel, Spindelende, Spindelkopf
spindle quill Spindelpinole, Spindelhülse
spindle speed Spindeldrehzahl
spindle taper Spindelkegel
spiral Spirale, Wendel
spiral end mill Schaftfräser mit Spiralverzahnung
spiral flute Spiralnut
spiral gear schrägverzahntes Zahnrad
spiral groove Spiralnut
spiral mill spiralverzahnter Fräser
spiral milling Spiralfräsen
spiral milling head Spiralfräskopf
spirit level Wasserwaage
splined shaft Keilwelle
spline milling Fräsen von Keilwellen
split, to spalten, schlitzen

split nut Mutterschloss
split pin Splint
spoil, to verderben
spray coolant system Ölnebel-Kühlanlage
spray cooling Ölnebelkühlung
spreading the load Verteilung der Belastung
spring Feder
spring collet Spannbüchse, Federspannfutter
spring-loaded federgespannt, federbelastet
spring operated federbetätigt
spring plunger Federstift
spring returned mit Rückführung durch Feder
spring tension Federspannung
sprocket wheel Kettenrad
spur gear Stirnrad
square Vierkant, Quadrat, Vierkanteisen
square quadratisch, rechtwinklig, gerade
square-base U-ring Nutringmanschette
squared im Quadrat, hoch zwei
squareness Rechtwinkligkeit
square-nosed tool Flachmeissel
square root Quadratwurzel
square-threaded screw Spindel mit Flachgewinde
square up, to auf rechten Winkel vorbearbeiten
square with each other rechtwinklig zueinander
squaring Bearbeiten auf rechten Winkel
stable stabil
stage Stufe, Stadium
stagger, to auf Lücke stehen, gegeneinander versetzen
staggered arrangement (of teeth) Kreuzverzahnung
staggered teeth Kreuzverzahnung
staggered tooth side mill kreuzverzahnter Scheibenfräser
stainless rostfrei

submersible pump
Tauchpumpe

stall, to zum Stehen bringen, zum Stehen kommen
stamina Kraft, Stärke
standard Norm, Ständer
standard equipment Normalausrüstung
star handle Handkreuz
start, to starten, in Gang setzen, einschalten
starting end Anschnittseite
starting hole Führungsbohrung, Vorbohrung
starting torque Anlaufdrehmoment
start of the cut Anschnitt
state Zustand
static discharge head statische Förderhöhe
static head statische Höhe
static suction lift statische Saughöhe
station Spannstelle, Arbeitsstelle, Station
stationary ortsfest, unbeweglich, fest, feststehend, ruhend
stay put, to in Stellung bleiben
steady Setzstock, Lünette
steady gleichmässig, stetig
steady flow stationäre Strömung
steady rest Setzstock
steel cutter blank Stahlfräserrohling
steel plate Stahlblech
steel scrap Stahlschrott
steel sheet Stahlblech
steep steil
steep angle taper Steilkegel
step Stufe
step cone Stufenscheibe
stepless stufenlos
steplessly variable stufenlos veränderlich
steplessly variable speed stufenlos veränderliche Drehzahl
stepless speed changing stufenlose Drehzahländerung
step milling Stufenfräsen
stepped gestuft, abgestuft, überhöht
stepped cone pulley Stufenscheibe
stepping Abstufung, Überhöhung

stick, to steckenbleiben, festsitzen, festfahren, haften
stiffness Steifigkeit
stock Material, Werkstoff
stock, from vom Lager
stock allowance Werkstoffzugabe
stocking Vorfräsen von Zahnlücken
stocking Lagerung
stocking cutter Zahnformvorfräser
stocking gear milling cutter Zahnformvorfräser
stock left for Werkstoffzugabe für, Materialzugabe für
stock removal Werkstoffabnahme, Spanleistung, abgenommenes Spanvolumen, Zerspanung, Spantiefe
stone, to (mit Ölstein) abziehen
stop, to anhalten, stillsetzen
stop Anschlag
stop button Ausschaltknopf
stoppage time Stillstandszeit
storage Speicherung, Lagerung
store, to speichern, lagern
straddle broaching gleichzeitiges Räumen paralleler, gegenüberliegender Flächen
straddle mill Scheibenfräser
straddle milling Fräsen zweier paralleler Flächen mit zwei Scheibenfräsern
straight bore zylindrische Bohrung
straight broaching Räumen ebener Flächen
straight edge Lineal
straighten, to ausrichten, richten
straight flute milling Fräsen gerader Nuten
straight length gerades Rohrstück
straight line Gerade
straight lined geradlinig
straight oil Öl ohne chemische Zusätze
straight toothed geradverzahnt
straight tooth side and face cutter geradverzahnter Scheibenfräser
strain Formänderung, Verformung, Dehnung

tailstock arrangement
Anordnung des Reitstockes

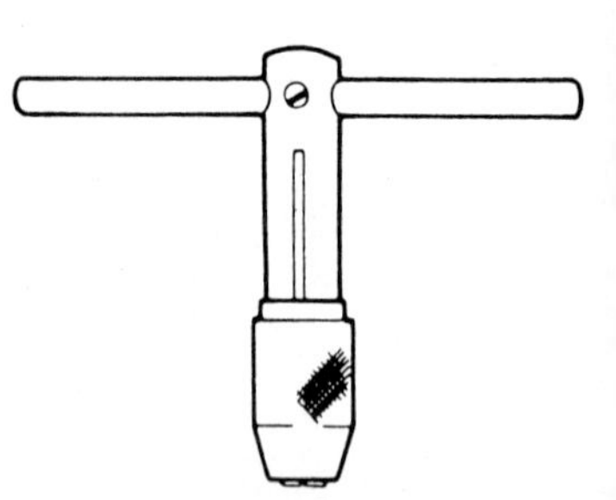

tap wrench
Lochschlüssel; sonst auch: Windeisen

tap wrench
Windeisen

One dowel for endwise location
Tenon slot
Tenon

tenon and slot for positioning
Zapfen und Schlitz zum Positionieren

strainer Saugkorb
strap (clamp) Spanneisen
stream Strahl, Strömung
stream in, to einströmen
streamline Stromlinie, Strömungslinie
strength Festigkeit
strengthen, to verstärken, versteifen, festigen
strength of materials Werkstoffestigkeit, Festigkeitslehre
strength test Festigkeitsprüfung
stress Spannung, Beanspruchung
stretch, to dehnen, strecken
stroke Hub
structure Bau, Konstruktion, Struktur
strut Strebe, Versteifung, Druckstab
stub angular cutter Aufsteckwinkelfräser
stub arbor kurzer Dorn, Aufsteckdorn
stub expansion arbor fliegend aufgespannter Spreizdorn
stuck broach festgefahrenes Räumwerkzeug
stud Stiftbolzen, Stiftschraube
stuff, to ausstopfen, abdichten
stuffing box Stopfbuchse
stuffing box gland Brille der Stopfbuchse
stuffing box packing Stopfbuchsendichtung
stylus Fühlerstift, Fühler, Taster
sub-assembly Teilmontage, Vormontage, Baugruppe
subdivision Unterteilung
subject to, to unterwerfen, aussetzen
submerge, to eintauchen
substitute, to austauschen, ersetzen
substitute Ersatz
stubstitute for, to einsetzen an Stelle von
successive aufeinanderfolgend
suck, to saugen
suction Ansaugung, Saugwirkung, Sog
suction boost Verstärkung der Ansaugmenge

suction dust remover Staubabsauger
suction filter Saugfilter
suction lift Saughöhe
suction port Saugkammer, Ansaugkanal
suggestion Verbesserungsvorschlag
suggestion award Prämie für Verbesserungsvorschlag
suggestion for improvement Verbesserungsvorschlag
suit, to passen, sich eignen
superior (to) überlegen
supply, to versorgen, liefern, zuführen
support, to unterstützen, abstützen, tragen
support Auflage, Unterstützung
support clamp Halteklemme
supporting bracket Tischstütze
surface Oberfläche
surface below the cutting edge Freifläche
surface broach Aussenräumwerkzeug
surface broach bar Schneideneinsatz für Räumzeug
surface broaching Aussenräumen
surface broaching machine Aussenräummaschine
surface cratering Auskolkung der Spanfläche
surface finish Oberflächengüte
surface flaw Oberflächenfehler
surface gauge Parallelreisser, Höhenreisser
surface hardening Oberflächenhärten
surface hardness Oberflächenhärte
surface milling Flächenfräsen
surface plate Anreissplatte, Tuschierplatte
surface speed Schnittgeschwindigkeit
surface texture Oberflächenfeingestalt
surplus Überschuss
surplus metal überschüssiges Metall
suspend, to unterstützen (einseitig), frei tragen
swan-necked finisher gekröpfter Schlichtmeissel
swan-necked rougher gekröpfter Schruppmeissel

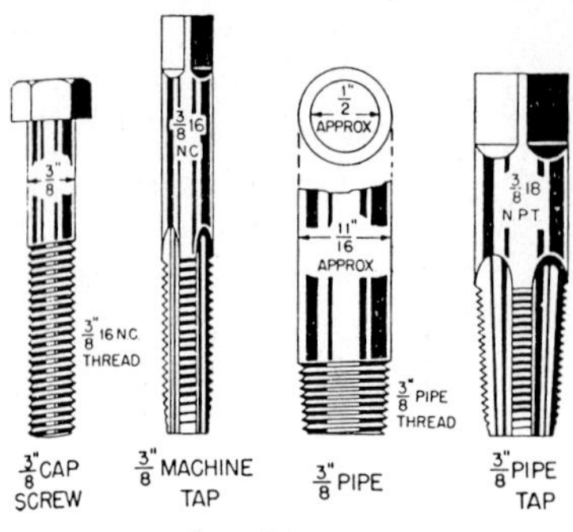

thread taps
Gewindebohrer

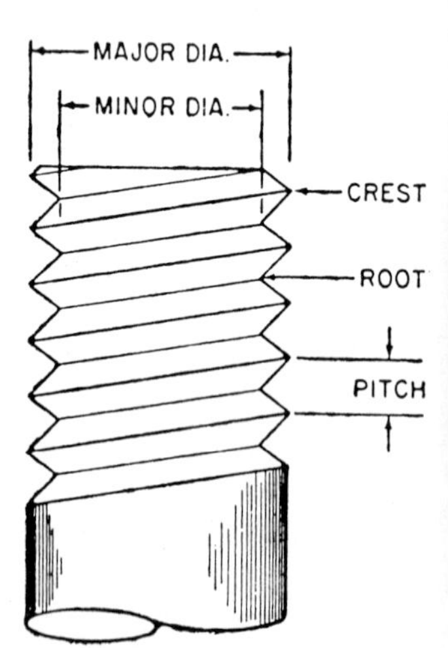

thread terminology
Bezeichnungen am Gewinde

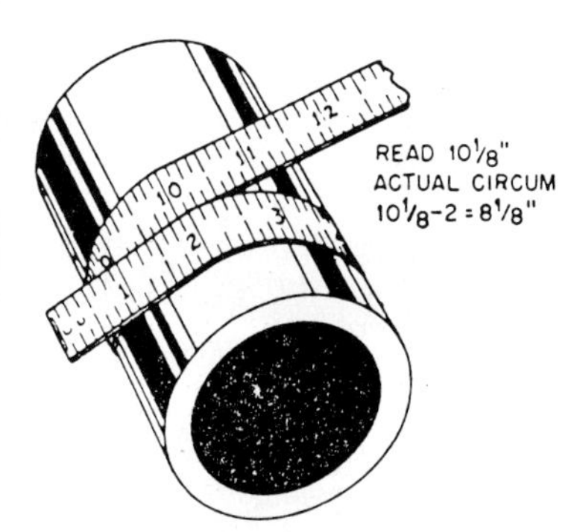

tape for measuring circumference
Band zum Ermitteln des Umfanges

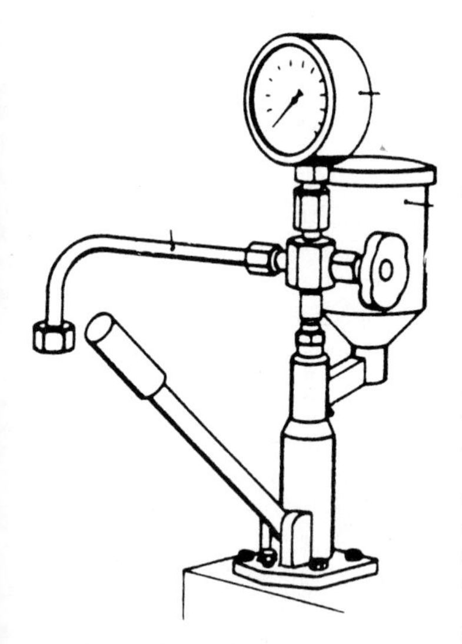

testing device for injectors
Prüfvorrichtung für Einspritzventile

swarf ölige Späne
swash-plate Taumelscheibe
swing, to schwenken, schwingen
swinging arm schwingende Kurbelschleife
switch, to schalten
switch Schalter
switch lever Schalthebel
switch off, to abschalten, ausschalten
switch on, to einschalten
switch position Schalterstellung
swivel, to schwenken, drehen
swivel Drehteil, Drehscheibe
swivel head Drehsupport, Kippsupport, schwenkbarer Spindelkopf
swivel head miller Fräsmaschine mit drehbarem Spindelkopf
swivelling attachment Universalspindelkopf
swivel pipe coupling Gelenkrohrverbindung
swivel plate Drehteil
swivel table Schwenktisch
swivel vice drehbarer Schraubstock
symmetrical angle cutter Prismenfräser

T

table Tabelle, Tisch
table drive Tischantrieb
tailstock Reitstock
tailstock spindle sleeve Reitstockpinole
take a cut, to einen Span abnehmen
take-up strip Nachstelleiste
tang Austreiberlappen, Mitnehmerlappen
tap, to Gewinde bohren
tap Gewindebohrer
tape-controlled bandgesteuert
tape-controlled milling machine bandgesteuerte Fräsmaschine
taped kegelig, konisch, abgeschrägt
taper Verjüngung, Abschrägung, Kegel

taper adapter Kegeleinsatz, Konuseinsatz
tape reader Bandlesegerät
tapered kegelig, konisch, abgeschrägt
tapered gib Keilleiste
tapered roller bearing Kegelrollenlager
taper milling Kegelfräsen
taper parallels Keilstücke
taper pin Kegelstift
taper plug Hahnkücken
tapped mit Innengewinde (versehen)
tappet Anschlag, Stössel
tapping cutter Gewindefräser
tear chip Reissspan
tee-piece T-Stück
teeth on the end Stirnzähne
teeth on the periphery Umfangszähne
telescopic teleskopisch, ausziehbar
telescopic coolant return ausziehbare Kühlmittel-Rückflussleitung
temperature rise Temperaturanstieg
template Schablone
tend, to neigen zu
tendency Tendenz, Neigung
tender Kostenvoranschlag
tensile strength Zugfestigkeit
tensile testing machine Zerreissmaschine
tensile test specimen Prüfstück für Zugfestigkeitsprüfung
tension Spannung (mech.)
tension spring Zugfeder
test cutting Zerspanungsprobe
test piece Prüfstück, Probestück
thermal conductivity Wärmeleitfähigkeit
thicken, to dickflüssiger machen
thickness Dicke
thin walled dünnwandig
thread, to Gewinde schneiden
thread Gewinde
threaded mit Gewinde (versehen)
threaded hole Bohrung mit Gewinde
thread miller Gewindefräsmaschine
thread milling Gewindefräsen
three-jaw chuck Dreibackenfutter
throttle, to drosseln

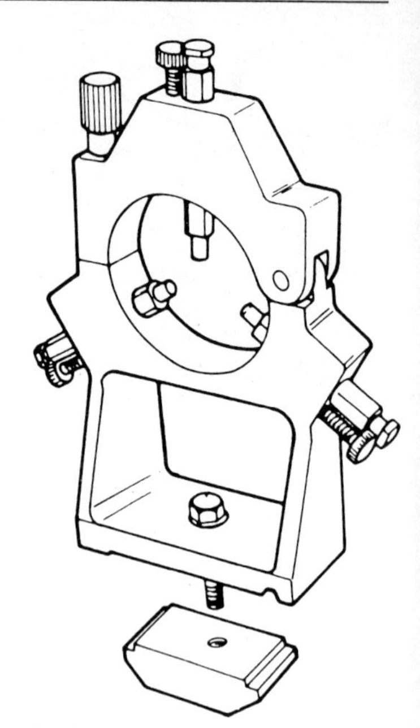

three-point steady
feststehender Dreipunkt-Setzstock

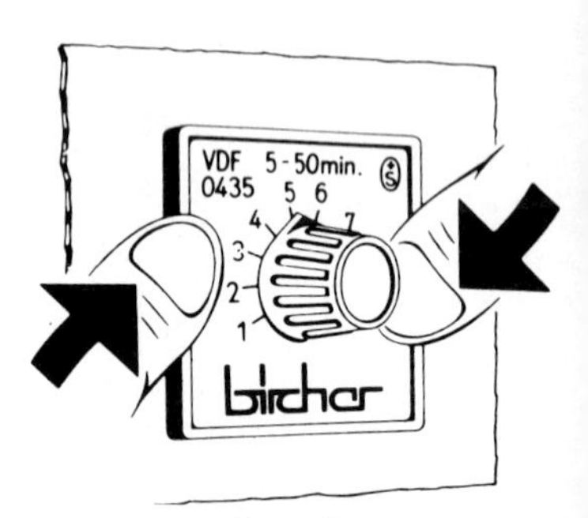

time relay
Zeitrelais

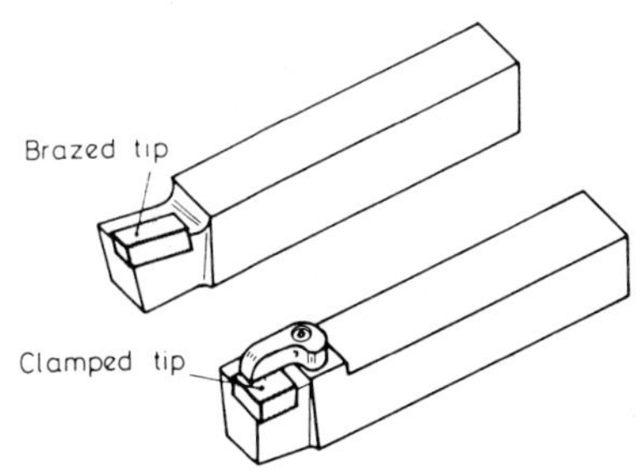

tipped lathe tools
bestückte Drehmeissel

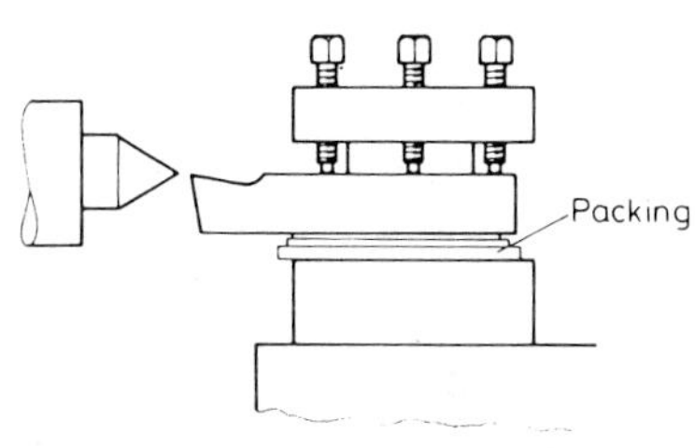

tool setting to centre
Einrichten des Werkzeugs auf Mitte

throttling valve Drosselventil
through piston rod durchgehende Kolbenstange
throw-away carbide insert blade Hartmetall-Wegwerfmesser
throw-aways Wegwerfwerkzeuge
throw in, to einrücken
throw out, to ausrücken
thrust Druck, Schub
tight fest, dicht
tighten, to anziehen
tilt, to kippen, schrägstellen
tilt Schrägstellung, Neigung
tiltable kippbar, schwenkbar, neigbar
tilting table Kipptisch
time of dwell Stillstandszeit, Verzögerungsdauer, Bewegungspause
time of operation Nutzungszeit
tip, to kippen
tip, to mit Schneide versehen, bestükken
tipped solid cutter massiver Fräser mit bestückter Schneide
tipped with sintered carbide mit Sinterhartmetallschneide (versehen) bestückt
titanium Titan
toe dog Niederhalter, Spannfinger
toggle Kniegelenk
tolerance Toleranz
tool Werkzeug, Meissel, Werkzeugmaschine
tool and die miller Werkzeug- und Gesenkfräsmaschine
tool apron Meisselhalterklappe
tool bit Meisseleinsatz, Schneidplatte
tool bit holder Meisseleinsatzhalter
tool box Support, Werkzeugträger
tool box on cross slide Quersupport
tool ejector Werkzeugauswerfer
tool face Spanfläche
tool failure Unbrauchbarwerden des Werkzeugs
tool geometry Geometrie der Schneide
tool grind Nachschliff des Werkzeugs
tool head Support, Werkzeugträger, Stösselkopf

tool head slide Schieber (des Supports)
tool holder Werkzeughalter, Meisselhalter
tooling Einrichten, Aufspannen der Werkzeuge, Auswahl der Werkzeuge
tooling diagram Werkzeuganordnung
tooling time Rüstzeit
tool interchange Werkzeugwechsel
tool life Lebensdauer des Werkzeugs, Standzeit (between grinds)
tool lifter Meisselabheber
toolmaking Werkzeugmacherei
tool point Schneidenkopf
tool rake angle Spanwinkel
tool releasing Ausspannen des Werkzeugs
toolroom Vorrichtungsbau, Werkzeugmacherei
tool section Meisselquerschnitt
tool setter Einrichter
tool set-up time Aufspannzeit der Werkzeuge
tool shank Werkzeugschaft
tooth back Zahnrücken
toothed gezahnt, verzahnt
tooth gullet Zahngrund
tooth pitch Zahnteilung
tooth space Zahnlücke
top Oberteil, Oberfläche, Oberseite
top coming cutter rechtsdrehender Fräser
top edge Oberkante
top face Spanfläche (Drehmeissel)
top going cutter linksdrehender Fräser
top land Fase an Umfangsschneide (Fräser)
topping cutter Gewindefräser (der auch Aussendurchmesser fräst)
top position oberste Stellung
top rail Querhaupt, Traverse
toroidal ringförmig
torque Drehmoment
total head Gesamthöhe, gesamte Energie
total length Gesamtlänge

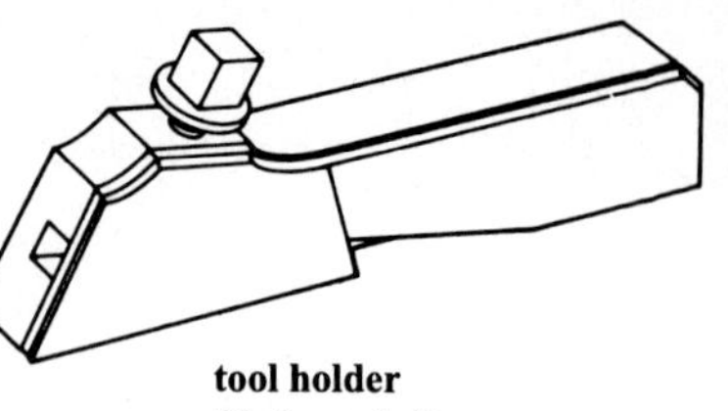

tool holder
Werkzeughalter

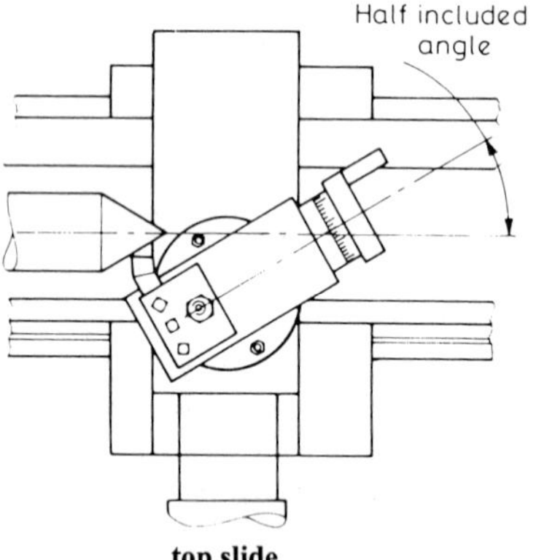

top slide
Oberschlitten

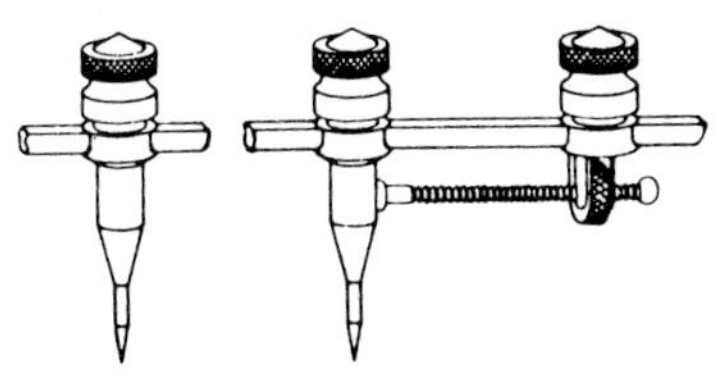

trammel
Stangenzirkel

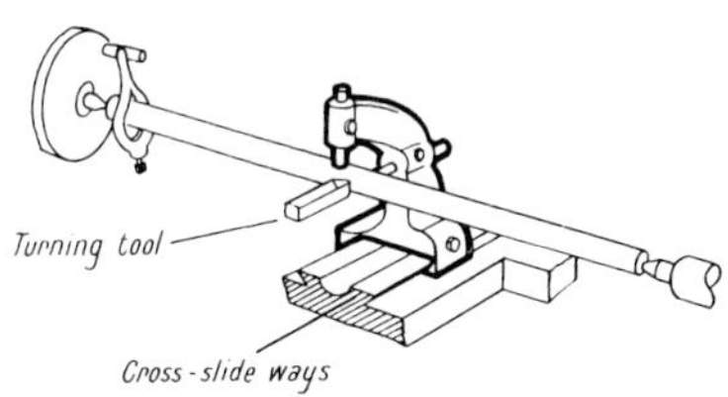

travelling steady
mitlaufender Setzstock

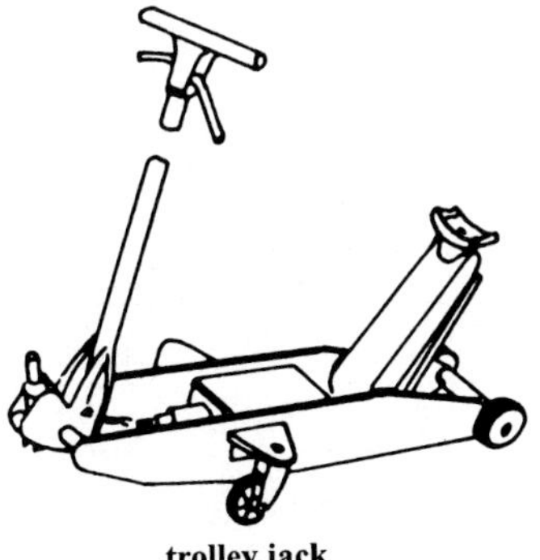

trolley jack
Wagenheber

total table travel gesamter Fräsweg
tote box Förderkasten, Transportkasten
tough zäh
toughness Zähigkeit
towards the right nach rechts
towards the left nach links
trace, to abtasten, der Form folgen
tracer Fühler, Taststift
tracer controlled fühlergesteuert
tracer finger Taststift
tracer-mill, to fräsen (nachformen) mit Fühlersteuerung
tracer pin Taststift
tracer roll Fühlerwalze
track Laufbahn, Führungsbahn
track ring Laufring
trade school Berufsschule
train, to ausbilden (Lehrlinge)
trainee machine tool operator Anzulernender für Werkzeugmaschinenbedienung
trammel point Stangenzirkel
transmission Übertragung
transmission ratio Übersetzung, Übersetzungsverhältnis
transmit, to übertragen
transverse quer
transverse dog Anschlagnocke zum Ausschalten des Quervorschubs
transverse planing Querhobeln
travel, to sich bewegen, sich verschieben
travel Bewegung, Verschiebbarkeit
travelling bridge miller Fräsmaschine mit beweglicher Doppelständerbrücke
travelling head beweglicher Spindelstock
traverse, to bewegen, fahren
traverse shaper traversierende Waagrechtstossmaschine
traversing head miller Fräsmaschine mit verstellbarer Spindel
treadle Fusshebel
treat, to behandeln
treatment Behandlung
trial Versuch

trial period Probezeit
trigger Auslöser
trim, to beschneiden, entgraten
trip, to auslösen
trip Anschlag, Knagge
trip dog Anschlag (für die Auslösung)
triplex milling machine Dreispindelfräsmaschine
trip-out Auslösung
tripping mechanism Ausklinkeinrichtung
trouble-free störungsfrei
trough Schale, Behälter, Trog, Spänetrog
true, to zentrieren, ausrichten, abrichten
true massgerecht, rundlaufend, in Waage
true running genauer Rundlauf
true to size massgerecht, massgenau
true within genau auf eine Toleranz, innerhalb
trunnion Zapfen, Drehzapfen
tube Rohr
tubing Rohre, Rohrleitung
tubular rohrförmig, Rohr . . .
tungsten-carbide face milling cutter Wolframkarbid-Messerkopf
tungsten-carbide insert Wolframkarbideinsatz
tungsten-carbide tipped blade wolframkarbidbestücktes Messer
tungsten high-speed steel Wolframschnellstahl
turn, to drehen
turn Umdrehung
turning tool Drehmeissel
turn out, to ausstossen, fertigen
turnstile Drehkreuz
turret Revolver
twin Zwillings . . ., Doppel . . .
twist Verdrehung, Drehung
twist drill Spiralbohrer
twisting force Verdrehungskraft
two-column vertical broaching machine Zweiständer-Senkrechträummaschine
type Typ, Art, Bauart, Ausführung

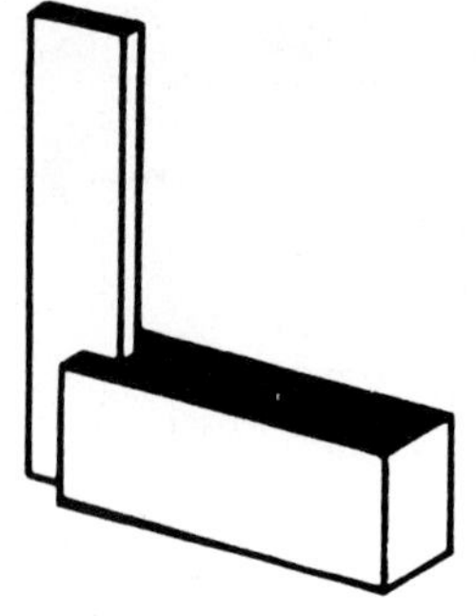

try-square
Anschlagwinkel

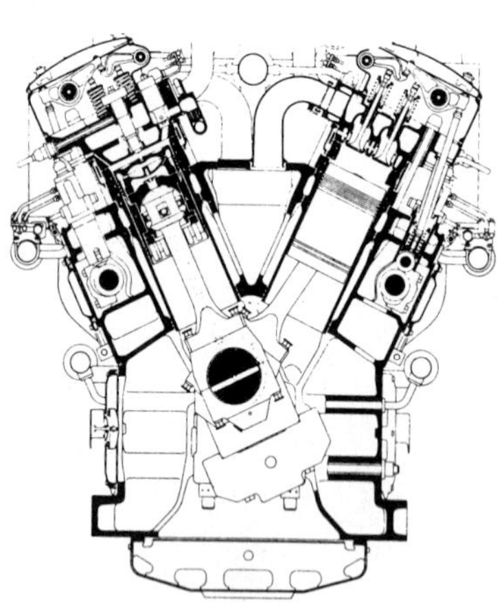

V-type engine
V-Motor (Diesel)

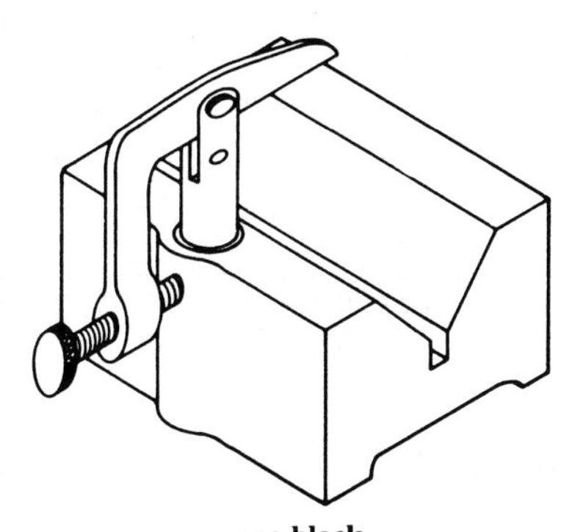

vee-block
Prismenblock,
Prismenaufsatz

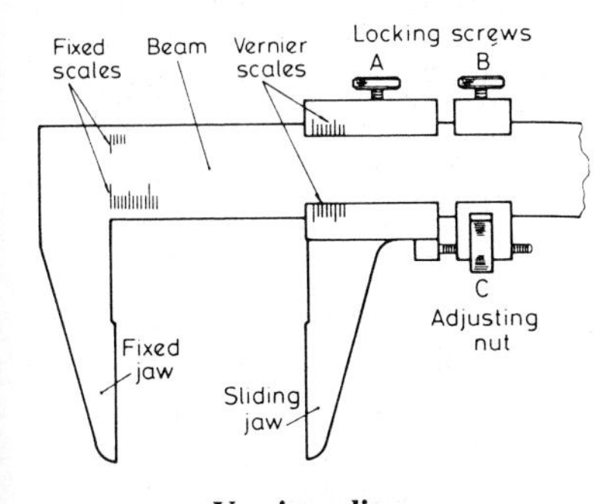

Vernier caliper
Schieblehre

U

ultimate load Grenzlast
ultrasonic testing Überschallprüfung
unbalance Unwucht
unbalanced unausgeglichen, nicht ausgewuchtet
unclamp, to ausspannen
unclamping Ausspannen
unclutch, to entkuppeln
undercut, to unterschneiden
undercut Unterschnitt, Untergriff
undercut angle Spanwinkel
underrate, to zu niedrig auslegen
undersize Untermass
undertighten, to zu locker anziehen
undo, to aufmachen, lösen, losschrauben
undue unzulässig, unangemessen
unequal spacing ungleiches Teilen
unfinished unbearbeitet, nicht fertig bearbeitet, nicht geschlichtet
uniform indexing Durchführung gleicher Teilungen
unilateral einseitig
union Anschlussstück
union nut Überwurfmutter
union socket Verbindungsmuffe
unite, to verbinden, vereinigen
unit production Einzelfertigung
universal chuck Universalfutter
universal head milling machine Fräsmaschine mit schwenkbarem Spindelkopf
universal joint Kardangelenk
universal shaft Gelenkwelle
unlatch, to ausklinken, lösen
unloading Abnehmen des Werkstücks, Abspannen, Entnahme
unlock, to lösen, Festklemmung lösen, entriegeln, ausspannen
unmachinable unzerspanbar
unmachined unbearbeitet
unrelieved nicht hinterschliffen, ohne Freiwinkel
unscrew, to abschrauben, lösen
unskilled ungelernt
unstable unbeständig

unsupported length of column freie Knicklänge
untrue ungenau, unrund
unwearable unverschleissbar
unwieldy sperrig
up-cut milling Gegenlauffräsen
up-cutting Gegenlauffräsen
up-feed method of milling Gegenlauf-Fräsverfahren
up-milling Gegenlauffräsen
upright housing Ständer
usable nutzbar
use, to verwenden, benutzen
use Gebrauch, Verwendung, Nützlichkeit
utilise, to ausnutzen, nutzbar machen

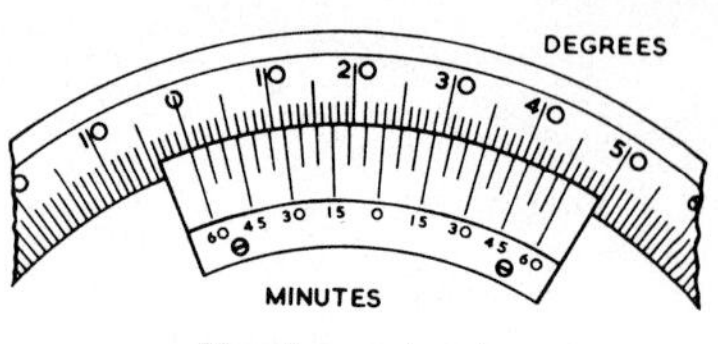

Vernier protractor
Winkelmesser

V

vacuum Vakuum
value Wert
valve Ventil, Schieber
vane Flügel
vane pump Flügelpumpe
variable veränderlich
variable delivery pump Pumpe mit veränderlicher Fördermenge
variation in dimension Abweichung vom Mass
vary, to sich ändern
vee block Prismenstück
vees (Pl.) V-Bahnen, Prismenführung
velocity head Geschwindigkeitshöhe
vent valve Entlüftungsventil
vernier caliper Schublehre
vernier scale Nonienteilung
versatile vielseitig
vertical adjustment Senkrechtverstellung
vertical milling Senkrechtfräsen
vibrate, to vibrieren, schwingen
vibration Vibration, Schwingung
vibration-free schwingungsfrei
vibrationless erschütterungsfrei, schwingungsfrei
vice Schraubstock

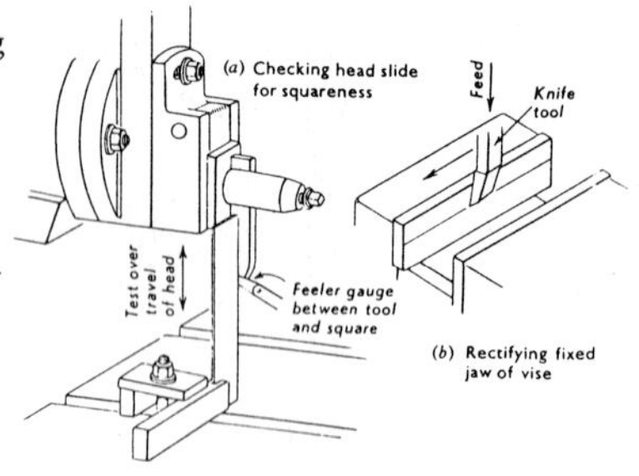

vertical facing
Senkrechtflächenbearbeitung

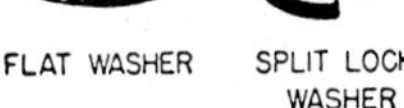

washers
Unterlegscheibe und Federringe

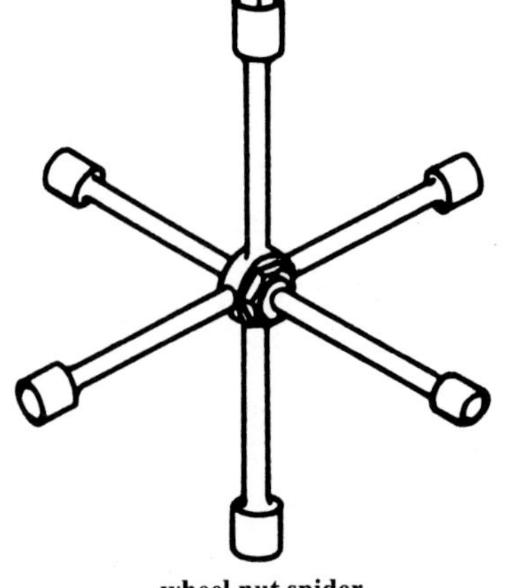

wheel nut spider
Rad(dreh)kreuz

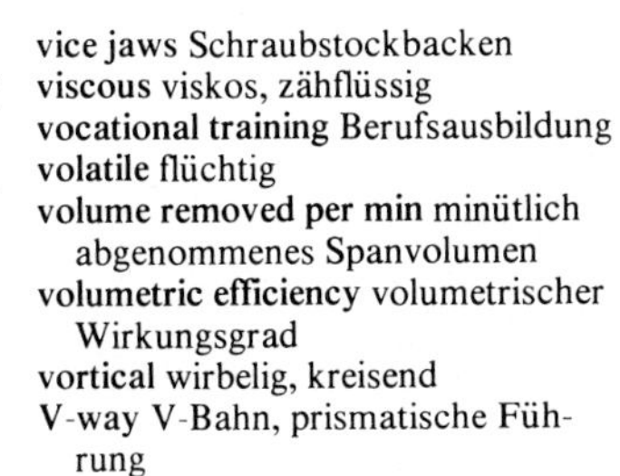

vice jaws Schraubstockbacken
viscous viskos, zähflüssig
vocational training Berufsausbildung
volatile flüchtig
volume removed per min minütlich abgenommenes Spanvolumen
volumetric efficiency volumetrischer Wirkungsgrad
vortical wirbelig, kreisend
V-way V-Bahn, prismatische Führung

W

wabbling disc Taumelscheibe
wake Sog
wall Wand, Wandung
wall of pipe Rohrwand
wall thickness Wandstärke, Wanddicke
warp, to verkrümmen, verziehen
warpage Verkrümmung
wash, to waschen, spülen
washer Unterlegscheibe, Dichtungsscheibe
waste quantity per unit time Verlustmenge je Zeiteinheit
water hammer Wasserschlag
water-in-oil emulsion Emulsion von Wasser in Öl
water repellent wasserabstossend, wasserabweisend
watertight wasserdicht
wavy wellig
waviness Welligkeit
ways (Pl.) Führungen
weaken, to schwächen
wear, to tragen, abtragen, abnutzen, verschleissen
wear Verschleiss, Abnutzung
wear land Verschleissfase
wearless verschleissfrei
wear life Standzeit
wear mark Verschleissmarke
wear out, to verschleissen
wear resistant verschleissfest

wear resistance Verschleissfestigkeit
wear take-up Ausgleich für Verschleiss
web Rippe, Steg, Seele
wedge, to verkeilen
wedge Keil
wedge-shaped keilförmig, kommaförmig (Span)
weight distribution Gewichtsverteilung
weight loading Gewichtsbelastung
weld, to schweissen
weld on, to anschweissen, aufschweissen
welded joint Schweissverbindung
wet, to benetzen
wet nass
wet grinding Nassschleifen, Nassschliff
wetting Benetzung
wide breit, weit
wide finishing Breitschlichten
wide-pitched mit grosser Teilung
wide-range divider Teilkopf für weite Bereiche
width Breite, Weite
width of cut Spanbreite, Schnittbreite
width of land Fasenbreite
wing nut Flügelmutter
withdraw, to zurückziehen
without operator attention wartungsfrei
withstand, to widerstehen, standhalten, aushalten
wobble plate Taumelscheibe
work, to arbeiten, bearbeiten
work Arbeit, Werkstück
work blank Rohling
work cycle Arbeitsablauf
work ejector Werkstückauswerfer
work feed Werkstückvorschub
work handling Transport des Werkstücks
work harden, to kaltverfestigen, sich härten bei der Bearbeitung
workhardening Kaltverfestigung
workholder Aufspannvorrichtung
work horn Aufnahmedorn

worm gear
Schneckengetriebe

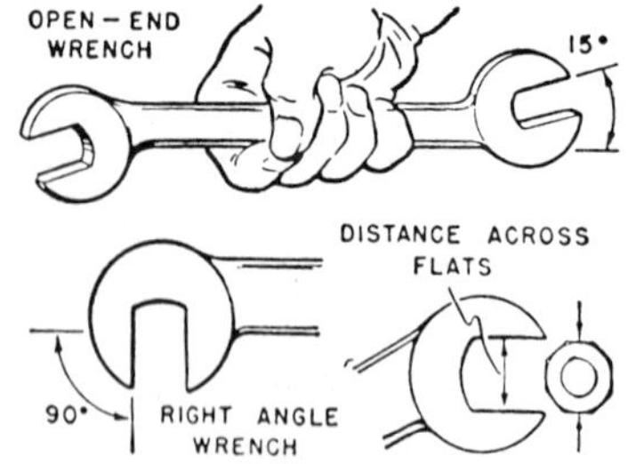

wrenches
Schraubenschlüssel

working edge Hauptschneide
working life Nutzungszeit
work in hand vorzunehmende Arbeit; Werkstück, das in Bearbeitung ist
work layout Anreissen
work light Arbeitsleuchte
workpiece Werkstück
work platen Arbeitstisch
works (Pl.) Werk (Fabrik)
work setting Werkstückaufspannung
worm Schnecke
worm gearing Schneckengetriebe
worn out abgenutzt
wrinkle washer federnde, gewellte Unterlegscheibe
wrist Kurbelzapfen

Y

yield, to abgeben, ausbringen, nachlassen, nachgeben
yield Fertigung, Ausbeute, Ausstoss
yield point Fliessgrenze, Streckgrenze
yoke Traglager, Gabel, Joch

Z

zero, to auf Null stellen

Werkzeugmaschinen -Vokabular

Deutsch - Englisch

A

abändern to alter, to modify, to change
abätzen to etch, to remove by etching
Abätzen etching off, removing by etching
Abbau demounting, dismantling, reduction
abbauen to demount, to dismantle, to reduce, to relieve (stress)
abbeizen to pickle, to cauterise
Abbeizen pickling, cauterising
Abbeizmittel pickling agent
abbiegen to bend off
abblasen to blow off, to bleed, to blast
abblättern to flake off, to peel off, to scale off
Abbrand melting loss by oxidation, smelting loss
abbrechen to break away, to crack away, to break off
Abbrennlängenverlust flash-off
abbrennschweissen to flash-weld
Abbrennschweissen flash-welding
abbrennstumpfschweissen to flash butt-weld
Abbrennstumpfschweissen flash butt-welding
Abbrenn-Stumpfschweissmaschine flash-butt welding machine
abbröckeln to spall, to crumble (cutting edge)
Abbröckeln der Schneide spalling (crumbling) of the cutting edge
Abdeckblech flashing, cover plate
abdecken to cover, to mask
abdichten to seal (off), to pack, to stuff, to make tight
Abdichtmittel sealant
Abdichtung seal, sealing packing
Abdrehdiamant diamond dresser (grinding wheel)
abdrehen to turn off, to shut off, to dress (grinding wheel)
abdrücken to force away, to force off, to push off
Abdrückschraube forcing screw
Abdrückversuch proof test (tubes)
abfahren to follow the outline, to trace, to track
Abfall waste, refuse, discard, loss, scrap (metal), drop (pressure)
Abfallbehälter waste bin
Abfallbeseitigung waste disposal
Abfallprodukt by-product, waste product
abfasen to chamfer, to bevel, to cant
Abfasen chamfering, bevelling, canting
Abfasmeissel chamfering tool
Abfasung chamfer, bevel, canting
abfedern to cushion, to spring
Abfederung cushion, cushioning, spring suspension, springing
Abfläch- und Zentriermaschine face-milling and centring machine
abflächen to spot-face (bores), to end-face
Abflächmaschine für Walzenstirnseiten roll-end milling machine
Abflächmesser spot-facing cutter
Abflachung truncation, flattening
abflanschen to notch
Abflanschen notching
abfliessen to flow off
Abfluss drain, discharge
Abflussrinne spout
Abfrage retrieval
abfragen to retrieve

abfräsen to mill off, to cut off
abfühlen to scan, to sense (punched card), to trace
Abfuhr removal, clearance, disposal (chips), unloading (work), dissipation (heat)
abführen to remove (chips), to dispose of, to unload (work), to conduct away, to dissipate (heat), to lead off, to carry away
Abführen removing, disposing, unloading, conducting away, dissipating, leading off, carrying away
Abführrinne unloading chute (work)
abgeben to deliver, to yield, to dispense, to release, to give off
abgenutzt worn out
Abgleich balance, alignment, gauging, trimming to value
abgleichen to balance, to align, to gauge, to trim to value
abgraten to burr, to deburr, to flash, to snag, to trim
Abgraten burring, deburring, flashing, snagging, flash trimming (die)
Abgratmaschine deburring machine, burr-removing machine
Abgratmatrize trimming die
Abgratmeissel burring chisel
abhacken to chop off, to cut off
abhämmern to peen
Abhämmern peening
abhauen to part off
abheben to lift, to relieve (planer tool), to rise, to raise, to bring clear off, to remove (chip)
Abhebung lifting, relieving (tool), unloading (work), removal (chip)
abhobeln to plane off
abkanten to bevel, to fold, to fold over, to fold on the edge, to brake (sheet metal)
Abkanten bevelling, folding, folding-over, folding on the edge, braking (sheet metal)
Abkantmaschine folding machine
Abkantpresse brake, press brake, folding press
Abkantung fold
abklopfen to tap, to sound out (grinding wheel), to knock off, to scale
abkneifen to pinch off, to nip off
abkratzen to scrape off
abladen to unload
ablagern to deposit
Ablagerung deposit
Ablagetisch finish table (workpiece)
ablängen to cut off to length
Ablängen cutting off to length
ablassen to drain (liquid), to vent (gas)
Ablasshahn drain cock
Ablassventil drain valve
Ablauf drainage, outlet, cycle
Ablauf der Arbeitsgänge sequence of operations, work cycle
ablenken to deflect, to divert, to deviate
Ablenkung deflection, deviation
Ablesefehler reading error
ablesen to read off, to take readings
Ablesung reading, readout
ablösen to strip, to detach, to separate
Abmessungen dimensions, size, data
abmontieren to demount, to detach, to dismantle, to remove
Abnahme acceptance (machine), removal (chip), demounting, unloading (work), reduction, decrease, drop, fall
Abnahmeprotokoll test record, test certificate
Abnahmeprüfung acceptance test
abnehmen to accept (test), to remove (chip), to take off, to demount, to detach, to unload (work), to decrease, to drop, to fall
abnutzbar wearable
abnutzen to wear
Abnutzung wear, abrasion
Abnutzungsausgleich wear compensation
Abnutzungsversuch abrasion test
abplanen to remove metal by facing, to face

abplatten to flatten
Abplattung oblateness, flattening
abplatzen to peel off, to flake off, to spall off
Abprall rebound, ricochet
abprallen to rebound, to ricochet
abputzen to clean, to polish, to wipe
abquetschen to squeeze off, to pinch off
abreissen to tear off, to break down (oil film), to break
Abrichtdorn truer mandrel
abrichten to true, to dress, to redress, to level, to plane
Abrichter abrasive-wheel dresser
Abrichthobel planer
Abrieb abrasion, abrasive wear
abriebbeständig resistant to abrasion
abrollen to uncoil, to unwind, to unreel
abrufen to retrieve, to extract (data), to request
abrunden to round off
abrutschen to slip off
Absatzdrehen shoulder turning
absaugen to suck away
Abschätzung evaluation, estimation, appraisal
abscheiden to precipitate, to separate, to deposit, to settle
Abscheidung precipitation
abscheren to shear off
Abscheren shearing off, shear cutting
Abscherfestigkeit shearing resistance, shearing strength
abscheuern to abrade
abschirmen to screen off, to shield
Abschirmung screening off, shielding
abschlacken to slag off, to remove the slag
Abschlackung slag removal, deslagging
abschmelzbar fusible
abschmelzen to melt off, to fuse, to flash, to consume (electrode)
abschneiden to cut away, to cut off, to chop off
abschrauben to unscrew, to loosen, to screw off, to unbolt
abschrecken to quench, to chill
Abschrecken quenching, chilling
abschreckgehärtet quench-hardened
abschreckhärten to quench-harden
abschreiben to depreciate
Abschreibung depreciation
Abschrot bottom chisel, anvil cutter
abschroten to part off (by means of a chisel)
absenken to lower
absetzen to step, to shoulder (shaft)
absondern to separate, to segregate
absorbieren to absorb
Absperrventil gate valve, shut-off valve, stop valve
abspülen to rinse, to cleanse, to wash
Abstand distance, spacing, space, clearance, interval, pitch (rivets, holes)
abstandsgleich equally spaced
Abstandsring spacer, collar
Abstechdrehmaschine parting-off lathe
abstechen to cut off, to part off
abstellen to stop (machine), to shut down, to cut off, to switch off, to disconnect
abstemmen to chisel off
Abstich tapping (furnace)
abstossen to repel, to repulse, to push off
abstossend repulsive
abstrahlen to emit, to radiate
abstufen to step, to grade
Abstufung stepping, graduation, progression (speeds)
abstützen to support, to back up
Abszisse abscissa
abtasten to scan, to trace, to follow (copying)
Abtasten scanning
Abtaststift stylus, tracer
Abteilung section, department, division
abtragen to erode, to abrade
Abtragung erosion, abrasion

abtrennen to sever, to detach, to separate, to part off, to remove
Abtrennung severance, separation, parting off, removal
Abtrieb output, driven side
abwandeln to modify
Abwärtshub downstroke
abweichen to deviate, to deflect
Abweichung deviation, deflection, error
abweisen to reject
Abweiser rejector
abwürgen to stall (motor)
abzundern to scale off
Abzunderung scaling off
abzweigen to branch
Abstand centre-to-centre distance
Achsbund axle collar
Achse axle (vehicle), axis (math.)
achsenfluchtend axially aligned
Achsenfluchtung axial alignment
Achsenkreuz system of coordinates
achteckig octagonal
Achtkant octagon
Adressenteil address part
addressierbar addressable
Aggregat set, unit, assembly, package
Ahle awl, bodkin, reamer
altern to age, to age-harden (light metal)
Altern ag(e)ing, age-hardening (light metal)
Altöl used oil, waste oil
Amboss anvil
amorph amorphous
Analyse analysis
analysieren to analyse
anbauen to attach, to fit
anbohren to spot-drill, to start a hole
Anbohrer start drill, spotting drill
Anbohrmaschine centring lathe
anbringen to fix, to fit, to attach, to mount
ändern to change, to alter, to modify, to vary
Änderung change, alteration, variation, modification
andrücken an to force against, to press on
aneinanderfügen to join together, to fit together
aneinanderpassend mating (surfaces), fitting together
aneinanderstossen to butt
anfahren to start (up)
anfasen to chamfer, to break the corners
Anfasen chamfering, breaking the corners
anfertigen to make, to manufacture, to produce
Anfertigung manufacture, production, making
anfordern to demand, to require, to order, to claim, to ask for
anfressen to pit, to corrode, to erode
angegossen cast integrally with
angekörnt punch-scribed
angelegt applied (voltage)
angelernt semi-skilled
angeordnet arranged, located, positioned
angeschlossen connected
angezogen von Hand manually tightened
angiessen to cast on
angrenzend adjacent
anhaften to adhere, to stick
anhalten to stop, to bring to rest
anheben to hoist, to elevate, to lift, to raise
anheften to pin
Ankerbolzen anchor bolt, foundation bolt
ankleben to glue on, to paste on, to agglutinate
ankörnen to punch, to centre-punch, to mark
Ankörnung centre mark
Anlage plant, installation, equipment, unit
Anlagelineal contact rule
anlassen to temper (steel), to start
Anlassknopf start button
Anlassmittel tempering medium (steel)

Anlassofen tempering furnace
anlassversprödet temper-brittle
Anlauf start, starting, run-up
Anlerner für Werkzeugmaschinenbedienung trainee machine tool operator
anlöten to solder (to)
anmelden (Patent) to apply for a patent
annullieren to cancel
anordnen to arrange, to place, to group, to lay out, to dispose
Anordnung arrangement, array, disposition, placing, lay-out, mounting, assembly
anpassen to adapt, to adjust, to fit, to match
anpassend adaptive
anpunkten to spot-weld
anquetschen an to crimp to
anreichern to concentrate, to enrich
Anreicherung concentration, enrichment
anreissen to line out, to mark, to scribe (lines)
Anreisser marker, liner, layout man
Anreissnadel scriber
Anreissplatte marking-out plate, marking-off table
Anreissschablone marking stencil
Anreisswerkzeug marking-off tool, layout tool
ansammeln to accumulate
Ansammlung accumulation
Ansatzstück lateral fitting, attached piece
ansaugen to prime (pump), to draw in
Ansaugen priming (pump), suction, drawing-in, taking-in
Ansaugkanal suction port
anschärfen to scarf, to sharpen, to resharpen
Anschlag stop, trip dog, dog
anschlaggesteuert dog-controlled
Anschlagknagge stop dog
Anschlagleiste stop bar
Anschlagwalze stop roll
Anschluss connection, junction
Anschlussrohr joining pipe
anschmieden to forge to (or: on)
Anschnitt starting cut, start of the cut
anschrauben to bolt to
ansenken to spotface
Ansenken spotfacing
Ansicht view, elevation (drawing)
Anspitzen pointing
Ansprechbereich range of sensitivity
ansprechen to respond, to react, to function
anstauchen to upset (metal forming), to upset-forge
Anstauchen upsetting, upset-forging
ansteigen to rise, to increase
Anstich initial pass, first pass
Anstieg rise, increase
anstiften to peg
Anstiften pegging
anstossen to butt, to abut
anstreichen to paint, to paint-coat, to coat with paint
Anstreichen painting, coating with paint, paint-coating, brushing
Antrieb drive
Antriebselement driving member
Antriebsritzel driving pinion
anwenden to apply, to use, to employ
Anwender user
Anwendung application, use
anzapfen to tap, to bleed (fluid)
anzeichnen to mark, to sign, to scribe
Anzeige indication, readout, reading
Anzeigelampe tell-tale light, indicating light
anzeigen to indicate, to display, to read
anziehen to tighten (nut), to attract (physics), to fasten, to snug
Apparat device, apparatus, appliance, unit, set
Arbeit work, job, task, operation
arbeiten to work, to operate
Arbeitsablauf machining cycle, sequence of operations
Arbeitsaufwand expenditure of work
Arbeitsbeispiel case study, typical application

Arbeitsfolge sequence of operations, operational sequence
Arbeitsgang operation, process of manufacture, procedure, cycle
Arbeitsgebiet field of activity
Arbeitsgruppe working party
Arbeitsplattform working platform
Arbeitsplatz site of work, work site, workplace
Arbeitsprinzip operating principle
Arbeitsradius coverage
Arbeitsspiel work cycle
Arbeitsstück workpiece
Arbeitsstudie work study
Arbeitstisch work-table, machine table
Arbeitsverfahren working method, manufacturing process
Arbeitsversuch trial operation
Arbeitsweise operating principle, technique, operation, performance, procedure
Arm arm, bracket, lever, support
Armatur instrument, armature, fitting
arretieren to arrest, to block, to impede, to lock
Arretierung arrest, detent, blocking, locking
Asbest asbestos
Asche ash, cinders
asynchron asynchronous
Atemschutz respiratory protection
ätzen to etch
Ätzen etching
ätzend etching, caustic, corroding, mordant
Ätzlauge caustic lye
Aufbau einer Maschine construction of a machine, assembly of a machine
aufbauen to erect, to build up, to construct, to assemble, to set up
aufbereiten to dress, to purify, to treat, to make up, to reclaim
aufbocken to jack up
aufbohren to bore
Aufbohren tiefer Bohrungen deep-hole boring
aufbüchsen to bush on
aufdecken to uncover
aufdrehen to screw open
Auflagedruck support pressure
Auflagefläche locating surface, bearing surface, seating surface
Auflager support, bearing
Aufnahmedorn mounting mandrel
Aufprall bouncing, impact, impingement
aufprallen to bounce on, to impinge
aufpunkten to spot-weld
aufrauhen to roughen
aufrecht upright
aufreiben to ream
aufrichten to erect
Aufriss front view (elevation)
aufsaugen to absorb, to soak
aufschichten to pile up, to stack
aufschrauben to bolt on, to screw on
aufschrumpfen to shrink on
aufschweissen to weld on, to attach by welding
Aufschweissen welding-on, attaching by welding
aufsetzen to put on, to place on, to mount on, to load
Aufspannblock planer jack
Aufspanndorn work arbor, mouting mandrel
aufspannen to clamp, to mount, to fix, to chuck
Aufspannen clamping, mounting, fixing, chucking
Aufspanntisch clamping table
Aufspannwerkzeug holding tool, clamping tool
Aufspannwinkel angle plate
Aufspannzeit clamping time, loading time
Aufsprühmethode spray-up technique
aufspulen to reel up, to wind up
aufstapeln to stack, to pile up
Aufstecken fitting-on, slipping over
aufstecken auf to fit on, to slip over
Aufsteckfräser arbor cutter, hole-type cutter

Aufsteckgewindebohrer shell tap
Aufstellung installation, erection, assembly, setting-up, building-up
Aufstellungsfläche floor space (machine)
Aufstieg rise, ascent, climb
Auftragsschweisselektrode hard-facing electrode
auftragsschweissen to weld-face, to weld-surface, to pad-weld, to deposit-weld
Auftragsschweissung weld facing, hard facing, pad-welding, surfacing by welding
Auftragswalze spreader roll
Auftragung deposition, coating, padding
Auf- und Abbewegung vertical reciprocating motion, raising and lowering, rise and fall
aufweiten to widen, to expand, to enlarge, to bulge
aufwickeln to wind up
aufzeichnen to record, to plot
Aufzugsfeder power spring
«Aus»-Stellung off-position
ausbauen to dismantle, to demount
ausbessern to repair, to mend, to recondition, to patch
ausbeulen to bulge, to buckle, to planish (sheet forming)
ausbilden to train, to instruct
Ausbilder instructor, apprentice trainer
Ausbildung training instruction
Ausbohr- und Stirndrehmaschine boring and facing mill
ausbrechen to tear loose, to break away, to break out, to chip (cutting edge)
ausbröckeln to crumble (cutting edge)
ausdehnen to expand, to stretch, to extend, to dilate
ausdrehen to recess by turning, to turn inside diameters, to turn with single-point tool
Ausdrehmeissel internal turning tool, internal boring tool
Ausdrückplatte ejector plate
Ausfall loss, deficiency, failure, stoppage
ausfugen to groove
Ausfugen grooving
ausführen to perform, to carry out, to accomplish, to practice
ausfüttern to line, to bush, to pad
Ausfüttern lining, bushing, padding
Ausgaberinne unloading passage, chute, channel
Ausgangsmetall parent metal
ausgebrochen torn loose (abrasive)
Ausgleich compensation, counterbalance, equalisation, correction
ausgleichen to compensate, to counterbalance, to equalise, to correct
Auslegerbohrmaschine radial drilling machine
Auslegerkran jib crane
auslösen to release, to disengage, to trip
ausschalten to switch off, to disconnect, to disengage, to stop
Ausschaltknopf stop button
Ausschuss scrap, scrap work, waste, rejects
Ausschussquote reject rate, scrap percentage
Aussendurchmesser outside diameter, outer diameter, major diameter (thread)
Aussengewinde external thread, male thread
Aussentaster outside caliper(s)
aussenverzahnt with external teeth, externally toothed
ausserachsig eccentric
aussermittig drehen eccentric turning, turning off centre
aussetzen to interrupt, to intermit, to fail
aussortieren to sort, to select
ausspannen to unclamp, to dechuck, to release
auswechseln to replace, to interchange, to substitute

auswerfen to eject
Auswerfer ejector
Automatenstahl free cutting steel, automatic steel
Azetylen acetylene
Azetylenflasche acetylene cylinder
Azetylenschweissen acetylene welding

B

Backe jaw (chuck, vice), block, die (upsetting, die-head)
bahngesteuert continuous-path controlled
Bahnsteuerung continuous-path control
Balken beam
Ballenpresse baling press, baler
Bandförderer belt conveyor, band conveyor
bandgesteuert tape-controlled
Bandpoliermaschine band polishing machine
Bauelement component, construction element
bauen to build, to construct
Bauform design, structural form
Baukasteneinheit standard unit, module, building block
Baustahl mild steel
beanspruchen to load, to stress, to impose stress (or: load) on
bearbeitbar machinable, workable
bearbeiten to machine, to work
Bearbeitungsfolge machining sequence, sequence of operations
beaufschlagen to admit
Beaufschlagung admission
bedienen to operate, to attend, to control, to handle
Bedienungspult operating desk
Bedienung operation, attendance, operator attention
Bedienungs- und Wartungshandbuch operating and maintenance manual
Bedienungsanweisung operating instructions, operator's manual
Bedienungspersonal operators
beeinflussen to affect, to influence
beeinträchtigen to impair, to interfere with
Befehl instruction, command, order
befestigen an to attach to, to fasten to, to fix to, to secure to, to mount to
Befestigung attachment, fastening, fixing, securing, mounting
befeuchten to dampen, to humidify, to moisten
Befeuchtung dampening, humidification, moistening
befördern to convey, to transport, to handle, to carry
Beförderung conveyance, conveying, transport, handling
begrenzen to limit, to bound
begrenzen, den Weg to limit the travel
behalten to retain, to maintain
Behälter container, bin, case, receptacle, receiver, tank
behandeln to treat, to process, to handle
Behandlung treatment, processing, handling
beharren to persist, to remain, to stay
Beharrungsvermögen inertia
Behelfs . . . auxiliary, temporary, makeshift
behindern to interfere with, to obstruct
beibehalten to maintain, to retain
beidseitig on both sides, both-way
Beilage supporting block, shim
Beilegering packing
Beilegescheibe shim
beistellen, von Hand to feed by hand
Beistellung feed
Belag coating, covering, lining, facing, plating
belastbar loadable
Belastbarkeit loadability, load-carrying capacity
belasten to load, to stress

Belastung load, stressing
Belastungsstoss shock load
Belastungsverlauf load cycle
bemassen to dimension
Bemassung dimensioning
benachbart adjacent, adjoining, neighbouring
benennen to denominate, to term, to designate
Benennung denomination, designation
benetzen to wet, to moisten
benutzen to use, to employ, to utilise
Benzin petrol, gasoline (am.)
beranden to rim, to bound
beraten to advise, to consult
berechnen to compute, to calculate
Berechnung computation, calculation
Bereich range, field, scope
berichtigen to correct, to rectify, to adjust, to set right
Berichtigung correction, rectification, adjustment
berieseln to sprinkle, to spray, to wet
Berieselung sprinkling, spraying, wetting, flooding (coolant)
Berufsausbildung vocational (craft) training
Berufsschule apprentice training school
berühren to touch
Berührung contact, tangency
berussen to soot
besäumen to edge, to trim, to square up
beschädigen to damage, to mar (surface)
Beschädigung damage, defect
beschaffen to supply, to provide
beschäftigen to employ, to occupy
beschichten to coat, to face (with)
Beschichtung coating, facing
beschleunigen to accelerate
Beschleunigung acceleration
beschneiden to trim (sheet metal)
beschränken to limit, to restrict, to confine
beschriften to mark (tool), to letter (drawing)
Beschriftung marking (tool), lettering (drawing)
beschweren to weight
beseitigen to dispose of, to remove, to eliminate
Beseitigung disposal, removal, elimination
besetzen to occupy, to man
Besetzung occupation, manning (operators)
Bessemerstahl acid bessemer steel
beständig constant, continuous, stable, resistant, durable
Bestandteil component, constituent, ingredient
bestimmen to determine, to analyse, to specify
bestücken to face, to tip (tool)
betätigen to actuate, to operate
Betätigung actuating, actuation, operating, operation
Betätigungsorgan actuating element, operating element
Betrag amount, rate
Betrieb factory, mill (steel, rolling), workshop
Betriebsanleitung operating instructions, instruction to operators
Betriebsart method of operation, mode of operation
betriebsbereit ready for service, ready for operation
Betriebsdaten operating data, ratings, performance
betriebsfähig in good working order, ready for service
betriebssicher safe in operation, reliable, reliable in service, dependable in service
Betriebssicherheit operating safety, reliability, reliability in operation
Betriebsstörung breakdown, stoppage, failure
Betriebsüberwachung production control, operation control

Betriebsunfall factory accident
Bett bed (machine)
bewegen to move, to agitate, to stir
Bewegung movement, motion, travel (slide), traverse
bewerkstelligen to accomplish
bewirken to cause, to effect
bezeichnen to mark, to denote, to designate, to specify
Bezeichnung marking, designation, specification
Bezugsachse reference axis
Bezugsgrösse reference magnitude
Biege- und Richtmaschine bending and straightening machine
Biegebeanspruchung bending load
Biegefestigkeit bending strength
biegen to bend
Biegepresse bending press
biegsam flexible, bendable
Biegung bending, flexure
Bild image, figure, pattern, graph
bilden to form, to shape
Bildung formation, generation, production
binär binary
Bindemittel binding agent, bonding material (grinding wheel)
binden to bind, to tie, to bond, to cement, to fix
Blasebalg smith's bellows
blasenfrei non-porous, without blisters, without flaws
Blatt blade (tool), leaf (spring)
Blättchen lamina, foil
blauglühen to open-anneal
Blauglühung open-annealing
Blech plate, sheet, sheet metal, metal sheeting
Blei lead
bleiben to remain, to stay, to be permanent
Bleibronze lead bronze
Blende diaphragm, slit, aperture (optics)
Block-Abstechdrehmaschine ingot slicing lathe
Blockbild block diagram
blockieren to lock, to jam, to interlock (machine)
Blockierung locking, jamming, interlocking (machine)
Bockkran gantry crane
Boden bottom, ground, floor, earth
Bodenabstand ground clearance
Bogen arc (math.) bend, bow, curve, curvature
Bohr- und Fräswerk horizontal boring and milling machine
Bohr- und Plandrehmaschine boring and facing lathe
Bohrausleger radial drill arm
Bohrautomat automatic drilling machine (or: boring machine)
bohren to drill (from the solid), to bore (enlarging an existing hole)
Bohren drilling, boring
Bohrer drill, drilling (or: boring machine operator)
Bohrer- und Gewindebohrer-Schärfmaschine drill and tap grinder
Bohreranschliff drill grind
Bohrerausspitzen drill pointing
Bohrerstandzeit drill life
Bohrfutter drill chuck
Bohrknarre ratchet brace
Bohrkopf drill (or: boring) head
Bohrmaschine drilling machine, boring machine
Bohrmeissel boring tool
Bohrpinole boring sleeve
Bohrschneide bit
Bohrstange boring bar
Bohrtisch drilling machine table, worktable of drilling machine
Bohrung bore hole, hole
Bohrwerk boring mill
Bohrwinde breast drill
Bolzen bolt
Bolzenanschweisspistole stud welding gun
Bolzenaufschweissen stud welding
Bolzenschere bolt cutter
Bor boron
Borax borax

Bördeleisen bordering tool
Bördelmaschine flanging (or: bordering, clinching) machine
bördeln to flange, to bead, to border, to clinch
Borkarbid boron carbide
brauchbar suitable, applicable, serviceable
Brecheisen crowbar
brechen to break, to fracture, to crush, to crack
Brechstange jimmy, jim-crow
Breite width
breitschlichten to finish with broad cut
Bremse brake
bremsen to brake, to slow down
brennbar combustible, inflammable
brennen to burn, to calcine
Brenner blowpipe, burner, torch
brennschneiden to flame-cut, to torch-cut, to gas-cut
Brennschneiden flame cutting, torch cutting, gas cutting
Brennschneider flame cutting torch
Brennstoff fuel
Brinellhärte Brinell hardness
Bronze bronze
Bruch breakage, fracture, failure, rupture
Bruchspan discontinuous chip, segmental chip
brünieren to burnish, to bronze, to blue, to brown
Brustbohrmaschine breast drill
Buchse bush, bushing, collar, sleeve
buckelschweissen to projection-weld
Buckelschweissen projection welding
Bügelsäge hacksaw
Bund collar, flange, shoulder (shaft)
bündig flush
Buntmetall non-ferrous metal
Bürste brush
bürsten to brush
Butan(gas) butan

C

C-Stahl carbon steel
Chrom chromium
chromieren to chromise

D

dämpfen to absorb, to damp, to muffle, to attenuate (electricity)
Dämpfer absorber, damper, muffler, cushion
Daten data
Datenspeicher data storage unit
Datenverarbeitung data processing
Daube stave
Dauer duration, period
Dauerbetrieb continuous duty, continuous operation
dauernd, Dauer ... permanent, continuous, constant, sustained
Deckel cover, lid, cap
Deckplatte cover plate
defekt defect, flaw, fault
deformieren to deform
dehnen to strain, to expand, to extend, to stretch
Dehnung strain, expansion, extension, stretching
Demontage dismantling, disassembling, dismounting, demounting
demontieren to dismantle, to dismount, to disassemble, to demount
Diagonalschaben diagonal shaving
Diagramm diagram, chart, graph, plot
Diamant diamond
Diamantabrichten diamond truing
Diamantabrichter diamond wheel dresser
diamantbestückt diamond-tipped
Diamantbohrkrone diamond drill bit
diamanthart adamantine
Diamantschneide diamond edge

Diamantspitze diamond point, diamond tip
Diamantsplitter diamond particle
dicht tight, dense, compact, close, proof
dichten to seal, to pack
Dichtung seal, sealing, packing
Dicke thickness
dickflüssig viscous, consistent
dickwandig thick-walled, heavy-walled
dienen to serve
Dieselaggregat diesel-driven generating set
Dieselmotor diesel engine
dimensionieren to determine safe dimensions, to size, to proportion
dispergieren to disperse
Dispersion dispersion
Distanzbüchse spacer bushing
Distanzstück spacing piece, distance piece, spacer
Docht wick
Dochtschmierung wick-feed lubrication
doppeln to double, to fold, to duplicate
doppelseitig bilateral, double-ended
Doppelspindelmaschine duplex machine, double-spindle machine
Dorn arbor, mandrel
dosieren to proportion, to dose, to batch, to measure out
Draht wire
Drahtstrasse wire mill
Drahtwalzwerk wire-rod mill
Drahtziehmaschine wire drawing machine
drallförmig helical
Drallnut helical flute, helical groove
Drallnutenräumen helical broaching
Drallrichtung hand of helix
Drallsteigung lead of helix
Draufsicht top view, plan view
Drehachse rotation axis, axis of rotation
Dreharbeit turning operation, lathe work
Drehausleger hinged cantilever
Drehautomat automatic lathe, automatic screw machine, automatic
Drehbank lathe, turning machine
Drehbohren rotary drilling
Drehbohrer rotary drill
Drehdorn lathe mandrel, lathe arbor
drehen to rotate, to swivel, to revolve, to swing, to turn
Dreher lathe operator, metal turner
Drehherz lathe dog, driving dog
Drehkreuz turnstile, capstan
Drehling tool bit, insert tool
Drehmaschine lathe, turning machine
Drehmaschine in Tischausführung bench lathe
Drehmaschinenleitspindel lathe leadscrew
Drehmaschinenreitstock lathe tailstock
Drehmaschinenschlossplatte lathe apron
Drehmaschinensupport lathe rest, lathe steady, lathe carriage
Drehmeissel turning tool, lathe tool, lathe cutting tool
Drehmeisselstandzeit turning tool life
Drehmoment torque
Drehmomentschlüssel torque spanner
Drehrichtung direction of rotation, hand of rotation
Drehtiefe turning tool mark
Drehschablone strickle tackle
Drehspindelkasten lathe headstock
Drehstab torsion bar
Drehtisch rotary table, revolving table
Drehung rotation, revolution, turning
Drehvollautomat fully-automatic lathe
Drehzahl speed, rotational speed, rpm (revolutions per minute)
Drehzahlstufe speed increment
Drehzahlverhältnis speed ratio
Drehzapfen swivel pin, central pivot, trunnion
Dreibackenfutter three-jaw chuck

Dreibein tripod
Dreieck triangle
dreieckig triangular
Dreikantfeile triangular file, three-cornered file
Dreikantschaber triangular scraper
dreireihig three-row, triple-row
Dreischichtbetrieb three-shift working (or: operation)
dreiseitig trilateral
dreistufig three-step, three-stage
Dreiwegehahn three-way cock
drosseln to choke, to restrict, to throttle
Druck pressure, compression, thrust, head
Druckbehälter pressure vessel, pressure tank
druckbetätigt pressure-operated
druckdicht pressure-tight
drücken to press, to force, to depress, to push
druckgeschmiert pressure-lubricated
druckgesteuert pressure-controlled, pressure-responsive
druckgiessen to pressure die-cast
Druckgiessen pressure die casting
Druckknopf push-button
druckknopfbetätigt push-button-operated, push-button-actuated
Drucklager thrust bearing, spigot
Druckluft compressed air
druckluftbetätigt pneumatically operated, air-operated, air-actuated
Druckluft-Kleinbohrmaschine air-power drill
druckschmieren to force-lubricate
Dübel dowel, peg, plug
dübeln to dowel, to peg
dünn thin
Dünnblech thin sheet metal
dünnwandig thin-walled
Duowalzwerk duo mill, two-high mill
durchbiegen to deflect, to sag
durchbohren to through-drill
durchbrechen to pierce, to break through
durchbrennen to burn out, to blow (fuse)
durchdringen to penetrate, to permeate, to intersect (engineering drawing)
durchfahren im Eilgang to rapid-traverse
durchfliessen to flow through, to pass (current)
durchführbar feasible, practicable
Durchführbarkeit feasibility, practicability
durchführen to carry out, to perform, to accomplish, to execute
Durchgang passage, pass (workpiece)
Durchhang sag, dip
durchhängen to sag, to dip
durchkonstruiert, gut well-designed
Durchlass passage
durchlassen to admit, to pass, to transmit
durchlaufen to pass through, to travel through, to flow through
durchlochen to punch, to pierce
Durchmesser diameter
durchqueren to cross
Durchschlag puncture, breakdown, rupture (insulation), leakage (liquid), punch (tool)
durchschlagen to puncture, to break down, to punch
durchschleifen to cut off (by abrasive cutting)
Durchschleifen abrasive cutting
Durchschnitt average, mean, intersection
durchspülen to flush
durchstossen to push through, to punch, to pierce
durchweichen to soak
Düse nozzle, orifice, tip

E

eben flat, even, smooth, plane
Ebene plane

Ebenheit flatness, evenness, smoothness
ebnen to flatten, to smooth, to planish
Ecke corner
eckig cornered, angular
Eckstoss corner joint
Edelgas-Lichtbogenschweissen inert-gas arc welding
Edelmetall precious metal, noble metal
Edelstahl high-qualitiy steel, special steel
Eichblock calibration block
eichen to calibrate, to gauge, to measure
Eichung calibration
Eigenfrequenz naturel frequency
Eigenschaft property
eignen, sich to suit, to be suited
Eilgang rapid traverse, quick traverse
Eilrücklauf rapid return, quick return
einbauen to install, to build in, to insert, to mount, to assemble, to place, to fit in, to incorporate, to include
einbetonieren to embed in concrete
einbetten to embed
einbohren, Ölkanäle to drill oil holes
einbrennen to burn in
eindrehen to turn a neck (or: flute, groove) into
einengen to narrow, to restrict, to neck down, to reduce
Einengung narrowing, restriction, necking down, reduction
Einfachfräsmaschine plain milling machine
einfassen to line, to set, to border
einfügen to fit in, to insert
einführen to introduce, to feed into, to let in, to insert
einfüllen to fill in, to pour in, to feed, to charge
Eingabe input, feeding, delivery, filling-in
Eingabedaten input data
Eingabespeicher input memory, input storage
Eingriffsstelle meshing point, point of engagement
einkehlen to channel
einkerben to notch, to indent, to score, to nick
einkuppeln to engage the clutch, to throw in
Einlass intake, inlet, admission
einordnen to classify
Einordnung classification
einpassen to fit in
Einpassung fitting-in
einrasten to engage, to catch in, to click in
einrichten to adjust, to set, to adapt, to position, to tool up, to arrange
einrücken to throw in, to start, to slide into
Einsatz application, utilisation, use, employment, service, operation, insert (tool)
einsatzbereit ready for operation, ready-to-use
einsatzgehärtet case-hardened
Einsatzmeissel insert tool, bit-type insert
einschalten to switch on, to start
einschneidig single-edged, single-point, single-ended
einschränken to restrict, to limit, to restrain, to reduce
einseitig one-sided, unilateral, single-ended
einsetzen to insert, to install, to charge (ingot), to case-harden
einspannen to clamp, to chuck, to hold
Einspeisung feed, supply
Einspindelmaschine single-spindle machine
Einspritzdüse injection nozzle
einspritzen to inject
Einständerbauart single-column construction (or: design)
einstechen to plunge-cut, to recess, to neck, to groove
Einstechmeissel recessing tool
Einsteckschlüssel face spanner

einstellbar adjustable
einstellen to adjust, to set, to position, to locate
Einstellschraube adjusting screw, setscrew
Einstellung adjustment, setting, positioning, location
eintauchen to immerse, to dip
einteilen to divide, to subdivide, to grade
Einteilung division, subdivision, grading
Eintritt intake, entrance, inlet
einwirken auf to act upon, to affect
Einzelstückfertigung one-off production
Einzelteil one-off part individual component (or: part, piece)
einziehen to retract, to pull in, to draw in
Eisen iron
Eisenschrott iron scrap, ferrous scrap
elastisch elastic, flexible, resilient, springy
Elektrofeile rotary file
Elektromagnet electromagnet
elliptisch elliptic
eloxieren to anodise
Endanschlag limit stop, end stop
Endausschalter limit switch
Endform final shape
Endkontrolle final inspection
Endmass end measure
eng narrow, close (tolerance), closely spaced
entdecken to discover
entfernen to remove, to clear (chips)
Entfernung removal, clearing, distance, range
entgegenwirken to counteract
entgraten to deburr, to burr
entladen to unload, to discharge
Entladung unloading, discharging
entlangfahren an to trace (copying)
entlasten to unload, to relieve of load
Entlastung unloading, load relieving
entleeren to empty, to drain (liquid)
entlüften to vent, to ventilate, to deaerate
Entlüfter ventilator, deaerator, air exhauster
entriegeln to unlock, to unclamp
entrosten to remove the rust, to derust, to unrust
Entroster rust remover
entstören to clear faults
entwerfen to design
entwickeln to develop
entwirren to disentangle, to untwist
entzinnen to detin, to untin
Entzundern descaling, scaling-off, removal of scale
Erdung earthing, grounding
erfinden to invent
erfüllen to meet, to satisfy, to fulfil
ergänzen to complement, to supplement, to complete, to replenish
ergreifen to grip
erhalten to receive, to obtain, to maintain
erhitzen to heat up
erhöhen to raise, to increase, to elevate
Erhöhung raise, increase, elevation
erholen to recover
erkalten to cool down, to chill
erleichtern to facilitate
ermitteln to determine, to detect, to find out
Ermittlung determination, detection
Ermüdung fatigue
erneuern to renew, to recondition, to renovate
Erneuerung renewal, reconditioning, renovation
erodieren to erode
erproben to try, to test
errichten to erect, to set up, to build
Ersatz spare, replacement
Ersatzteil spare part
Ersatzteillager spare part stock, replacement stock
Ersatzteilliste list of spare parts
erschütterungsfrei vibrationfree, vibrationless, smooth

ersetzen to replace, to exchange
erwärmen to heat up, to warm up
Erwärmung heating, generation of heat
erweitern to widen, to enlarge, to expand, to extend
Erweiterung widening, enlargement, expansion, extension
erwerben to acquire, to obtain, to gain
Erz ore
erzeugen to generate, to manufacture, to produce, to make
erzielen to obtain, to attain
Evolvente involute, involute curve
Extremwert extreme value, peak value
Exzenter eccentric cam, eccentric
exzentrisch eccentric, off-centre

F

Fabrik factory
fabrizieren to produce, to manufacture
Fach field, subject, branch, line
Facharbeiter skilled worker, craftsman
Fachgebiet subject field
fahrbar mobile, movable
fahren to travel, to traverse, to move, to run, to ride, to drive
Fahrzeug vehicle
fallen to fall, to drop, to decrease
Falte wrinkle
Falz seam, hem, flange, welt (sheet metal working)
Farbe colour, paint
Farbspritzanlage spray-painting unit
Fase chamfer, land, margin, bevel, bevelling
fasen to chamfer
Fasen chamfering
Fassondrehen form turning
Faustregel rule of thumb
Feder spring, key (in slot), feather
federbelastet spring-loaded
fehlen to lack, to be absent
Fehler error, fault, defect, failure, flaw (material)
Fehlerbestimmung trouble shooting, determination of errors, localisation of faults
fehlerfrei error-free, without error, faultless, flawless
fehlerhaft defective, faulty
Feile file
feilen to file
Feilen filing
Feilenheft file handle
fein fine, minute, pure
Feinbearbeitung finishing, metal finishing
Feinbohren fine boring, diamond boring, fine-hole boring
Feineinstellung fine adjustment, fine setting, inching
Feingewinde fine thread, fine-pitch thread
feinstufig fine-step, in fine steps
Feinvorschub fine feed
Fernbedienung remote control
fertigen to manufacture, to produce
fertiggestellt completed
Fertigmass finish size, finished size, finishing dimension
Fertigung manufacture, production, yield, output
Fertigungstechnik production engineering
fest solid, strong, fixed, stationary, stable
festbinden to tie
festdrehen to tighten (nut)
festfahren to stick, to get stuck, to become jammed, to stall out (spindle)
festgefahren stuck (tool), jammed
festhalten to hold in place, to arrest, to retain
festigen to strengthen, to compact, to stabilise
Festigkeit strength (material), compactness, stability, solidity, firmness

Festigkeitslehre strength theory, strength of materials, theory of the strength of materials
festklemmen to clamp in place, to clamp in position, to lock, to jam, to stick
festmachen to fix, to fasten
festschrauben to bolt down, to attach by bolts (or: screws)
festschweissen to attach by welding
Fett grease
Fettbüchse grease cup, greaser
Fettschmierung grease lubrication
feucht humid, damp, moist
Feuchte humidity, moisture
feuchten to wet, to moisten, to damp
feuerfest fire-proof, fire-resistant, refractory
Filzscheibe felt wheel
Fingerfräser slotting cutter, slot drill
Finne peen (hammer)
Fläche area, surface, face
flächen to spot-face
Flächen spot-facing
Flächenfräsen surface milling
Flächenschleifen surface grinding
Flachgewinde square thread, flat thread
flammspritzen to flame-spray
Flanke flank, side (thread)
Flansch flange
Flasche bottle, flask
Flaschenzug pulley block, set of pulleys, block and tackle
Fleck spot, stain, blotch (oil)
fleckenfrei stainless
Fliehkraft centrifugal force
Fliessarbeit flow line production
Fliessband flow line
fliessen to flow
Fliessspan flow chip, continuous chip
Flucht alignment
fluchten to be aligned, to be in line, to align
Fluchten aligning, lining-up, alignment
Flügelmutter wing nut, butterfly nut, thumb nut
Flügelschraube thumb screw
Fluss flow, flux (electricity)
Flussstahl mild steel, low-carbon steel, ingot steel
Folge sequence, succession, progression
folgen to follow
Folgeschaltung sequence control
Folie foil, sheet
Förderband conveying belt, conveyor
fördern to convey, to handle, to deliver, to transfer
Fördertechnik materials handling engineering
Form shape, form, profile, contour
Form- und Schraubenautomat single-spindle automatic screw machine
formabrichten to form-true
formdrehen to contour-turn, to form-turn
Formlehre form tool gauge, profile gauge
Formlineal former plate
Formmaschine moulding machine
Formmeissel forming tool, contouring tool, formed tool
formrichtig correctly shaped, geometrically true
Formscheibe formed wheel
formschleifen to contour-grind, to form-grind
Formschleifen contour grinding, form grinding
Formschmieden precision grinding
Formstanze stamping die, stamping machine
formstanzen to stamp
formstossen to shape with formed tool
Formstossen shaping with formed tool
Formteil machined part, shaped part
Formung forming, shaping
Formwälzwerkzeug circular broach
Formzahnradschleifen form-tooth grinding
Forschung research

fortbewegen to move away, to move on
fortlaufend continuous, progressive
fortschreiten to progress, to make progress
Fortschritt progress, advance
Franzose monkey wrench
Fräsarbeit milling work, milling job, milling operation
Fräsautomat automatic milling machine
Fräsdorn milling arbor, cutter arbor
Fräsdornmutter cutter arbor nut
fräsen to mill, to cut
Fräsen milling, cutting
Fräser milling cutter, mill, cutter
Fräserstandzeit cutter life, milling cutter life
Fräskopf milling head
Fräsleistung milling capacity
Fräspinole milling quill, milling sleeve
Freihandschleifmaschine off-hand grinder
freikommen to clear, to come clear of
freisetzen to liberate, to set free
freitragend self-supporting, unsupported, cantilevered
Freiwinkel clearance angle, back-off angle (broach)
Frequenz frequency
fressen to fret (corrosion), to score, to seize (moving parts)
Front front, face
fühlen to feel, to sense, to trace (copying)
Fühler sensor, tracer (copying) stylus
Fühlerdruck force on stylus
fühlergesteuert tracer-controlled
Fühlerkopf tracer head
Fühlerlehre feeler gauge
Fühler-Nachformeinrichtung tracing attachment
Fühler-Nachformsteuerung tracer duplicator control
Fühlersteuerung tracer control, contouring control, tracing control
Fühlerstift stylus
Fühlerwalze roller follower, tracer roll
führen to lead, to carry, to guide, to pilot, to conduct
Führung leading, carrying, guiding, piloting, conducting
Führungsbahn guideway, slideway, track
Führungsbohrstange line boring bar, piloted boring bar
Führungsbohrung pilot hole, starting hole
Führungsstück pilot segment, alignment section (broach)
füllen to fill, to pour, to charge
Füllstandsmessung level gauging, level control
Fundament foundation
Fundamentplatte bottom plate, base plate
Fundamentschraube foundation bolt, anchor bolt
Funke spark
Funkenerodieren spark eroding
funkenfrei sparkless, nonsparking
funktionieren to function, to work, to operate, to be serviceable
furchen to riffle
Fuss foot, leg (machine), base (tooth), root, bottom (vice)
fussbetätigt foot-operated, foot-actuated, foot-controlled
Fussboden floor
Futter chuck (for workpiece), lining, packing
Futterbacken chuck jaws

G

Galvanisation electrodepositing, electroplating
galvanisieren to electrodeposit, to electroplate
Gang running, motion, gear, speed, thread
ganz integral, whole, intact, complete

Gasbrenner gas torch
Gasentwicklung gassing, gas formation
Gasflammenhärtung flame hardening
Gaslöten gas brazing
gasschweissen to gas-weld
Gasschweissen gas welding
Gebläse fan, blower, blowing fan
gebläsegekühlt fan-cooled
Gebrauch use, operation, application, utilisation
Gebrauchsanweisung instruction for use, operation instructions
gebraucht second-hand, used, reconditioned
geeignet für adapted for
geerdet earthed, grounded
Gefüge structure, grain
gegenhalten to hold up, to back up, to dolly (riveting)
Gegenkraft counterforce
Gegenlauffräsen conventional milling, upcut milling
Gegenlaufschleifen up-grinding
Gegenmutter lock nut, jam nut, check nut, binding nut
gegenwirken to react, to act against
Gehalt content
Gehäuse casing, case, housing, enclosure
Gehrungsschnitt mitre cut, diagonal cut
Gehrungsstoss mitred joint
gekröpft offset, swan-necked, cranked
gekrümmt crooked, curved
Gelenk joint, hinge, articulation
Gelenkbolzen joint bolt
genau accurate, exact, precise, true, correct
Genauigkeit accuracy, exactness, precision
gerade straight, even, direct
Gerät device, instrument, appliance, implement, equipment, outfit
Geräusch noise, sound
Geräuschdämpfung sound-absorbing, silencing
geräuschlos noiseless
Geräuschpegel noise level
geräuschvoll noisy
geruchlos odourless
gesamt total, entire, whole, overall
Geschwindigkeit speed, velocity
Gesenk die, cavity block
gesenkdrücken to swage
Gesenkfräser die-sinking cutter, die mill
Gesenkfräsmaschine die-milling machine, die-sinking machine
Gesenkpresse die stamping press
Gesenkschmiede drop forge, drop-forging shop
gesenkschmieden to drop-forge, to die-forge
Gestänge leverage
Gestell frame, fixed link, base, rack
Getriebe gear
Gewicht weight
Gewinde thread
Gewinde- und Schneckenschleifmaschine thread and worm grinding machine
Gewindebohreinheit tapping unit
Gewindebohren tapping
Gewindebohrer tap
Gewindebohrmaschine tapping machine
Gewindedrehen thread cutting
Gewindedrehmeissel threading tool
Gewindeflanke thread flank
Gewindefräsen thread milling
Gewindefräser thread cutter, milling cutter
Gewindefräsmaschine thread milling cutter
Gewindegang thread
Gewinde(gang)lehre screw-pitch gauge
Gewindekernbohren tap drilling
Gewindekluppe screw stock
Gewinderollen cylindrical die thread rolling
Gewindeschleifen thread grinding
Gewindeschleifmaschine thread grinding machine

Gewindeschneidautomatik automatic thread cutting device
Gewindeschneiden thread cutting
Gewindesteigung thread lead
Gewindestrehlen thread chasing
giessen to cast (process), to pour, to teem
Giessen casting
Giessereimodell foundry pattern
Gitter grid, grating, lattice (structure)
glänzend glossy, bright
glatt smooth, even, plain
glattdrücken to burnish
glätten to smooth, to planish, to burnish (metal surfaces)
gleichlauffräsen to climb-cut
Gleichlauf-Fräsmaschine climb-cut milling machine
Gleichlaufschleifen down-grinding
Gleichung equation
Gleitbahn slideway
gleiten to slide, to slip, to glide, to chute (work)
Gleitsitz sliding fit
Glied link, member
glühen to anneal, to glow
Glühen annealing, glowing
Grauguss cast iron
Gravierfräser engraving cutter
greifen engraving cutter
greifen in to mesh in, to engage into
Grenzschalter limit switch
Griff handle, grip
griffbereit easy-to-reach
grob coarse, rough
Grossdrehmaschine large lathe, heavy-duty lathe
Grösse size, magnitude, amount
Grossserienfertigung long-run (or: large-batch) production
Grundfläche base surface, floor area
Grundform basic shape, basic design
Grundmetall parent metal, base metal
Grundplatte baseplate, bottom plate, bedplate
Gummi rubber
Gurt belt, band
Guss casting, founding
Gusseisen cast iron

H

Haarriss hair crack
haften to adhere, to stick, to hold
Haken hook, bracket
Hakenmeissel hook tool, recessing tool
Halbautomat semi-automatic lathe, semi-automatic machine
Halbkugel hemisphere
halten to hold, to keep, to maintain, to retain, to last
handbedient manually operated, hand-operated
Handbedienung manual control, manual operation
handbetätigt hand-actuated, manually actuated, manually operated
Handschutz hand guard
Handvorwahl manual preselection
Härte hardness
härten harden to
Härteöl quenching oil
Härtung hardening
hartlöten to braze, to hard-solder
Hartlöten brazing, hard soldering
Hartmetall cemented carbide, sintered carbide
Hartmetall-Aufbohrwerkzeug cement-carbide boring tool
Hartmetallauflage carbide facing
hartmetallbestückt carbide-tipped
Hartmetallbestückung carbide tipping, carbide facing
Hartmetallbohrer carbide drill
Hartmetalleinsatz carbide insert, cement-carbide insert, carbide-lined
Hartmetallplättchen carbide tip
Hartnickel solid nickel
Hauen cutting (file)
Hauptschneide primary cutting edge, principal cutting edge

Hauptständer main column (machine tool)
Hebebock lifting jack
Hebel lever, dolly
Hebelarm lever arm, moment arm
Hebelschere alligator shear, lever shear
Hebemagnet lifting magnet
heben to elevate, to raise, to lift, to hoist, to jack
Heben elevating, raising, lifting, hoisting, jacking
Heber jack
Hebezeug hoisting unit
heftschweissen to tack-weld
heisslaufen to get hot, to heat
Heizanlage heating plant
hemmen to stop, to block, to arrest, to stem
herabdrücken to press down, to force down
herablassen to lower
herabsetzen to reduce, to decrease, to lower
heranfahren to approach (tool)
heranführen to advance, to approach
herauslösen to extract, to dissolve out, to leach
herausragen aus to project from, to protrude from
herausreissen to break out, to tear out
herausschrauben to screw out, to unscrew
herausziehen to extract, to retract, to withdraw, to back out, to project clear of (dimensions in drawing)
hereinziehen to draw in (reamer)
hergestellt manufactured, made, produced
herleiten to derive, to deduce
herrichten to prepare
herstellen to manufacture, to make, to produce
Hersteller manufacturer, maker, producer
Herstellung manufacture, making, production
herumschalten to index around
herunterfahren to lower, to move down
Hilfsantrieb auxiliary drive, accessory drive
Hilfspumpe booster pump
hin und her to and fro
Hin- und Herbewegung reciprocation, reciprocating movement, to-and-fro movement
hin- und herfahren to move back and forth, to shuttle
hin zu towards
Hindernis obstacle, obstruction
hindurchgehen to pass through, to penetrate, to traverse
hineinreichen to extend into
hinterarbeiten to relieve, to back off
hinterdrehen to relief-turn, to relieve by turning
Hinterschliff relief grinding, relieving by grinding
hinterstechen to undercut
hitzebeständig heat-resistant, heat-resisting
HM-bestückt carbide-tipped
Hobel plane
Hobelmaschine planing machine, planer
Hobelmeissel planer tool, planing tool
Hobeln planing
Hobler planer (operator)
hochbeansprucht highly stressed
Hochfahren acceleration, raising (slide etc.)
Hochfrequenzschweissen high-frequency welding, electronic sewing
hochglanzpolieren to finish bright, to mirror-finish
hochheben to lift, to raise, to elevate, to hoist
hochlegiert high-alloy, high-alloyed
Hochleistungsautomat high-duty automatic
Hochofen blast furnace
Höchstbelastung maximum load
Höchstdrehzahl top speed, maximum speed

hochwertig high-grade (material), high-quality
Höhe level, height, head, lift
Höhenreisser mit Teilung vernier height gauge, surface gauge
höhenverstellbar vertically adjustable
hohl hollow, tubular, concave, female
Hohlbohr- und Ausbohrbank trepanning and reboring lathe
Hohlbohrbank trepanning lathe
hohlbohren to trepan, to hollow-drill, to hollow-bore
Hohlbohren trepanning, hollow boring, hollow drilling
Hohlbohrmaschine trepanning lathe
Hohlfräser hollow milling cutter
Hohlkehle fillet, round corner
Hohlkörper hollow part, tubular part, vessel
Hohlkugel hollow sphere
Hohlmeissel hollow chisel
Hohlmeisselbohrer mortising tool
hohlprägen to emboss
Holz wood, timber, lumber
Honahle honing tool
honen to hone
Honen honing
Honmaschine honing machine
horizontal horizontal
Horizontalausführung horizontal design
Horn horn, beak (anvil)
Hub stroke, travel, motion
Hubhöhe stroke height
Hubkraft hoisting capacity, hoisting power, lifting power
Hutmutter acorn nut, domed nut
Hydraulikaggregat hydraulic power pack
hydraulisch hydraulic

I

ideal ideal, perfect, pure
Impuls pulse, impulse, momentum
inaktiv non-activated, inert
Inbetriebnahme putting into operation, setting to work, starting, commissioning (acceptance)
Indexstift index pin
Indexfehler index error
indizieren to indicate, to index
Induktionserwärmung induction heating
Induktionshärtemaschine induction hardening machine
Induktionshärten induction hardening
Induktionsofen induction furnace
Induktionsschweissen induction welding
Industrieanlage industrial plant
ineinandergreifen to mesh, to engage, to gear
Ineinanderpassen mating (gear), fitting together
Inertgasschweissung inert arc welding
infrarot infrared
Ingangsetzen starting, initiating
Inhalt contents
innen inside
Innenabmessung internal dimension
Innendrehen turning internal surfaces
Innendrehmeissel internal turning tool, boring tool
Innendurchmesser internal diameter, inside diameter, minor diameter (thread)
Innenflächenfräsen internal milling
Innengewinde internal thread, female thread
Innengewindeschneiden internal thread cutting, internal tapping
innenseitig on the interior
Innenspannfutter internal chuck
Innentaster internal caliper gauge, inside calipers
Innenverzahnung internal toothing
Innenvierkant square hole
Innenwandung inside wall
Inneres interior
innerhalb inside, within
Installateur fitter

instandhalten to maintain, to service, to keep in order
Instandhaltung maintenance, servicing
instandsetzen to repair, to recondition, to reservice
Instandsetzung repair, reconditioning, reservicing
Instrument instrument, implement, device
Inventar stock, inventory
ISA-Einheitsbohrungssystem ISA basic-hole system
ISA-Einheitswellensystem ISA basic-shaft system
ISA-Toleranzsystem ISA system of tolerances
Isolator insulator
Isolierband insulating tape
isolieren to insulate, to isolate (separation)
Istanzeige actual indication
Istgrösse actual dimension, actual size
Istwert actual value

J

justierbar adjustable
justieren to adjust, to true, to set, to level
Justierfehler maladjustment

K

Kabel cable
Kabelverseilmaschine quadding machine, cable stranding machine
Kadmiumlot cadmium solder
Käfig cage (bearing)
Kaliber roll opening, groove, roll pass, calibre, bore
Kaliberpresse sizing press
Kaliberring female gauge
Kaliberdorn mandrel, broach, plug
Kalibrierwalze sizing roll
Kalilauge caustic potash solution
Kalotte cap, spherical surface, ball indentation
kaltgestaucht cold-headed
kaltbearbeiten to cold-work
kaltbiegen to cold-bend
kältefest cold-resisting
kaltrichten to cold-straighten
kaltschmieden to cold-forge
kaltverarbeiten to cold-process, to cold-form, to cold-work
kaltverformbar cold-formable, cold-workable
kaltwalzen to cold-roll, to roll cold
kämmen to mesh (gears), to engage
Kanal duct, conduit, channel, passage
Kanonenbohrer cylinder bit
Kante edge
kanten to edge, to cant, to tilt
Kantenabschrägen bevelling of the edges
Kantenbrechen chamfering, breaking of the corners
Kapazität capacity, efficiency
Kappe cover, cap, bonnet, hood
Kapsel capsule, casing, can
kapseln to encapsulate, to can
Kapselung encapsulation, canning
Karabinerhaken snap hook
Karbid carbide
karburieren to carburise
Kardangelenk cardan joint, Hooke's joint
Kardanwelle cardan shaft
Karren truck, trolley (tool storage for machining centres)
Kartei card index, card file
Kartenleser card reader
Kartenlochen card punching, card perforating
Kartenlocher card punch, card perforator
Karusselldrehmaschine vertical turning and boring mill
Karussell-Revolverdrehmaschine vertical turret lathe

kaschieren to coat, to apply coatings
Kassette cassette
Kasten case, box, bin
Kastenbett box-section bed
Kastenfuss cabinet leg (lathe)
Kathodenstrahlröhre cathode-ray tube
Käufer customer, purchaser
Kegel cone, taper
Kegeldrehen taper turning
kegelig conical, tapered
Kegelrad bevel gear
Kegelreibahle taper reamer
Kegelstift taper pin
Kehle throat, neck, groove
kehlen to groove, to flute
Keil wedge, key, spline
keilförmig wedge-shaped
Keilnut keyway, key groove
Keilnutenfräsen keyway milling, splining
Keilriemen vee-belt, V-belt
Keilsitz key seat
Kenndaten specification(s), data
Keramik ceramics
Kerbe notch, groove, nick
Kern core, centre
kernbohren to trepan
Kernbohren trepanning
Kessel boiler
Kette chain
Kippachse tilting axis
kippen to tilt, to cant
Kiste chest, box, case
klammern to brace, to clip
klappbar tiltable, hinged
Klaue claw, dog, jaw
Klebeband adhesive tape
kleben to glue, to adhere, to stick
Kleinbohrmaschine power drill
Kleinmengenfertigung small-quantity production
Kleinserie small batch, short run, small lot
Klemmbacke clamping jaw
klemmen to clamp, to jam
Klempner plumber, pipe fitter
Klinge blade
klopfen to knock (engine), to beat
Kluppe die-stock (thread)
Knagge dog, catch, trip
knapp short, scarce, insufficient
Knarre ratchet
Kneifzange pincers, nipper
Knick bend, kink
Knickung buckling
Knickversuch buckling test, collapsing test
Knickzahl buckling coefficient
Knie knee, angle, elbow
Knopfdruck, durch by pressing a button
Knüppel billet (metal)
Kobaltstahl cobalt steel
Kohlendioxid carbon dioxide
Kohlenstoff carbon
kohlenstoffarm low-carbon
Kohlenstoffgehalt carbon content
Kokille chilled mould, ingot mould
Kolben piston, plunger
Kolbenpresse ram-type press
Kombinationszange combination pliers
Kompaktbauweise compact design
kompensieren to compensate, to neutralise, to balance
Komponenten der Schnittkraft cutting force components
komprimieren to compress
konisch conical, taper
Konkurrent competitor
Konkurrenz competition
konkurrieren to compete
Konsole knee (machine tool)
Konsolfräsmaschine column-and-knee type milling machine
konstruieren to design
Konstrukteur designer, design engineer
Konstruktion design, designing, structure
Konstruktionsabteilung design department
Konstruktionsbüro design office, design bureau

Konstruktionsleiter chief designer
Kontermutter lock nut, jam nut
Kontrollampe telltale light, control light, indicating light
Kontrolle control, check, checking, inspection
kontrollieren to control, to check, to inspect
Kontur contour, outline
Konus cone, taper
konzentrisch concentric
Koordinatenbohren coordinate boring, coordinate drilling
Kopf head, top layer, upper end
Kopfhöhe addendum (gear)
Kopfmeissel end-cut tool
Kopier- und Gesenkfräser copy and die milling cutter
Kopierautomat automatic copier
kopierdrehen to copy-turn
Kopierdrehen copy turning, contour turning
Kopierdrehmaschine copying lathe, copy-turning lathe
kopieren to copy, to duplicate, to contour, to trace
Kopieren copying, duplicating, contouring, tracing
kopierfräsen to copy-mill
Kopierfräsen copy milling
Kopierfühler copying tracer
Kopierlineal master plate, form plate, guide plate, template
Kopiermaschine copying mill, duplicating machine, contouring machine
Kopierschiene template
Kopierstift guide pin, tracer point
koppeln to couple, to link
Kordel diamond pattern, diamond-shaped knurling
Kordelmutter knurled nut
kordeln to knurl
Körner centre punch, punch mark
Körnermarke punch mark
Korngefüge grain structure
korrigieren to correct
korrosionsfest corrosion-resistant
Korund corundum
Kostenanschlag estimation, tender
Kraft force, power, strength
Kran crane
Kranportal gantry
kratzen to scratch, to mar, to scrape
Kratzer scratch, mar, scraper
Kreide chalk
Kreis circle
Kreislauf circuit, cycle
Kreuz cross
Kreuzkopf crosshead
Kreuzschlitten compound-rest slide, compound slide
kröpfen to offset, to double-bend
Kröpfung offset
krumm curved, bent, crooked
krümmen to curve, to bend, to warp
Krümmung curvature, bow, bend
Kücken plug
Kugel sphere, ball, globule, pellet
Kugeldrehen ball turning, spherical turning
Kugeldrehmaschine ball turning lathe
kugelförmig spherical, ball-shaped, globular
Kugelfräsen cherrying
Kugelfräser cherry
Kugellager ball bearing
kühlen to cool
Kühlflüssigkeit coolant, cooling liquid
Kühlmittel coolant, cooling fluid
Kühlmittelfluss coolant flow
Kupfer copper
kuppeln to couple, to clutch, to connect, to engage
Kupplung coupling, clutch
Kurbel crank
Kurbelwelle crankshaft
Kurve curve, graph, plot
kurz short
kürzen to shorten
kurzhobeln to shape
Kurzhobeln shaping
kurzhubig short-stroke
Kurzspan short chip, finely broken chip

L

labil instable
laden to load (machine), to charge
Ladung loading (machine), charging
Lage position, location
Lager bearing (machine), stock, store
Lagerhaltung stocking, stock keeping, storage
lagern to mount, to house, to seat, to stock, to store
Lagerschale bearing bush, bearing shell
Lamelle lamina
Langdrehautomat sliding-head type single-spindle automatic machine, Swiss-type automatic, traversing-head bar machine
langdrehen to slide, to turn cylinder surfaces
Länge length
längen to lengthen, to extend
Längenanschlag length stop, longitudinal stop
Langfräsmaschine plano-milling machine, planer-type milling machine
Langhobelmaschine planer miller
Langloch- und Keilnutenfräsmaschine slot and keyway milling machine
Langlochfräsmaschine slot milling machine
Längsabrichten traverse truing
Langvorschub sliding feed, longitudinal feed
läppen to lap
Läppen lapping
Lärm noise
lärmfrei noiseless
Last load, weight
Lauf run, running, working
laufen to run, to work, to move
Laufkatze trolley
Lauge lye, leach, caustic
Lebensdauer life
lecken to leak
leer empty
Leerlauf idle running, idling
leerlaufen to run idle
legen to lay, to place, to put down
Legierung alloy
Lehre gauge, jig, template
Lehrenbohren jig boring
Lehrwerkstatt training workshop, apprentice workshop
leicht light in weight
Leichtmetall light metal
leisten to perform, to carry out, to effect
Leistung power, performance, efficiency, capacity
Leistungsaufnahme power input
Leistungsbedarf power demand, power requirement
leistungsfähig efficient, productive
Leit- und Zugspindeldrehmaschine sliding, surfacing and screwcutting lathe, lathe with leadscrew and feed rod, regular engine lathe
Leiter conductor (current), manager (factory)
Leiterplatte printed circuit board, circuit card
Leitkanal conduit
Leitlineal guide bar, former plate
Leitlinie directrix
Leitschiene guide rail
Leitspindel leadscrew
Leitung line, mains, management (factory)
Leonard-Antrieb Ward-Leonard drive
Lesegerät reading head, reader
Lesekopf reading head
leuchtend bright, luminescent
Leuchtmelder indicating light
Libelle air level
Lichtblitz light flash (stroboscope)
Lichtbogen arc
Lichtbogenschweissen arc welding
Lichtpause print
Lichtschranke light barrier, photo-electric guard
lichtundurchlässig opaque
Lieferant supplier

liefern to deliver, to furnish
Liefertermin delivery date
Lieferung supply, delivery
liegen auf to rest on, to lie on, to be placed on
Lineal ruler, straight edge
linear linear, straightline
Linie line
links left, left-hand, on the left
Linksdrall left-hand twist
Linksgewinde left-hand thread
Linkslauf counter-clockwise (ccw)
linksschneidend left-hand cutting
linkssteigend left-hand helical
Loch hole, aperture, pit
Lochband punched tape
lochbandgesteuert punched tape controlled
Locheisen hollow punch
lochen to punch, to pierce
Locher punch, perforator
Lochkarte punched card, perforated card
Lochkreis circle of holes, index circle, hole circle
Lochmitte hole centre
Lochstreifen punched tape, perforated tape
lochstreifengesteuert punch tape controlled
locker loose, slack
lockern, sich to slacken, to come loose
Lockern slackening off, loosening, releasing, easing, coming loose
lösbar unscrewable, removable, disconnectable
losbinden to untie
löschbar erasable (data storage)
löschen to delete, to erase, to cancel, to extinguish (arc)
lose loose, free
lösen to unscrew, to remove, to disconnect, to release, to untie, to unclamp (work)
loslassen to release
losreissen to tear off, to pull off
losschrauben to unscrew, to unbolt
Lötbad solder bath, dipping bath, pool of solder
Lötdraht solder wire
löten to solder
Lötfahne soldering tag
Lötkolben soldering iron, soldering bit, copper bit for soldering
Lötlampe blowtorch for soldering, soldering blowlamp
Lötmittel solder, soldering flux
Luftabzug air escape, air vent
luftbetätigt air-operated, air-actuated
lüften to vent, to ventilate
Lüfter blower, fan, ventilator
luftleer evacuated
Luftspalt air gap
Lüftung ventilation, venting
Lunker cavity, shrinkage, pipe, void
lunkerfrei pipeless, sound
Lupe magnifying glass, loupe

M

machen to make, to produce, to manufacture, to accomplish, to render
Magnetfutter magnetic chuck
Magnetventil solenoid valve
mahlen to grind, to crush, to pulverise
Makroschliff macrosection
Malteserkreuz Geneva cross, Maltese cross
Mangel lack, shortage, deficiency
mangelhaft defective, imperfect, incomplete, faulty
manövrierfähig manoeuvrable
Manschette collar, sleeve, lip-type of packing
Mantel jacket, outer casing, sheat, coat
markieren to mark, to sign, to label
Masche mesh
maschinell by machine
Mass dimension, measure, size, gauge
Masse dimensions, size
Masse mass

Massenfertigung mass production
massiv solid, one-piece, heavy
Massstab scale, rule
massstäblich full scale, full to scale
Material material, stock
Maul jaw, mouth
Maulschlüssel jaw wrench, open-end wrench
Maulweite wrench opening
Mechanikerdrehmaschine precision lathe
mechanisch mechanical
Mehrkurvensteuerung independent cam control for each operation
Mehrmeisselausstattung multiple tooling
Mehrspindelbauart multiple-spindle design
mehrstufig multiple-step, multi-stage
Meissel tool, chisel (hand tool)
Meisselabheber tool lifter
Meisseleinsatz tool bit
Meisselhalter toolholder
Meisselhalterklappe tool apron, toolholder flap
meisseln to chip, to chisel (by hand)
Meisselschneide tool tip, cutting edge
Meisterschalter master controller
Meldeleuchte indicating light, indicating lamp
melden to indicate, to report
Menge amount, quantity
messbar measurable
messen to measure, to gauge, to check, to determine
Messen measuring, gauging, checking, determining
Messing brass
Messkopf measuring head, gauging head
Messstift feeler pin, test pin
Messuhr dial gauge
Messung measurement, measuring, gauging, determination
Messwert measured value, indicated value, recorded value
Messwertgeber transducer
Metall metal
Metalldrücken metal spinning
Metallkleben metal gluing
Mikrometer micrometre, micron, micrometre screw
mindern to decrease, to reduce
Mindesthärte minimum hardness
mischen to mix, to blend, to compound
Mischung mixture, mixing, blend
mitgehen to follow
mitlaufend revolving, rotating
Mitnahmestift driving pin
Mitnehmer driver, work driver, driving carrier, dog
Mitte centre
Mittel medium, average (value), means
mittelgross medium-size, medium-sized
Mittellinie centre line
Mittelteil centre portion
Modellbau patternmaking
Modelltischler patternmaker
Modul module
Molybdän molybdenum
momentan instantaneous, immediate
Montage assembling, assembly, assemblage, mounting, erection, setting up, fitting
montagefertig ready-to-mount, ready-to-fit, ready-to-assemble
Monteur millwright, fitter, assembler
montieren to assemble, to mount, to fit, to erect, to install
Motor motor (electric), engine (internal combustion)
Motorsäge power saw
Muffe sleeve, bush, muff
Mühle mill, crusher, pulveriser
Mundstück tip, nozzle
Mündung mouthpiece, aperture
Muster sample, specimen, pattern, prototype
Mutter nut
Mutterabkantmaschine nut-bevelling machine
Muttergewinde internal thread, nut thread

Muttergewindebohren nut tapping
Muttergewindebohrer nut tap
Muttergewinde-Schneideautomat nut-tapping automatic
Mutterschloss leadscrew nut
Mutterschlüssel wrench

N

Nabe hub, boss
nachaltern to afterage
Nachalterung afteraging
Nacharbeit rework, re-machining
nacharbeiten to rework, to re-machine
nacharbeitsfrei without reworking, without re-machining
nachbauen to reproduce, to duplicate
Nachbehandlung additional treatment, retreatment
nachbestellen to reorder
nachbilden to copy, to reproduce
Nachbildung copy, reproduction
nachbohren to rebore, to finish-bore
nachdrehen to re-turn, to copy-turn
nacheichen to re-calibrate
Nacheichung re-calibration
nacheinander successively, progressively
nachfahren to trace
nachfolgend following, subsequent
Nachformarbeit copy machining, duplication
Nachformdrehen copy turning, duplicate turning
nachformen to copy, to duplicate, to contour, to post-form
Nachformfräsmaschine copy-milling machine
Nachformfühler form tracer
Nachführung follow-up
nachfüllen to refill, to replenish
nachgeschliffen reground
nachgeschnitten recut
nachlassen to slacken, to loosen, to weaken
Nachlauf lag
nachprüfen to verify, to reinspect, to check
nachrechnen to re-calculate, to check the calculation
nachrichten to restraighten, to redress, to re-align
Nachrichten restraightening, redressing, re-alignment
nachschärfen to resharpen, to regrind
Nachschub feeding
nachspannen to reclamp, to restretch
Nachspannung reclamping, restretching
nachstellbar readjustable
nachstellen to readjust, to reset
Nachstellung readjustment, resetting
nachsteuern to follow up
nachwärmen to postheat
Nachwärmen postheating
nachweisen, experimentell to verify experimentally
Nadel needle, stylus, broach
Nadellager needle bearing
Nagel nail
nageln to nail
nahe near, close
Nähe proximity, closeness
nähern, sich to approach, to approximate
Naht weld seam, seam, joint
nahtlos seamless
nahtschweissen to seam-weld
Nahtschweissen seam welding
Nasenkeil gib-head key
nass wet, moist, damp
nassschleifen to grind wet
natriumgekühlt sodium-cooled
Nebenschneide trail edge, secondary-cutting edge
neigen to incline, to tilt
Neigung inclination, tilt, slope
Neigungswinkel angle of inclination, inclination angle
Nennbelastung nominal load, basic load, nominal stress
Nenndaten ratings

Nenndrehzahl nominal speed, rated speed
Netz mains, network, supply system
netzen to wet
Nibbelmaschine nibbling machine
nibbeln to nibble
Nichteisenguss non-ferrous casting
Nickelstahl nickel steel
niedergehen to descend, to move downward
Niederhaltedruck stripper pressure
niederhalten to keep down, to hold down
Niederhalter hold-down, blank holder, toe dog
Niet rivet
nieten to rivet
Nietmaschine riveter, riveting machine
Nietpresse riveting press
Nippel nipple
Nitrierstahl nitrided steel
Niveau level
nivellieren to level
Nocke cam, disk cam
Nonius vernier
Noniusablesung vernier reading
Norm standard
Normalausrüstung standard equipment
normen to standardize
normgerecht corresponding to the standard
normieren to standardize, to normalize
Normschrift standardized lettering
Normung standardization
Nullanzeige zero indication
numerieren to number
Numerik numerical control system
Numerikfräsmaschine numerically-controlled milling machine
Nut groove, flute, slot
nuten to groove, to flute, to slot
Nutendrehen recessing, grooving, turning necks
Nutenfräsen flute milling
Nutenfräser grooving cutter, keyway cutter, slot drill
Nutenziehmaschine pull-type keyseating machine, groove-drawing machine
nutzbar utilisable, usable
nutzbringend useful, profitable

O

obenliegend overhead
Oberfläche surface
Oberflächenbearbeitung surface machining
oberflächenbehandelt surface-treated, finished, coated
Oberflächengüte surface quality, surface finish
Oberflächenhärte surface hardness, superficial hardness
Oberteil upper part, top
Ofen furnace, kiln (ceramics), oven, stove
offen open, non-enclosed, uncovered
öffnen to open, to unlock, to uncover
Öffnung opening, aperture
Ölablasshahn oil-drain cock
Ölabscheider oil separator
ölanlassen to oil-temper
Ölbad oil bath
ölen to oil
Öler oil cup, oiler
ölfest oil-resistant
Ölkanne hand oiler
Ölnebel oil mist
Ölumlauf oil circulation
ortsbeweglich portable, movable
ortsfest stationary, fixed
Öse eyelet, ear, loop, grommet
oxidieren to oxidize

P

Palette pallet
Palettenhubwagen pallet-lift truck

palettieren to palletize
Panne breakdown, failure
Panzerschlauch armoured hose
parallelgeschaltet connected in parallel
Parallelverschiebung parallel translation
partiell partial
Passarbeit fitting work
pàssen to suit
passend suitable, fitting, well-fitting
Passfeder fitting key, feather, spline
passieren to pass
Passschraube precision bolt, reamed bolt, fitting screw
Passstift fitting pin, dowel, alignment pin
Passstücke mating parts, mating members
Patentanmeldung patent application
Patrone collet (workholding), sleeve, cartridge
pausen to trace, to copy, to print
Pendelbetrieb shuttle service
Pendelfräsen reciprocal milling, pendulum milling
Pendellager self-aligning bearing
pendeln to oscillate, to reciprocate, to shuttle, to float
perforieren to perforate, to punch
pfeilverzahnt with double-helical teeth, with herringbone gear teeth
Pfeilverzahnung herringbone gear teeth, double-spiral gear teeth
Pferdestärke horse-power
Pflege attendance, maintenance, service, treatment
pflegen to maintain, to service, to tend
Pfusch slipshod work, bungle
Pfuscharbeit shoddy work
pfuschen to bungle, to botch
Pilgerdorn pilger mandrel, piercer
Pinole spindle sleeve, quill
plan plane, flat
Planarbeit facing operation
Planbearbeitung surfacing, facing
plandrehen to face-turn, to face, to surface
Plandrehen face turning, facing, surfacing
Plandrehmaschine facing machine, surfacing machine
planen to plan, to project, to schedule
planfräsen to slab-mill
Planfräsen plain milling, slab milling
Planfräsmaschine fixed-bed miller, fixed-bed milling machine
plankopieren to contour-face, to copy-face
Planscheibe faceplate
planschleifen to face-grind
Planschleifen face-grinding, surface grinding
Planschleifmaschine face grinder, face grinding machine
planschlichten to finish-face
Planschlichten finish facing
plansenken to spotface
Plansenken spotfacing
Plansenker spotfacer, spotfacing tool
Plansupport cross-slide
Plasmabeschichtung plasma deposition
Plasmabrenner plasma torch
Platin platinum
Platine sheet bar, plate bar
platinieren to platinize, to slab
Platte plate, slab, sheet
plattieren to clad
Platz space, seat, spot, point, position
Platzbedarf space required, space requirement
platzen to burst, to crack
Platzersparnis space saving
Platzleuchte built-in spot light
platzsparend space-saving, room-saving
Pleuel connecting rod
plombieren to seal
pneumatisch pneumatic
polierdrücken to burnish
Polierdrücken die burnishing
polieren to polish
polierläppen to buff
Polierläppen buffing
Poliermaschine polishing machine

reissen to crack, to fracture, to tear, to rupture
Reissen cracking, fracturing, tearing, rupturing
Reissspan tear chip
Reitstock tailstock
Rentabilität profitability
Reparatur repair, overhaul
reparieren to repair, to overhaul, to recondition
Resultat result
Resultierende resultant
Revolverautomat automatic turret lathe
Revolverdrehen turning in turret lathe
Revolverdrehmaschine turret lathe
richten to align, to straighten, to planish
richtig correct, true, proper, right
Richtung direction, hand, sense
Richtwert reference value, recommended value
Riefe ridge, flute, channel, groove
riefen to ridge, to flute, to channel, to groove
riefenfrei without tooling marks
riefenlos ridgeless
Riegel latch, tie, beam
Riemen belt
riffeln to riffle, to serrate
Rille groove, ridge, scratch
Rillentiefe scratch depth
Ring ring, annulus, washer, ferrule
ringförmig ring-shaped, annular, toroidal
Rinne channel, chute, spout, groove
Rippe rib, ridge, web, stem
rippen to rib, to fin
Riss crack, fissure, fracture
Rissbildung crack formation, cracking
rissig cracky
ritzen to scratch, to score
roh raw, rough, crude, blank (workpiece), undressed
Roheisen pig iron
Rohling blank, raw piece
Rohr pipe, tube, conduit, duct
Rohrgewinde pipe thread
Rohstoff raw material
Rolle roller, roll, pulley, reel, coil
rollen to roll, to curl
Rollengang rolling mill table
röntgen to X-ray
Röntgenbild X-ray image
Rostbildung rust formation, rusting
rosten to rust
Rostschutz rust prevention, rust-proofing
Rotmetall wrought copper-tin-zinc alloy
Ruck jerk
ruckartig jerky, jerkily
rucken to jerk
rückführen in den Umlauf to recirculate
Rückführung return, recirculation, feed-back
Rücklauf reverse stroke, non-cutting stroke, reverse motion, retraction, relieving stroke, backward movement
Rücklauf zum Ausspänen swarf relieving stroke
rücklaufen to return, to retract, to reverse
rückmelden to feed back
Rückmeldung feed-back
Rückprall rebound, bouncing, recoil
Rückschlag flashback, backfiring, recoil, rebound
Rückschlagventil non-return valve, check valve
Rückseite rear side, back, reverse
Rückstand remainder, residue
Rückstellung resetting, returning, restoration
Rückwand back wall, rear wall
rückwärtig at the rear
rückwärts backwards, reverse
ruckweise jerky, jerkily
rückziehbar rectractable
Rückzugfeder release spring
Ruhe rest, dwell, quietness, silence
Ruhestellung position of rest, neutral position

ruhig quiet, silent, smooth
rühren to agitate, to stir
rund round, circular, cylindrical, spherical, rounded
Rundlauf true running, concentricity
Rutsche gravity chute, slide
rutschen to slip, to slide, to skid
rütteln to jolt, to jar, to vibrate
rüttelsicher vibration-proof, shake-proof

S

Säge saw
Sägeblatt saw blade
sägen to saw
Sägegewinde saw-tooth thread, buttress
Sägespäne saw waste
Sammelrohr header
sandstrahlen to sand-blast
Sandstrahlen sand blasting
Saphir sapphire
sättigen to saturate
Sättigung saturation
Satzfräsen gang milling
sauber clean, neat
Sauberkeit cleanliness, neatness
Sauerstoff oxygen
saugen to suck, to absorb
Säure acid
schaben to scrape
Schaber scraper, scraping tool
Schablone templet, template, master plate, stencil
Schaden damage
schadhaft damaged, defective, faulty
Schaftfräser shank cutter, end mill cutter
schallabsorbierend sound-absorbing
schalldicht sound-proof
Schallpegel sound level
Schalter switch, control lever
Schaltpult control desk, control console
Schaltung circuit, circuitry, control, switching, actuation, indexing
scharf sharp, keen
Schärfe sharpness, keenness
schärfen to sharpen, to regrind, to swage, to point, to scarf
Scharnier hinge
schätzen to estimate
Schätzung estimation
Scheibe disk, plate, pane, washer, dial
scheiden to separate, to sort, to part
Schelle clip, clamp, bracket
Schenkel side, leg, angle, arm
Schere shear
Schermeissel shear tool
Scherspan shear chip
scheuern to chafe, to gall
Schicht layer, lamina, film
schieben to push, to shift, to slide
Schieber toolholder, slide, slider, face slide
Schieberad sliding gear, shifting gear
schief oblique, inclined
Schiene rail, beam
Schlag blow, impact, percussion
Schlagbohren percussion drilling
schlagen to blow, to beat, to strike
schlagfräsen to fly-cut
Schlagfräser fly cutter
Schlauch hose
Schleichgang creep feed, inching
Schleife loop
schleifen to grind
Schleifen grinding
Schleifscheibe grinding wheel
Schleifscheiben-Abrichtdiamant grinding wheel dressing diamond
Schleifscheibenabrichter grinding wheel dresser
Schlichtarbeit finish-machining operation
schlichtdrehen to finish-turn
schlichten to finish, to finish-turn
Schliff grind
Schlitten slide, carriage, saddle
Schlitz slot, nick, groove
schlitzen to slot-mill
Schlitzfräsen slot milling gashing

Schlosskasten apron housing
Schlossmutter leadscrew nut
Schlossplatte lathe apron, apron
Schlupf slip, slipage, slipping
Schlüssel key, spanner, wrench
Schlüsselweite wrench opening, width across flats
Schmelze melt, molten metal
schmelzen to melt
schmieden to forge
Schmieden forging
Schmierbüchse oil cup
schmieren to lubricate
Schmiermittel lubricant
Schmieröl lubricating oil
Schmierung lubrication
Schmirgel emery
Schmutz dirt
schmutzig dirty
schnappen to catch
Schnecke worm
Schneckenrad worm wheel
Schneidarbeit cutting operation, cutting action
Schneidbacke cutting die
schneidbrennen to flame-cut
Schneidbrenner cutting torch, flame cutter
Schneide cutting edge, bit
Schneideinsatz insert
Schneiden cutting, shearing, blanking, trimming
Schneidenausbruch edge chipping
Schneidfähigkeit cutting capacity
Schneidkluppe stock and die, hand die-stock
Schneidkopf die head
Schneidlippe cutting edge
Schneidplatte cutting tip, tool bit, insert
Schneidspäne cuttings, chips
schnell fast, rapid, high-speed, quick, speedy
Schnelldrehmaschine high-speed lathe
Schnellzugriff immediate access
Schnitt cut, cutting, notch, incision
Schnittbau punch and die making
Schnittbreite width of cut
Schnittfläche area of cut, shoulder of cut
Schnittflanke kerf wall, face of cut, cut surface
Schnittiefe cutting depth
Schnittkraft cutting force
Schnittmatrize cutting die
Schnittrichtung hand of the cut, direction of the cut
Schräglager inclined roller bearing
schrägverzahnt helically toothed
Schrägverzahnung helical teeth
Schränkmaschine saw-setting machine
Schränkung tooth set
Schraube screw, bolt
schrauben to screw, to tighten, to loosen
Schraubenanziehmaschine nut runner
Schraubenautomat automatic screw machine
Schraubenfutter screw chuck, cat-head
Schraubenschlüssel wrench, spanner
Schraubenzieher screw driver
Schraubstock vice
schreiben to plot, to record
Schritt step, pitch, increment
Schrittmotor stepping motor
Schrott scrap
Schrotthaufen scrap pile, scrap heap
schrumpfen to shrink
Schrumpfung shrinkage, shrinking, contraction
Schrupparbeit roughing operation
Schruppdrehen rough turning
schruppen to rough, to rough-machine
Schruppspan roughing cut
Schub thrust
Schublehre caliper rule, vernier caliper
Schulter shoulder, collar
schütteln to shake, to vibrate
Schutz guard, safeguard, protection, cover
Schutzbrille goggles

schützen to protect, to guard, to shield
Schutzkleidung protective clothing
Schutzüberzug protective coating
Schwalbenschwanz dovetail
schwanken to vary, to pulsate, to fluctuate
Schwankung variation, pulsation, fluctuation
Schwefel sulphur
schweissen to weld
Schweissen welding
schwenkbar swivelling
schwerbearbeitbar difficult to machine
schwierig difficult, intricate
Schwimmerschalter float switch
schwingen to oscillate, to vibrate, to rock
Schwingung oscillation, vibration, rocking
Schwund shrinkage
Schwungmasse gyrating mass
Schwungrad flywheel
Sechskant hexagon
sechskantig hexagonal
Sechskantmutter hexagon nut, hexagonal nut
Sechskantschlüssel hex wrench, hexagonal wrench
Seite side, end, hand
Seitendrehmeissel side-turning tool
Seitensupport side tool-head, horizontal head
Selbstausgleich self-compensation
selbsthemmend self-locking
selbsttragend self-supporting
senken to lower, to descend
Senker counterbore, countersinking cutter
senkrecht vertical, perpendicular
Serienfertigung series production
serienmässig in series, serial
Sicherheit safety
Sichtanzeige visual indication, visual display, readout
Sichtfeld field of view
Sieb sieve, screen, strainer
sieben to sieve, to screen, to strain
Silberlot silver solder
Silizium silicon
Sinterhartmetall cemented carbide
Sinterkarbid sintered carbide
sintern to sinter, to cement, to cake
Skala scale, dial
Skizze sketch
Sinuskurve sine curve, sinusoidal curve
Sollwert required value, nominal value, rated value, reference value
Sonde probe
Sonderausführung special design
sondieren to probe
Sorte kind, grade
sortieren to sort, to classify, to separate, to size-grade
Spalt gap, clearance
Spaltfilter gap-type filter
Span chip, cut, cutting
Spanabnahme chip removal, metal removal
spanbar machinable, cutable
Spanbrecher chip breaker
Spänestauung chip clogging, chip congestion
Spanfläche tool face, rake face, top face
Spannbüchse work-holding bushing
Spanndorn mandrel, arbor
Spanneisen holding strap, clamp
spannen to clamp, to chuck, to grip, to hold, to set up, to load, to mount
Spannfutter chuck
Spannung stress, tension, voltage (el.)
Spannweite span, chucking capacity, capacity of vice jaws, opening capacity
Spanwinkel rake angle, cutting rake
Speicher rack (tool), memory, storage, store
Speiseleitung feed line, supply line
sperren to lock, to arrest, to retain, to hold, to interlock, to shut
Spiel clearance, play, allowance, backlash
Spindel spindle, screw

Spindelstock headstock, spindle head
Spiralbohrer twist drill
Spirale spiral, helix
spitz pointed, acute (angle), sharp-pointed
Spitze point, tip, peak, top, apex, crest
spitzen to point, to sharpen
Spitzendreharbeit between-centres turning operation
Spreizbüchse split bushing
spreizen to expand, to spread
springen to crack, to burst, to jump
spritzen to spray
spritzwassergeschützt splash-proof
spröde brittle, short
sprühen to spray, to sputter
Sprung increment (step), crack, fissure
Sprungschaltung jump feed
Spule coil, reel
spülen to rinse, to flush, to wash, to cleanse
Spur channel (tape), trace, gauge
stabil stable, rugged, robust
Stadium stage
Stahlbeton reinforced concrete
Stand level, state, position
Standardausführung standard design
Ständer column, upright housing, upright standard
Ständerbohrmaschine box-column drilling machine
Ständerführung column way
Standort position, site, location
Standzeit tool cutting life, tool life
Stange bar, rod
Stangenautomat automatic bar machine
Stanze stamping press, blanking machine, punching machine
stanzen to stamp, to blank, to punch
stapeln to pile, to stack, to tier
starr rigid
stationär stationary, fixed, steady
staubdicht dust-tight
stauen to stagnate, to choke, to jam, to clog
stechen to lance, to louvre
Stechmeissel parting-off tool, cut-off tool
Steckschlüssel box spanner, socket wrench
Stehbolzen staybolt
stehenbleiben to stall, to stop
steif stiff
Steife stiffness
steigen to climb, to ascend, to rise, to increase
steigern to increase
steil steep
Stelle place, site, spot, point, digit
Stellglied control element
Stellschraube expansion screw (reamer), setting screw, setscrew
Stellung position, setting
Stempel punch, male die, stamp, puncheon
steuern to control
Steuerpult control desk, control console
Steuertafel control panel
Steuerung control
Stickstoff nitrogen
Stift pin, stud
Stiftbolzen stud
Stiftschlüssel pin-type wrench
stillsetzen to stop, to shut down, to bring to rest
Stillstand rest, standstill, stoppage
Stirn front, face, end
Stirnschneide front cutting edge, face cutting edge
Stopfbüchse stuffing-box
stören to disturb, to interfere with
Störung breakdown, disturbance, failure, fault
Stoss impact, shock, percussion, push, impulse
Stossdämpfer shock absorber, dashpot
Stössel ram, tappet
stossen to push, to strike, to knock, to shape
Stossnaht butt joint
stossräumen to push-broach

straff tight, taut, tensioned
strählen to chase
Strählen chasing
strangpressen to extrude
Strebe brace, bracing
streifengesteuert tape-controlled
Streifenleser tape reader
Strom stream, flow, current (el.)
Stück piece, workpiece, component, part
Stückliste part list, bill of materials
stumpf blunt, dull, obtuse (angle)
stumpfschweissen to butt-weld
stumpfwinklig obtuse-angled
stützen to support, to back
suchen to look for, to search, to find (calculation)
Support slide rest, carriage, tool box, head (planer), tool slide, cross-slide (boring mill)

T

Tabelle table chart
Taktzeit cycle time
Taste key, push-button
tasten to feel, to trace, to scan
Taster tracer, stylus, feeler, feeler pin
tauchen to dip, to immerse
taumeln to wabble, to nutate
Taumelscheibe wabbling disc, wobble plate
Teil part, section, portion, piece, element
teilen to divide, to index, to space
Teilkopf indexing head
Teilkreis pitch circle
Teilung division, indexing
Temperaturanstieg temperature rise
Tiefe depth
Tisch table, platen, desk
Tischpresse bench press
Titan titanium
Toleranz tolerance, limit
Topffräser cup-chaped cutter
Totpunkt dead centre
Totzeit dead time, non-cutting time
Tragbalken girder
träge inert, inactive
tragen to carry, to bear, to support
Träger beam, girder, carrier, support
Trägheit inertia
Trägheitsmoment moment of inertia
Transport transfer, conveyance, handling, transport
Traverse top beam, cross beam, cross girth
treffen to hit, to meet
Treiber driver
trennen to part off, to cut off, to separate, to disconnect
Trennsäge cut-off saw
Trockenabrichten dry truing
Trog trough, pan, tray
Trommel drum, reel
Trommelfräsmaschine drum-type milling machine
Trommelmagazin drum magazine
Trommelschalten drum indexing
tropfen to drop, to drip
Tropfen drop

U

überdimensionieren to oversize, to overdimension
Überdrehzahl overspeed
überholen to overhaul, to recondition
Überholen overhauling, reconditioning
überlasten to overload
Überlast overload
überprüfen to check, to verify, to inspect
überschalten to override
Übersetzung transmission ratio
übertragen to transmit
Übertragung transmission
überwachen to monitor, to supervise
umarbeiten to rework, to modify
Umdrehung revolution, turn

Umdrehungen je Minute revolutions per minute (rpm, rev/min)
umgehen to by-pass
Umgehung by-pass
Umkehr reverse, reversal
umkehren to reverse, to invert, to return
umkippen to tilt over, to upset, to overturn
umlaufen to rotate, to revolve, to circulate, to spin
Ummantelung jacket, encasing, encasement
umrissfräsen to contour-mill
Umrissfräsen contour milling
umrüsten to reset, to retool
Umrüsten resetting, retooling, tool changeover
Umrüstzeit changeover time
Umspannen reclamping, rechucking, relocating, changing the set-up
umsteuern to reverse
unbehandelt untreated
unbeweglich immovable, stationary
undicht leaky, leaking
undurchführbar unfeasible, impracticable
uneben uneven
Unfall accident
ungleichmässig irregular, non-uniform
Universalfutter universal chuck
Unrundheit lack of roundness, out-of-roundness, non circularity
Unrundlauf untrue running, runout
unrundlaufend untrue-running
unterbauen to bolster, to pack, to jack, to shim
unterbringen to accomodate, to house, to place, to store
unterdrücken to suppress
Unterlage backing, support, back-up
Unterlegscheibe washer, grommet, packing shim
unterschneiden to undercut
Unterseite bottom side, underside
unterstützen to support, to carry
untersuchen to investigate, to examine, to analyse
Unterteil lower part, base, bottom
Unwucht unbalance, imbalance
unzerspanbar unmachinable, non-machinable
unzulässig inadmissible, undue

V

V-Bahn vee, wee way, V-way
Vakuum vacuum
Ventil valve
Ventilator ventilator, fan, blower
verarbeiten to process, to fabricate, to work
verbinden to connect, to tie, to couple
verblitzen to flash
verblitzt flashed
Verbrauch consumption
verbrauchen to consume
verbrennen to burn, to combust
verdichten to compress, to compact, to pack
verdunsten to volatise, to vaporise, to evaporate
veredeln to refine, to finish
verengen to narrow, to neck, to reduce, to contract
verfügbar available
Verhältnis ratio
verhindern to prevent
Verkleidung cover, lining
verklemmen to jam, to stick, to get stuck
Verlust loss
Vermögen power, ability
versenken to counterbore, to countersink
versorgen to supply, to provide
verstärken to strengthen, to reinforce, to boost, to amplify
Versuchsanordnung test set-up
verteilen to distribute, to spread
Verteilung distribution, spreading
verursachen to cause

verwenden to use, to employ
Verwendung use, employment
verzogen warped, distorted, crooked
verzögern to retard, to lag, to delay
Verzögerung retardation, lag, delay
Vielspindelbohrmaschine multi-spindle drilling machine
Vierkant square, squared
Vollast full load
Vollautomat fully automatic lathe, fully automatic machine
vollelektrisch all-electric
Vor- und Rücklaufhobeln two-way planing
vordrehen to rough-turn
vorspannen to prestress, to preload
Vorspannung prestressing, preloading
Vorwärtsbewegung forward movement

W

Waagrechtbohrmaschine horizontal boring machine
Waagrechtbohrwerk horizontal boring mill, boring and facing mill
wackeln to totter
wählen to select
Wählschalter selector switch
wälzen to roll, to generate
Walzenfräser cylindrical milling cutter, cylindrical slab mill
wälzfräsen to hob
Wälzfräsen hobbing
Wälzfräsmaschine hobbing machine
Wälzlager antifrication bearing
Wanddicke wall thickness
Wandler transducer, converter
Warmbearbeiten hot working
Wärmeabfuhr heat dissipation, heat removal
Wärmeausdehnung thermal expansion
warten to service, to attend, to maintain
Wartungshandbuch service manual
Wasserstoff hydrogen
wechseln to change, to reverse
Wechselrad change gear
Wechselrädergetriebe change-gear train
Weg travel, distance, traverse
weissglühend white-hot
Welle shaft
Wende reversal, turn
wenden to reverse, to turn
Wenden von Schneidplatten indexing of inserts
Wendeschneidplatte throw-away cutting tool tip
werfen to throw, to cast
Werkstatt workshop
Werkstoff material, stock
Werkstück workpiece, work, part, component
Werkstückzubringung work feeding, part feeding
Werktisch work bench
Werkzeug tool, implement
Wert value
Widerlager abutment
Widerstand resistance
wiederholen to repeat
Wiederholung repetition
wiegen to weigh, to scale
willkürlich random, arbitrary
Winkel angle
Winkeltrieb bell-crank drive
winklig angular
wirken to act, to effect
Wirkung action, effect
Wirkungsgrad efficiency
Wismut bismuth
Wolframkarbid tungsten carbide
wuchten to balance
Wuchten balancing

X

X-Achse X-axis
X-Koordinate X-coordinate

X-Naht double vee weld

Y

Y-Achse Y-axis

Z

zäh tough
Zähigkeit toughness
Zähler counter, numerator
Zahn tooth
Zahneingriff tooth engagement
Zahnfuss tooth dedendum
Zahnrad toothed gear, gear
Zahnformfräsen gear milling
Zahnteilung tooth spacing
Zapfen journal, pivot
zeigen to indicate, to show
Zeilenfräsen line-by-line milling
zentrieren to centre
zerfressen to corrode, to pit, to cauterise
zerlegen to decompose, to disassemble, to dismantle
zerreissen to break, to tear, to rupture
zerspanbar machinable
zerspanen to machine, to cut
Zerspankraft cutting force
zerstören to destroy
Zerstörung destruction
ziehen to draw, to pull
Ziehen drawing, pulling
Ziehräumen pull broaching
Zink zinc
Zubehörteile accessories
zubringen to feed, to load, to handle
Zubringer feeding device, feeder loader
Zug-Druck-Versuch push-pull test
zugeben to add, to charge
Zugriff access
Zugspindel feed shaft
zulässig permissible, admissible
Zustand condition, state
zweiarmig double-armed
zweischneidig two-edged, double-edged
Zylinder cylinder

Anhang/Appendix

Formulae - how to express in words

1. x^0 x to the power of zero
2. x^1 x to the power of one
3. x^{-1} x to the power of minus one
4. x^{-2} x to the power of minus two
5. x^2 x squared
6. x^3 x cubed
7. x^4 x to the fourth

Bei Potenzen auch: x^n = x to the power n, x to the nth power, x raised to n.

8. $\sqrt[2]{x}$ Square root of x
9. $\sqrt[3]{x}$ Cube root of x
10. $\sqrt[4]{x}$ Fourth root of x
11. $(a+b)^2$ a plus b all squared (or: a plus b in parentheses squared)
12. $R_t = \frac{u_t U^2 10^4}{S_n}$ (Capital) R sub (small) t equals (small) u sub t times (capital) U squared times 10 to the fourth, all over (capital) S sub (small) n.
13. $\frac{dy}{dx}$ dy by dx
14. $\int ax\,dx = a\int x\,d\,x = \frac{a\,x^2}{2} + C$

    Integral of ax dx equals a integral of x dx equals a x squared over two, plus C.

Verbs used in describing electric circuits

abfallen to release, to drop out
Contact d 4-2/4 opens, releasing contactor c2.
Der Kontakt d4-2/4 öffnet, wobei das Schütz c2 abfällt.

abschalten, durch Relais to switch off, to disconnect, to cut off, to break
Motor m 1 is switched off by relay d 2.
Motor m 1 wird durch Relais d 2 abgeschaltet.

anlegen, Spannung to apply voltage
Mains voltage is applied to terminals 1, 2 and 3.
Netzspannung liegt an den Klemmen 1, 2 und 3 an (oder: ist an den Klemmen... vorhanden).

anziehen, Relais to pick up, to pull in, to operate, to actuate
When voltage Ug has reached its rated value, relay d2 picks up.
Wenn die Spannung Ug ihren Nennwert erreicht hat, zieht Relais d 2 an.

aufleuchten to light (up), to go on
Indicating lamp h 9 lights up (or: goes on).
Anzeigeleuchte h 9 leuchtet auf.

ausgehen to go out, to switch off
When the reset button is operated, firing is interrupted,
and the lamp goes out.
Mit dem Betätigen des Rückstellknopfes wird das Zünden unterbrochen, und die Leuchte geht aus.

auslösen, Schnellschlussventil to trip, to release
When the emergency valve has been tripped...
Nachdem das Schnellschlussventil ausgelöst hat...
(Hinweis: Es muss heissen «when the...»; nicht: «after the...»)

beenden, Anlassvorgang to complete starting sequence
Motor m2 does not start (up) until m1 has completed its starting sequence.
Motor m2 läuft nicht an, solange m1 seinen Anlassvorgang nicht beendet hat.

betätigen to actuate, to operate, to energise
Contactor c2 is actuated (here also: activated) by relay d3.
Schütz c2 wird durch Relais d3 betätigt.

betreiben to operate
The alarm system operates from a 24 V d.c. source (or: operates at 24 V d.c.).
Die Alarmanlage wird mit 24 V Gleichspannung betrieben.

bewirken to result in, to cause
Simultaneously contact d 2-6/8 opens resulting in the indication 'oil pressure too low'.
Gleichzeitig öffnet der Kontakt d 2-6/8 und bewirkt die Anzeige «Öldruck zu niedrig».

blockieren, verriegeln to block, to interlock
In such cases the interlock is blocked by relay d 7.
In solchen Fällen ist die Umschaltung durch Relais d 7 blockiert.

drücken, betätigen to depress, to press, to actuate, to operate
Depressing the reset button interrupts the firing and causes the lamp to go out.
Durch Drücken des Rückstellknopfes wird das Zünden unterbrochen, und die Lampe geht aus.

einleiten to initiate
If the change-over is initiated by a generator fault, relay d 7 will block the phase comparison relay.
Sollte eine Umschaltung durch einen Generatorfehler eingeleitet werden, so wird das Phasenvergleichsrelais durch Relais d 7 blockiert.

einschalten to switch on, to switch in
Before commissioning, the device should be switched on without driving motors connected.
Vor dem Inbetriebsetzen sollte das Gerät ohne Antriebsmotoren eingeschaltet werden.

erhalten to obtain, to receive
Relay d 2 receives the signal from m 10.
Relais d 2 erhält das Signal von m 10.

erlöschen to go out
Warning light h 1 goes out.
Warnleuchte h 1 erlischt.

erregen to energise, to excite (machines)
Coil k 1 is energised by closing contacts d 2-4/6.
Spule k 1 wird durch Schliessen der Kontakte d 2-4/6 erregt.

fliessen, Strom to flow
Current Iw flows through coils k 1 and k 2 simultaneously.
Der Strom Iw fliesst gleichzeitig durch die Spulen k 1 und k 2.

gleich sein to be equal
The voltages are now equal.
Die Spannungen sind nun gleich.

halten, sich selbst to retain, to hold itself, to maintain in the closed position
Relay d 5 retains itself via contacts d 5-2/4.
Relais d 5 hält sich selbst über Kontakt d 5-2/4.

herausführen to take out to terminals, to lead out, to connect to terminals
The contacts of relay d 4 taken out to the terminals can be used for signalisation purposes.
Die an die Klemmen herausgeführten Kontakte des Relais d 4 können für die Signalisation verwendet werden.

löschen to turn off, to switch off
The thyristor is turned off by opening contact d 2-2/4.
Der Thyristor wird gelöscht, indem der Kontakt d 2-2/4 geöffnet wird.

schliessen, Stromkreis to complete, to close, to make a circuit
The control circuit is completed through coil k 1.
Der Steuerstromkreis wird durch die Spule k 1 geschlossen.

schliessen Stromkreis, wieder to re-establish a circuit
The monitoring circuit is re-established through time-lag relay d 9, contacts 6/8.
Der Überwachungsstromkreis wird durch das Zeitrelais d 9, Kontakt 6/8 wieder geschlossen.

setzen, in Betrieb to start, to operate, to put into operation, to commission
The installation is started by switch b 1.
Die Inbetriebnahme der Anlage erfolgt durch den Schalter b 1.

stellen, Wahlschalter to set, to position, to turn

Set selector switch to 'automatic'.
Wahlschalter auf «Automatik» stellen.

überbrücken to bridge
The open contacts are bridged by time switch d 7.
Die offenen Kontakte sind durch den Zeitschalter d 7 überbrückt.

übereinstimmen to correspond, to be in phase
If the two voltages being compared correspond, thyristor th 2 fires (or: turns on) at once.
Wenn die zu vergleichenden Spannungen übereinstimmen, so zündet der Thyristor th 2 sofort.

Verbs and nouns used in describing electronic circuits

alter (ver)ändern
amplify verstärken
apply anlegen (Spg.)
bias vorspannen
cause bewirken, verursachen
compensate ausgleichen
conduct leiten
control steuern, schalten
divert ableiten
drive (aus)steuern
emit emittieren, ausstrahlen
fire zünden
flow fliessen
forward-bias in Vorwärtsrichtung vorspannen
gain zunehmen, gewinnen, erlangen
release freigeben
reverse umkehren
saturate sättigen
supply speisen, versorgen
switch off abschalten
switch on einschalten
transfer überleiten, übertragen
turn off abschalten, ausschalten
turn on einschalten

alteration Veränderung
amplification Verstärkung
arrangement Anordnung
array Anordnung, Reihe
barrier Grenzschicht, Sperre
base Basis
bias Vorspannung
blocking capability Sperrvermögen
charge Ladung
collector Kollektor
condition Zustand
conduction Leitung
divider Teiler
drive (Aus-)Steuerung
driver Treiber
drop Abfall
emitter Emitter
firing Zünden
flow Fliessen
forward bias Durchlassspannung, Vorwärtsvorspannung
gain Verstärkung
gate Tor
input Eingang
module Modul
'off' state «Aus»-Zustand
'on'-state «Ein»-Zustand
output Ausgang
reverse-biasing Betrieb in Sperrichtung
reverse bias Sperrvorspannung
saturation Sättigung
secondary breakdown Sekundärdurchbruch
source Quelle
state Zustand
status Zustand
unity Eins, Einheit
unity gain Verstärkungsfaktor Eins

Description of a basic circuit diagram

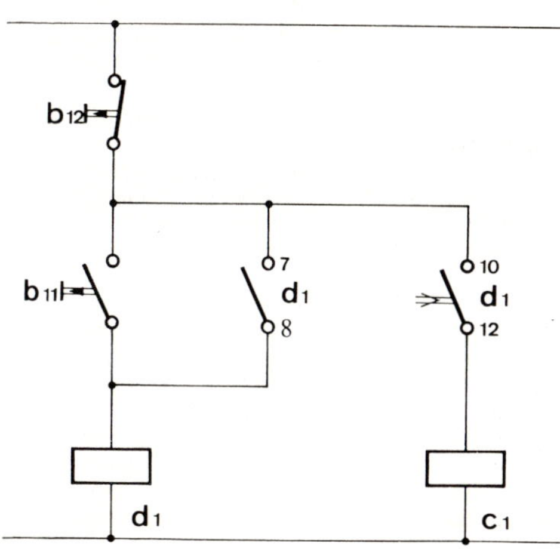

Description 1
On pressing push-button b11 time-lag relay d1 is energised and retained by contact d1-7/8. After lapse of a preset time contactor c1, controlled by d1-10/12, picks up. Pressing b12 causes relay d1 and contactor c1 to drop out.

Description 2
Depressing push-button b 11 causes time-lag relay d 1 to operate, holding itself by contact d 1-7/8. Contactor c 1 picks up after a preset time, controlled by contact d 1-10/12. Depressing button b 12 puts relay d 1 and contactor c 1 back into its normal position.

Description 3
Pressing push-button b 11 initiates the operation of time-lag relay d 1. Once activated, the relay is maintained in the closed position by contact d 1-7/8. Contactor c 1, controlled by d 1-10/12, is activated after a preset period. Depressing push-button b 12 resets relay d 1 and contactor c 1 to their normal position.

Description 4
Upon pressing push-button b 11, time-lag relay d 1 is energised. It is held in the closed position via contact d 1-7/8. After a preset period has elapsed, contactor c 1, controlled by d 1-10/12, is activated. Depressing push-button b 12 causes relay d 1 and contactor c 1 to drop out.

Verbs and nouns used for voltage fluctuation

abfallen to drop, to fall, to diminish
ansteigen to rise, to increase
aufbauen, sich to build up
bewirken, Spannungsanstieg to cause a voltage rise
erhöhen, sich to increase, to rise
gleich sein to be equal
konstant halten to keep constant, to maintain constant
schwächer werden to diminish, to decrease
schwanken to fluctuate, to oscillate
überschreiten to exceed, to overshoot, to overrun
unterschreiten, Wert to fall below a value
verringern to reduce, to diminish, to decrease
verschieben, sich to become out of phase, to shift out of phase
verursachen, Spannungsabfall to cause a voltage drop
voreilen to lead
voltage between lines verkettete Spannung
v. drop Spannungsabfall
v. deviation Spannungsabweichung
v. fluctuation Spannungsschwankung
v. gain Spannungsverstärkung
v. level Spannungspegel
v. peak Spannungsspitze
v. pulse Spannungsimpuls
v. range Spannungsbereich
v. rise Spannungsanstieg
v. spike Spannungsspitze
v. surge Spannungsstoss
v. variation Spannungsschwankung
fall in voltage Absinken der Spannung
line voltage Leiterspannung

Verbs used for mounting and fitting equipment

accommodate unterbringen, aufnehmen
affix anhängen, anbringen
apply anwenden, auftragen, eintragen, anlegen (Spannung)
arrange anordnen, aufstellen, beiordnen, einrichten, abmachen
assemble zusammenbauen, zusammensetzen, montieren, einbauen
attach anbauen, befestigen, anbringen, aufsetzen, anhängen, vorsetzen
bolt verschrauben, festschrauben, anschrauben, befestigen, verbolzen
build bauen, errichten, aufbauen
embed einbetten, einlassen, einhülsen
encapsulate einkapseln, kapseln, einhülsen
enclose einkapseln, kapseln, einschliessen, umgrenzen
equip ausrüsten, ausstatten, einrichten, versehen (mit)
erect errichten, aufbauen, aufrichten, montieren, setzen
fasten befestigen, festmachen
fit einpassen, zusammenpassen, versehen, passen, anbringen, montieren
fix anbringen, befestigen, festmachen, einsetzen, aufspannen
house unterbringen, aufnehmen, einbauen
incorporate einbauen, unterbringen
insert einsetzen, aufstecken, einstecken, einlegen
install installieren, einbauen, montieren, errichten, aufstellen
lay legen, verlegen
locate plazieren, anordnen, in Stellung bringen
mount montieren, aufstellen, zusammensetzen, einbauen
place anordnen, plazieren, verlegen, auflegen, anlegen
provide vorsehen, liefern, sorgen für
screw festschrauben, verschrauben
set setzen, einrichten, einstellen, anziehen

anbringen to affix
A warning sign must be affixed to the unit housing warning the operator of the presence of high voltage within.
Ein Warnschild ist auf dem Geräteschrank anzubringen, das den Bedienungsmann (oder das Personal) davon in Kenntnis setzt, dass Hochspannung im Schrank ist.

anbringen to fit
This equipment may even be fitted where space is at premium.
Diese Geräte können selbst dort angebracht werden, wo der Platz äusserst knapp bemessen ist.

aufnehmen to house
The secondary windings of the current transformers must be earthed at the panel which houses the associated relays.
Die Sekundärwicklungen der Stromwandler müssen bei der Tafel geerdet werden, welche die dazugehörigen Relais aufnimmt.

aufstellen to install
The battery charger and panel are to be installed in the emergency diesel generator room.
Das Batterieladegerät mit der Ladetafel ist im Notdiesel(generator)-raum aufzustellen.

einbauen to mount
Where steam and oil gauges are mounted on the main propulsion console, care is to be taken that the steam or oil cannot come into contact with the energised parts.
Wo Dampf- und Oelmanometer in das Hauptfahrpult eingebaut werden, ist dafür Sorge zu tragen, dass Dampf oder Oel nicht in Kontakt mit spannungsführenden Teilen kommen kann.

einlassen to embed
Foundation bolts that are to be embedded in concrete must not be painted.
Fundamentschrauben, die in Beton eingelassen werden, dürfen nicht angestrichen werden.

unterbringen to locate
Control cabinets should be located in easily accessible positions.
Steuerschränke sollten an gut zugänglichen Stellen untergebracht werden.

zusammenbauen to assemble
Each switchboard must be completely assembled at the factory and tested by the manufacturer.
Jede Schalttafel ist in der Fabrik vollständig zusammenzubauen und vom Hersteller zu prüfen.

Verbs used for removing and stripping equipment

detach abnehmen, lostrennen, lösen, loslösen
disconnect Verbindung lösen, unterbrechen, abschalten, abstellen, abkuppeln, abhängen
dismantle abbauen, abbrechen, abmontieren, ausbauen, auseinandernehmen, demontieren
remove entfernen, abnehmen, abbauen, abheben
strip abmontieren, abbauen, auseinandernehmen, demontieren, von Umhüllung befreien, abziehen
take apart auseinandernehmen, zerlegen
take off abnehmen, wegnehmen, abheben
unbolt losschrauben, abschrauben, lösen
unfasten losmachen, lösen
unscrew abschrauben, aufschrauben, losdrehen, losschrauben

A microcomputer glossary

interface (Schnittstelle): as a noun, a physical circuit or subsystem which allows two different types of circuits or systems to be connected together. As a verb, to perform the above-mentioned function.

index register (Indexregister): a register whose function is to store a number which is used as a pointer to reference a parameter.

instruction (Befehl): the machine language words in a computer program that tell the computer what to do.

I/O device (Ein/Aus-Gerät): (input/output device) an input device or an output device. Examples of input devices are keyboards, sensors, and switches. Examples of output devices are displays, audio indicators, and X-Y-plotters.

interrupt (Unterbrechung): a request to the computer to service an external device. The external device can get the attention and the service of the computer by sending an interrupt signal to the computer. The computer will first finish the execution of the current instruction and it will service the external device.

keyboard (Tastatur): portion of a terminal that is used to input information to the computer. A terminal has two separate logical entities; one is the keyboard which is used to input information to the computer. The other is the printer or display.

LED: (light emitting diode) an electronic device which sends out light when it is turned on.

micro-code: the logic-level definition of the instruction set a bit-slice microcomputer or similar type of machine. Some microprocessors have a micro-coded architecture.

microcomputer: a computer whose CPU is a microprocessor.

microprocessor: a micro-electronic chip which contains an ALU, registers, input/output ports, and control and timing circuits. It is capable of performing arithmetic and logical operations.

mnemonic (Gerätekurzbezeichnung, mnemotechnischer Gerätename): the short-hand symbolic names or abbreviations which have predefined meaning, and which represent instructions in assembly language.

multiplexer (Datenübertragungs-Steuereinheit): a circuit which performs path selection function so that a computer can «talk» to several external devices one at a time.

object codes (Objektcode): the binary codes which are obtained as a result of the translation done by an assembler.

octal number (Oktalzahl): a number with a value ranging from 0 to 7.
operand (Rechengrösse): the number which follows an instruction. Example: *LDA 5. LDA* is a mnemonic which represents an instruction that tells the computer to load the operand into the accumulator. The operand in this case is 5. When this instruction is executed, the number 5 is loaded into the accumulator.
page (Seite, Bogen): a number of consecutive memory locations, nominally 256. If the memory of a computer has 4096 locations, the computer has 4096 ÷ 256 = 16 pages of memory.
parity bit (Prüfbit): a bit which is used to detect a transmission error.
pointer (Hinweisadresse, Zeiger): an address which corresponds to the beginning address of a program or table.
PROM: (**p**rogrammable **r**ead-**o**nly **m**emory) a random-access type of memory which can be programmed once only with a PROM programming machine.
RAM: (**r**andom **a**ccess **m**emory) memory devices which can store and retrieve information in any location in an amount of time which is independent of the memory location selected.
ROM: (**r**ead **o**nly **m**emory) memory devices which allow only retrieval of information. The information in the ROM is stored during manufacture before the ROM is put in operation. The computer can only read information out of the memory device, therefore, altering the information by the program is not possible.
register: an electronic device for storing information. In addition, a register may perform other functions such as shifting of bits, selectively clearing some bits, and selectively setting some bits.
resident assembler (Residentassemblierer): an assembler that runs

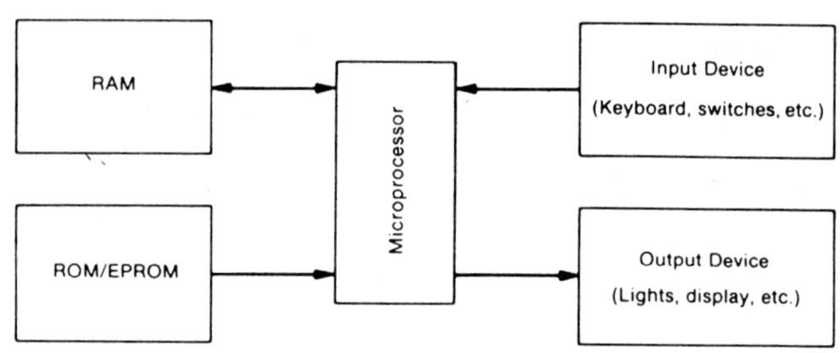

A basic microcomputer system

in a system that executes the object codes which are generated by that assembler.

sequential memory (sequentieller Speicher): memory in which information storage and retrieval is done in a sequential manner. For example, to read the contents of lcoation 50, the computer must search through locations 00 through 49. It cannot jump from location 00 to location 50.

software (immaterielle Ware, Programmausrüstung): computer programs which are written using instructions, operands, and labels in a logical fashion to accomplish the required task.

source code (Quellcode): the program as it appears in high-level language or in assembly language (as opposed to machine language).

stack (Stapel, Belege): storage and retrieval of information in a first-in-last-out fashion, i.e., the word that is stored first will be the word which is read out last. The term also refers to a collection of sequential memory cells which store and retrieve information in a first-in-last-out fashion.

stack pointer (Stapelzeiger): a register whose contents correspond to the address of the top of the stack.

table-driven (tabellengesteuert): a technique of branching to different subroutines. The starting address of a subroutine is contained in a table and in order to branch to a subroutine, the computer must first go to the appropriate location in the table.

terminal (Datenstation): an input/output device through which a person can «talk» to a computer.

tri-state (mit drei Stellungen): a type of logic circuit whose output can assume three states: a logical zero, a logical one, or a high impedance stated. In the high impedance state, the operation of other tri-state devices connected to the same point is uneffected.

word (Wort): a group of bits (the most common are 8, 16, 32, or 64, depending on the computer) which comprises a single unit of information.

access time (Zugriffszeit): the time between a request for information from a storage medium and the time the information is available.

accumulator (Akkumulator): register(s) which contain results of the arithmetic/logic unit operations.

address (Adresse): a computer word used for designating a specific location in memory.

A/D converter (A/D-Umsetzer): (**a**nalog-to-**d**igital converter) an electronic device for changing a DC voltage to a binary-coded-value. Computers cannot process a continuous voltage waveform but can process binary value.

addressing mode (Adressiermodus): techniques for specifying memory locations for the purpose of storing and/or retrieving information.

ALU: (**a**rithmetic/**l**ogic **u**nit) the hard-ware portion of a computer which performs arithmetic functions, such as addition and substraction and logic functions such as shift, AND and OR.

architecture (Aufbau): the logical organization of the hardware portion of the CPU.

assembler (Assemblierer): a program which translates application programs written in English-like symbolic language (assembly language) into machine language (binary).

assembly listing (Übersetzungsprotokoll): a listing which shows the assembly-language program and the translated machine-language program.

ASCII: (American standard code for information interchange) a 7-bit code for representing the English alphabet, numbers, and special symbols, such as $, £, and §.

baud: a data transmission-rate unit. For most applications, a baud is equal to one bit per second.

BCD: (binary coded decimal) a coding scheme for representing the ten decimal numbers.

binary number: a number whose digits are either zero or one. Computers can «read» and «write» only binary numbers.

bit: a binary digit.

bit-slice microprocessor (Bitteil-Mikroprozessor): an *n*-bit wide processing element usually connected in parallel to implement a microcomputer of *n*-bit word length. The instruction set is customer defined (by a micro-code).

buffer (Zwischenspeicher): a register for holding temporary data.

bus (Übertragungsweg, Bus): a set of conductors which carries all necessary computer signals, such as data, address, and control signals.

byte: (Zelle): a computer word of 8 bits.

cell (Zelle): a memory bit.

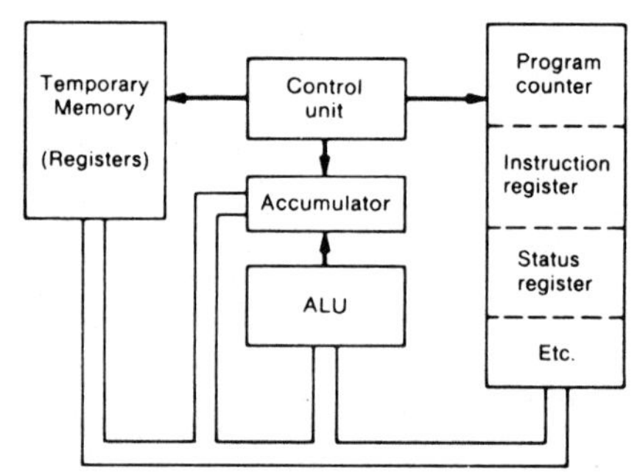

Block diagram of a basic microprocessor

CMOS: (**c**omplementary **m**etal **ox**ide **s**ilicon) a very long power logic technology.

CPU (**c**entral **p**rocessing **u**nit) (Zentraleinheit): the arithmetic/logic unit (ALU), registers, and control circuits of a computer. The CPU decodes instructions, issues timing signals, and performs all control functions.

cross assembler (Querassemblierer): an assembler for use in a computer with an instruction set other than the one which the application program is written for.

cycle time (Zykluszeit): the shortest period of time at the end of which a sequence of events repeats itself. Some of the events may be retrieving a word from a memory cell, interpreting the meaning of the word, and executing the instruction.

D/A-converter (D/A-Umsetzer): (**d**igital-to-**a**nalog converter) a device which changes a binary signal into a dc voltage level.

decode (entschlüsseln, decodieren): to translate or interpret a computer word into something familiar or useful for performing tasks.

display (Anzeige): a device which shows a letter or a number or simply emits light; as a miniature light bulb.

DMA (**d**irect **m**emory **a**ccess) (Direktspeicher Zugriff): a technique for receiving information from or transferring information to the main memory of a computing system without having the CPU involved.

drivers (Treiber): circuits which increase the driving capability of an output circuit.

encoder (Geber, Codierer): a device for translating a one-bit signal into a multibit signal.

editor (Arbeitsvorbereiter, Editor): a computer program which aids a programmer in his source program creation. An editor helps to perform the following functions: typing in program, making corrections, assigning line numbers to all lines, resequencing of line numbers, locating selected characters, and listing of partial or entire program.

EPROM: (**e**rasable **p**rogrammable **r**ead **o**nly **m**emory) a type of non-volatile random-access memory chip which can be programmed in the field.

execute cycle (Ausführungszyklus): the amount of time for an instruction, that has been fetched, to be executed.

fetch cycle (Abrufzyklus): the amount of time for an instruction or data to be retrieved from a memory location.

flag (Kennzeichen, Kettung): a bit in a register which is used to keep tract of the state of an input/output device, the state of a register, or the state of the microprocessor. For example, a flag can be used to keep tract of the state of the interrupt feature by setting the flag to 0 for an uninterrupted state and to 1 to denote an interrupt state.

firmware (Fest[speicher]wert):

software that is stored in random-access read-only memory.

handshaking (Austausch von Synchronisationsimpulsen): a predefined procedure for sending information from a terminal (sender) to a computer (receiver) or vice versa that informs the sender that the receiver is ready to receive.

hardware (physikalische Ausrüstung [und Bauteile]): computer circuits and peripheral devices.

hexadecimal (hexadezimal [zur Basis 16]): a number with a digit value ranging from 0 to 15. The 16 hexadecimal numbers are: 0, 1, 2, 3, 4, 5, 6, 7, 8, 9, A, B, C, D, E, and F.

high-level language (Höhere Programmiersprache): a computer language such as FORTRAN or BASIC which resembles English. One high-level language instruction is translated by a compiler or interpreter program into several machine-language instructions.

IC: (**i**ntegrated **c**ircuit) very small electronic circuits contained in a single package which perform sophisticated functions.

"Now we have solved our fly-ash disposal problem."
(Power, New York)

At the meeting

Let's arrange a meeting with the microprocessor section.
Wir werden eine Besprechung mit der Mikroprozessor-Gruppe vorbereiten.

Mr. Aubrey wants to see us at ten fifteen.
Mr. Aubrey möchte uns um zehn Uhr fünfzehn sprechen.

It won't take very long.
Es wird nicht sehr lange dauern.

I've tried to contact you all over the place.
Ich habe überall versucht, Sie zu erreichen.

I've arranged to see some customers this evening.
Ich habe abgemacht, mich heute abend mit Kunden zu treffen.

Well, we mustn't keep you.
In Ordnung, wir wollen Sie nicht aufhalten.

I recall a talk with Mr. White on those TTL circuits half a year ago.
Ich erinnere mich an ein Gespräch vor einem halben Jahr mit Mr. White über diese TTL-Schaltkreise.

We shouldn't get confused over these findings.
Wir sollten uns wegen dieser Feststellung nicht verwirren lassen.

I felt all along that there was something wrong.
Die ganze Zeit dachte ich schon, dass da etwas nicht richtig sein könnte.

We've got to allow for depreciation.
Wir müssen die Abschreibung berücksichtigen.

We should draw up a capsule appraisal.
Wir sollten eine kurzgefasste Abschätzung der Lage entwerfen.

First of all we shall tackle the control problem.
Zuerst müssen wir mit dem Steuerungsproblem fertig werden.

This is only a very rough-and-ready guide to our controls.
Dieses ist nur eine sehr grobe Vorstellung von unserer Steuerung.

Mr. Summers has put forward a number of good ideas which make good sense.
Mr. Summers hat eine Reihe guter Vorschläge vorgelegt, die sich als (durchaus) sinnvoll erweisen.

I'll tell you off the record.
Ich sage Ihnen nebenbei (nicht offiziell).

We shouldn't interfere too much with the engineers' work.
Wir sollten uns nicht so sehr in die Arbeiten der technischen Abteilung einmischen.

The most recent modification on these diagrams are marked in red.
Die letzten Änderungen auf diesen Plänen sind in Rot eingetragen.

The progress report for Salamanca power station will be issued tomorrow.
Der Montagebericht für das Kraftwerk Salamanca wird morgen herausgegeben.

Has Mr. Jones got enough people working with him?
Hat Mr. Jones genügend Leute zu seiner Unterstützung?

Can we now revert to the components to be supplied by our subcontractors?
Können wir nun auf die Teile zurückkommen, die von unseren Unterlieferanten zugeliefert werden?

We've got to order change-over devices und control panels ourselves.
Umschaltgeräte und Steuertafeln müssen wir selbst bestellen.

Now to the last but one item of this meeting.
Wir kommen jetzt zu dem zweitletzten Punkt dieser Besprechung.

We would be very competitive if it weren't for those ECL circuits as per specification.
Wir könnten sehr konkurrenzfähig sein, wenn da nicht, laut Spezifikation, die ECL-Schaltkreise wären.

There is still something else that bothers me.
Da ist noch etwas anderes, was mir Kummer bereitet.

Now to something more promising.
Nun zu etwas Erfreulicherem.

That really surprises me. Why didn't you tell us earlier?
Das überrascht mich wirklich. Warum hast du uns das nicht schon eher erzählt?

I wanted some proper results first.
Ich benötigte erst einige konkrete Ergebnisse.

Machine tool phrases
Fachliche Redewendungen aus dem Werkzeugmaschinenbau

A Commissioning of equipment
Inbetriebsetzung von Maschinen

1. **See that instructions under lubrication Sheet X have been carried out**
   Überzeugen Sie sich davon, dass alle Schmiervorschriften nach Blatt X befolgt sind

2. **Note which speed is engaged and set the range change gears in neutral. Engage the feed brake, Fig. Y, Sheet X**
   Die gewählte Drehzahlreihe feststellen und die Wechselräder für den Bereich auf Null stellen. Die Vorschubbremse betätigen, siehe Blatt X, Bild Y

3. **Check the direction of rotation of the motor (clockwise looking on pulley)**
   Die Drehrichtung des Motors prüfen (muss auf die Riemenscheibe gesehen, im Uhrzeigersinn sein)

4. **Run the motor to check that the lubrication system (Sheet X) is working. Observe the pressure gauge, sight feeds for the spindle drum, and all drip points by removing covers. Regulate the oil feed where necessary**
   Motor laufen lassen und beobachten, ob die Schmierung richtig arbeitet (Blatt X). Das Manometer, die Schaugläser für die Spindeltrommel und alle Tropfschmierpunkte beobachten; hierbei sind die entsprechenden Deckel abzunehmen. Falls erforderlich, ist der Ölfluss nachzuregeln

5. **Disengage the index clutch and handwind the machine through a complete cycle to check that all parts operate freely**
   Die Schwenkkupplung ausrücken und die Maschine von Hand durch einen vollständigen Arbeitstakt drehen, um zu prüfen, ob alle Teile frei arbeiten

6. **Engage the index clutch, disengage the collet operating shoe, and handwind through index. If excessive pressure is required on the handcrank, locate and correct the cause**
   Die Schwenkkupplung einrücken, den Klotz für die Spannzangenbetätigung ausrücken und von Hand schwenken. Wenn eine übermässige Kraft für Betätigung der Handkurbel erforderlich ist, so ist die Ursache festzustellen und zu beheben

7. **Engage the low range change gears and run the machine under power feed**
   Die Wechselräder für den unteren Bereich einrücken und die Maschine mit Arbeitsvorschub laufen lassen

B Your special attention please
Bitte besonders beachten

1. **Check the oil level in the sump. This oil level must not fall below the level indicated on the dipstick**
   Den Ölstand in der Wanne prüfen. Er darf nicht unter den auf dem Tauchstab angezeigten Stand sinken

2. **Regular checks of the spindle bearing and spindle gear sight feeds are necessary**
   Die Sichtschmierung zu den Spindellagern und Antriebsrädern ist regelmässig zu überprüfen

3. **Turn the Purolator knob daily and apply oil gun to all nipples**
   Den «Purolator»-Filter täglich durchdrehen und mit der Ölpresse alle Nippel abschmieren

4. **Whenever the machine is stopped for adjustment, ensure, as a safety measure, the front-rear selector switch is set for the side at which the operator is working**
   Sobald die Maschine zwecks Einstellung angehalten wird, ist sicherheitshalber zu kontrollieren, ob der Wahlschalter «Vorn-Hinten» auf jene Seite eingestellt ist, auf welcher der Bedienungsmann beschäftigt ist

5. **Ensure the feed clutch is disengaged before starting the machine. When starting from cold, use the jog button several times to allow the oil to circulate before running the machine**
   Kontrollieren Sie, ob die Vorschubkupplung ausgerückt ist, bevor man die Maschine anlässt. Wird die Maschine im kalten Zustand in Betrieb gesetzt, so ist die Taste «Schrittbetrieb» mehrmals zu drücken, damit das Öl in Umlauf gesetzt wird, bevor die Maschine auf Touren kommt

6. **For normal starting press start button and allow control gear to change from star to delta connections before engaging the clutch**
   Beim normalen Anlassen Starttaste drücken und dem Anlasser Zeit lassen, von Stern auf Dreieck zu schalten, bevor die Kupplung eingerückt wird

7. **Should it be necessary to engage the clutch before starting the motor, start the motor on the jog button**
   Sollte es notwendig sein, die Kupplung vor dem Starten des Motors einzurücken, ist der Motor durch die Taste für den Schrittbetrieb anzulassen

8. **Ensure that tools are clear of work before using the fast motion lever**
   Achten Sie darauf, dass sich alle Werkzeuge frei vom Werkstück bewegen, bevor der Hebel für Eilvorschub geschaltet wird

9. **If a slipping clutch disengages, locate the cause of overload before re-engaging the clutch and restarting the machine**

Wenn eine Rutschkupplung ausrückt, die Ursache der Überlastung feststellen, bevor man die Kupplung wieder einrückt und die Maschine anlässt

10. **Additional plungers and springs can be fitted to the slipping clutch to increase torque; spares are included in the machine equipment kit for this purpose. They should only be applied in exceptional circumstances**
Auf der Kupplung lassen sich zusätzliche Stifte und Federn zur Erhöhung des Drehmoments anbringen; entsprechende Ersatzteile sind der Maschinenausrüstung beigelegt. Man sollte sie jedoch nur in Ausnahmefällen verwenden

11. **Do not handwind the machine through the feed part of the cycle when threading attachments are in use, unless the diehead is opened**
Die Maschine nicht von Hand durch den Vorschubabschnitt des Taktes drehen, wenn eine Gewindeschneideinrichtung eingesetzt ist, es sei denn, der Gewindeschneidkopf ist geöffnet

12. **Regularly check adjustment of all multiple plate clutches by hand. They should be neither so slack as to cause slip, nor so tight that excessive pressure is required to engage them**
Die Einstellung aller Lamellenkupplungen ist regelmässig mit der Hand zu prüfen. Sie dürfen weder so schlaff sein, dass sie rutschen, noch so straff, dass eine übermässige Kraft zum Einrücken nötig ist

13. **Remember to release the slide stop before changing stroke or slide adjustments and to reset afterwards**
Daran denken, den Schlittenanschlag vor einer Hubänderung oder Schlitteneinstellung zu lösen und anschliessend neu einzustellen

14. **After adjusting cross slides by means of the micrometer knob, remember to clamp the locking screw**
Nach einer Verstellung der Querschlitten durch die Mikrometerschraube nicht vergessen, die Stellschraube zu sichern

15. **Gib strip adjustment needs great care. Strips should not be so tight as to bind or allowed to become too slack**
Das Einstellen der Führungsleisten erfordert grosse Sorgfalt. Die Leisten dürfen nicht so straff sein, dass sie Schwergang verursachen, jedoch auch nicht übermässiges Spiel aufweisen

16. **Avoid whenever possible the use of soluble oils which could cause rust**
Möglichst keine Emulsionen verwenden, die Rosterscheinungen hervorrufen könnten

17. **Keep the coolant level as high as possible**
Den Kühlmittelstand so hoch wie möglich halten

18. **Clean the tray of swarf and sediment at regular intervals**

Die Kühlmittelwanne in regelmässigen Zeitabständen von Spänen und Ablagerungen reinigen

C Controls
Bedienungselemente

1. **Principal controls are the main motor push-buttons, feed clutch lever, fast-slow clutch lever, handwind crank, index clutch lever, hand collet lever, bar feed shoe lever and bar stop lever**
   Die Hauptbedienungselemente sind die Tasten (oder: Druckknöpfe) für den Spindelmotor, der Vorschubhebel, der Eilgang-Langsam-Hebel, die Handkurbel, der Schwenkhebel, der Handhebel für die Spannzange, der Hebel für den Nachschubklotz und jener für den Stangenanschlag

2. **Subsidiary controls are the auto-stop knob on the auto-stop (stock exhaustion) mechanism, the manual trip switch and conveyor push-buttons**
   Die Hilfsbedienungselemente sind der Knopf für die automatische Abschalteinrichtung (wenn eine Stange aufgebraucht ist), der Handabschalter und die Tasten für den Späneförderer

3. **Main motor push-buttons and front-rear selector switches. The three motor controls: stop button, front-rear selector switch and start-jog button are duplicated on the front and rear of the machine**
   Die Taster für den Spindelmotor und der Wahlschalter «Vorn-Hinten». Die drei Steuerungselemente für den Motor: Halttaster, Wahlschalter «Vorn-Hinten» und Taster für den Schrittbetrieb sind vorn und hinten auf der Maschine angeordnet

4. **The front-rear selectors are two-position switches and both must be at the same setting, either front or rear, before the appropriate start-jog button can be used to start the motor**
   Die Wahlschalter «Vorn-Hinten» haben zwei Stellungen, und beide müssen sich in der gleichen Stellung befinden, entweder vorn oder hinten, bevor der Taster für den Betrieb des Spindelmotors betätigt wird

5. **The machine is arranged so that the main motor will stop when bar stock is exhausted. If this occurs, the motor cannot be restarted until the selector switch on the control panel is set to «off» position and the bar feed trip plungers are pulled out**
   Die Steuerung ist so ausgelegt, dass der Antriebsmotor stillgesetzt wird, wenn der Stangenvorrat aufgebraucht ist. Ist dies der Fall, so kann man den Motor nicht wieder anlassen, bevor der Wahlschalter auf der Steuertafel in die «Aus»-Stellung gebracht ist und die Abschaltstifte des Stangennachschubs herausgezogen sind

6. **The stop button, on either side of the machine, will stop the motor irrespec-**

tive of the setting of any other switch. As a safety precaution a front-rear selector switch should always be set to the side of the machine on which the operator is making adjustments
Die Stopptasten, von denen sich je eine auf jeder Maschinenseite befindet, setzt den Motor unabhängig von der Stellung aller anderen Schalter still. Aus Sicherheitsgründen sollte der «Vorn-Hinten»-Wahlschalter immer auf jene Seite der Maschine eingestellt werden, auf welcher der Bedienungsmann Einstellungen vornimmt

7. Feed clutch levers are fitted at front and rear of the machine. They have three positions: «up» to engage the feed clutch, «down» to engage the feed brake (used to prevent over-run when disengaging the feed clutch) and a neutral position in which the handwind gear can be engaged
Die Vorschubkupplungshebel sind an der Front- und der Hinterseite auf der Maschine angebracht. Sie haben drei Stellungen: «Oben» zum Einrükken der Vorschubkupplung, «unten» zum Betätigen der Vorschubbremse (um ein Überlaufen beim Ausrücken der Vorschubkupplung zu vermeiden) und eine Nullstellung, in welcher man die Handkurbel betätigen kann

8. Fast-slow clutch levers are fitted at front and rear for use when setting up. Each lever is free on the cross shaft and must be pulled outwards to engage the slot in the shaft end
Die Eilgang-Langsam-Kupplungshebel zum Rüsten der Maschine sind vorn und hinten angebracht. Jeder Hebel kann frei auf einer Querwelle gleiten und muss nach aussen gezogen werden, damit er in die Nut am Ende der Welle eingreift

9. These levers should be used with care, and only when all tools are clear of the work
Diese Hebel sind mit Vorsicht zu betätigen, und zwar nur dann, wenn alle Werkzeuge frei am Werkstück vorbeigehen können

10. They have two positions: «up» to engage slow feed and «down» to engage fast motion. There is no neutral because a freewheel or roller over-running clutch is in the feed drive
Sie haben zwei Stellungen: «Oben» zum Einrücken des Arbeitsvorschubs und «unten» zum Einrücken des Eilgangs. Es gibt keine Nullstellung, da ein Freilauf oder eine die Rollen überlaufende Kupplung im Vorschubantrieb eingebaut ist

11. The handwheel crank can be fitted to the handwind pinion shaft at either front or rear
Die Handkurbel kann man vorn oder hinten auf die Verstellwelle aufsetzen

12. Handwind can only be engaged with the feed clutch lever in neutral, as an interlock prevents the pinion being slid into mesh
Eine Betätigung durch die Handkurbel ist nur möglich, wenn der Kupp-

lungshebel in seiner Nullstellung liegt, da eine Verriegelung es verhindert, dass das Antriebsritzel in den Eingriff gleitet

13. **Index clutch levers are provided at front and rear of the machine and an interlock is provided so that the clutch can be disengaged only during part of the slow feed period of the cycle**
Die Schwenkkupplungshebel sind vorn und hinten auf der Maschine angebracht und so verriegelt, dass man die Kupplung nur in der Vorschubperiode des Taktes ausrücken kann

14. **Because the index clutch is spring-loaded into engagement, latches are provided adjacent to the index clutch levers to hold the clutch out against the springs**
Da die Schwenkkupplung durch Federkraft in Eingriff gehalten wird, sind Klinken neben den Schwenkhebeln vorgesehen, um die Kupplung gegen die Federkraft ausgerückt zu halten

15. **These latches must be disengaged to allow the indexing clutch to be engaged**
Diese Klinken müssen freigegeben sein, bevor man die Schwenkkupplung einrücken kann

16. **The hand collet lever is loose and is fitted into its boss for checking collet adjustments**
Der Handhebel für die Spannzange ist abnehmbar und zum Prüfen der Zangeneinstellung auf den Wellenstumpf aufsteckbar

17. **The bar feed shoe lever operates the bar feed shoe and allows the bar feed to be disengaged for setting up or bar loading purposes**
Der Hebel für den Nachstellklotz betätigt den Klotz für den Stangennachschub, so dass man den Nachschub beim Rüsten oder Laden von Stangen ausrücken kann

18. **The bar stop lever is used to retract the bar stop against its operating spring, to allow the removal of bar ends from the collet**
Der Hebel für den Stangenanschlag wird zum Zurückziehen des Anschlags gegen die Betätigungsfeder verwendet, damit man die Stangenreste aus der Spannzange entfernen kann

19. **The auto-stop knob is pulled out to release the latch holding the limit switch after the latch has engaged when a bar is exhausted**
Der Knopf der automatischen Abschalteinrichtung wird zur Freigabe der Klinke, die den Endschalter hält, nach dem Aufarbeiten einer Stange betätigt

20. **The manual trip switch on the main control panel is used when the operator wishes to stop the machine just before the next index for checking components or making adjustments. This avoids the operator having to stand by ready to disengage the feed clutch**
Der handbetätigte Stoppschalter im Hauptsteuerschrank wird betätigt,

wenn der Bedienungsmann die Maschine kurz vor dem nächsten Schwenken anhalten will, um die Werkstücke nachzumessen oder eine Neueinstellung vorzunehmen. Hierbei ist es nicht nötig, dass der Bedienungsmann bereitsteht, um den Vorschubhebel auszurücken

D Main drive
Hauptantrieb

1. **The motor is mounted on a platform**
Der Motor ist auf einer Spannplatte montiert

2. **This platform pivots on a shaft mounted in a bracket which is bolted to the side of the tray**
Diese Spannplatte ist auf einer Welle schwenkbar, die in zwei auf der Seite der Wanne angeschraubten Lagern aufgehängt ist

3. **Two screws are provided to adjust the platform for belt tension and should be adjusted to clamp the platform securely**
Die beiden Schrauben, welche die Spannplatte zwecks Riemenspannung halten, müssen gut angezogen sein

4. **The constant speed pulley shaft drives the second shaft by means of the range change gears which provide high and low speed ranges**
Die mit konstanter Drehzahl laufende Riemenscheibenwelle treibt die zweite Welle über die Schieberäder für den Drehzahlbereich, die für den Bereich der hohen und niedrigen Drehzahlreihe bestimmt sind

5. **The sliding double gear is located on the second shaft by a spring loaded plunger engaging a V notch in one of the splines**
Das doppelte Schieberad wird auf der zweiten Welle durch einen Federstift in einer Einkerbung in einer der Keilnuten aufgenommen

6. **A neutral position is provided for running the oil pump independently**
Eine Nullstellung ist vorgesehen, damit die Ölpumpe unabhängig laufen kann

7. **Access to the range change gears is obtained by removing the right-hand front cover and the gears are moved by levering with a bar against the housing**
Die Schieberäder sind durch Abnehmen des vorderen rechten Deckels erreichbar und durch die Hebelwirkung einer gegen das Gehäuse gedrückten Stange bewegbar

8. **Speed pick-off gears give the different spindle speeds in each range and carry the drive from the second shaft to the centre drive shaft**
In jedem Bereich werden die verschiedenen Drehzahlen durch Wechsel-

räder bewirkt, die den Antrieb von der zweiten auf die mittlere Welle übertragen

E Electrical equipment
Elektrische Ausrüstung

1. **The standard control panel includes an isolator and main fuses**
   Der Schaltschrank in der Normalausführung ist mit einem Trennschalter (oder: Hauptschalter) und den Hauptsicherungen versehen

2. **It is necessary to connect the line and earth**
   Es ist erforderlich, die Erdleitung anzuschliessen

3. **Direction of main motor rotation should be clockwise**
   Der Spindelmotor soll sich, auf die Riemenscheibe gesehen, im Uhrzeigersinn drehen

4. **Careful attention should be given to see that all main and control fuses, overload tripping devices and transformer tappings are correct for the customer's electrical supply**
   Es ist sorgsam darauf zu achten, dass alle Sicherungen der Haupt- und Steuerstromkreise, Auslöser für Überlast und Transformatoranschlüsse (Anzapfung) richtig für das Netz des Kunden ausgelegt sind

5. **These details are checked before dispatch by our company, but should be rechecked in case of any change in the customer's supply**
   Diese Einzelheiten werden von uns vor dem Versand geprüft, sollten jedoch nochmals überprüft werden, wenn irgendwelche Änderungen am Kundennetz vorgenommen werden sind

6. **Wiring diagram, spare parts and instruction sheets are included in the pocket inside the control panel door**
   Schaltplan, Ersatzteile und Betriebsanweisungen werden in der an der Innenseite der Tür befindlichen Tasche mitgeliefert

F Spindle drum
Spindeltrommel

1. **The spindle drum carries the work spindles and the centre guide on which the centre tool block slides**
   Die Spindeltrommel trägt die Arbeitsspindeln und die Mittenführung, auf welcher der Hauptschlitten gleitet
2. **This arrangement ensures consistent alignment of work spindles with the centre block**

Diese Konstruktion gewährleistet die stets gleichbleibende Ausrichtung der Arbeitsspindeln mit dem Hauptschlitten

3. **The end thrust of the tools on the spindle drum is taken by the flange of the drum on thrust blocks bolted to the rear face of the drum housing**
   Der Axialdruck der Werkzeuge auf die Spindeltrommel wird durch auf die Fläche des Trommelgehäuses verschraubten Druckstücke gegen den Trommelflansch aufgenommen

4. **The spindle drum is geared to the four-slot Geneva wheel and is indexed by the Geneva arm on the camshaft**
   Die Spindeltrommel hat eine Zahnradübersetzung zum Malteserrad mit vier Nuten und wird durch den Malteserhebel auf der Hauptkurvenwelle geschwenkt

G Workspindles and feed drive
Arbeitsspindeln und Vorschubantrieb

1. **The workspindles are mounted in extra precision preloaded ball journal bearings at the front and an extra precision parallel roller bearing at the rear**
   Die Arbeitsspindeln laufen vorn in vorgespannten Höchstgenauigkeits-Kugellagern und hinten in einem zylindrischen Genauigkeitswälzlager

2. **The front bearing end caps also carry the cross slide stop screws which are used to cancel out slight variations in spindle position and drum indexing**
   Die vorderen Lagerdeckel sind weiterhin mit den Anschlagschrauben für die Querschlitten versehen, die dem Ausgleich kleiner Ungenauigkeiten zwischen der Lage der Trommel nach dem Schwenken und der Spindel dienen

3. **Collets are of the drawback type with internal thread and are screwed up to the shoulder on the collet tube**
   Die Spannzangen (Selbstrückzug) mit Innengewinde sind gegen den Absatz auf dem Zangenrohr verschraubt

4. **The feed drive is taken from the centre shaft by gears to the first feed pick-off gear shaft and then through the pick-off gears to the second pick-off gear shaft**
   Der Vorschubantrieb wird von der Mittelwelle durch eine Übersetzung zu den ersten Wechselrädern und von dort zu einem zweiten Wechselräderpaar geleitet

5. **Feed pick-off gears are mounted on taper shafts and an extractor is supplied in the tool kit**
   Die Wechselräder sitzen auf Kegelwellen; eine Abziehvorrichtung gehört zur Werkzeugausrüstung der Maschine

H Feed trip mechanism
Vorschubabschaltung

1. **The feed trip mechanism is provided to disengage the feed clutch and engage the brake when the trip solenoid is energised by switches operated by:**
   a) disengagement of the feed slipping clutch
   b) manual trip switch
   c) the auto-stop mechanism which operates when bar stock is exhausted in any spindle
   d) special tooling safety devices
   Die Vorschub-Abschaltvorrichtung ist vorgesehen, um die Vorschubkupplung auszurücken und die Bremse zu betätigen, wenn der Abschaltmagnet wegen der folgenden Ursachen erregt wird:
   a) Ausrücken der Rutschkupplung für den Vorschub
   b) Handabschalter
   c) Betätigung der automatischen Abschalteinrichtung, wenn die Stange in einer Spindel aufgebraucht ist
   d) Besondere Sicherheitsschalter für die Werkzeugausstattung

P.C.M.
'And do you know, darling, ever since those lovely engineers came and installed this p.c.m. (I think that means postcode motivated or something), I haven't had a single wrong number...'

Technical phrases

Commissioning of equipment

Inbetriebsetzung von Anlagen

1 **Disconnect power (plug) before inserting or removing circuit cards**
Vor dem Einsetzen oder Herausnehmen der Leiterplatten ist der Stecker für die Einspeisung zu ziehen

2 **This device allows fully automatic operation**
Durch dieses Gerät wird eine vollautomatische Arbeitsweise erreicht

3 **The equipment is started by operating switch b1**
Die Inbetriebnahme der Anlage erfolgt durch Betätigung des Schalters b1

4 **Set selector switch b2 to (position) "automatic"**
Wahlschalter b2 auf «Automatik» stellen

5 **Turn switch b3 to (position) "X"**
Schalter b3 auf «X» stellen

6 **Set limit switch b4 to "0"**
Endschalter b4 auf «0» stellen

7 **Press lighted push-button b5**
Leuchttaster b5 betätigen (drücken)

8 **Warning light h1 goes out and after two seconds indicating light h2 lights up (or: comes on)**
Warnleuchte h1 erlischt und nach zwei Sekunden leuchtet Anzeige(leuchte) h2 auf

9 **Relay d1 monitors the supply line**
Relais d1 überwacht die Speiseleitung

10 **Contactor c1 is actuated (or: tripped) by relay d2**
Schütz c1 wird durch Relais d2 betätigt (abgeschaltet, ausgelöst)

11 **What happens if the emergency generator does not come on line?**
Was geschieht, wenn sich der Notgenerator nicht zuschaltet?

12 **A number of safety precautions have been taken against such a case**
Für einen solchen Fall sind mehrere Sicherheitsvorkehrungen getroffen worden

13 **The system was modified and is now operating satisfactorily**
Das System wurde abgeändert und arbeitet nun zufriedenstellend

14 **All diagrams are now in line with this modified system**
Alle Zeichnungen stimmen jetzt mit der abgeänderten Anlage überein

15 **The controls are simple and easy to operate**
Die Steuerungen sind einfach und leicht bedienbar

16 **The test equipment permits the testing of all the individual components taking part in changeover**
Die Prüfeinrichtung gestattet die Prüfung aller an der Umschaltung beteiligten Einzelgeräte

Designing a device

Konstruktion eines Gerätes

1 **This device provides a high degree of safety and convenience for operating personnel**
Dieses Gerät bietet ein hohes Mass an Sicherheit und ist leicht zu bedienen (wörtlich: . . . Erleichterung für das Betriebspersonal)

2 **These control devices lend themselves to modular designs**
Diese Steuergeräte eignen sich für eine Modularkonstruktion

3 **Sufficient space should be allowed between the components on the rack**
Genügend Platz sollte zwischen den Apparaten auf dem Gestell vorhanden sein

4 **You may even fit this equipment where space is at premium**
Sie können diese Geräte selbst dort anbringen, wo der Platz äusserst knapp bemessen ist

5 **Corrosion-resisting steel will be used for all bolts and nuts**
Korrosionsbeständiger Stahl wird für alle Schrauben und Muttern verwendet

6 **All external and internal surfaces shall be painted, except where painting will interfere with satisfactory operation of the equipment**
Alle äusseren und inneren Flächen sind zu streichen, ausgenommen wo die Farbe den ordnungsgemässen Betrieb der Geräte beeinträchtigen könnte

7 **Internal surfaces will preferably be finished glossy white**
Innenflächen erhalten vorzugsweise einen weissglänzenden Endanstrich

8 **Foundation bolts and other steel foundation members for embedding in concrete shall not be painted**
Fundamentschrauben und andere Stahlteile für Fundamente, die in Beton eingelassen werden, sind nicht zu streichen

9 **Where ventilation is necessary, louvred covers of non-ferrous material shall be employed**
Wo Ventilation notwendig ist, sind mit Kühlschlitzen versehene Abdeckungen aus NE-Metall zu verwenden

10 **Mechanisms shall be designed to prevent sticking due to corrosion**
Die Vorrichtungen sind so zu konzipieren, dass ein Festsetzen durch Rost verhindert wird

11 **Each device is suitable for manual and remote control**
Jedes Gerät ist für Hand- und Fernsteuerung geeignet

12 **All switches must be of the plug-in type**
Alle Schalter müssen steckbar sein

13 **Not more than two wires shall be connected to one terminal**
Nicht mehr als zwei Leiter sind an einer Klemme anzuschliessen

14 **No equipment shall be shipped until it has been inspected and released for shipment by the consulting engineers**
Geräte dürfen erst dann spediert werden, wenn sie vom Konsulenten inspiziert und freigegeben worden sind

In the drawing office (or: designing office)

Im Konstruktionsbüro

1 **We reduce the one-line diagrams to half their original size in order to obtain handy plans**
Wir verkleinern die Übersichts(schalt)pläne um die Hälfte ihrer Originalgrösse, damit wir handliche Unterlagen erhalten (d. h. Verkleinerung mittels Photokopiermaschine; z. B. mit «Repromaster»)

2 **The first thing to do when changes have been made to the systems is to modify the appropriate one-line diagrams (accordingly)**
Wenn Änderungen an der Anlage vorgenommen worden sind, müssen als erstes die entsprechenden Übersichtspläne abgeändert werden

3 **It is important for these diagrams to contain as much detail as possible since it eases the work of all concerned**
Es ist wichtig, so viele Einzelheiten wie möglich auf diesen Plänen anzugeben, denn dadurch wird allen, die damit zu tun haben, die Arbeit erleichtert

4 **If you have a customer on the phone, and he has the same reduced (or: scaled down) one-line diagram in front of him, you can avoid misunderstandings (or: you know you are talking about the same thing)**
Wenn Sie einen Kunden am Telefon haben, und er hat denselben verkleinerten Übersichtsplan vor sich, so vermeiden Sie Missverständnisse

5 **Sometimes people call up who have difficulty in expressing themselves clearly in English; our comprehensive plans here help to overcome those problems**
Manchmal rufen Leute an, die Schwierigkeiten haben, sich im Englischen klar auszudrücken; hier sind unsere ausführlichen Pläne eine grosse Hilfe

6 **In this file you will find all the circuit diagrams (wiring diagrams) of the equipment on which our group is working**
In diesem Ordner (Dossier) finden Sie alle Stromlaufpläne (Verdrahtungspläne) für Anlagen, die von unserer Gruppe bearbeitet werden

7 **The engineer in charge of the project is responsible for updating all diagrams**
Der Projektleiter ist dafür verantwortlich, dass alle Pläne dem tatsächlichen Stand entsprechen

8 **For key plans great care must be taken of the size of lettering used; otherwise the information might become illegible when they are reduced**
Bei wichtigen Plänen (Schlüsselplänen) muss man streng auf die Schriftgrösse achten; bei der Verkleinerung von Plänen können die Angaben sonst unleserlich werden

9 **When an arrangement drawing is referred to in our specification, a drawing is meant that gives three separate views, namely front view, side elevation and plan**
Wo in unserer Spezifikation von Anordnungsplänen die Rede ist, sind darunter Zeichnungen mit drei verschiedenen Ansichten zu verstehen, nämlich Frontansicht, Seitenansicht und Draufsicht

10 **When a schematic diagram is referred to, a drawing that explains the operation and control of devices is meant**
Wo von Prinzipschaltbildern die Rede ist, so sind Zeichnungen gemeint, aus denen die Funktion und Steuerung von Geräten zu ersehen ist

Rating of equipment

Bemessung (Auslegung) von Maschinen und Geräten

1 **As per ANSI the rating of electrical equipment in general is expressed in VA, HP (h.p.), kW or appropriate units**
Laut ANSI (American National Standards Institute) werden die Nenndaten für elektrische Maschinen und Geräte im allgemeinen in VA, PS, kW oder in den entsprechenden Einheiten angegeben

2 **The rated continuous current of the largest "Biggs Switchgear" circuit-breaker is 3000 A at 4.16 kV**
Der grösste Leistungsschalter der Firma «Biggs Switchgear» ist für einen Nennbetriebsstrom (Nenndauerstrom) von 3000 A bei 4.16 kV ausgelegt

3 **All circuit-breakers of the same current rating are fully interchangeable one with another**
Alle Leistungsschalter mit den gleichen Nennstromdaten sind untereinander voll auswechselbar

4 **The current rating of main buses is not less than that of the largest incoming breaker**
Die Nennstrombemessung der Hauptsammelschienen ist nicht kleiner als die des grössten Eingangsschalters

5 **Each of the three "QE 2" generators is rated 5.5 MW at 3.3 kV, 60 c/s, 0.833 power factor**
Jeder der drei Generatoren der «QE 2» (Queen Elizabeth 2) hat eine Nennleistung von 5,5 MW bei 3,3 kV, 60 Hz, Leistungsfaktor 0,833

6 **The rated output of each Kariba hydro generator is 100 MW at 18 kV, 50 c/s, 166.7 rev/min**
Die Nennleistung jedes Generators des Wasserkraftwerkes Kariba beträgt 100 MW bei 18 kV, 50 Hz, 166,7 U/min

7 **The A 300 B Airbus is equipped with three 400 Hz generators, each rated 60 kVA at 28 V**
Der Airbus A 300 B ist mit drei 400 Hz-Generatoren ausgerüstet, von denen jeder eine Nennleistung von 60 kVA bei 28 V abgibt

8 **A diesel generator station rated 1.28 MW, operated 96 m above ground level, supplies power to a Goliath crane at the Belfast shipyard of Harland & Wolff**
Eine Diesel-Generator-Station (oder: ein Dieselaggregat) mit einer Nennleistung von 1,28 MW, betrieben in einer Höhe von 96 m, liefert die Energie für einen Riesenbrückenkran auf der Schiffswerft von Harland & Wolff in Belfast

Telecommunication Problems

1. Nachrichtentechnische Probleme
2. Verständigungsschwierigkeiten

Applying and measuring voltage

Spannung anlegen und messen

1. **This device operates on (or: at) 24 V d.c.**
   Dieses Gerät wird mit 24 V Gleichstrom betrieben
2. **We usually employ (or: use) + 5 V for our TTL (transistor-transistor logic) circuits**
   Wir verwenden gewöhnlich + 5 V für unsere TTL-Schaltungen (oder: Schaltkreise)
3. **First of all, we check the input voltage**
   Als erstes (über)prüfen wir die Eingangsspannung
4. **The voltage leads the current**
   Die Spannung eilt dem Strom vor
5. **Voltages U_1 and U_2 are in phase**
   Die Spannungen U_1 und U_2 sind phasengleich (oder: gleichphasig)
6. **The quantities are 15° out of phase (or: the quantities have a phase displacement of 15°)**
   Die Grössen sind um 15° phasenverschoben
7. **The voltages are now equal**
   Die Spannungen sind jetzt gleich
8. **U_1 is kept (or: maintained) constant**
   U_1 wird konstant gehalten
9. **The voltage drops (or: diminishes) by 10%**
   Die Spannung fällt um 10% ab (oder: geht zurück, sinkt)
10. **The voltage increases (or: rises, is increasing, is rising)**
    Die Spannung steigt an (oder: erhöht sich, wird grösser)

Choosing logic equipment

Wahl der logischen Schaltungen

1. **We must carefully consider whether we shall go on using TTL logic (transistor-transistor logic) for our new contract**
   Wir müssen es uns gründlich überlegen, ob wir auch weiterhin TTL-Logik für unseren neuen Auftrag verwenden sollen
2. **Nowadays TTL logic is most widely employed but here are some points which are not in favour of this design**
   Heutzutage wird weitgehendst TTL-Logik verwendet, jedoch gibt es einige Punkte, welche nicht (unbedingt) für diese Ausführung sprechen

3 **This type of logic requires a rather high power consumption**
Diese Art Logik erfordert einen ziemlich hohen Speisestrom(verbrauch)

4 **Furthermore, TTL logic is liable (or: susceptible) to malfunction if zero leads are not clean (which means a drop in voltage)**
Weiterhin ist TTL-Logik störanfällig, wenn die Null-Leitungen nicht sauber sind (was Spannungsabfall bedeutet)

5 **We could use HTL logic (high threshold logic) as this logic is scarcely liable to malfunction**
Wir könnten HTL-Logik verwenden, da diese Logik kaum störungsanfällig ist

6 **However, HTL logic is very slow in operation and needs a high power input**
Jedoch ist HTL-Logik sehr langsam im Betrieb und benötigt viel Speiseleistung

7 **We would operate our TTL systems at +5 V**
Wir würden unser TTL-System mit +5 V betreiben

8 **CMOS (complementary metal oxide semiconductors) is an up-to-date design and requires hardly any power, except during change-over**
CMOS ist eine moderne Ausführung und benötigt fast keine Leistung, ausgenommen beim Umschalten

9 **CMOS is slower than TTL but switching frequencies in the range of 1 MHz are quite sufficient for our case**
CMOS ist langsamer als TTL, jedoch sind Schaltfrequenzen im Bereich von 1 MHz stets ausreichend für unseren Fall

10 **Compared with TTL, this type of logic is three times as dear**
Verglichen mit TTL ist diese Art Logik dreimal so teuer

11 **The extra expense for CMOS will partly be made good by employing smaller supply units**
Der Mehraufwand für CMOS wird teilweise durch die Verwendung von kleineren Speisegeräten gutgemacht

12 **We have found out that the manufacturers furnish CMOS integrated circuits with different technical data**
Wir haben festgestellt, dass die Hersteller CMOS-Prints (oder: Leiterplatten) mit verschiedenen technischen Daten liefern

Everyday phrases

Agreement

Zustimmung

1 **I agree with you entirely**
Ich stimme Ihnen vollkommen zu

2 **I fully agree**
Ich stimme (Ihnen) voll zu

3 **It looks all right to me**
Es scheint mir in Ordnung (zu sein)

4 **You can certainly say that again**
Das kann man wohl sagen

5 **It's quite right with us**
Bei uns ist alles in Ordnung

6 **That's just what I was going to say**
Genau das wollte ich gerade sagen

7 **You are absolutely right**
Sie haben vollkommen recht

8 **That will do fine**
Das trifft sich gut (oder: das geht gut)

9 **Quite so**
Das stimmt

10 **Sounds good**
Das lässt sich hören

11 **I quite see that**
Ich stimme hier zu

12 **I think so**
Ich denke ja (oder: ich denke schon)

13 **I see the point**
Ich sehe schon (oder: ich verstehe, was Sie meinen)

14 **Right you are**
Sie haben recht

15 **He thought so as well**
Er dachte das gleiche

16 **I'm sure you know the problem better than I do**
Ich bin sicher, dass Sie sich hier besser auskennen als ich

17 **An attractive enquiry, certainly**
Eine interessante Anfrage, sicher

18 **We agreed on this subject very soon**
Bei diesem Thema wurden wir uns schnell einig

Praise

Lob

1. **You've done a fine job**
   Das haben Sie gut gemacht
2. **I'm very grateful, you've done all you can**
   Ich bin Ihnen sehr dankbar, Sie haben getan was Sie konnten
3. **You've put it very nicely**
   Das haben Sie gut gesagt
4. **I can't but say a word of praise to Mr Jones**
   Ich komme nicht umhin, Mr Jones besonders zu erwähnen (zu loben)
5. **It's very wise of you**
   Das ist sehr klug von Ihnen
6. **I declare you've done fine work**
   Ich muss schon sagen, Sie haben gute Arbeit geleistet
7. **You can take pride in that design**
   Auf diese Konstruktion können Sie stolz sein
8. **I'm glad you've been looking after that business**
   Ich bin froh, dass Sie sich um diese Angelegenheit gekümmert haben
9. **You've got a good command of the German language**
   Sie beherrschen die deutsche Sprache ausgezeichnet
10. **I'm happy you'll assist me with this job**
    Ich bin froh, dass Sie mich bei dieser Arbeit unterstützen wollen
11. **I've never seen a layout like this before**
    Ich habe bisher noch keinen Entwurf wie diesen gesehen
12. **I'm sure you'll manage it somehow**
    Ich bin sicher, dass Sie es (schon) irgendwie meistern werden

Disagreement

Unstimmigkeit

1. **I can't possibly agree to your proposal**
   Ihrem Vorschlag kann ich unmöglich zustimmen
2. **Things are different with us**
   Bei uns ist das anders
3. **It's not quite so simple as you think**
   Es ist nicht ganz so einfach wie Sie denken

4 **It's not as simple as that**
So einfach ist das (auch) nicht

5 **I won't consider that**
Das kommt nicht in Frage

6 **I'm not going to argue about something so absurd**
Ich werde mich über so alberne (lächerliche) Sachen nicht streiten

7 **I don't like the sound of that**
Das gefällt mir gar nicht

8 **What's the use of going on if you don't agree on this item**
Was hat es für einen Zweck weiterzumachen, wenn Sie in diesem Punkt nicht zustimmen

9 **I felt all along that something was wrong here**
Ich dachte schon die ganze Zeit, dass hier etwas nicht stimmen könnte

10 **That's not at all clear to me**
Das ist mir durchaus nicht klar

11 **One really oughtn't to say that**
Das sollte man wirklich nicht sagen

12 **You can never hope to achieve the same results again**
Diese Ergebnisse werden Sie nicht noch einmal erreichen

13 **I don't think we should leave it as it is**
Ich meine, wir sollten es so nicht lassen

14 **I see absolutely no reason why I shouldn't go on like this**
Ich sehe absolut keinen Grund, warum ich nicht so weitermachen sollte

15 **I've told him again and again that we can't go on like this**
Ich habe ihm immer wieder gesagt, dass wir so nicht weitermachen können

16 **Nothing doing, we must turn down your request**
Nichts zu machen, wir müssen Ihr Gesuch ablehnen

17 **Catch me ever admitting that**
Da können Sie lange warten, bis ich das zugebe

18 **You don't say**
Was Sie da nicht sagen

19 **Please, don't side-track**
Weichen Sie bitte nicht vom Thema ab

20 **What's that to you?**
Was geht Sie das an?

21 **That's not your business**
Das geht Sie gar nichts an

22 **That's too much**
Das geht zu weit

23 **Don't drag me into that affair**
Lassen Sie mich aus dem Spiel

24 **It's all over and done with**
Es ist alles aus

25 **Don't think about it any longer**
Schlagen Sie es sich aus dem Sinn

26 **We have not seen the last of it yet**
Es ist noch nicht aller Tage Abend

27 **I declare, I don't know where to put it**
Ich weiss wirklich nicht, wo ich es unterbringen soll

28 **It doesn't pay (or: It's not worth while)**
Es lohnt sich nicht

29 **We are arguing at cross-purposes**
Wir reden aneinander vorbei

30 **You'd better leave that alone**
Sie sollten lieber die Finger davon lassen

On the telephone

Am Telefon

1 **This is Faver speaking from Baden, Switzerland**
Hier spricht Faver aus Baden in der Schweiz

2 **I should like to speak (or: talk) to Mr Evans**
Ich möchte gern mit Herrn Evans sprechen

3 **Just a moment, I'll see if Mr Evans is available (or: free)**
Einen Moment, ich werde mal nachsehen, ob Herr Evans da ist (oder: frei ist)

4 **Mr Evans, a Mr Zürcher is on the line for you**
Herr Evans, ein Herr Zürcher möchte Sie am Telefon sprechen

5 **Put him through, please**
Stellen Sie ihn bitte durch (oder: verbinden Sie mich bitte mit ihm)

6 **Mr Zürcher, I'm putting you through now**
Herr Zürcher, ich stelle Sie jetzt durch (oder: ich verbinde Sie jetzt)

7 **I'll transfer you to Mr Robertson now**
Ich stelle Sie jetzt zu Herrn Robertson durch

8 **Please give me Alan again**
Geben Sie mir bitte noch einmal Alan (oder: stellen Sie mich bitte noch einmal zu Alan durch)
9 **Biggs & Walker here, my name is Baker. I'd like to speak to Mr Hunter**
Hier ist Biggs & Walker, mein Name ist Baker. Ich möchte gern mit Herrn Hunter sprechen
10 **Cheerio Alan. Thanks for calling**
Mach's gut Alan. Vielen Dank für den Anruf.

Letter beginnings

Briefanfänge

1 **Thank you very much for your immediate reply to our questions**
Wir danken Ihnen sehr für die sofortige Beantwortung unserer Fragen
2 **We are very pleased to note from your letter of 2nd May . . .**
Wir freuen uns, aus Ihrem Brief vom 2. Mai zu erfahren . . .
3 **Thank you very much for your letter of 20th May regarding . . .**
Vielen Dank für Ihren Brief vom 20. Mai betreffend . . .
4 **In reply to your letter of 3rd April, we have pleasure in offering you . . .**
Wir freuen uns, Ihnen gemäss Brief vom 3. April . . . anbieten zu können
5 **We regret to learn from your report . . .**
Wir bedauern, Ihrem Bericht entnehmen zu müssen . . .
6 **In response to your request of 19th May, I would advise you . . .**
Entsprechend Ihrer Anfrage (Ihres Gesuches) möchte ich Ihnen mitteilen (auch: empfehlen, raten) . . .
7 **I am delighted to tell you . . .**
Es freut mich, Ihnen mitteilen zu können . . .
8 **This is to let you know . . .**
Hiermit möchten wir Sie informieren . . .
9 **I am sorry to have to tell you . . .**
Es tut mir leid, Ihnen mitteilen zu müssen . . .
10 **I must write you a few words of thanks for . . .**
Ich muss Ihnen ein paar Worte des Dankes wegen (für) . . . schreiben

11 **Thanking you for your above mentioned enquiry, we take pleasure in quoting you for . . .**
Wir danken Ihnen für die oben erwähnte Anfrage und möchten Ihnen folgendes Angebot unterbreiten: . . .

12 **In accordance with your request, we submit the following quotation: . . .**
Gemäss Ihrer Anfrage übersenden wir folgende Offerte: . . .

13 **We have much pleasure in attaching . . .**
Es freut uns, Ihnen als Beilage . . . übersenden zu können

14 **Please submit additional data providing the background information and parameters which went into this calculation**
Übersenden Sie bitte weitere grundsätzliche Angaben und Parameter, die dieser Berechnung zugrunde gelegt worden sind

15 **This letter confirms our verbal acceptance given to you on the 5th May at your offices**
Mit diesem Brief bestätigen wir unsere mündliche Zusage vom 5. Mai in Ihrem Hause

16 **Following a discussion with your (esteemed) Mr Jones on . . . at . . .**
Gemäss einer Diskussion mit Ihrem Herrn Jones am . . . in . . .

17 **Would you be so kind as to allow me to bring this matter to your notice**
Gestatten Sie bitte, dass ich mich in dieser Angelegenheit an Sie wende (oder: dass ich Ihnen diese Angelegenheit zur Kenntnis bringe)

18 **It is understood that you have been favoured with an order for (the delivery of) two 1200 MW generators**
Uns wurde bekannt, dass Sie den Auftrag für zwei 1200 MW-Generatoren erhalten haben . . .

Letter endings

Briefabschlüsse

1 **We trust that we have been of some assistance (or: have been able to help you) in this matter**
Wir hoffen, dass wir Ihnen in dieser Angelegenheit etwas Unterstützung geben konnten (oder: behilflich sein konnten)

2 **If you require any further information, please do not hesitate to contact us (or: should you require any further information on this matter, please let us know)**
Falls Sie weitere Angaben (oder: Unterlagen) benötigen, so setzen Sie sich bitte mit uns (unverzüglich) in Verbindung

3 **We shall be glad to supply any further information which may be required**
Gern übersenden wir Ihnen weitere Angaben (oder: Unterlagen), falls erforderlich

4 **Thanking you in anticipation for . . .**
Vielen Dank im voraus für . . .

5 **We should be glad if you would accept our proposals and hope to hear from you soon**
Wir würden uns freuen, wenn Sie unsere Vorschläge annehmen würden und hoffen bald wieder von Ihnen (etwas) zu hören

6 **We should be very obliged if you could furnish us the . . .**
Wir wären Ihnen sehr dankbar, wenn Sie uns die . . . übersenden könnten

7 **We trust that our explanation will convince you...**
Wir hoffen, dass unsere Erklärung Sie (davon) überzeugt . . .

8 **We shall let you know the result of this affair**
Wir werden Sie über das Ergebnis (dieser Angelegenheit) unterrichten

9 **When you have received the information, we naturally would be interested in getting the benefit of your opinion**
Wenn Sie die Angaben erhalten haben, so hätten wir gern Ihre Meinung hierüber erfahren

10 **We look forward to hearing from you again**
Wir würden uns sehr freuen, von Ihnen wieder etwas zu hören

◁ How to ease the job.
Wie man sich die Arbeit erleichtert.

Calming down people

Leute beruhigen

1 **Never mind, it's a natural error (or: it can happen to anybody)**
Macht nichts, (das) kann jedem passieren

2 **Think nothing of it**
Macht nichts (mach dir keine Gedanken)

3 **Put yourself at ease, we have not lost yet**
Beruhige dich, noch haben wir nicht(s) verloren

4 **Unbend a little, it isn't his fault either**
Sei ein bisschen nachgiebig, seine Schuld ist es auch nicht (wörtlich: lockere deine starre Haltung)

5 **Take it easy, we still have got time to start from scratch**
Nimm es leicht, noch haben wir Zeit, ganz von vorn zu beginnen

6 **No use moaning about it, we can't do any more in this case**
Es hat keinen Zweck herumzujammern, wir können in diesem Fall nichts mehr tun

7 **Calm down, Alan, I really didn't mean it**
Beruhige dich, Alan, ich habe es wirklich nicht so gemeint

8 **I wouldn't dream of blaming you for that, we should have seen it, too**
Ich denke nicht daran, dir die Schuld zu geben, wir hätten es auch sehen sollen!

9 **I'm afraid you must have thought me extremely rude but I was very tired, I haven't had any sleep for two days**
Ich fürchte, Sie haben mich für sehr grob (unhöflich) gehalten, aber ich war sehr müde, ich hatte zwei Tage keinen Schlaf gehabt

10 **Don't take it amiss, we all make mistakes in a situation like this**
Nimm es (mir) nicht übel, wir alle machen Fehler in solchen Situationen

Phonetic alphabeths / Buchstabiertabelle

	GB	USA	D (A)	CH
A	Andrew	abel	Anton	Anna
B	Benjamin	Baker	Berta	Bertha
C	Charlie	Charlie	Cäsar	Cäsar
D	David	Dog	Dora	Daniel
E	Edward	Easy	Emil	Emil
F	Frederick	Fox	Friedrich	Friedrich
G	George	George	Gustav	Gustav
H	Harry	How	Heinrich	Heinrich
I	Isaac	Item	Ida	Ida
J	Jack	Jig	Julius	Jakob
K	King	King	Kaufmann (Konrad)	Kaiser
L	Lucy	Love	Ludwig	Leopold
M	Mary	Mike	Martha	Marie
N	Nellie	Nan	Nordpol	Niklaus
O	Oliver	Oboe	Otto	Otto
P	Peter	Peter	Paula	Peter
Q	Queenie	Queen	Quelle	Quelle
R	Robert	Roger	Richard	Rosa
S	Sugar	Sugar	Samuel (Siegfried)	Sophie
T	Tommy	Tare	Theodor	Theodor
U	Uncle	Uncle	Ulrich	Ulrich
V	Victor	Victor	Viktor	Viktor
W	William	William	Wilhelm	Wilhelm
X	Xmas	X	Xanthippe (Xaver)	Xaver
Y	Yellow	Yoke	Ypsilon	Yverdon
Z	Zebra	Zebra	Zacharias (Zürich)	Zürich
Ä			Ärger	
Ö			Ökonom (Österreich)	
Ü			Übermut	
Ch			Charlotte	
Sch			Schule	

Bildquellennachweis / Illustration credits

Nachschlagewerke

Collier's Standard® Dictionary, Funk & Wagnalls Comp., New York
Bildwörterbuch, VEB Verlag Enzyklopädie, Leipzig
The Oxford Illustrated Dictionary, Oxford University Press
(Book Club Associates, London)

Lehrbücher/-hefte

About a Motor Car, Penguin Books Ltd, Hermondsworth/UK
Basic Electronics, The Technical Press Ltd, New York/USA
Engineering Report, Nussbaumen/Schweiz
Fischer Bücherei, Technik 3, Frankfurt/Main, BRD
Metals in the Service of Man, Pelican Books, London/UK
The Incredible Illustrated Electricity Book,
Pathfinder Publications Inc. , Boston/USA
The Incredible Illustrated Tool Book,
Pathfinder Publications Inc., Boston/USA
Metalwork, Edward Arnold Ltd., London

Zeitschriften

Iron Age
Iron and Steel Engineer
Stahl und Eisen

Firmen-Veröffentlichungen

Allen-Bradley Ltd, Milton Keynes/UK
Asea Brown Boveri AG, Baden/Schweiz
BICC Wiring Systems Ltd/UK
DEMAG/BRD
Hawker Siddely Group, Staines/UK
Heyl & Patterson, Pittsburgh/USA
H.K. Porter Comp. Inc., New York/USA
INDUMAT, Bergisch Gladbach/BRD
North American Mfg. Co., Cleveland/USA
Vogel-Bucher SA, Morges/Schweiz

Springer und Umwelt

Als internationaler wissenschaftlicher Verlag sind wir uns unserer besonderen Verpflichtung der Umwelt gegenüber bewußt und beziehen umweltorientierte Grundsätze in Unternehmensentscheidungen mit ein. Von unseren Geschäftspartnern (Druckereien, Papierfabriken, Verpackungsherstellern usw.) verlangen wir, daß sie sowohl beim Herstellungsprozess selbst als auch beim Einsatz der zur Verwendung kommenden Materialien ökologische Gesichtspunkte berücksichtigen.
Das für dieses Buch verwendete Papier ist aus chlorfrei bzw. chlorarm hergestelltem Zellstoff gefertigt und im pH-Wert neutral.

Druck: Mercedesdruck, Berlin
Verarbeitung: Buchbinderei Lüderitz & Bauer, Berlin